Volumes already published

Volume 1: Trees I (1986)
Volume 2: Crops I (1986)
Volume 3: Potato (1987)
Volume 4: Medicinal and Aromatic Plants I (1988)
Volume 5: Trees II (1989)
Volume 6: Crops II (1988)
Volume 7: Medicinal and Aromatic Plants II (1989)
Volume 8: Plant Protoplasts and Genetic Engineering I (1989)
Volume 9: Plant Protoplasts and Genetic Engineering II (1989)
Volume 10: Legumes and Oilseed Crops I (1990)
Volume 11: Somaclonal Variation in Crop Improvement I (1990)
Volume 12: Haploids in Crop Improvement I (1990)
Volume 13: Wheat (1990)
Volume 14: Rice (1991)
Volume 15: Medicinal and Aromatic Plants III (1991)
Volume 16: Trees III (1991)
Volume 17: High-Tech and Micropropagation I (1991)
Volume 18: High-Tech and Micropropagation II (1992)
Volume 19: High-Tech and Micropropagation III (1992)
Volume 20: High-Tech and Micropropagation IV (1992)
Volume 21: Medicinal and Aromatic Plants IV (1993)
Volume 22: Plant Protoplasts and Genetic Engineering III (1993)
Volume 23: Plant Protoplasts and Genetic Engineering IV (1993)
Volume 24: Medicinal and Aromatic Plants V (1993)
Volume 25: Maize (1994)
Volume 26: Medicinal and Aromatic Plants VI (1994)
Volume 27: Somatic Hybridization in Crop Improvement I (1994)
Volume 28: Medicinal and Aromatic Plants VII (1994)
Volume 29: Plant Protoplasts and Genetic Engineering V (1994)
Volume 30: Somatic Embryogenesis and Synthetic Seed I (1995)
Volume 31: Somatic Enbryogenesis and Synthetic Seed II (1995)

Volumes in preparation

Volume 32: Cryopreservation of Plant Germplasm I (1995)
Volume 33: Medicinal and Aromatic Plants VIII
Volume 34: Plant Protoplasts and Genetic Engineering VI
Volume 35: Trees IV
Volume 36: Somaclonal Variation in Crop Improvement II
Volume 37: Medicinal and Aromatic Plants VIII
Volume 38: Plant Protoplasts and Genetic Engineering VII

Biotechnology in Agriculture and Forestry 31

Somatic Embryogenesis and Synthetic Seed II

Edited by Y.P.S. Bajaj

With 208 Figures and 79 Tables

Springer-Verlag
Berlin Heidelberg New York
London Paris Tokyo
Hong Kong Barcelona Budapest

Professor Dr. Y.P.S. BAJAJ
A-137
New Friends Colony
New Delhi 110065, India

ISBN-13: 978-3-642-78645-7 e-ISBN-13: 978-3-642-78643-3
DOI: 10/1007.978-3-642-78643-3

Library of Congress Cataloging-in-Publication Data Somatic embryogenesis and synthetic seed/edited by Y.P.S. Bajaj. p. cm. – (Biotechnology in agriculture and forestry; <>-31 Includes bibliographical references and index. (v. 2: Berlin: acid-free paper) – (v. 2: New York: acid-free paper) 1. Somatic embryogenesis. 2. Synthetic seeds. 3. Crops. I. Bajaj, Y.P.S., 1936- . II. Series. SB 118.45.S66 1995 631.5′21 – dc20 94-28237

Softcover reprint of the hardcover 1st edition 1995

SPIN: 10127472 31/3130/SPS–5 4 3 2 1 0 – Printed on acid-free paper

Dedicated to
Professor Dr. Jacob Reinert
with whom I had the privilege of working
at the Freie Universität, Berlin, Germany
from 1974–1976

Preface

While working in the laboratory of Professor Dr. Jacob Reinert at the Freie Universität Berlin (1974–1976), I had the opportunity to become deeply involved in studying the intricacies of the fascinating phenomenon of somatic embryogenesis in plant cells and protoplasts. In numerous stimulating discussions with Professor Reinert on this subject, I was fully convinced that somatic embryogenesis would become one of the most important areas of study, not only regarding basic and fundamental aspects, but also for its application in crop improvement. During the last decade, we have witnessed tremendous interest and achievements in the use of somatic embryos for the production of synthetic seeds, for micropropagation, genetic transformation, cryopreservation, and conservation of germplasm. The en masse production of somatic embryos in the bioreactors has facilitated some of these studies. Somatic embryos have now been induced in more than 300 plant species belonging to a wide range of families. It was therefore felt that a compilation of literature/state of the art on this subject was necessary. Thus, two volumes on *Somatic Embryogenesis and Synthetic Seed* have been compiled, which contain 65 chapters contributed by International experts.

Somatic Embryogenesis and Synthetic Seed I comprises 31 chapters, arranged in 3 sections:

Section I. Commitment of the cell to somatic embryogenesis; early events; anatomy; molecular basis; gene expression; role of polyamines; machine vision analysis of somatic embryos.

Section II. Applications of somatic embryos; technology of synthetic seed; fluid drilling; micropropagation; genetic transformation through somatic embryos; cryopreservation.

Section III. Somatic embryogenesis in various tree species of *Aesculus, Betula, Carica, Citrus, Cocos, Corylus, Elaeis, Hevea, Juglans, Larix, Liriodendron, Magnolia, Olea, Picea, Populus,* and *Theobroma.*

Somatic Embryogenesis and Synthetic Seed II contains 34 chapters, arranged in 4 sections:

Section I. Cereal and Grasses – wheat, rice, maize, rye, oat, ryegrass, fescue, orchardgrass, bluestem grasses, sugarcane.

Section II. Vegetables and Fruits – asparagus, chicory, cucurbits, cucumber, okra, carrot, banana.
Section III. Legumes and Oilseed Crops – peanut, soybean, cotton, white mustard, meadowfoam.
Section IV. Ornamental, Medicinal and Miscellaneous Plants – daylily, freesia, celery, coriander, *Coptis*, *Panax*, *Rauwolfia*, *Arabidopsis*, *Bellevalia*, *Brimeura*, *Dendrophthoe*, *Rumex*, *Fagopyrum*, and *Ranunculus*.

These books will be of interest to students, teachers, and research workers in the field of botany, horticulture, forestry, tissue culture, general plant biotechnology, and to those involved in micropropagation.

New Delhi, January 1995

Professor Y.P.S. Bajaj
Series Editor

Contents

Section IV Ornamental, Medicinal, and Miscellaneous Plants

List of Contributors

ALIBERT, G., Laboratoire de Biotechnologie et Amélioration des Plantes (BAP), INP-ENSAT, Unité Associée à l'INRA, 145 Avenue de Muret, F-31076 Toulouse Cedex, France

ALTMAN, A., Department of Horticulture, Faculty of Agriculture, The Hebrew University of Jerusalem, P.O. 12, Rehovot, 76–100 Israel

BAILEY, M.A., Department of Crop and Soil Sciences, The University of Georgia, 3111 Plant Sciences Building, Athens, GA 30602–7272, USA

BAJAJ, Y.P.S., A-137 New Friends Colony, New Delhi 110 065, India

BREGITZER, P.P., National Small Grains Germplasm Research Facility, United States Department of Agriculture, Agricultural Research Services, P.O. Box 307, Aberdeen, ID 83210, USA

BROWN III, A., Department of Biological Sciences, Clark Atlanta University, Atlanta, GA 30314, USA

CARMAN, J.G., Plants, Soils, and Biometerology Department, Utah State University, Logan, UT 84322-4820, USA

CAVALLINI, A., Department of Agricultural Plant Biology, Genetics Section, University of Pisa, via Matteotti 1/B, I-56124 Pisa, Italy

CHEE, P.P., Discovery Research, Molecular Biology Division, The Upjohn Company, 301 Henrietta St., Kalamazoo, MI 49007, USA

CHEN, R.R., Department of Biology, Shanxi University, Taiyuan 030006, People's Republic of China

CHENIEUX, J.C., Plant Biotechnology Laboratory,
Faculty of Pharmacy, University of Tours, 31 Avenue Monge,
F-37200 Tours Cedex, France

CONGER, B.V., Department of Plant and Soil Science,
University Tennessee, Knoxville, TN 37901-1071, USA

COOK, D.A., Department of Biological Sciences,
Clark Atlanta University, Atlanta, GA 30314, USA

COUILLEROT, J.P., Physiologie Cellulaire et Morphogénèse Végétales,
Universite des Sciences et Technologies de Lille, Bt SN2, F-59655
Villeneuve d'Ascq Cedex, France

CULAFIC, Lj., Institute of Botany, Faculty of Science
and Institute for Biological Research "S. Stankovic",
University of Belgrade, Takovska 43, 11060 Belgrade, Yugoslavia

DUBOIS, J., Physiologie Cellulaire et Morphogénèse Végétales,
Université des Sciences et Technologies de Lille, Bt SN2,
F-59655 Villeneuve d'Ascq Cedex, France

DURHAM, R.E., Department of Plant and Soil Sciences,
Texas Tech University, Box 42122, Lubbock,
TX 79409-2122, USA

EKSOMTRAMAGE, T., CIRAD-CA, BIOTROP-GERDAT Laboratory,
2477 Avenue du Val de Montferrand, BP 5035,
F-34032 Montpellier Cedex 01, France
(Present address: Department of Plant Science,
Faculty of Natural Resources, Hatyai, Songkhla, 90112,
Thailand)

EMONS, A.M.C., Department of Plant Cytology and Morphology,
Wageningen Agricultural University, Arboretumlaan 4,
6073 BD Wageningen, The Netherlands

FELDMANN, P., CIRAD-CA, Station de Roujol,
97170 Petit Bourg, Guadeloupe, French West Indies

FRANKLIN, C.I., Plant Biology Division,
The Samuel Roberts Noble Foundation, P.O. Box 2180,
Ardmore, OK 73401, USA

GAWEL, N.J., Cooperative Agriculture Research Programm,
Tennessee State University, 3500 John A. Merritt Blvd.,
Nashville, TN 37209, USA

Glaszmann, J.C., CIRAD-CA, BIOTROP-GERDAT Laboratory,
2477 Avenue du Val de Montferrand, BP 5035,
F-34032 Montpellier Cedex 01, France

Grand, C., RAGT, 18 ruede Seguret-Sainerie,
F-12003 Rodez Cedex, France

Gray, D.J., Central Florida Research and Education Center,
University of Florida, I.F.A.S. 5336 University Avenue,
Leesburg, FL 34748, USA

Gui, Y., Institute of Botany, Academia Sinica,
Beijing 100044, People's Republic of China.

Guiderdoni, E., CIRAD-CA, BIOTROP-GERDAT Laboratory,
2477 Avenue du Val de Montferrand, BP 5035,
F-34032 Montpellier Cedex 01, France

Guo, S.S., Institute of Botany, Academia Sinica,
Beijing 100080, People's Republic of China

Hao, J.P., Department of Biology, Shanxi University,
Taiyuan 030006, People's Republic of China

Hilbert, J.L., Physiologie Cellulaire et Morphogénèse Végétales,
Université des Sciences et Technologies de Lille, Bt SN2,
F-59655 Villeneuve d'Ascq Cedex, France

Huang, B., Institute of Genetics and Cytology, Northeast
Normal University, Changchun 130024, People's Republic of China

Jones, T.J., Agricultural Biotechnology, Dupont Company,
Experimental Station 402/2250, Wilmington,
DE 19880-0402, USA

Kann, R.P., Department of Biochemistry and Cell Biology,
State University of New York at Stony Brook, Stony Brook,
NY 11794-5215, USA

Ke, S., Wuhan Institute of Botany, Academia Sinica,
Wuhan 430074, People's Republic of China
(Present address: Department of Crop and Weed Sciences,
North Dakota State University, Fargo, ND 58105, USA)

Kieft, H., Department of Plant Cytology and Morphology,
Wageningen Agricultural University, Arboretumlaan 4,
6073 BD Wageningen, The Netherlands

Kishira, H., Nippon Mining Co., Ltd.,
Bioscience Research Laboratories, 3-17-35, Niizo-Minami,
Toda, Saitama 335, Japan

Komatsuda, T., Department of Cell Biology,
National Institute of Agrobiological Resources, Tsukuba,
Ibaraki 305, Japan

Krikorian, A.D., Department of Biochemistry and Cell Biology,
State University of New York at Stony Brook, Stony Brook,
NY 11794-5215, USA

Kwiatkowski, S., Department of Biology,
Southern Oregon State College, Ashland, OR 97520-5071, USA

Ladyman, J.A.R., Plant Cell Research Institute, Dublin,
CA, USA
(Present address: JnJ Associates, 1701 Father Sky CT. N.E.,
Albuquerque, NM 87112, USA)

Levi, A., USDA-ARS, PSI, Fruit Laboratory, BARC-West,
Beltsville, MD 20705, USA

Li, B.P., Department of Biochemistry, High Training School,
Yuancheng 044000, People's Republic of China

Linacero, R., Department of Genetics, Faculty of Biology,
University Complutense of Madrid, Madrid 28040, Spain

Matsunaga, T., Department of Biotechnology,
Tokyo University of Agriculture and Technology, 2-24-16 Naka-cho,
Koganei, Tokyo 184, Japan

Matsushita, H., Faculty of Pharmaceutical Sciences,
Kyushu University, Higashiku, Fukuoka 812, Japan

McDaniel, J.K., Department of Plant and Soil Sciences,
University of Tennessee, Knoxville, TN 37901-1071, USA

McKently, A.H., Science and Technology, The Land,
Epcot Center P.O. Box 10,000, Lake Buena Vista,
FL 32830, USA

Merot, B., Laboratoire d' Amélioration des Plantes, Bat 360,
Université d' Orsay, F-91405 Orsay Cedex 05, France
(Present address: BIOCEM, Campus Universitaire des Cezeaux,
24 Avenue des Landais, F-63170 Aubiere, France)

Meynet, J., Station d'Amélioration des Plantes Florales,
INRA, La Gaudine-Fréjus, F-83370 Saint-Aygulf, France

Milach, S.K.C., Department of Agronomy and Plant Genetics,
University of Minnesota, St. Paul, MN 55108, USA

Nadel, B.L., AGECOTEC,
Agricultural and Ecological Biotechnologies, Ltd., Kfar Pinnes,
Post Karkur 37920, Israel

Natali, L., Department of Agricultural Plant Biology,
Genetics Section, University of Pisa, via Matteotti l/B,
I-56124 Pisa, Italy

Neskovic, M., Institute of Botany, Faculty of Science
and Institute for Biological Research "S. Stankovic",
University of Belgrade, 29 novembra 142, 11060 Belgrade,
Yugoslavia

Parrott, W.A., Department of Crop and Soil Sciences,
The University of Georgia, 3111 Plant Sciences Building,
Athens, GA 30602-7272, USA

Paulet, F., CIRAD-CA, BIOTROP-GERDAT Laboratory,
2477 Avenue du Val de Montferrand, BP 5035,
F-34032 Montpellier Cedex 01, France

Rines, H.W., Plant Sciences Research Unit,
United States Department of Agriculture,
Agricultural Research Service, St. Paul, MN 55108, USA

Robacker, C.D., Department of Horticulture,
Georgia, Agricultural Experiment Station, University of Griffin,
GA 30223, USA

Rost, T.L., Section of Botany, University of California,
Davis, CA 95616-8537, USA

Sallandrouze, A., Laboratoire de Biotechnologie et Amélioration
des Plantes (BAP), INP-ENSAT, Unité Associée à l'INRA,
145 Avenue de Muret, F-31076 Toulouse Cedex, France

Scott, M.E., Department of Biochemistry and Cell Biology,
State University of New York at Stony Brook, Stony Brook,
NY 11794-5215, USA

Shoyama, Y., Faculty of Pharmaceutical Sciences,
Kyushu University, Higashiku, Fukuoka 812, Japan

SINK, K.C., Department of Horticulture, Michigan State University, East Lansing, MI 48824, USA

SKIRVIN, R.M., University of Illinois, Plant and Animal Biotechnology Laboratory, 1201 Gregory Drive, Urbana, IL 61801, USA

SMITH, D.L., Department of Embryology, Carnegie Institution of Washington, 115 W. University Parkway, Baltimore, MD 21210, USA

SOMERS, D.A., Department of Agronomy and Plant Genetics, University of Minnesota, St. Paul, MN 55108, USA

SOUTHWORTH, D., Department of Biology, Southern Oregon State College, Ashland, OR 97520-5071, USA

TREMOUILLAUX-GUILLER, J., Plant Biotechnology Laboratory, Faculty of Pharmacy, University of Tours, 31 Avenue Monge, F-37200 Tours Cedex, France

TRIGIANO, R.N., Department of Ornamental Horticulture and Landscape Design, University of Tennessee, Knoxville, TN 37901-1071, USA

UMETSU, H., Department of Bioscience and Biotechnology, Aomori University, 2-3-1 Kobata, Aomori, Aomori 030, Japan

VASSEUR, J., Physiologie Cellulaire et Morphogénèse Végétales. Université des Sciences et Technologies de Lille, Bt SN2, F-59655 Villeneuve d'Ascq Cedex, France

VAZQUEZ, A.M., Department of Genetics, Faculty of Biology, University Complutense of Madrid, Madrid 28040, Spain

VUJICIC, R., Center for Multidisciplinary Studies and Institute for Biological Research "S. Stankovic", University of Belgrade, 29 Novembra 142, 11060 Belgrade, Yugoslavia

WAKE, H., Biotechnology Section, Central Research Laboratory, Pentel Co. Ltd., 4-1-8 Yoshi-cho, Soka, Saitama 340, Japan

WANG, L., Institute of Genetics and Cytology, Northeast Normal University, Changchun 130024, People's Republic of China

Yeoman, M.M., Institute of Cell and Molecular Biology,
The University of Edinburgh, The King's Buildings,
Edinburgh EH9 3JH, United Kingdom

Zhang, J.T., Department of Biology, Shanxi University,
Taiyuan 030006, People's Republic of China

Zhou, X.M., Department of Biology, Shanxi University,
Taiyuan 030006, People's Republic of China

Zhu, X.X., Faculty of Pharmaceutical Sciences,
Kyushu University, Higashiku, Fukuoka 812, Japan

Ziv, M., Department of Agricultural Botany,
Faculty of Agriculture, The Hebrew University of Jerusalem,
P.O. Box 12, Rehovot 76-100, Israel

Section I
Somatic Embryogenesis in Cereal and Grasses

I.1 Somatic Embryogenesis in Wheat

J.G. CARMAN[1]

1 Introduction

Wheat is the most intensively bred species in the world. It is second to rice in world production, which in recent years has approached 550 million metric tonnes per year (Young et al. 1990). The bread wheats (*Triticum aestivum* L.) are divided into four main categories, based on the protein content of the grain. Hard red spring (11 to 18% protein) and winter (9 to 15% protein) wheats are used primarily for bread. Soft red winter (8 to 12% protein) and white (8 to 11% protein) wheats are used primarily for muffins, noodles, cakes, crackers, and cookies. Durum wheat (*Triticum durum* L.) is used primarily for macaroni and pasta. While an extremely adaptive crop, wheat is still subject to many diseases, which makes genetic engineering an appealing biotechnology for wheat improvement.

Somatic embryogenesis is used in wheat improvement programs via (1) transformation protocols (Vasil et al. 1992; Weeks et al. 1993); (2) generating somaclonal variants (Hashim et al. 1990); (3) cloning sterile derivatives from wide hybridization (Tabaeizadeh and Campeau 1992); and (4) encouraging chromosomal translocations, eliminations, and other aberrations in sterile or fertile derivatives obtained from wide hybridization (Larkin et al. 1990). The use of somatic embryogenesis in transformation protocols is appealing because thousands of unique transgenic wheat plants (obtained by insertion of genes at different loci) could potentially be produced. Such plants could then be screened for agronomically important temporal and/or spatial variation in transgenic expression. While transformation has been accomplished (Vasil et al. 1992; Weeks et al. 1993), poor efficiency in maturing and germinating somatic embryos is a serious impediment to the mass production of somatic embryogenesis-based transgenic plants.

Aberrant somatic embryos of wheat were first recognized in tissue cultures by Ahloowalia (1982), and Ozias-Akins and Vasil (1982). Numerous reports followed these early studies and have been reviewed (Carman and Campbell 1990; Scott et al. 1990). This chapter summarizes early findings, reviews recent studies, presents concepts amenable to further experimentation, and provides a brief protocol for somatic embryo production and germination.

[1] Plants, Soils and Biometeorology Department, Utah State University, Logan, Utah 84322-4820, USA

Biotechnology in Agriculture and Forestry, Vol. 31
Somatic Embryogenesis and Synthetic Seed II (ed. by Y.P.S. Bajaj)

2 Somatic Embryogenesis

The frequency of somatic embryogenesis in wheat is affected by a number of factors discussed below.

2.1 Donor Plants and Explants

Genotypic variation in frequency and intensity of embryogenic responses from various wheat explants is well documented (Carman and Campbell 1990; Scott et al. 1990). It is perhaps best described as variation in genetic plasticity for embryogenic competence such that poorly responding genotypes respond favorably under a narrow range of explant and medium conditions, while superior genotypes respond favorably under a broad range of explant and medium conditions (Carman et al. 1988; Carman 1990; Redway et al. 1990b).

Kaleikau et al. (1989a,b) crossed highly regenerative wheat lines and monosomic, ditelosomic, and compensating nullisomic-tetrasomic wheat lines to study the genetic basis for embryogenic competence. Segregation ratios suggested that a major regeneration gene is located on chromosome 2*D* and that minor regeneration genes are found on 2*A* and 2*B*. Previously, Galiba et al. (1986) showed that genes on wheat chromosomes 7*B*, 7*D*, and 1*D* are important to regeneration, and Lazar et al. (1987) reported that regeneration from callus cultures was enhanced when wheat addition lines contained rye chromosomes 6 or 7.

Breadth of genotypic plasticity for embryogenic response is more fully experienced when donor plants are grown under favorable conditions. Variation in growth conditions may cause embryogenic responses from immature embryo explants to range from 5 to 50% for individual, poorly responding genotypes and from 60 to 100% for individual, superior genotypes. Growth conditions that decrease the frequency of embryogenic responses include heat or water stress, fungal or insect infestations, poor lighting, or poor plant nutrition. In general, we have found that the percentage of favorably responding explants from greenhouse-grown plants occurs in the winter when plants are produced under supplemental lighting. In contrast, our poorest results occur when such plants are produced in the summer, when it is difficult to maintain moderate daytime greenhouse temperatures (J.G. Carman, unpubl.).

Donor plant growth conditions affecting explant competence are poorly understood. We have evaluated effects of temperature. Two sets of plants from each of two wheat lines, a poorly responding line (Yaqui 50) and a superior line (PCYT 10), were grown in growth chambers at constant temperatures of 15 and 25 °C, when donor plants were grown at 25 °C, Yaqui 50 and PCYT 10 calli produced 105 and 309 somatic embryos per g (fr. wt.), respectively. These numbers were 258 and 676, respectively, when the donor plants were grown at 15 °C. Hence, continuous, moderately high temperatures during donor plant growth contribute to poor embryogenic response. Endogenous hormone levels in wheat caryopses during embryo differentiation were not strongly correlated with the donor plant temperature treatments or subsequent embryogenic responses (Hess 1992).

Embryogenic wheat callus is readily obtained from immature embryos, about 0.8 to 2 mm in length (10 to 13 days postanthesis, DPA), and young inflorescences, about 5 to 20 mm in length. Embryogenic callus can also be obtained from coleoptilar nodes, intercalary leaf sheath tissues, or shoot primordia of mature embryos or young seedlings. Favorable results from the use of these latter explants are more dependent on genotype, occur less frequently, and usually occur only after one or more selective subcultures (Barcelo et al. 1992, and reviews by Carman and Campbell 1990; Scott et al. 1990). After careful screening of genotypes and optimization of procedures, the use of non-conventional explants, such as shoot tips (Wernicke and Milkovits 1986) or basal leaf sections (Rajyalakshmi et al. 1991), may yield results nearly comparable to those obtained with immature embryos or inflorescences.

2.2 Embryogenic Callus Cultures

Many factors influence the induction of embryogenic callus. The best explants are immature embryos (0.8 to 2 mm long, 10 to 13 DPA). These are generally cultured with the scutellum facing upward and the embryo axis pressed down into semisolid medium. Greater quantities of embryogenic callus and improved plant regeneration efficiencies were obtained by culturing wheat embryo explants on filter paper bridges (Mohmand and Nabors 1991). Regardless of the technique, embryos damaged or excessively exposed to dry air (desiccated) during excision often fail to produce embryogenic callus. Favorable results occur with incubation temperatures ranging from 22 to 28 °C. Total darkness is preferred in many labs, though favorable results for some genotypes have been reported when cultures are incubated in the light.

In addition to providing favorable growing conditions for donor plants, the potential breadth of genotypic plasticity for embryogenic response can be more fully experienced with improved media. The use of MS medium (Murashige and Skoog 1962), containing twice the concentration of inorganic salts, has improved the frequency of embryogenic callus formation for many genotypes (Ozias-Akins and Vasil 1983; Carman et al. 1988; Redway et al. 1990b). Redway et al. (1990b) hypothesized that this improvement is due to increased N. They observed similar improvements by adding 100 mg/l casein hydrolysate and 500 mg/l glutamine to single-strength MS. Likewise Liu et al. (1992) observed significant improvements in callus formation and plantlet regeneration with several wheat genotypes when the tissue culture medium was amended with 4 g/l casein hydrolysate.

He et al. (1989) altered levels of individual macroelements of MS medium. Formation of white embryogenic callus from epiblasts of immature embryos of the cultivar Oxley was enhanced when NH_4NO_3, KH_2PO_4, and $CaCl_2$ levels were individually elevated substantially above MS concentrations. In contrast, embryogenic callus formation from scutella of this same cultivar was more frequent when levels of N, P, or K were about 50% of MS levels or when levels of $CaCl_2$ or $MgSO_4$ were substantially higher (up to 24-fold) than those of MS. The authors hypothesized that scutella generally required lower salt concentrations because of their well-developed vascular systems, which are absent in epiblasts.

Proliferation of established embryogenic callus occurred on media that spanned a wide range of salt concentrations but was most favorable for this genotype when $CaCl_2$ and $MgSO_4$ were elevated. Differentiation of plantlets via somatic embryogenesis also occurred over a wide range of salt concentrations but was generally more favorable when cultures were incubated on media containing low levels of macrosalts (half-strength MS).

He et al. (1991) conducted a study of the effects of minor elements similar to that described above for major elements. They found that a complete elimination of zinc from the medium promoted the formation of white embryogenic callus from the explants of the two cultivars tested. In general, the literature indicates that improvements in embryogenic response are possible through media modifications, but these are frequently genotype or environmentally dependent. The results of such optimizations have not been so impressive as to find their way into general use.

2.3 Embryogenic Suspension Cultures

The production of embryogenic wheat suspension cultures has recently been reported by numerous labs. The fundamental requirement identified in these reports is the inclusion of a procedure that enriches concentrations of disassociated embryogenic cells and cell clusters. This unique cell and tissue type forms only after several to many months of selective subculture in which nonembryogenic cells, tissues, and organs are repeatedly eliminated from the suspension or callus culture.

Wang and Nguyen (1990) initiated highly embryogenic cell suspensions from immature embryo callus that had not been selected for its embryogenic nature prior to the initiation of suspension culture. They hypothesized that proliferating embryogenic cells and cell clusters in highly heterogeneous suspensions would be dense (settle rapidly to the bottom upon termination of shaking) and would not form roots upon exposure to root-forming conditions (low auxin concentrations or prolonged subculture interval). They repeatedly replaced nonembryogenic free-floating cells with new MS-based medium and challenged the denser, settled cells with root-forming, low-auxin conditions. Cell clusters that formed roots (or shoots) were manually removed and nonrooty, highly proliferative embryogenic suspensions eventually formed. These contained mostly small disassociated compact cells and cell aggregates with dense cytoplasms. Individual cells and cell clumps containing < 10 cells were plated on a semisolid MS-based regeneration medium. Approximately 0.1% of the plated cells formed embryogenic (shoot-competent) callus.

Yang et al. (1991) enriched the embryogenic cell component of their suspensions in a manner similar to that of Wang and Nguyen. However, embryogenic callus, obtained from immature embryos, was used to initiate the suspensions. Subcultures were made weekly and involved the transfer of embryogenic callus clumps (solid, bright yellow callus, 2 to 8 mm in diameter, with very small proembryo-like protuberances) to fresh liquid medium every week. Nonembryogenic callus and callus that had differentiated shoots or roots were discarded.

Enrichment was improved with one cultivar by challenging suspension-derived callus clumps with regeneration conditions (low or no auxin) on semisolid medium. After 1 month, those calli that remained embryogenic (without root or plantlet regeneration) were returned to auxin-containing suspension culture. Highly proliferative, fine suspensions of embryogenic aggregates (2 to 200 cells) eventually formed. These produced embryogenic callus and green plants upon plating on semisolid hormone-free regeneration medium.

Wang and Nguyen (1990) and Yang et al. (1991) selected in suspension culture embryogenic tissue and cell morphotypes that eventually (several to many months) disassociated into fine suspensions of embryogenic cells and cell clusters. Redway et al. (1990a) found that such tissue and cell morphotypes could largely be identified and selected for on semisolid medium prior to the initiation of suspension culture. Upon repeated selective subculture (5 to 8 months), they identified two types of "aged" embryogenic callus: type C, compact and nodular, which was composed of embryogenic protuberances surrounded mostly by densely cytoplasmic (embryogenic) cells, and type D, soft and friable, which was composed of groups of densely cytoplasmic cells interspersed with nonembryogenic cells (Redway et al. 1990a,b). Use of either type resulted in suspensions that, upon repeated subculture, without selection, proliferated and disassociated into fine embryogenic suspensions within 2 to 3 months. Plants were regenerated on semisolid media only from those suspensions that were initiated from type C callus (Redway et al. 1990a). The identification of this type of embryogenic callus has played an important role in achieving stable transgenic wheat plants (Vasil et al. 1992).

Calli similar to that described by Redway et al. (1990a) were derived after repeated selection and subculture of immature, embryo-derived calli (Li et al. 1992b; Qiao et al. 1992) and from immature, inflorescence-derived calli (Li et al. 1992a). In each case highly regenerable embryogenic suspensions were obtained. Embryogenic calli derived by Li et al. (1992a,b) appeared after about 8 months of selection and were described as small granular calli composed of granules measuring 0.2 to 1 mm in size. This starting material produced highly regenerable suspensions of finely disassociated embryogenic cells and cell clusters within 3 to 6 weeks.

2.4 Embryogenic Protoplast Cultures

Plants have been regenerated from wheat protoplasts isolated from embryogenic anther-derived suspension cultures (Harris et al. 1988), callus obtained from immature embryos (Hayashi and Shimamoto 1988; Guo et al. 1990), callus obtained from young inflorescences (Ren et al. 1989), cell suspensions obtained from mature embryo callus (Wang et al. 1989), cell suspensions obtained from immature embryo callus (Vasil et al. 1990; Chang et al. 1991; Yang et al. 1991; He et al. 1992; Li et al. 1992b; Qiao et al. 1992), and cell suspensions obtained from young Inflorescence callus (Li et al. 1992a). In most cases, the regenerated plants were reported as being albino or sterile, being highly aneuploid, or having died prior to reaching maturity. While no comment on fertility was made, regenerated plants with a normal chromosome number, $2n = 42$, were reported by He et al. (1992).

Rather remarkable percentages of plated protoplasts have been reported to have formed somatic embryos directly or to have formed cell colonies that later produce somatic embryos. Li et al. (1992a) induced the direct formation of distinguishable somatic embryos from 0.06 to 0.1% of plated protoplasts (20% of the total number of protoplast-derived embryos). He et al. (1992) induced colonies from 0.1 to 5% of plated protoplasts and Qiao et al. (1992), using conditioned medium (from embryogenic cell suspensions), induced colonies from 2 to 8% of plated protoplasts. In the latter study, 4.8 to 45% of the colonies produced plants. Calli from these colonies were used to reestablish cell suspensions from which plant-competent calli formed upon plating on regeneration medium. The frequency of protoplasts ultimately giving rise to plants in this study often exceeded 0.1%.

Several factors have led to improved regeneration from protoplasts. Starting materials consisting of highly embryogenic suspensions of finely disassociated, dispersed cells and small cell clusters are essential for high yields of regenerable protoplasts. He et al. (1992) found that the optimal size of embryogenic cell clusters for protoplast release was between 185 to 750 μm.

The growth phase of suspensions is also critical to protoplast isolation. Conflicting stages have been reported. Qiao et al. (1992) obtained superior results (high frequencies of protoplast division and subsequent colony formation) when 7- to 8-day old suspensions, which had reached the lag phase, were used. In contrast, superior results were obtained by Li et al. (1992a,b) when 2-day-old suspensions were used. This discrepancy may be due to the use of conditioned media by Qiao et al. (1992). Fresh media were used by Li et al. (1992a,b).

He et al. (1992) observed a decline in regeneration competence of protoplast-derived cell colonies over time. Seventeen percent of 40-day-old colonies produced plants when transferred to regeneration medium, while only 0.5% of 54-day-old colonies produced plants.

2.5 Maturing and Germinating Somatic Embryos

In each of the cases reviewed above, frequencies of regeneration from distinguishable somatic embryos were extremely low. This might be due to an inability to provide somatic embryos with environments conducive to normal development, maturation, and germination.

The occurrence of well-formed mature somatic embryos in wheat tissue cultures, complete with coleoptile, coleorhiza, epiblast, and a normal scutellum, is extremely rare, as evidenced by the few published photomicrographs thereof (Magnusson and Bornman 1985; Carman et al. 1988). Out of tens of thousands of wheat somatic embryos observed by the authors, only a few have had a reasonably normal, mature morphology. In contrast, the development of well-formed early stage globular somatic embryos is common, especially in suspension cultures (Harris et al. 1988; Redway et al. 1990a; Li et al. 1992a). Common developmental abnormalities of wheat somatic embryos as they begin to mature beyond the globular/ovoid stage include; (1) failure to develop a normal coleorhiza or coleoptile; (2) fusion with other somatic embryos; (3) development of grossly abnormal scutella; (4) precocious germination prior to complete histodifferentiation; and (5) formation of secondary embryos and embryogenic callus. Apparently, normal development beyond the globular stage requires conditions that are not being met by current tissue culture procedures.

Germination of somatic embryos is usually induced on hormone-free culture medium or medium that contains low levels of an auxin or low levels of both an auxin and a cytokinin. Under these conditions, those somatic embryos developmentally capable of germinating do so without maturing. Unlike mature zygotic embryos, which germinate vigorously into sizeable plants within several days, germinating somatic embryos generally lack vigor and may require many days to develop into healthy, rapidly growing plants.

Germination frequencies of somatic embryos have been increased by a procedure where embryogenic callus is incubated in low O_2 tensions on medium containing low levels of abscisic acid (ABA). Somatic embryos embedded in calli are then matured in low O_2 tensions on medium containing high levels of ABA and indole-3-acetic acid (IAA), which prevents precocious germination and permits continued enlargement and development of the somatic embryos. Embryo-containing callus is then desiccated. Vigorous germination (up to 44% of recognizable somatic embryos) occurred synchronously within 3 days of incubation on hormone-free medium under normal O_2 tensions (Carman 1988).

It has been hypothesized that developmentally normal somatic embryos will form from embryogenic cells if they are exposed to temporally dynamic simulations in vitro of the physical, nutritional, and hormonal conditions present in ovules during zygotic embryogenesis (Carman 1989, 1990). Published studies concerning the in ovulo environments in which wheat embryos develop are sketchy at best. We have characterized dissolved oxygen (dO_2) tensions, nutrients, hormones, pH, and osmolarity in the in ovulo apoplastic solution that nourishes developing wheat embryos.

Diurnal fluxes in in ovulo dO_2 were measured in situ at 50- to 100-μm intervals through greenhouse-grown wheat kernels using an O_2 microelectrode. Levels of dO_2 decreased nearly linearly with the depth of microelectrode penetration in kernels. At midday at 18 DPA, dO_2, levels at the embryo surface were about 75% of saturation. These levels dropped to 60% of saturation at night (after 8 h in the dark). Oxygen was nearly undetectable (< 1% saturation) in the center of kernel, both day and night (Carman et al. 1993).

We excised numerous wheat caryopses at specific time intervals from anthesis through embryo development to characterize pH, osmolarity, minerals, free amino acids, hormones, and sugars in the ovular apoplastic solution. Caryopses were gently ruptured, microcentrifuged, and the supernatant collected for analyses. The pH in this solution increased slightly from 6.5 at 3 DPA to 6.8 at 20 DPA and the osmolarity declined from 820 mM/kg at 0 DPA to 400 mM/kg at 12 DPA and then rose to 560 mM/kg at 20 DPA. In ovulo levels of nitrogen were nearly equivalent to those of MS medium; however, 85% was in the form of free amino acids (Carman et al. 1993). Concentrations of apoplastic in ovulo cytokinins, ABA, and IAA were elevated at different times from 0 to 6 DPA. Thereafter cytokinin levels declined while ABA and IAA levels increased (Hess 1992).

The development of cultured, 12-DPA zygotic embryos in media that simulated ovular levels of dO_2 and ABA approached normalcy in that precocious germination was prevented. Endogenous ABA levels were elevated after a 6- to 12-day exposure of 12-DPA embryos to elevated levels of ABA. Such levels did not prevent precocious germination when embryos were cultured in atmospheric (supraoxic) levels of O_2. Embryos did not germinate when exposed to low (normoxic) O_2 tensions (Hess and Carman 1993). Young wheat embryos cultured in media that simulates ovular fluids closely resembled embryos sequentially cultured in extracted ovular apoplastic sap of the appropriate DPA (Carman, unpubl. data). Studies are currently underway to evaluate the usefulness of simulating ovular environments with respect to maturing and germinating somatic embryos from callus and suspension cultures.

3 Summary and Conclusions

Embryogenic callus can be induced from a variety of explants, including immature embryos and inflorescences, mature embryos, shoot tips, and intercalary meristems of young leaves. Induction frequencies depend on explant source, the physiological condition of the explant, and genotype. Genotypic variation in the rate of conversion from embryogenic to root-forming or other nonregenerative tissues necessitates careful subculture procedures for some lines. As our understanding of the physiology of competence increases, we should be able to utilize economically important wheat cultivars, regardless of genotype, in biotechnologies involving suspension and protoplast culture.

Major procedural improvements have been made recently with respect to producing highly embryogenic suspension and protoplast cultures. These include the development of subculture procedures for callus or suspension cultures that selectively enrich the quantity of disassociated granular embryogenic callus. Production of this tissue morphotype generally requires 5 to 8 months of repeated subculturing. Protoplasts isolated from such suspensions divide at very high frequencies to produce somatic embryos either directly or from embryogenic cell colonies. The identification of this type of embryogenic callus has played an important role in achieving stable transgenic wheat plants.

Somatic embryos of wheat are seldom physiologically, biochemically, or even morphologically normal, and they frequently become nongerminative or even nonregenerative. Somatic embryos that germinate often do so precociously. Careful control of culture nutrition, O_2 tensions, and exogenous growth regulator concentrations has moderated these abnormalities, yet much improvement remains to be made. Future improvements in normalizing somatic embryo maturation should occur by simulating in vitro those factors that permit such development in ovulo.

4 Protocol

Kernels are excised from wheat spikes collected 10 to 14 DPA, surface sterilized in 1% sodium hypochlorite for 12 min, and rinsed three times in sterile distilled water. Excised embryos 0.8 to 1.5 mm long are cultured with the embryo axis pressed into MS medium containing double the concentration of mineral salts, 20 g/l sucrose, 200 mg/l casein hydrolysate, 5 mmol/l 2,4-dichlorophenoxyacetic acid, and 1% agar, pH 5.8. Embryogenic callus, which appears within 14 to 28 days is selectively subcultured to the same medium every 3 to 4 weeks. Regeneration from somatic embryos occurs upon transfer of embryogenic callus to MS medium without hormones. Alternatively, callus may be incubated in low O_2 tensions (8%) and low temperatures (16 °C) on MS medium containing 5 mmol/l ABA and 5 mmol/l IAA for 2 weeks. The callus may then be desiccated to 12% moisture (slowly at 40 to 60 % RH) or transferred directly to hormone-free regeneration medium. Rooted plantlets 5 cm long are established directly in soil and grown in a mist tent for the first 3 days.

Acknowledgments. Research supported in part by NASA-Ames Coop. agreement no. NCC2-139; Nat. Res. Initiative, Competitive Grants Program, USDA, award no. 91-37300-6457; Utah Dept. of community and Econ. Dev., contract no. 92-0768; and the Utah Agric. Exp. Stn., Utah State University, Logan, Utah 84322-4810. Approved as technical paper no. 4352.

References

Ahloowalia BS (1982) Plant regeneration from callus culture in wheat. Crop Sci 22: 405–410

Barcelo P, Lazzeri PA, Martin A, Lörz H (1992) Competence of cereal leaf cells. II. Influence of auxin, ammonium and explant age on regeneration. J Plant Physiol 139: 448–454

Carman JG (1988) Improved somatic embryogenesis in wheat by partial simulation of the in-ovulo oxygen, growth-regulator and desiccation environments. Planta 175: 417–424

Carman JG (1989) The in ovulo environment and its relevance to cloning wheat via somatic embryogenesis. In Vitro 25: 1155–1162

Carman JG (1990) Embryogenic cells in plant tissue cultures: occurrence and behavior. In Vitro 26: 746–753

Carman JG, Hess JR, Bishop D, Hole DJ (1993) In ovulo environments and embryo dormancy in wheat. In: Walker-Simmons MK, Ricd JL (eds) Pre-harvest sprouting in cereals 1992. Am Assoc Cereal Chemists, St Paul, Minnesota, pp 163–170

Carman JG, Campbell WF (1990) Factors affecting somatic embryogenesis in wheat. In: Bajaj YPS (ed) Biotechnology in agriculture and forestry, vol 13, Wheat. Springer, Berlin Heidelberg New York, pp 68–87

Carman JG, Jefferson NE, Campbell WF (1988) Induction of embryogenic *Triticum aestivum* L. calli. I. Quantification of genotype and culture medium effects. Plant Cell Tissue Organ Cult 12: 83–95

Chang YF, Wang WC, Warfield CY, Nguyen HT, Wong JR (1991) Plant regeneration from protoplasts isolated from long-term cell cultures of wheat (*Triticum aestivum* L.). Plant Cell Rep 9: 611–614

Galiba G, Kovacs G, Sutka J (1986) Substitution analysis of plant regeneration from callus culture in wheat. Plant Breed 97: 261–263

Guo GQ, Xia GM, Li ZY, Chen HM (1990) Direct somatic embryogenesis and plant regeneration from protoplast-derived cells of wheat (*Triticum aestivum* L.). Sci Sin 9: 970–974

Harris R, Wright M, Byrne M, Varnum J, Brightwell B, Schubert K (1988) Callus formation and plantlet regeneration from protoplasts derived from suspension cultures of wheat (*Triticum aestivum* L.). Plant Cell Rep 7: 337–340

Hashim ZH, Campbell WF, Carman JG (1990) Morphological analysis of spring wheat (CIMMYT cv PCYT-10) somaclones. Plant Cell Tissue Organ Cult 20: 95–99

Hayashi Y, Shimamoto K (1988) Wheat protoplast culture: embryogenic colony formation from protoplasts. Plant Cell Rep 7: 414–417

He DG, Yang YM, Scott KJ (1989) The effect of macroelements in the induction of embryogenic callus from immature embryos of wheat (*Triticum aestivum* L.). Plant Sci 64: 251–258

He DG, Yang YM, Scott KJ (1991) Zinc deficiency and the formation of white structures in immature embryo cultures of wheat (*Triticum aestivum* L.). Plant Cell Tissue Organ Cult 24: 9–12

He DG, Yang YM, Scott KJ (1992) Plant regeneration from protoplasts of wheat (*Triticum aestivum* cv. Hartog). Plant Cell Rep 11: 16–19

Hess JR (1992) Wheat kernal hormone levels during development and their relevance to zygotic and somatic embryogenesis. PhD Diss, Utah State University, Logan, UT

Hess JR, Carman JG (1993) Normalizing development of cultured *Triticum aestivum* L. Embryos. I. Low oxygen tensions and exogenous ABA. J Exp Bot 44: 1067–1073

Kaleikau EK, Sears RG, Gill BS (1989a) Control of tissue culture response in wheat (*Triticum aestivum* L.). Theor Appl Genet 78: 783–787

Kaleikau EK, Sears RG, Gill BS (1989b) Monosomic analysis of tissue culture response in wheat (*Triticum aestivum* L.). Theor Appl Genet 78: 625–632

Larkin PJ, Spindler LH, Banks PM (1990) The use of cell culture to restructure plant genomes for introgressive breeding. In: Kimber G (ed) Proc 2nd Int Symp Chromosome engineering in plants. University of Missouri Press, Columbia, pp 80–89

Lazar MD, Chen THH, Scoles GJ, Kartha KK (1987) Immature embryo and anther culture of chromosome addition lines of rye in Chinese Spring wheat. Plant Sci 51: 77–81

Li ZY, Xia GM, Chen HM (1992a) Somatic embryogenesis and plant regeneration from protoplasts isolated from embryogenic cell suspensions of wheat (*Triticum aestivum* L.). Plant Cell Tissue Organ Cult 28: 79–85

Li ZY, Xia GM, Chen HM, Guo GQ (1992b) Plant regeneration from protoplasts derived from embryogenic suspension cultures of wheat (*Triticum aestivum* L.). J Plant Physiol 139: 714–718

Liu H, Misoo S, Kamijima O, Sawano M (1992) Effects of casein hydrolysate on the response in immature embryo culture of wheat (*Triticum aestivum* L). Jpn J Breed 42: 367–373

Magnusson I, Bornman CH (1985) Anatomical observations on somatic embryogenesis from scutellar tissues of immature zygotic embryos of *Triticum aestivum*. Physiol Plant 63: 137–145

Mohmand AS, Nabors MW (1991) Comparison of two methods for callus culture and plant regeneration in wheat (*Triticum aestivum*). Plant Cell Tissue Organ Cult 26: 185–187

Murashige T, Skoog F (1962) A revised medium for rapid growth and bioassays with tobacco tissue cultures. Physiol Plant 15: 473–497

Ozias-Akins P, Vasil IK (1982) Plant regeneration from cultured immature embryos and inflorescences of *Triticum aestivum* L. (Wheat): evidence for somatic embryogenesis. Protoplasma 110: 95–105

Ozias-Akins P, Vasil IK (1983) Improved efficiency and normalization of somatic embryogenesis in *Triticum aestivum* (wheat). Protoplasma 117: 40–44

Qiao YM, Cattaneo M, Locatelli F, Lupotto E (1992) Plant regeneration from long term suspension culture-derived protoplasts of hexaploid wheat (*Triticum aestivum* L.). Plant Cell Rep 11: 262–265

Rajyalakshmi K, Grover A, Maheshwari N, Tyagi AK, Maheshwari SC (1991) High frequency regeneration of plantlets from the leaf-bases via somatic embryogenesis and comparison of polypeptide profiles from morphogenic and non-morphogenic calli in wheat (*Triticum aestivum*). Physiol Plant 82: 617–623

Redway FA, Vasil V, Vasil IK (1990a) Characterization and regeneration of wheat (*Triticum aestivum* L.) embryogenic cell suspension cultures. Plant Cell Rep 8: 714–717

Redway FA, Vasil V, Lu D, Vasil IK (1990b) Identification of callus types for long-term maintenance and regeneration from commercial cultivars of wheat (*Triticum aestivum* L.). Theor Appl Genet 79: 609–617

Ren JG, Jia JF, Li MY, Zhen GC (1989) Plant regeneration from protoplasts isolated from callus initiated from immature inflorescences of wheat (*Triticum aestivum* L.). Sci Sin B 9: 693–695

Scott KJ, He DG, Yang YM (1990) Somatic embryogenesis in wheat. In: Bajaj YPS (ed) Biotechnology in agriculture and forestry, vol 13, Wheat. Springer, Berlin Heidelberg New York, pp 46–67

Tabaeizadeh Z, Campeau N (1992) Embryogenic cell suspensions of *Triticum aestivum* x *Leymus angustus* F_1 hybrids: characterization and plant regeneration. Plant Cell Rep 11: 81–85

Vasil V, Redway F, Vasil IK (1990) Regeneration of plants from embryogenic suspension culture protoplasts of wheat (*Triticum aestivum* L.). Bio/Technology 8: 429–434

Vasil V, Castillo A, Fromm M, Vasil IK (1992) Herbicide resistant fertile transgenic wheat plants obtained by microprojectile bombardment of regenerable embryogenic callus. Bio/Technology 10: 667–674

Wang HB, Li XH, Sun YY, Chen J, Zhu Z, Fang R, Wang P (1989) High frequency of callus formation and plant regeneration from protoplasts of wheat (*Triticum aestivum* L.). Sci Sin 8: 828–834

Wang WC, Nguyen HT (1990) A novel approach for efficient plant regeneration from long-term suspension culture of wheat. Plant Cell Rep 8: 639–642

Weeks JT, Anderson OD, Blechl AE (1993) Rapid production of multiple independent lines of fertile transgenic wheat (*Triticum aestivum*). Plant Physiol 102: 1077–1088

Wernicke W, Milkovits L (1986) The regeneration potential of wheat shoot meristems in the presence and absence of 2,4-dichlorophenoxyacetic acid. Protoplasma 131: 131–141

Yang YM, He DG, Scott KJ (1991) Establishment of embryogenic suspension cultures of wheat by continuous callus selection. Aust J Plant Physiol 18: 445–452

Young E, Allen E, Evans S (1990) The wheat program in the 1990s: issues for decision makers. USDA Econ Res Ser, Agric Inf Bull 606: 1–16

I.2 Somatic Embryogenesis in Rice (*Oryza sativa* L.)

T.J. Jones[1] and T.L. Rost[2]

1 Introduction

Cultivated rice belongs to the genus *Oryza*, in the grass family Poaceae (Gramineae). There are 22 species within the genus, 20 wild species and 2 cultivated species, *O. glabberima* (African rice) and *O. sativa* (common, or Asian rice). *Oryza sativa* is commonly divided further into two subspecies, *indica* (or tropical rice) and *japonica* (or temperate rice) with distinct ranges of eco-geographic distribution.

Rice is arguably the most important food crop in the world, accounting for 20% of the calories and 13% of the protein of the human diet (Chang 1985). In Asia, where 90% of the world's rice is consumed, rice represents roughly 50% of all food consumption (Juliano 1985). During the 1960s, improvements in the agronomics and yield of rice, due in large part to the breeding efforts of the International Rice Research Institute (IRRI) in Manila, resulted in the release of several high-yielding, enhanced rice varieties spawning what became known as the "Green Revolution". Nevertheless, it is estimated that rice production must increase approximately 65% within the next 30 years to keep pace with estimated population growth (Bajaj 1991). Obviously, further improvements, both agronomic and nutritional, must be made in both *japonica* and *indica* varieties, utilizing conventional breeding techniques and biotechnology (now possible using embryogenic cultures) to exploit new sources of variation. In this chapter literature on rice somatic embryogenesis is reviewed, and our work on the induction of somatic embryogenesis through the culture of mature rice embryos is summarized.

2 Somatic Embryogenesis

The first documented reports of rice somatic embryogenesis were from cultures initiated from immature rice leaves (Bhattacharya and Sen 1980; Wernicke et al. 1981). Since that time there have been numerous accounts of somatic

[1] Agricultural Biotechnology, Dupont Company, Experimental Station 402/2250, Wilmington, Delaware 19880-0402, USA
[2] Section of Botany, University of California, Davis, California 95616-8537, USA

Biotechnology in Agriculture and Forestry, Vol. 31
Somatic Embryogenesis and Synthetic Seed II (ed. by Y.P.S. Bajaj)

Table 1. Somatic embryogenesis from different source tissues

Explant	Reference
Mature embryo	Heyser et al. (1983; Siriwardana and Nabors (1983); Jones and Rost (1989)
Immature embryo	Lai and Liu (1982); Heyser et al. (1983)
Roots	Abe and Futsuhara (1985); Sticklen (1991)
Leaves	Bhattacharya and Sen (1980); Wernicke et al. (1981)
Inflorescences	Ling et al. (1983); Chen et al. (1985)
Anthers	Ling et al. (1983); Chen et al. (1985)
Microspores	Genovesi and Magill (1982)
Protoplasts	Fujimura et al. (1985); Abdullah et al. (1986) Yamada et al. (1986); Lee et al. (1989)

embryogenesis from virtually every conceivable source tissue (Table 1). Regardless of the source tissue, somatic embryo production was greatly dependent upon the genotype of the donor plant, and Peng et al. (1992) concluded that the regeneration capacity in rice is under the control of several nuclear and cytoplasmic genes.

There is no clear consensus among researchers as to the media salt formulation for optimum somatic embryogenic culture production. MS (Murashige and Skoog 1962), LS (Linsmaier and Skoog 1965), and N6 (Chu et al. 1975) media are utilized with nearly equal regularity. There is, however, a general consensus that 2,4-D is the preferred, and usually, only plant growth regulator needed, typically in the range of 0.5 to 3 mg/1. Once somatic embryos have been produced, plant regeneration is commonly achieved by transfer to medium lacking plant growth regulators.

The development of embryogenic suspension cultures and subsequent achievement of plant regeneration from suspension culture-derived protoplasts (Abdullah et al. 1986; Yamada et al. 1986; Lee et al. 1989) has led directly to the recovery of transgenic rice plants from both *japonica* and *indica* varieties (Toriyama et al. 1988; Zhang et al. 1988; Shimamoto et al. 1989). More recently transgenic rice plants have been obtained using microprojectile bombardment to introduce DNA into suspension culture cells (Cao et al. 1992) or directly into rice scutellar tissue (Christou et al. 1991) with subsequent production of somatic embryos. The latter technique exploits the direct development of somatic embryos from scutellar epithelial cells (Jones and Rost 1989). This particular transformation technique extends the range of genotypes amenable to transformation beyond those that can be regenerated from protoplasts.

2.1 Tissue Culture of Mature Embryos

We have characterized tissue cultures derived from mature rice embryos that develop somatic embryos directly from single scutellum epithelium cells (Jones and Rost 1989). Although this response appears to be broadly applicable to most rice genotypes, L-202 was particularly responsive.

Mature zygotic rice embryos were plated on agar-solidified MS medium supplemented with 2 mg/1 2,4-D and 3% sucrose. Within 7–10 days, somatic embryos could be observed developing from the surface of the scutellum epithelium (Fig. 1A). Most of the somatic embryos arose from epithelial cells from the basal (root) half of the scutellum, often clustered around the longitudinal midline of the scutellum. After 15 days, scutellar/coleoptilar-stage somatic embryos could be detected (Fig. 1B), and morphologically mature somatic embryos were present 28 days after culture initiation. The embryogenic capacity of the cultures could be maintained for months by transferring embryogenic callus to a medium with a reduced 2,4-D concentration of 1 mg/l.

2.2 Anatomy and Ultrastructure of Somatic Embryo Initiation

The ontogeny of the somatic embryos arising from mature rice scutella was examined at a light and electron microscope level. Light microscopy was performed on sectioned, plastic-embedded tissue fixed at intervals after culture initiation. We were able to determine that the somatic embryos arose directly from individual scutellum epithelium cells (Jones and Rost 1989). Electron microscopy was subsequently performed on material fixed at 12-h intervals spanning the time of the first embryogenic division of the epithelium cells. Rice scutella from embryos plated on the same medium but lacking 2,4-D were also fixed and observed by transmission electron microscopy. By comparing the ultrastructure of the epithelial cells from the two treatments, we could distinguish between changes due to somatic embryo induction from those due to rehydration and the onset of germination.

The scutellum epithelium of a mature zygotic embryo forms a unicellular layer of columnar cells abutting the endosperm (Fig. 2A). The epithelial cells are densely cytoplasmic with a prominent nucleus and numerous peripherally located lipid bodies and mitochondria (Fig. 3A). Amyloplasts and proplastids are also present but protein bodies are few. The initial step in somatic embryo formation from induced epithelial cells is a transverse division (Fig. 2B), usually preceded by a change in conformation of the nucleus from spherical to highly lobed. The resultant terminal cell becomes the embryo proper while the basal cell gives rise to a suspensor-like structure. This division is accompanied by an increase in the number of free ribosomes and a proliferation of rough endoplasmic reticulum (RER) and Golgi bodies. However, unlike the epithelial cells on germination medium (medium lacking 2,4-D), where the RER is present in compact, parallel arrays and long profiles (Fig. 3B), in the embryogenic epithelial cells it is swollen and distended, as are the nuclear envelope and the Golgi body membranes (Fig. 3C). This distention of the endomembrane system seems to be a transient event and is no longer obvious 96 h after plating. Interestingly, the embryogenic epithelial cells do not immediately catalyze their lipid bodies nor do they produce starch grains within their plastids, two features common to epithelial cells undergoing germination. This also appears to be a transient feature as lipid catabolism and starch accumulation are evident 96 h after initiation.

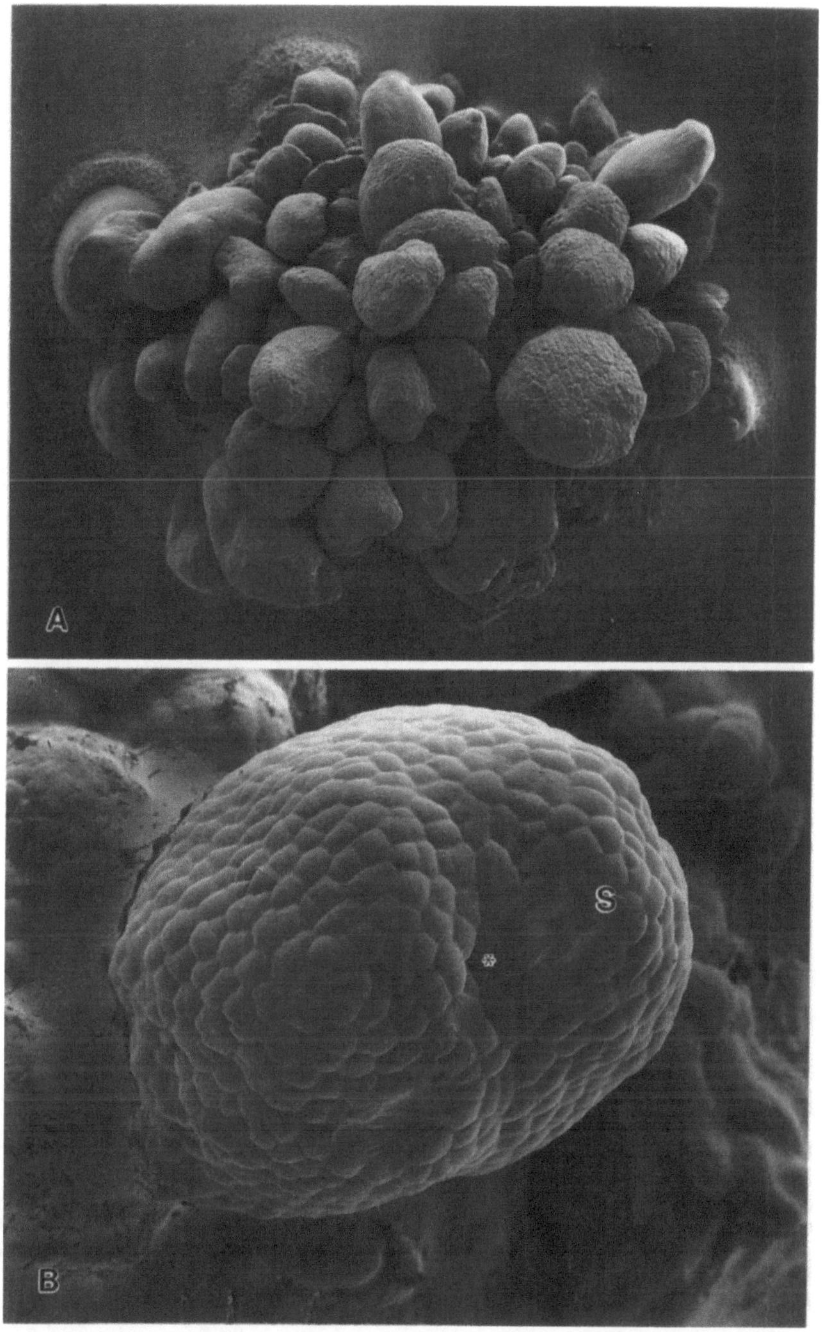

Fig. 1A,B. Scanning electron micrographs of rice somatic embryos. **A** Dense cluster of nascent somatic embryos originating from the scutellum epithelium within 7–10 days after initiation. **B** Somatic embryo 15 days after initiation with developing scutellum (*S*). Shoot apex will differentiate in the lateral notch (*)

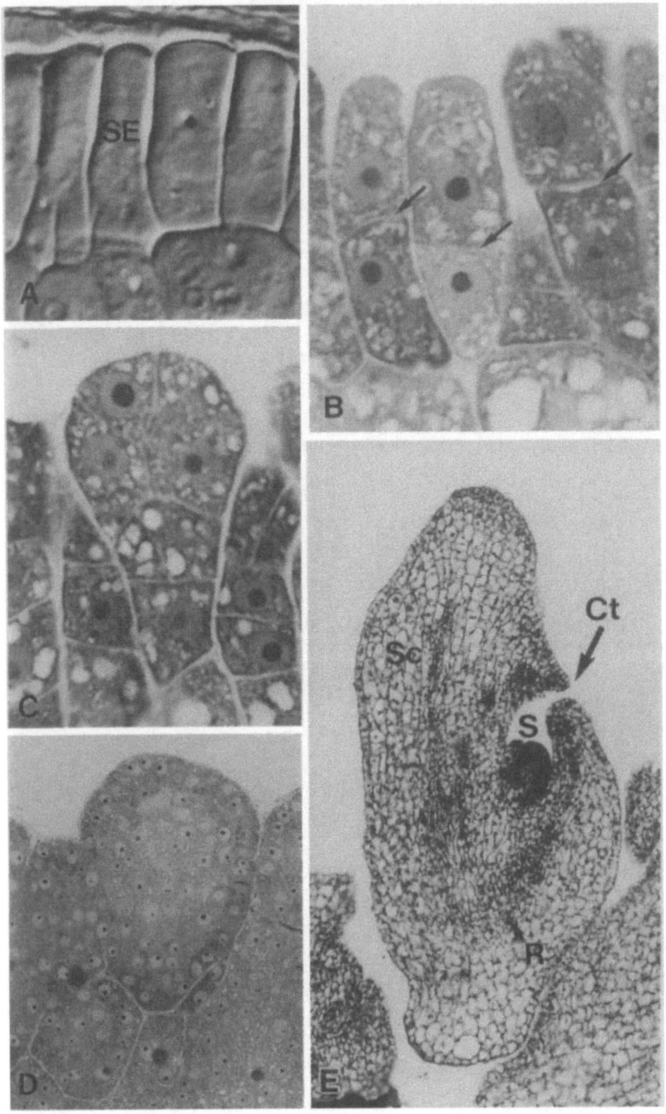

Fig. 2A–E. Development of a rice somatic embryo. **A** Differential interference contrast (DIC) micrograph of rice scutellum epithelium (*SE*) from a mature zygotic embryo. **B** Scutellum epithelium cells 72 h after culture initiation. The epithelial cells have divided transversely (*arrows*). One of the terminal, proembryonal cells is undergoing a longitudinal division. **C** Small globular somatic embryo. **D** Somatic embryo at the time of scutellum initiation. Protoderm has differentiated. **E** Nearly mature somatic embryo with a coleoptile (*Ct*), scutellum (*Sc*) and shoot (*S*) meristem. Radicle (*R*) meristem is beginning to develop

Fig. 3A–C. Ultrastructure of somatic embryo initiation. **A** Ultrastructure of a scutellum epithelium at the time of plating. The cell contains numerous mitochondria (*M*), amyloplasts (*A*), and lipid bodies (*L*); (*Nu*) nucleus. **B** Detail of epithelial cell after 60 h on germination medium. Endoplasmic reticulum

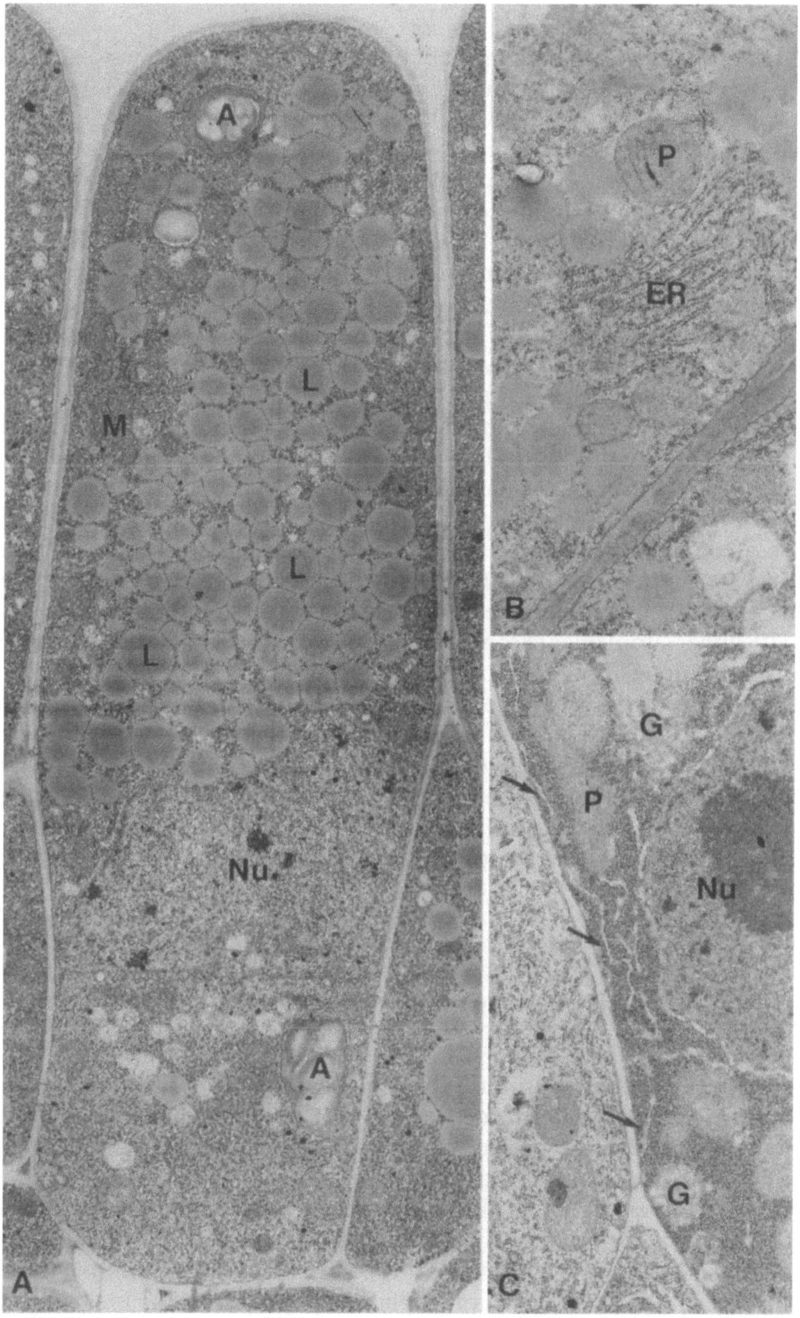

(*ER*) is found in parallel stacked arrays; *P* plastid. **C** Detail of terminal, proembryonal cell following division of the epithelium after 60 h on embryo induction medium. The nuclear envelope, ER (*arrows*), and Golgi body membranes (*G*) are distended and the cytoplasm is dense with numerous ribosomes; *Nu* nucleus; *P* plastid

2.3 Somatic Embryo Maturation

Somatic embryogenesis continues by a series of random divisions of the terminal cell to produce a small globular embryo (Fig. 2C). After the globular embryo has produced approximately 32 cells, a protoderm differentiates. As the embryo continues to grow, it becomes flattened at the apical end at the onset of scutellum development (Fig. 2D). The coleoptile and shoot apical meristem differentiate laterally (Fig. 2E) followed shortly by the differentiation of the root apical meristem, coleorhiza and, typically, several leaf primordia. Somatic embryos mature within 28 days of culture initiation and can be readily germinated upon transfer to medium lacking 2,4-D.

While fully capable of germination, the somatic embryos that mature on medium supplemented only with 2,4-D are lacking many of the storage proteins and lipids that accumulate during zygotic embryogenesis (Jones 1988). More biochemically complete somatic embryos can be obtained by supplementing the medium during the maturation phase of development with either abscisic acid or increased osmoticum. For example, rice lectin (a protein that accumulates during late zygotic embryogenesis) accumulation was examined in somatic embryos isolated from cultures matured on basal medium (2,4-D only) or basal medium supplemented with ABA (10 μM), or additional sucrose or mannitol (0.29 M). A nearly 100% increase in the lectin content of the somatic embryos was achieved with either supplement as measured by rocket immunoelectrophoresis (Fig. 4A), although none approached the 0.44 μg lectin/zygotic embryo as measured by the same assay or the 0.12 μg lectin/embryo reported in the literature (Stinissen and Peumans 1983). The lectin that accumulated in the somatic embryos was properly targeted to vacuoles (Fig. 4B) as it is in zygotic embryos (Stinissen et al. 1984; Jones 1988). Obviously, by manipulating the medium during somatic embryo development to more accurately represent the environment within which the zygotic embryo develops, it should be possible to obtain somatic embryos nearly biochemically identical to zygotic embryos.

3 Summary and Conclusions

Somatic embryogenesis in rice has been demonstrated in cultures derived from diverse explant tissues using a variety of different media. The description of scutellum epithelium-derived embryos in this chapter is a rare example of direct embryogenesis from single cells. The introduction of novel traits into well-adapted rice cultivars should soon be a reality as research extends the capability of producing embryogenic cultures to a broad range of agronomically important germplasm. The publication of protocols to regenerate plants from embryogenic suspension culture-derived protoplasts has made rice the first cereal crop to be amenable to routine genetic transformation. Somaclonal variants or mutants derived from directed in vitro selection of embryogenic cultures will also provide breeders with new genetic variability to exploit. Further, it should be possible to develop large-scale multiplication systems for elite germplasm, e.g., male sterile

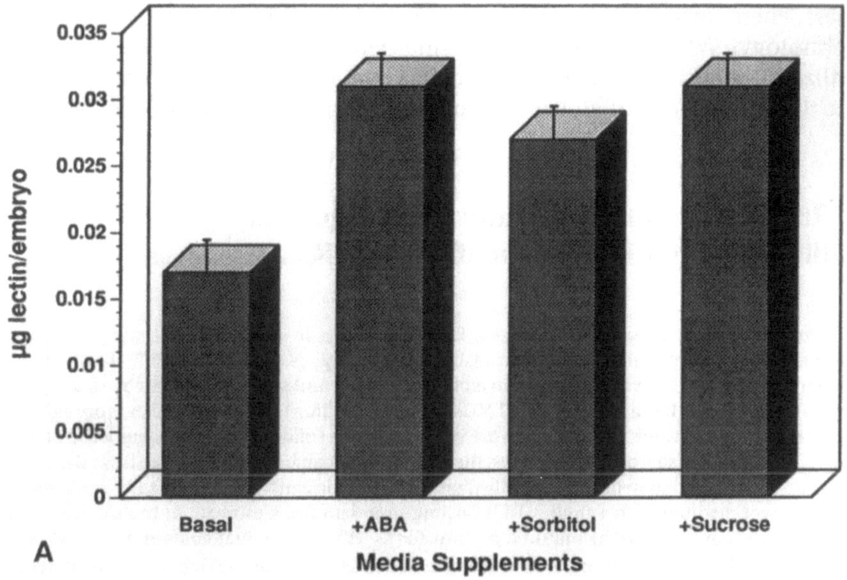

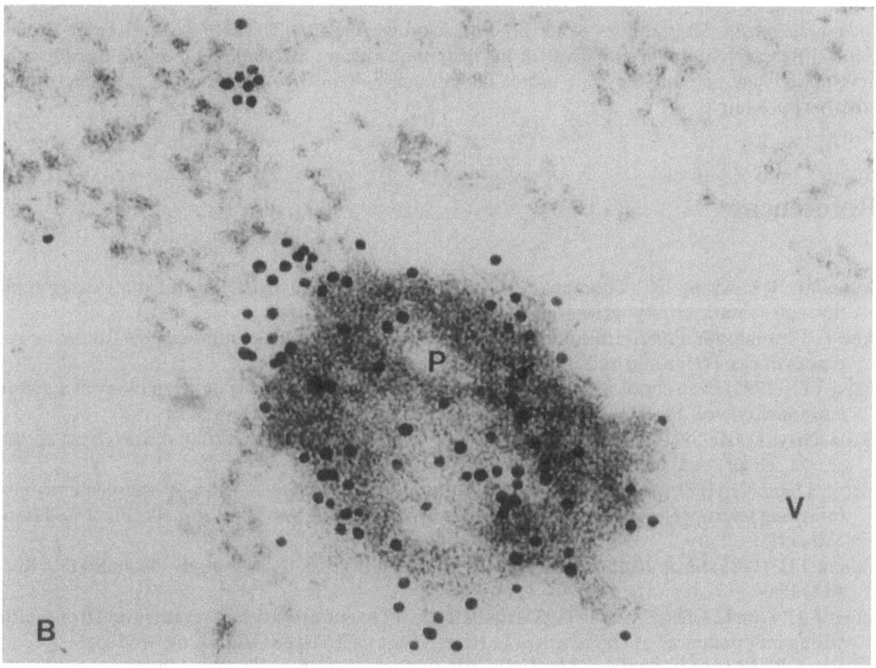

Fig. 4. A Lectin concentration in somatic embryos matured on a basal medium containing MS salts and 1 mg/l 2,4-D or supplemented with 10 µm ABA or 0.29 M sorbitol or mannitol. Rice lectin was extracted from somatic embryos and measured by rocket immunoelectrophoresis. Lectin content of zygotic embryos was measured to be 0.44 µg lectin/embryo. **B** Immunogold localization of rice lectin to a proteinaceous body (*P*) in the vacuole (*V*) of a somatic embryo matured in the presence of 10 µM ABA. Anti-rice lectin antibody was detected with a 15-nm gold-cojugated secondary antibody

lines, when embryogenic cultures are coupled with an efficient synthetic seed technology. While it will be some time before all the aforementioned can be realized, the advancements made in rice somatic embryogenesis have brought rice to the forefront of a new "green revolution" in crop improvement.

4 Protocol to Induce Somatic Embryogenesis in the Scutellum Epithelium of Mature Rice Embryos

Most of our work has been done using cv. L-202, however, it was found that many other cultivars respond to this protocol in a similar fashion, including cvs. M-101, S-202, Koshihikari, and Nipponbare. Mature caryopses are removed from the husk and surface sterilized by rinsing with 95% ethanol for 60 s and then agitating with 2.5% sodium hypochlorite for 5 min. After thoroughly rinsing with sterile distilled water, the caryopses are placed on agar-solidified medium composed of MS salts, 2 mg/l 2,4-D, and 3% sucrose. After 7 days, the rice embryos can be easily removed from the endosperm and the now enlarged scutella are carefully dissected from the embryos. The scutella are transferred to fresh MS salt medium with 1 mg/l 2,4-D for long-term culture. Cultures are transferred every 7–10 days. Somatic embryos can be matured by transfer to MS salt medium containing 10 μM ABA (2.6 mg/l) for 2 weeks and then germinated by transfer to MS salt medium lacking plant growth regulators.

Acknowledgment. The authors would like to acknowledge Dr. Rick Howard, Dupont Company Central Research and Development for his help in producing the scanning electron micrographs of developing somatic embryos. The use of his Hitachi S-4000 cryo-scanning electron microscope is greatly appreciated.

References

Abdullah R, Cocking EC, Thompson JA (1986) Efficient plant regeneration from rice protoplasts through somatic embryogenesis. Bio/Technology. 4: 1087–1090

Abe T, Futsuhara Y (1985) Efficient plant regeneration by somatic embryogenesis from root callus tissues of rice (*Oryza sativa* L.). J Plant Physiol 121: 111–118

Bajaj YPS (1991) Biotechnology in rice improvement. In: Bajaj YPS (ed) Biotechnology in agriculture and forestry, vol. 14. Rice. Springer, Berlin Heidelberg New York, pp 3–18

Battacharya P, Sen SK (1980) Potentiality of leaf sheath cells for regeneration of rice (*Oryza sativa* L.) plants. Theor Appl Genet 58: 87–90

Cao J, Duan X, McElroy D, Wu R (1992) Regeneration of herbicide resistant transgenic rice plants following microprojectile-mediated transformation of suspension culture cells. Plant Cell Rep 11: 586–591

Chang T-T (1985) Crop history and genetic conservation: rice – a case study. Iowa State J Res 59: 425–455

Chen T-H, Lam L, Chen S-C et al. (1985) Somatic embryogenesis and plant regeneration from cultured young inflorescences of *Oryza sativa* L. (rice). Plant Cell Tissue Organ Cult 4: 51–54

Christou P, Ford TL, Dofron M (1991) Production of transgenic rice (*Oryza sativa* L.) from agronomically important Indica and Japonica varieties via electric discharge particle acceleration of exogenous DNA into immature zygotic embryos. Bio/Technology 9: 957–962

Chu CC, Wang CC, Sun CS, Hsu C, Yin KC, Chu YC, Bi FY (1975) Establishment of an efficient medium for anther culture of rice through comparative experiments of the nitrogen sources. Sci Sin 18: 659–668

Fujimura T, Sakurai M, Akagi T, Negishi T, Hirosa A (1985) Regeneration of rice plants from protoplasts. Plant Tissue Cult Lett 2: 74–75

Genovesi AD, Magill CW (1982) Embryogenesis in callus derived from rice microspores. Plant Cell Rep 1: 257–260

Heyser JW, Dykes TA, DeMott KJ, Nabors MW (1983) High frequency, long term regeneration of rice from callus culture. Plant Sci Lett 29: 175–182

Jones TJ (1988) A comparative structural and immunocytochemical study of rice (*Oryza sativa* L.) zygotic and somatic embryogenesis. PhD Thesis, University of California, Davis

Jones TJ, Rost TL (1989) The developmental anatomy and ultrastructure of somatic embryos from rice (*Oryza sativa* L.) scutellum epithelial cells. Bot Gaz 150: 41–49

Juliano BO (1985) Rice. J Plant Foods 6: 129–145

Lai K-L, Liu L-F (1982) Induction and plant regeneration of callus from immature embryos of rice plants (*Oryza sativa* L.) Jpn J Crop Sci 51: 70–74

Lee L, Schroll RE, Grimes HD, Hodges TK (1989) Plant regeneration from indica rice (*Oryza sativa* L.) protoplasts. Planta 178: 325–333

Ling DH, Chen WY, Chen MF, Ma ZR (1983) Somatic embryogenesis and plant regeneration in an interspecific hybrid of *Oryza*. Plant Cell Rep 2: 169–171

Linsmaier E, Skoog F (1965) Organic growth factor requirements to tobacco tissue cultures. Physiol Plant 18: 100–127

Murashige T, Skoog F (1962) A revised medium for rapid growth and bioassays with tobacco tissue cultures. Physiol Plant 15: 473–497

Peng J, Kononowicz H, Hodges TK (1992) Transgenic indica rice plants. Theor Appl Genet 83: 855–863

Shimamoto K, Terada R, Izawa T, Fujimoto H (1989) Fertile transgenic rice plants regenerated from transformed protoplasts. Nature 338: 274–277

Siriwardana S, Nabors MW (1983) Tryptophan enhancement of somatic embryogenesis in rice. Plant Physiol 73: 142–146

Sticklen M (1991) Direct somatic embryogenesis and fertile plants from rice root cultures. J Plant Physiol 138: 577–580

Stinissen HM, Peumans WJ (1983) Lectin accumulation and synthesis in developing rice embryos. Plant Cell Rep 2: 277–280

Stinissen HM, Peumans WJ, Chrispeels MJ (1984) Subcellular site of lecin synthesis in developing rice embryos. EMBO J 3: 1979–1985

Toriyama K, Arimoto Y, Uchimiya H, Hinata K (1988) Transgenic rice plants after direct gene transfer into protoplasts. Bio/Technology 6: 1072–1074

Wernicke W, Brettell R, Wakizuka T, Potrykus I (1981) Adventitious embryoid and root formation from rice leaves. Z Pflanzenphysiol 103: 361–365

Yamada Y, Yang Z-Q, Tang D-T (1986) Plant regeneration from protoplast-derived callus of rice (*Oryza sativa* L.). Plant Cell Rep 5: 85–88

Zhang HM, Yang H, Rech EL, Golds TJ, Davis AS, Mulligan BJ, Cocking EC, Davey MR (1988) Transgenic rice plants produced by electroporation-mediated plasmid uptake into protoplasts. Plant Cell Rep 7: 379–384

Zhang Q, Saghai Maroof MA, Lu TY, Shen BZ (1992) Genetic diversity and differentiation of indica and japonica rice detected by RFLP analysis. Theor Appl Genet 83: 495–499

I.3 Somatic Embryogenesis in Maize (*Zea mays* L.)

A.M.C. Emons and H. Kieft[1]

1 Introduction

Maize is an herbaceous monocot with an annual cycle. Its embryo lies embedded in the endosperm at one side, toward the base of the caryopsis, with the primary root directed toward the attached end. The scutellum is large, lies directly against the endosperm, and partially encloses the embryonic axis. The primary root is enclosed by a coleorhiza. The suspensor persists during the maturation of the grain. Abbe and Stein (1954) have described eight stages of zygotic embryo development in maize by shoot apex development and this order of events has been confirmed in more detail by Van Lammeren (1986) in a scanning electron microscope study. The first stage is called the transition stage, in which a club-shaped embryo differentiates into a suspensor and the embryo proper. During the coleoptilar stage, the scutellum develops, a shoot apex becomes recognizable, and the almost circular coleoptilar primordium arises as a roll of meristematic tissue. The further six stages are distinguished by the number of leaf primordia. The scutellum becomes large and fleshy and initially has a groove from which the shoot meristem protrudes; the coleoptile covers more and more the shoot meristem and the developing leaf primordia; the first leaf primordium is situated opposite to the insertion of the scutellum and each subsequent leaf primordium forms opposite to the preceding one.

This introduction has primarily been written to better understand and compare the mode of development of zygotic embryos with that of somatic embryos of maize described in this chapter.

2 Somatic Embryogenesis

Somatic embryogenesis in maize has been reported by a number of workers (e.g. Armstrong and Green 1985; Vasil and Vasil 1986; Fransz and Schel 1991a,b; Emons and Kieft 1991).

Because plant regeneration via somatic embryogenesis starts with one or only a few cells, this type of regeneration is important for plant production and biotechnology, such as clonal propagation, multiplication of F1 hybrids, and especially genetic transformation (Gordon-Kamm et al. 1990).

[1] Department of Plant Cytology and Morphology, Wageningen. Agricultural University, Arboretumlaan, 4, 6703 BD Wageningen, The Netherlands

Biotechnology in Agriculture and Forestry, Vol. 31
Somatic Embryogenesis and Synthetic Seed II (ed. by Y.P.S. Bajaj)

2.1 Callus Types

Two types of embryogenic calli are distinguished in maize, type I and type II (Armstrong and Green 1985). Type II, also called friable callus, is the less differentiated type of regenerable callus. It consists of aggregates of undifferentiated cells, sometimes interspersed with vascular cells, and groups of small isodiametric meristematic cells on the outside of the aggregate (Emons and Kieft 1991), which are tightly packed, thin-walled, richly cytoplasmic, highly basophilic, and contain many small vacuoles, lipid droplets, and starch grains (Fransz and Schel 1991a, 1994). These groups of cells are comparable to the proembryogenic masses in carrot somatic embryogenesis and to the globular-stage zygotic embryo. After subculturing the calli for several years, these proembryogenic cells remain similar. In fact, they are the dividing cells giving rise to new embryogenic cells and to nondividing cells that will elongate. The nonembryogenic cells in the aggregate, i.e., the nondividing cells, have thicker cell walls, are enlarged, and have large vacuoles and few starch grains. In medium containing 2,4-D they form continuously from the embryogenic cells, but embryogenic cells remain present. High 2,4-D levels are not necessary to induce embryogenesis, but to preserve the undifferentiated nature of the culture. The yellowish, friable type-II embryogenic callus of maize can be propagated as such if maintained in or on medium containing a sufficient amount of 2,4-D. If the concentration of 2,4-D is decreased these globular-stage somatic embryos start

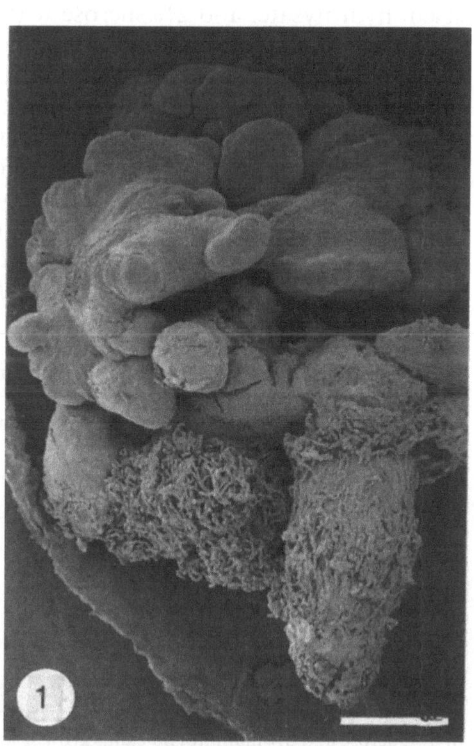

Fig. 1. Scanning electron micrograph showing rhizogenic callus on the lower side and developing somatic embryos on the upper side; *bar* = 500 μm. (Emons and de Does, Unpubl.)

differentiation and begin to mature, forming a scutellum. The white, opaque type-I embryogenic callus is no more than a further advanced differentiation step. The white opaque appearance stems from the starch containing scutellar-like structures. Type-I callus can be regenerated into plants, but cannot be propagated as such; it may produce friable callus. A third type of callus in monocots is the rhizogenic callus. Often this callus type is called nonembryogenic; however, the term rhizogenic is preferred. When the 2,4-D concentration is decreased, this callus forms only roots. The most important difference between embryogenic and rhizogenic callus is that the meristems in embryogenic calli are formed exogenously from a protodermal cell layer on the outside of the tissue, while in rhizogenic calli they are formed endogenously from a parenchymatous cell layer within the tissue (Emons et al. 1993b). Figure 1 shows a scanning electron micrograph with rhizogenic callus facing the agar medium and developing embryogenic somatic embryos on top of it. It is important to note that friable callus already contains the globular developmental stage of the zygotic embryo, as is the case in carrot proembryogenic masses.

2.2 Influence of Mannitol on the Maintenance of Embryogenesis

Table 1 shows the influence of mannitol in the growth medium on the embryogenic potential of callus. It compares calli of genotype 4C1 (Emons and Kieft 1991) grown on N6 medium containing 2 mg/1 2,4-D, 6 mM L-proline, 100 mg/l casein hydrolysate, and 2% sucrose with callus grown on medium of the same composition but with an additional 3% mannitol. Callus tissue of 4C1, proliferating on 2% sucrose, lost the ability to regenerate shoots after 6 months in culture. This ability was regained after growing on 2% sucrose with 3% mannitol. On 3% mannitol without sucrose the callus did not grow at all. Also, rice callus, which had lost the ability to regenerate shoots, regained this ability with a change in the osmolarity of the growth medium by the addition of metabolically inactive substances such as sorbitol and mannitol (Kishor and Reddy 1986). In the maize genotype 4C1 cultures the use of mannitol at the proliferation stage was crucial for later regeneration. It is not a mere effect of osmotic potential, because 6%

Table 1. Influence of mannitol in the proliferation medium on regeneration in hormone-free medium

Duration of experiment	Sucrose (%)	Mannitol (%)	Shoot/root ratio after regeneration
At start (embryogenic callus aggregates, 3–5 mm in diameter) N=1200	2	3	9:1
1 month	2	0	1:1
6 months	2	0	1:90
At start (rhizogenic callus aggregates, 3–5 mm in diameter) N=800	2	0	1:90
1 month	2	3	1:40
2 months	2	3	1:20

In 0% sucrose and 3% mannitol no callus growth was observed (Emons, unpubl.)

sucrose instead of 3% mannitol + 2% sucrose had a different effect. Mannitol acts as an agent that maintains the competence for embryogenesis rather than rhizogenesis. This means that exogenous meristems are preferably formed instead of endogenous ones. Generally, osmotic agents, such as mannitol, in the medium of embryo cultures lead to the inhibition of precocious germination. Somatic embryogenesis experiments show that these agents also play a role in determining the type of meristems formed. When isolated maize zygotic embryos in empty seed coats were incubated in a bathing medium, the rate of sugar and amino acid release depended on the osmolarity of the medium: net solute efflux decreased with increasing mannitol concentration in the medium (Wolswinkel and Ammerlaan 1989). These authors suggest that a net efflux of assimilates from isolated embryos may be controlled by the osmolarity of the medium either via an effect on efflux phenomena and/or an effect on uptake processes. Valine uptake into maize embryos in empty seed coats proved to be enchanced by a high osmolarity of the medium, showing that the balance between uptake and release of amino acids and sucrose by isolated maize embryos is controlled by the osmotic environment (Wolswinkel and Ammerlaan 1989). In our culture of somatic embryos high concentrations of L-proline and casein hydrolysate without mannitol kept the callus embryogenic, similar to a medium containing low concentractions of proline and casein hydrolysate but with 3% mannitol, suggesting that mannitol indeed increases the uptake of amino acids.

2.3 Embryogenic Suspension Cultures

Somatic embryogenesis from suspension culture is the morphological pathway that allows the study of embryogenesis at the cell and molecular biological level; it is also an elegant way of regenerating genetically manipulated single plant cells. To attain these goals, it is a prerequisite that suspension cultures be maintained and regenerated into fertile plants. In the reported attempts of regeneration of suspension cells of maize via somatic embryogenesis, regeneration was first via callus aggregates (Vasil and Vasil 1986; Rhodes et al. 1988; Emons and Kieft 1991), which was also the case for regeneration from protoplasts (Prioli and Sondahl 1989; Shillito et al. 1989). Suspension cultures of genotype 4C1 (Emons and Kieft 1991) were initiated by growing callus aggregates on agar plates in liquid media. Figure 2a shows a growth curve of the friable callus of maize genotype 4C1 that was used for suspension initiation, and grown in liquid proliferation medium. One to 3 weeks after transferring embryogenic friable callus aggregates from agar plates, callus aggregates grown in liquid medium produced only elongated cells. These elongated cells are sloughed off from the callus aggregates. When cultured alone, they elongate but do not divide. They die within 2 weeks and are not produced in medium lacking 2,4-D. After 1 to 3 weeks the culture with callus aggregates contains, apart from the elongated cells, isodiametric single cells, clusters of the latter, and globules, which develop into ovoid and polar structures. The globules are comparable to globular-stage somatic embryos attached to callus aggregates. By sieving through a 500-μm sieve and transferring the sieved units to maturation medium for 10 days and

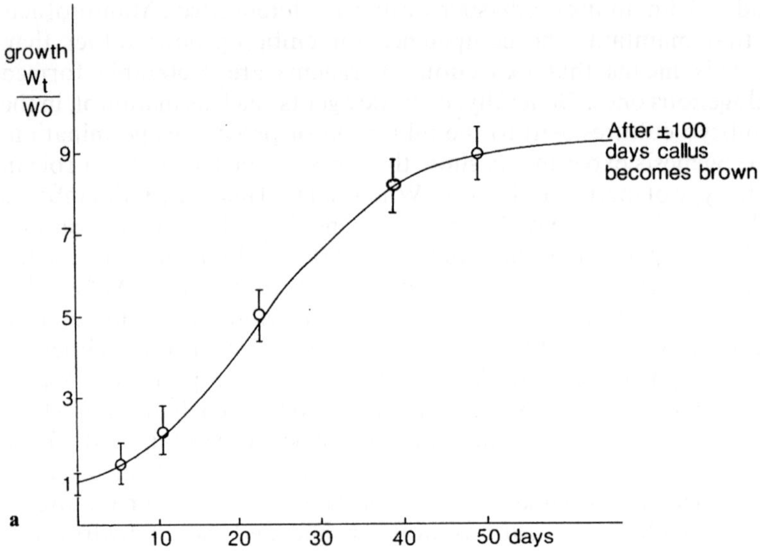
growth
$\frac{W_t}{W_o}$
9
7
5
3
1
10
20
30
40
50 days
After ±100
days callus
becomes brown

a

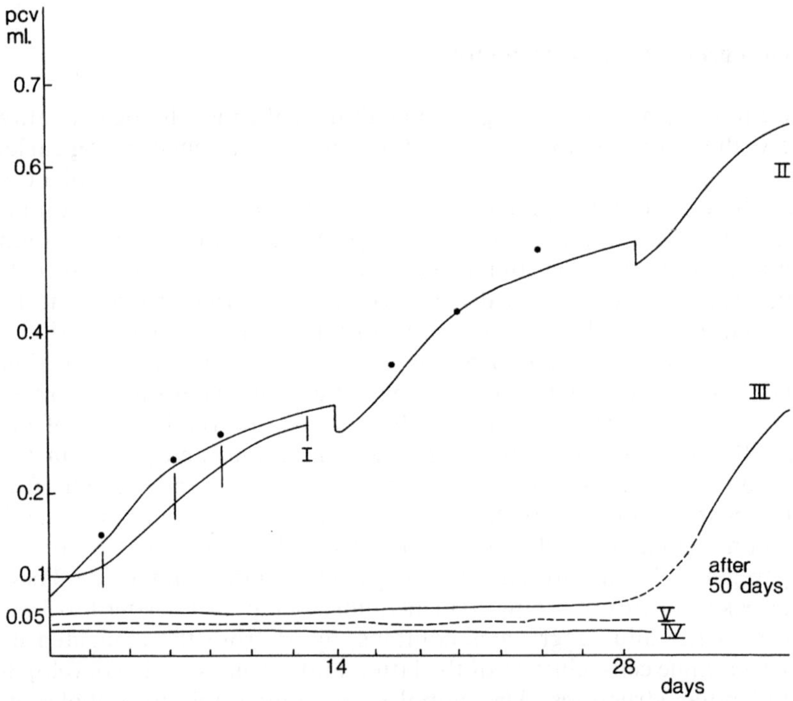
pcv
ml.
0.7
0.6
0.4
0.2
0.1
0.05
14
28
days
II
III
I
after
50 days
V
IV

b

then to regeneration medium, 100 to 1200 somatic embryos were counted per 50 ml volume of medium 4 weeks after the onset of regeneration. By refreshing the medium, the callus aggregates continued to produce the above-mentioned stages of somatic embryos for 3 to 4 months, after which they turned brown. New somatic embryo-producing cultures can be obtained by growing the decanted suspension in fresh proliferation medium. Figure 2b shows growth curves of suspension cultures, monitored by measuring packed cell volume (PCV). Each curve is the mean of three experiments. The PCV of elongated cells does not increase (Fig. 2b, V), nor do mixtures of elongated cells and cell clusters when below a critical cell concentration of 0.07 ml PCV/10 ml medium (Fig. 2b, IV). In cultures that contain elongated cells and cell clusters below the critical concentration (at 0. 05), but to which one or two globules have been added, growth starts after some months (Fig. 2b, III). The globules enrich the medium with elongated cells. Above the critical concentration of 0.07 ml PCV/10 ml medium, the cell volume doubles every 3 days (Fig. 2b, I) and growth slows down after 2 weeks. The culture starts growing again after the medium is refreshed (Fig. 2b, II). By refreshing the medium once every 2 weeks, growth continues until the flask is full of callus aggregates (2 to 3 months).

Somatic embryos from suspension cultures that are unattached to a callus aggregate show impaired shoot formation, precocious germination, and no scutellum formation. They resemble some of the DEK (defective endosperm) mutants described by Sheridan (1991) that develop a root primordium, but do not have a proper scutellum, and do not germinate. They also resemble immature zygotic embryos detached from the seed before scutellum and shoot meristem formation. In vitro maturation of these very young zygotic embryos was not possible and they were always abnormal in the scutellum and shoot (Van Lammeren 1988). In the attached somatic embryos the undifferentiated callus is a real prerequisite. Polyembryos that are attached to each other without intervening callus tissue either on the root or shoot side, which is the prevalent situation, do not develop scutella on maturation medium. If kept for a longer time in the proliferation medium, the globules themseleves develop into embryogenic calli and these regenerate into plants in regeneration medium. This process is more efficient if maturation and regeneration are carried out on solid media than in liquid.

Fig. 2. **a** Growth curve of friable callus aggregates in liquid callus maintenance medium (Emons et al. 1993b) Start: 1 (0.6–1.2) mg aggregates in 50 ml medium:

$$\text{growth: } \frac{\text{Weight after x days } (W_t)}{\text{Weight at beginning of experiment } (W_0)}$$

b Growth curves of suspension cultures (Emons, unpubl.).
I Embryogenic suspension culture;
II embryogenic suspension culture, medium refreshed once every weeks;
III embryogenic suspension culture (cell density too low, but containing one globule);
IV embryogenic suspension culture (cell density too low);
V suspension culture containing elongated cells only. * Globules produce elongated cells

2.4 Ultrastructure of Embryogenic Cells

The ultrastructural changes during embryogenic callus development from immature zygotic embryos of maize inbred line A188 have been studied and compared with changes occuring in the non-embryogenic inbred line A632 during the first 3 days of in vitro culture (Fransz and Schel 1987). On the first day of culture the reaction of both lines is the same and is characterized by an increase in the amount of organelles, and changes in vacuolation and nuclear morphology, comparable to a wound response in which the wounded tissue prepares for cell division. Under the same culture conditions, the two lines behaved differently after 1 day in culture. A188 embryos formed two meristems, one in the scutellum and one in the coleorhiza; A632, on the other hand, formed a secondary meristematic zone only in the coleorhiza. Embryogenic callus developed only in line A188; it originated from the scutellar meristematic areas. Thus callus contains globular-stage somatic embryos. Cells of the globules are small, rich in dense cytoplasm, mitotically active, and have small vacuoles and many starch grains (Fransz and Schel 1991b, 1994). The cells of embryogenic and nonembryogenic suspension cells of carrot are known to differ in division rate, elongation rate, size, shape, cell-to-cell contact, relative vacuole size, number of starch grains, amount of rough ER, ribosomes and polyribosomes, and nuclear activity (Halperin and Jensen 1967). All these characteristics were also found in maize friable calli (Franz and Schel 1994).

Freeze fracturing reveals a number of cell surface characteristics of embryogenic and nonembryogenic cells. A. freeze-fracture study on embryogenic and nonembryogenic suspension cells of maize confirmed the above-mentioned characteristics, and showed additional differences between the two cell types in cell wall texture and the number of plasmodesmata (Emons and Kieft 1990). In this study no differences were found in the number of plasma membrane rosettes. The cell wall texture (Fig. 3a) in embryogenic cells, which are able to divide and adhere to each other in cell clusters, is highly ordered, consisting of alternating lamellae of microfibrils with an angle of 20° to 40° between the microfibrils constituting the subsequent lamellae. Most wall facets of the isodiametric cells of the cell cluster observed have this texture, but also random textures occur, probably in recently divided cells. Elongated, nonembryogenic cells, which do not divide, have random wall textures with sweeping and curving microfibrils in the last deposited lamellae (Fig. 3b). Plasmodesmata, with an outer diameter of 80–100 nm including their collar, occur in groups in the embryogenic cells (Fig. 3c). In elongated cells hardly any plasmodesmata were observed. Putative microfibril synthesizing complexes, the particle rosettes (approx. 24 nm in diameter), of six (occasionally five) subunits (approx. 8 nm in diameter) occur infrequently in the PF face of the plasma membrane, and then in extremely low numbers ($0.3/\mu m^2$). The change in texture deposition, from ordered polylamellate in cell clusters to unordered in long, elongated cells, occurs mainly after the elongation of the single cells. These free cells seem to have lost a microfibril orientation coordinating factor. Also, the shape of these cells is highly variable, from balloon- to banana- and corkscrew-shaped. The two types of wall textures have also been observed in elongated cells of a carrot suspension culture (Emons et al. 1992). Elongated cells are formed if 2,4-D is present in the culture

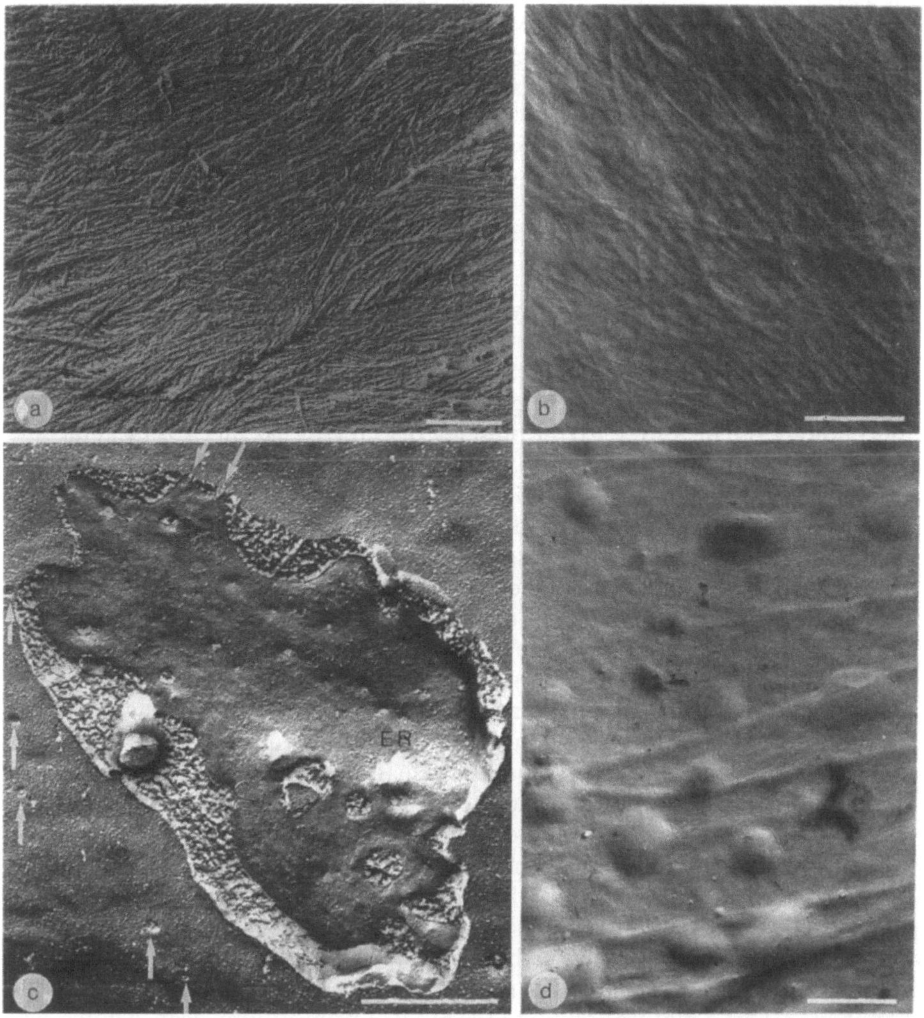

Fig. 3. Freeze fracture micrographs of the cell surface of embryogenic and nonembryogenic suspension cells (Emons and Kieft 1990). **a** Wall texture embryogenic cells; *bar* = 1 μm. **b** Last deposited wall texture of nonembryogenic cells, seen as imprints on the plasma membrane (EF face); *bar* = 1 μm. **c** Plasmodesmata on PF face of plasma membrane of embryogenic cells; ER endoplasmic reticulum; *bar* = 1 μm. **d** Exocytosis seen on the EF face of the plasma membrane of nonembryogenic cells; *bar* = 1 μm

medium. One of the effects of the presence of 2,4-D is the loosening of cells from each other. From the present data it is still unclear whether the change in wall texture is also caused by 2,4-D or due to the fact that the cells are no longer attached to other cells of a tissue. 2,4-D is a synthetic auxin, and auxins cause the elongation of plant cells by increasing the extensibility of the cell wall. Hager and coworkers (1991) used maize coleoptiles to study the action of this phytohormone and suggested that auxin activates genes responsible for (1) the induction and acceleration of exocytotic processes; (2) the synthesis of plasma membrane ATPases; and (3) the synthesis and exocytosis of wall precursors. Indeed, the cells in the carrot and maize suspension cultures that become elongated increase their exocytosis in the first days of culture when the 2,4-D concentration is high (maize: Fig. 3d) In carrot, density doubled after 1 day in fresh medium in the elongated cells and decreased in the next 4 days, decreasing almost to zero after 1 week (Emons et al. 1992). Figure 3b shows such a stage in maize.

2.5 Development of Somatic Embryos

The criteria indicating that regenerating structures are somatic embryos are: (1) direct development from typical globules; (2) a scutellum with characteristic morphology and histology; (3) the presence of a coleoptilar ring arising at the side of the scutellum; (4) a root-shoot axis without callus cells between them; (5) a closed vascular system; and (6) a single coleorhiza with a root meristem. Maize genotype 4C1 very efficiently produces friable embryogenic callus from which plants can be regenerated via the somatic embryogenesis pathway (Emons and Kieft 1991). Also, single somatic embryos, detached from the callus could be regenerated into mature plants, if they had been attached to a callus aggregate until scutellum and shoot meristem development. Scanning electron microscopy shows that embryo morphogenesis, as described by Abbe and Stein (1954) for zygotic embryos, is seen in the somatic embryos attached to callus aggregates, or detached from them after scutellum formation (Emons and De Does 1993). The somatic embryo in Fig. 4a represents the transition stage. The attachment side of the callus aggregate is equivalent to the suspensor. As reported also by Ho and Vasil (1983) for sugarcane, we agree that, in indirect somatic embryogenesis, a suspensor is indistinguishable from the callus tissue. In Fig. 4b the scutellum is large and fleshy and has a groove. In Fig. 4c the shoot meristem is visible and in Fig. 4d the coleoptile envelops shoot meristem and leaf primordia almost completely. The coleoptile lies on the same side as the scutellum. Cross-sectioned somatic embryos generally do not show more than two or three leaf primordia. Complete embryo maturation up to the 6th leaf primordium, as in zygotes, was not found. Somatic embryos can be recognized under binoculars as structures having a roundly lobed, white opaque structure during maturation, the scutellum, which stands more or less free from the callus aggregate. Just as in the zygotic embryo, the scutellum cells of the somatic embryo contain many large amyloplasts (Emons et al. 1993a) and the embryo axis lies at the base of the scutellum. However in callus aggregates not all of the regenerating structures are round, white opaque scutella, standing free from the callus. Many regenerating structures are more or less leafy and fasciated. The leafy structures may be hairy, green if maturation occurs in the light, and their cells do not contain an abundance of starch (Emons and De Does 1993). If the regenerating structures are leafy, the blade grows together with the callus cells, but they are not shoots standing free from the callus aggregate. A shoot meristem is present at their bases, and a root meristem may or may not be attached to it. Studying the order

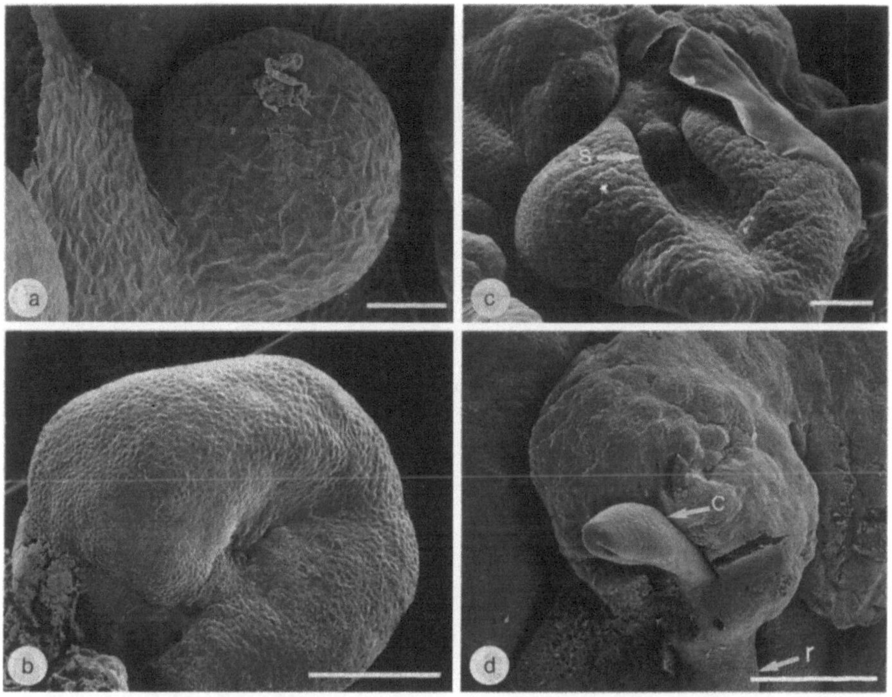

Fig. 4 a–d. Scanning electron micrographs of developmental stages of somatic embryos matured on solid medium (Emons and De Does.1993). **a** Globular stage; *bar* = 50 µm. **b** Somatic embryo with scutellar groove; *bar* = 500 µm. **c** Somatic embryo with scutellum and shoot meristem (*s*); *bar* = 100 µm. **d** Somatic embryo showing a root (*r*) and a coleoptile (*c*) enveloping the shoot meristem; *bar* = 500 µm. *sc* Scutellum; *cl* coleoptile

in which the primordia are formed on the shoot side, it becomes clear that the primordium that generally lies on the same side as the scutellum, the coleoptile, is missing; thus these structures are not comparable with embryos. Because these shoot primordia lie at the base of leafy structures, they should rather be regarded as homologous to axillary buds (Fig. 5a,b). The bud is perpendicular to the leafy structure as expected. Even under the rather stringent conditions of proliferation, maturation, and regeneration as described in the protocol of this chapter, somatic embryos as well as axillary buds develop on callus aggregates of maize. In tissue culture practice, using binoculars, one sees a whole range of structures in between the extremes, white opaque scutella and green leafy structures, and often one cannot be sure whether one is dealing with somatic embryos or with axillary, or even adventitious buds. This situation occurs not only in maize, but in most plant tissue culture systems.

2.6 Histology of Somatic Embryos

The first visible differentiation in a single globular somatic embryo from suspension culture is the formation of a protoderm. After that, at one side , a root

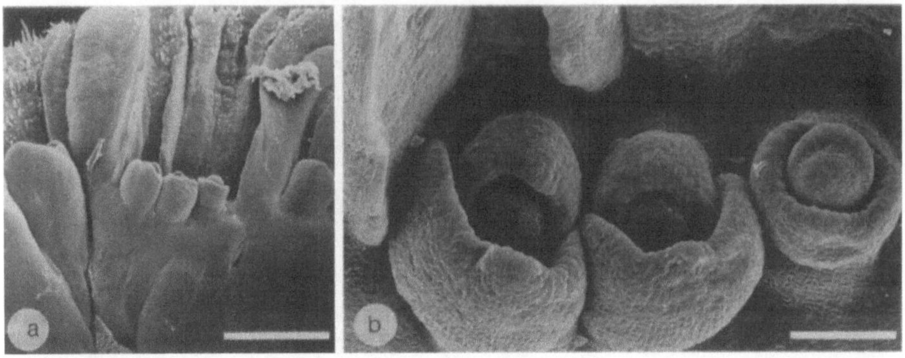

Fig. 5 a,b. Axillary buds developed on callus from solid medium. **a** Fasciated leafy structures with shoot meristems at their bases; *bar* = 500 μm (Emons and De Does 1993). **b** Detail of a, note the missing coleoptile; *bar* = 100 μm

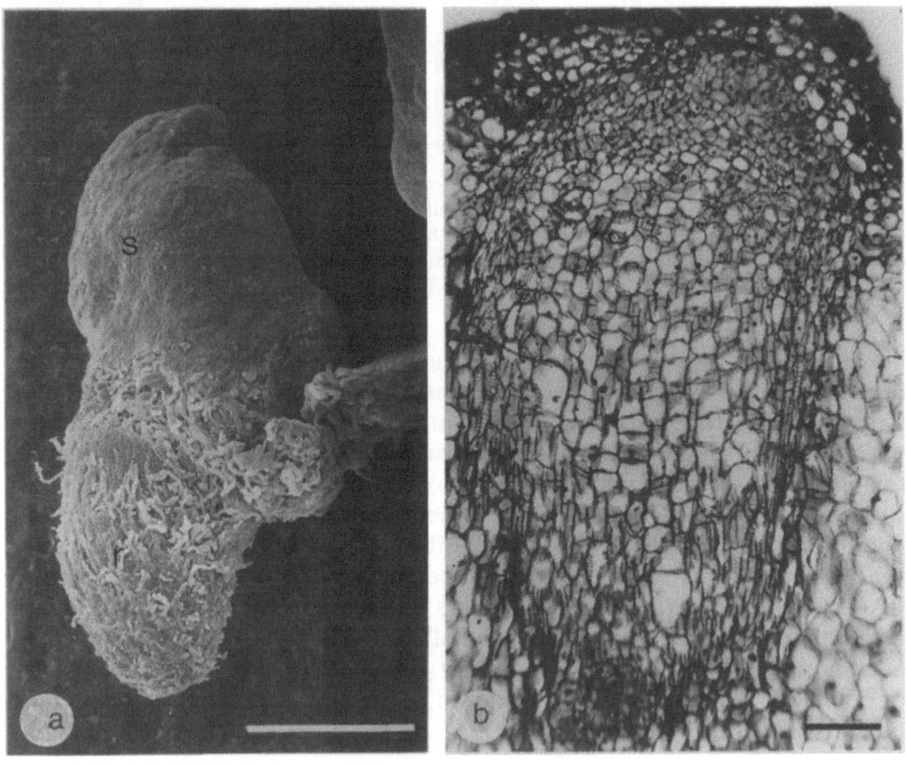

Fig. 6. a Morphology of an ovoid-shaped, single somatic embryo, showing **a** root(*r*) and a shoot (*s*) side; *bar* = 500 μm (Emons and De Does1993). **b** Histology of the meristematic zone on the shoot side of a polar, single somatic embryo; *bar* = 100 μm. (Emons and Kieft 1991)

meristem is formed. During this process, the unit attains a club-shaped, ovoid form (Fig. 6a). The shoot side consists of small cells, similar to a very young shoot apex, but serial sectioning shows that no leaf primordia are present (Fig. 6b). Vascular cells connect the root and shoot meristems. The single somatic embryo from suspension culture represents the embryo axis only. No scutellum is present. In regeneration medium only the roots grow out; the development of the shoot side is inhibited.

If a globular-stage somatic embryo grows while attached to a callus aggregate, a somatic embryo is formed that possesses the same histodifferentiation as a zygotic embryo (Emons and Kieft 1991). Figure 7a shows one of a series of sections through a somatic embryo that had grown while attached to callus cells. The histology of the somatic embryo closely resembles that of the zygotic embryo. The different tissues of the zygotic embryo of maize (Van Lammeren

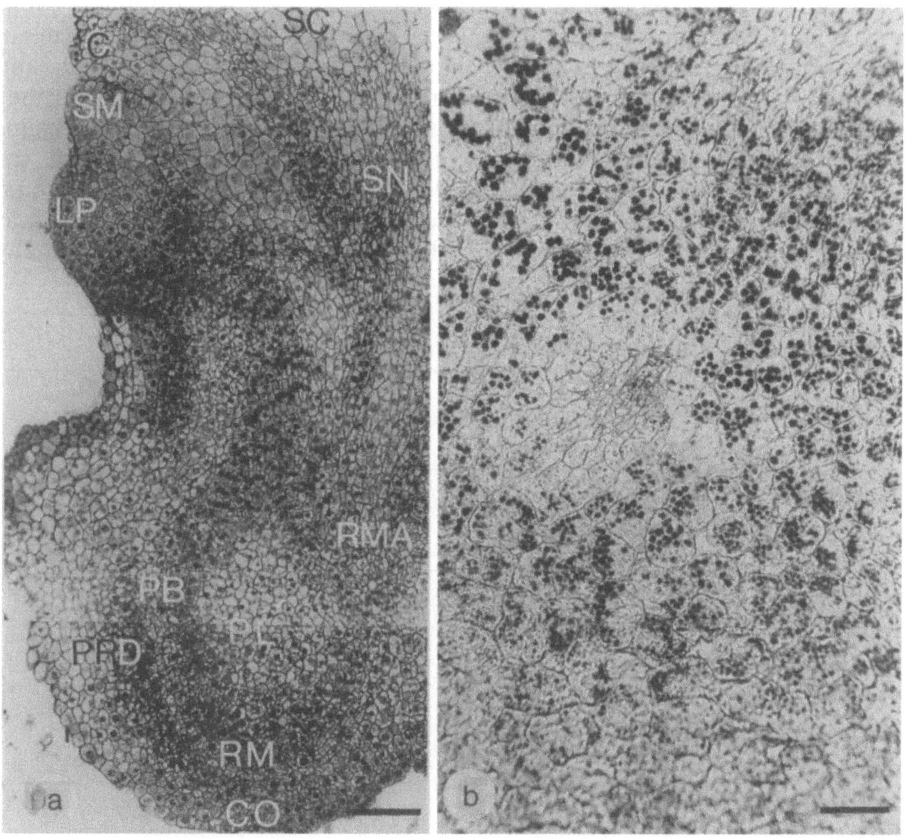

Fig. 7. a Section through complete somatic embryo, grown while attached to callus aggregate; *bar* = 100 μm (Emons and Kieft 1991). *C* Coleoptile; *CO* coleorhiza; *LP* leaf primordium; *PB* periblem; *PL* plerome; *PRD* root protoderm; *RM* root meristem, *RMA* adventitious root meristem; *SC* scutellum; *SM* shoot meristem; *SN* scutellar node; *bar* =100 μm. **b** Section through scutellum and part of embryo axis, showing numerous starch grains in the scutellum cells; *bar* = 100 μm. (Emons et al 1993a)

1988), e.g., coleoptile, calyptra, coleorhiza, embryo axis, leaf primordia, periblem, plerome, protoderm, root meristem, scutellum, shoot meristem, scutellar node, and adventitious root meristem, can be discerned. Just as in a zygotic embryo the scutellum of this somatic embryo contains many starch grains (Fig. 7b). In regeneration medium these somatic embryos develop into plants.

3 Summary and Conclusions

By using 2,4-D, many maize genotypes can produce embryogenic calli from immature zygotic embryos (Duncan et al. 1985). This callus type can be maintained in culture and can be regenerated. The callus consists of aggregated, globular-stage somatic embryos attached to a mass of more vacuolated cells. If suspension cultures can be produced from such calli, regeneration can take place in small cell clusters that first form globules and then develop into a callus aggregate. Cells with embryogenic potential differ from nonembryogenic cells in their ultrastructural features and the characteristics related to cell-cell adhesion appear to be important. Single globules are unable to regenerate into plants, they form only roots. The somatic embryos which grow while attached to a callus show the same histodifferentiation as zygotic embryos. Full maturation, i.e., the development of up to six leaf primordia, was not achieved. The maximum number of primordia before the structures germinate, seen in thin sections, was three. For such an embryo maturation, which includes the formation of a starchy

Table 2. Stages of somatic embryo formation from maize cell suspensions

Stages[a] Medium	I N6 Liquid	II N6 Liquid	III N6 Liquid	IV MS Liquid/solid	V MS Solid/liquid
Hormones:					
2,4-D	2 mg/l	2 mg/l	2 mg/l	0.25 mg/l	0
BAP	–	–	–	–	3.5 mg/l
Energy source:					
Sucrose	2 %	2 %	2 %	6 %	2 %
Osmoticum:					
Mannitol	3 %	3 %	3 %	–	–
Culture periods:					
Weeks	1–3	3–12	3–12	1.5	2–4
Subculture:					
Weeks	2	2	1	–	–
Light Condition	Dark	Dark	Light/dark	Light/dark	Light/dark

[a] I, Callus growth, formation of elongated cells; II as I, and formation of cell clusters, globules, ovoids, and polars; III, after sieving of aggregates, proliferating suspension; IV, maturation of somatic embryos, formation of scutella; V, regeneration of plantlets on agar plates (Emons and Kieft 1991)

scutellum, a maturation phase on abundant sucrose is necessary. Without this phase, calli regenerate via organogenesis. However, with such a maturation phase included in the regeneration cycle, not all regenerating structures are somatic embryos at the base of scutella, but in maize some are axillary buds at the base of leafy structures. Regeneration on agar plates is more efficient than in liquid culture.

4 Protocol for Induction of Somatic Embryos

Table 2 summarizes the main culture procedures and conditions (See also Emons et al. 1993a).

Callus Initiation

Zygotic embryos of maize genotype 4C1 (Mórocz et al. 1986) were cultured on agar-solidified N6 medium (Chu et al. 1975), containing 2 mg/1 2,4-D, 20 mM L-proline, 200 mg/1 casein hydrolysate, and 2% sucrose. From the three types of calli obtained, compact embryogenic callus, friable embryogenic callus, and rhizogenic callus, the friable embryogenic callus was used for suspension initiation. Callus initiation was in the dark. In all experiments culture temperature was 24–26 °C.

Callus Maintenance

Fast-growing friable embryogenic callus was selected and subcultured once every 2 weeks. It has been in culture for over 2 years on N6 medium containing 2 mg/l 2,4-D, 6 or 20 mM L-proline, 100 or 200 mg/l casein hydrolysate, 2% sucrose, and 3% mannitol, the proliferation, or callus maintenance medium. The medium was solidified with 0.7% agar. Growth was monitored by determination of fresh weight. Callus maintenance was under a 16-h light/8-h dark regime.

Suspension Initiation

Suspension cultures were initiated from friable embryogenic callus aggregates 1 week after subculture at a concentration of 2 g/50 ml callus maintenance medium without agar in 250-ml flasks. Also, callus maintenance medium with increased proline (20 mM) and casein hydrolysate (200 mg/l) was used. Flasks were placed in the dark and shaken at 100 rpm on a gyratory shaker. Over a 3-week period the whole biomass and medium of a flask were divided weekly among two flasks and equal volumes of fresh medium were added. Proliferation was in the dark.

Suspension Maintenance

After 3 weeks the suspension cells were poured weekly in other flasks such that only callus aggregates remained. Callus aggregates were given fresh callus maintenance medium to produce new suspension cells. The cell clusters were allowed to settle and 30 ml of the upper medium, primarily containing elongated cells was decanted; then 30 ml fresh callus maintenance medium was added to the culture. After 2 weeks the culture contained again a suspension of elongated cells, isodiametric cells, clusters of isodiametric cells, globular, ovoid and polar stages of somatic embryos and small callus aggregates. Suspensions were under a 16-h light/8-h dark regime, or in the dark.

Regeneration of Callus Aggregates

To obtain plants from callus aggregates grown on solid medium, the undifferentiated callus aggregates were transferred for 10 days to hormone-free solidified Murashige and Skoog (MS) medium (1962) supplemented with 6% sucrose, the maturation medium, and after that to medium with 2% sucrose, the regeneration medium. On this medium, plantlets developed.

Regeneration of Suspension Cells on Solid Medium

One week after subculturing, the suspension containing young somatic embryos was decanted into a sterile centrifuge tube, and centrifuged at 50 g. The supernatant was decanted and 2 ml PCV was spread onto a 9-cm Petri dish containing solid callus maintenance medium. In 2 to 4 weeks small callus aggregates formed. These aggregates were further regenerated as described above and plants developed.

Regeneration of Suspension Cells in Liquid Medium

Flasks were rotated and somatic embryos were allowed to settle. Thirty ml, containing primarily elongated cells was decanted carefully and 30 ml hormone-free MS medium containing 6% sucrose was added. Embryos were allowed to mature in this medium for 10 days. After 10 days 30 ml of the upper medium was replaced by MS medium containing no hormones and 2% sucrose. This medium was refreshed once every 2 weeks. The shoot meristem of these single somatic embryos is blocked and only roots develop.

Acknowledgments. The authors thank A. Samallo-Droppers for tissue culture work, H. De Does for scanning electron microscopy, S. Massalt for photography, A. Haasdijk for artwork, and Professor Dr. M.T. M. Willemse for helpful discussions.

References

Abbe EC, Stein OL (1954) The growth of the shoot apex in maize embryogeny. Am J Bot 4: 285–293

Armstrong CL, Green CE (1985) Establishment and maintenance of friable embryogenic maize callus and the involvement of L-proline. Planta 164: 207–214

Chu CC, Wang CC, Sun CS, Hsu C, Yin KC, Chu CY, Bi FY (1975) Establishment of an efficient medium for anther culture of rice through comparative experiments on the nitrogen sources. Sci Sin 18: 659–668

Duncan DR, Williams ME, Zehr BE, Widholm JM (1985) The production of callus capable of plant regeneration from immature embryos of numerous *Zea mays* genotypes. Planta 165: 322–332

Emons AMC, Kieft H (1990) Comparison of embryogenic and nonembryogenic suspension cells of maize by means of freeze -fracturing. Micron Microsc Acta 21: 255–256

Emons AMC, Kieft H (1991) Histological comparison of single somatic embryos of maize from suspension culture with somatic embryos attached to callus cells. Plant Cell Rep 10: 485–488

Emons AMC, De Does H (1993) Origin and development of embryo and bud primordia during maturation of embryogenic calli of *Zea mays* L. Can J Bot 71: 1349–1356

Emons AMC, Vos JW, Kieft H (1992) A freeze fracture analysis of the surface of embryogenic and non-embryogenic suspension cells of *Daucus carota*. Plant Sci 87: 85–97

Emons AMC, Mulder MM, Kieft H (1993a) Pyrolysis mass spectrometry of developmental stages of maize somatic embryos. Acta Bot Neerl 42: 319–399

Emons AMC, Samallo-Droppers A, Van der Toorn C (1993b) The influence of sucrose, mannitol, L-proline, abscisic acid and gibberellic acid on the maturation of somatic embryos of *Zea mays* L. from suspension cultures. J Plant Physiol 142: 597–604

Fransz PF, Schel JHN (1987) An ultrastructural study on early callus development from immature embryos of the maize strains A188 and A632. Acta Bot Neerl 36: 247–260

Fransz PF, Schel JHN (1991a) Cytodifferentiation during the development of friable embryogenic callus of maize (*Zea mays* L.). Can J Bot 69: 26–33

Fransz PF, Schel JHN (1991b) An ultrastructural study on the early development of *Zea mays* somatic embryos. Can J Bot 69: 858–865

Fransz PF, Schel JHN (1994) Ultrastructural studies on callus development and somatic embryogenesis in *Zea mays* L. In: Bajaj YPS (ed) Biotechnology in agriculture and forestry Vol 25. Maize. Springer, Berlin Heidelberg New York pp. 50–65

Gorden-Kamm WJ, Spencer TM, Mangano ML, Adams TR, Daines RJ, Strat WG, O'Brien JV, Chambers SA, Adams WR, Willetts NG, Rice TB, Mackey CJ, Krueger RW, Kausch AP, Lemaux PG (1990) Transformation of maize cells and regeneration of fertile transgenic plants. Plant Cell 2: 603–618

Hager A, Debus G, Edel HG, Strarsky H, Serrano S (1991) Auxin induces exocytosis and the rapid synthesis of a high turnover pool of plasma membrane ATP-ase. Planta 185: 527–537

Halperin W, Jensen WA (1967) Ultrastructural changes during growth and embryogenesis in carrot cell cultures. J Ultrastruct Res 18: 428–443

Ho WJ, Vasil JK (1983) Somatic embryogenesis in sugarcane (*Saccharum officinarum* L.). I. The morphology and physiology of callus formation and the ontogeny of somatic embryos. Protoplasma 118: 169–180

Kishor PBK, Reddy GM (1986) Retention and revival of regenerating ability by osmotic adjustment in long-term cultures of four varieties of rice. J Plant Physiol 126: 49–54

Mórocz S, Dudits D, Nemeth J (1986) Two approaches to rendering *Zea mays* L. applicable to tissue culture manipulation. In: Somers DA, Gengenbach BG, Biesboer DD, Hackett WP, Green CE (eds) IV Int Congr Plant tissue culture, Minneapolis pp 190

Murashige T, Skoog F (1962) A revised medium for rapid growth and bioassays with tobacco tissue cultures. Physiol Plant 15: 473–497

Prioli L, Sondahl MR (1989) Plant regeneration and recovery of fertile plants from protoplasts of maize (*Zea mays* L.). Biotechnology 7: 589–594

Rhodes CA, Lowe KS, Ruby KL (1988) Plant regeneration from protoplasts isolated from embryogenic maize cell cultures. Biotechnology 6: 56–60

Sheridan WF (1991) Isolation and characterization of 51 embryo-specific mutations in maize. Plant Cell 3:935-951

Shillito RD, Carswell GK, Johnson CM, DiMaio JJ, Harms CT (1989) Regeneration of fertile plants from protoplasts of elite inbred maize. Biotechnology 7: 581–578

Van Lammeren AAM (1986) Developmental morphology and cytology of the young maize embryo (*Zea mays* L.). Acta Bot Neerl 35: 169–188

Van Lammeren AAM (1988) Observations on the structural development of immature maize embryos (*Zea mays* L.) during in vitro culture in the presence or absence of 2, 4-D. Acta Bot Neerl 37: 49–61

Vasil V, Vasil IK (1986) Plant regeneration from friable embryogenic callus and cell suspension cultures of *Zea mays* L. J Plant Physiol 124: 399–408

Wolswinkel P, Ammerlaan A (1989) Effect of the osmotic environment on assimilate transport in isolated developing embryos of maize (*Zea mays*). Ann Bot 63: 705–708

I.4 Somatic Embryogenesis in Rye (*Secale cereale* L.)

A.M. Vázquez and R. Linacero[1]

1 Introduction

The genus *Secale* includes 14 species of perennials and annuals, but only one species, *Secale cereale* L., is extensively cultivated and has an important role in human and animal diets.

Rye is an allogamous crop, so it is difficult to keep rye varieties pure. It can be grown in areas that are generally unsuitable for growing other grain cereals, because it is extremely hardy and has the ability to grow in sandy soils having low fertility. Greatest production is in the cool temperate zones of the world; it can also grow in semiarid regions near deserts and at high altitudes (Evans and Scoles 1976).

Most rye species are annual. The so-called winter rye can be grown successfully in areas where the climate is too severe for winter wheat. Some spring rye is grown in areas where the winters are too severe for winter rye production. Rye still continues to be an important crop, because it has the sturdiness that other crop species do not have (Bushuk 1976).

Rye is mostly grown in the Czech Republic, France, Spain, Argentina, and Turkey; however, it has a special economic importance in Russia, Poland, and Germany. It has several uses: as flour for bread, as grain for livestock feed, and as a green plant for livestock pasture. Of the cereals, only wheat and rye produce flour that can be used for the production of bread. Rye flour alone can be used to produce black bread consumed in eastern Europe and in parts of Asia. Rye is also used as a pasture crop, and can be pastured both in the autumn and spring (Bushuk 1976). It is also a good rotational crop. In Argentina it is an important pasture crop, and in southern Australia it is planted to prevent wind erosion.

The development of molecular biological techniques and their application in in vitro genetic multiplication of rye would help to improve this crop. One of the most important uses of tissue culture may be in the multiplication of a particular genotype, especially if the desired plant is a hybrid. Embryogenic cultures have been shown to be a stable and highly efficient source of rapid clonal propagation of plants. The whole plant can be regenerated from somatic embryos, allowing easy transplantation to soil.

[1] Departmento de Génetica, Facultad de CC. Biológicas, Universidad Complutense de Madrid, Madrid 28040, Spain

Biotechnology in Agriculture and Forestry, Vol. 31
Somatic Embryogenesis and Synthetic Seed II (ed. by Y.P.S. Bajaj)

2 An Overview of Rye Tissue Culture

Compared to other cereals, not much work has been done on rye tissue culture. In 1955, Roberts and Street obtained callus from excised rye roots. The same response was obtained by Carew and Schwarting (1958) using different rye explants as mature embryos, endosperm, or leaves. Later, Rybczynski et al. (1980) induced callus from the rachis of a 7-mm-long inflorescence and subsequently this callus differentiated shoots. Eapen and Rao (1982) obtained, from 14–21 day postanthesis immature embryos, calli with the ability to regenerate plants.

In 1984, Kumbriegel-Schroeren et al. and Lu et al. described the formation of embryogenic calli from the scutellum of immature embryos of rye of different sizes. These calli were; yellow, compact, and nodular. They proliferated on the coleorhizal end and the surface of the scutellum. Kumbrigel-Schroeren et al. (1984) used five rye cultivars, and 1–3% of the cultures showed an embryogenic response. These authors also used 1–11 cm immature inflorescences as explants, reporting embryogenic calli in only one of the cultivars used. However, Lu et al. (1984) described the presence of somatic embryos in 40% of the immature embryos cultured in vitro.

Bebeli et al. (1988) also described the formation of embryogenic calli from immature embryos of different lines of rye. Linacero and Vázquez reported the formation of embryogenic calli from young leaves (1986), and immature inflorescences (1990) of four different rye cultivars.

The formation of somatic embryos has also been reported in in vitro cultures of immature embryos from other species of *Secale*, the so-called wild rye, *S. africanum*, *S. ancestrale*, *S. kuprianovii*, *S. segetale*, and *S. vavilovii* (Rybczynski and Zdunczyk 1986), and also in the in vitro cultures of other immature explants in *S. vavilovii* (Vázquez et al. 1991). A summary of morphogenetic studies in rye is given in Table 1.

3 Somatic Embryogenesis

The immature embryos of *Secale cereale*, cultured on Murashige and Skoog (1962) MS medium supplemented with different 2,4-D concentrations, showed the initiation of callus formation within 2–3 days. The proliferation on the scutellum produced a compact and yellow callus (embryogenic callus) in 2 or 3 weeks. The embryoids could be observed over the entire callus surface (Fig. 1A) and these embryoids developed into mature somatic embryos (Fig. 1B). In other cases the calli had a soft and white appearance (non-embryogenic callus), and sometimes this callus type showed adventitious root formation.

The immature inflorescences and young leaves produced embryogenic calli similar to that described previously. In these explants, the initiation of embryogenic calli occurred from the rachis in the inflorescence, and from the vascular tissues in the case of young leaves.

Table 1. Summary of morphogenetic (organogenesis/embryogenesis) studies in rye

Species	Nature of explant	Induction medium (mg/l)	Embryogenesis (E)/ organogenesis (O)	Best response frequency (%)	References
S. cereale					
	Immature embryos	MS+2,4,5- -Cl_3 phenoxypropionic	O	17.00	Eapen and Rao (1982)
	Immature inflorescences	MS+2,4-D	E	43.74	Krumbiegel-Schroeren et al. (1984)
	Immature embryos	1–2.5	E	1–3	
	Immature embryos	MS+ 2,4-D 2.5	E	40.00	Lu et al. (1984)
	Young leaves	MS + 2,4-D 0.5	E	52.94	Linacero and Vázquez (1986)
	Immature inflorescences	MS + 2,4-D 1.5	E	93.87	Linacero and Vázquez (1990)
Wild *Secale*					
S. segetale and others	Immature embryos	MS + 2,4-D 2.5–5	E+O	49.10	Rybczynski and Zdunczyk (1986)
S. vavilovii	Immature embryos and	MS + 2,4-D 1	E	33.33	
	inflorescences	0.5–2	E	100	Vázquez et al. (1991)
	Young leaves	2	E	25.64	

A

B

Fig 1. **A** 6-month-old embryogenic callus induced on Murashige and Skoog medium supplemented with 2 mg/l 2,4-D from immature embryos of *Secale cereale*. **B** Somatic embryos on Murashige and Skoog medium

When the embryogenic callus was cultured on an MS basal medium, most of the embryos germinated into plants. Usually, these plants had roots because the somatic embryos were bipolar structures with shoot and root apices.

The nonembryogenic calli sometimes produced plant regeneration by organogenesis. In this case, it was necessary to culture the regenerated shoot on a different medium to induce root formation.

The frequency of embryogenic callus formation was affected by nature of the explant, culture conditions, and the genotype of the donor plant.

3.1 Nature of the Explant

It is necessary to use initial explants from meristematic tissues to obtain embryogenic cultures. The period when an explant could proliferate in an embryogenic way was short. The explants which contained some cellular types in a competent state, in an adequate culture medium, divided and produced embryogenic calli. However, when the tissue started losing its competence, the frequency of embryogenic calli decreased. As they seem to be related to competence, the physiological condition of the plant from which the explant was taken and the condition of the explant itself are very important.

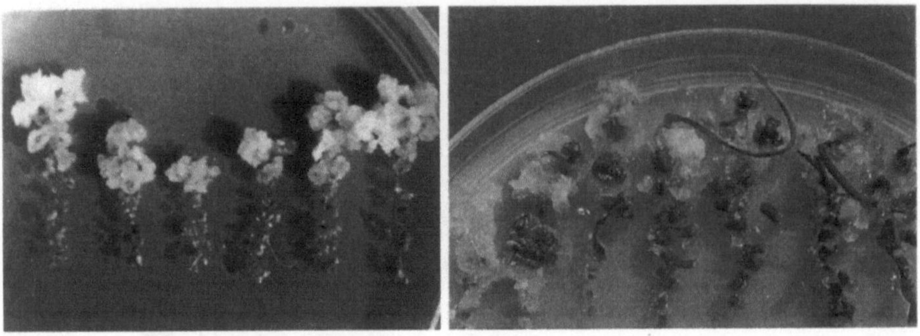

A B

Fig. 2A,B. Young leaf cultures. Calli are formed from the basal segments. **A** *S. vavilovii* cultured on media containing 1 mg/l 2,4-D (Vázquez et al. 1991) **B** *S. cereale* cultured on media containing 0.5 mg/l 2,4-D. (Linacero and Vázquez 1986)

When mature embryos of rye were used as the initial explant, embryogenic calli were not obtained. However, sporadically the regeneration of plants in *S. cereale* and *S. vavilovii* was observed (Vences et al. 1986).

Immature embryos have proved to be a more competent explant. The first researchers using immature embryos measured the developmental stage of the embryos in terms of days after pollination. However, this relation is not always correct, because development is affected by the environmental conditions such as temperature, rain, etc. We observed a negative effect of low temperatures; here, the development of embryos by the donor rye plant depended on the capacity of this explant to produce calli.

Embryos measuring 1 to 2 mm in length gave the best response in *S. vavilovii* (14–33%) (Vázquez et al. 1991) and *S. cereale* (11–26%). Other authors obtained their best responses from immature embryos of about 1 mm (Bebeli et al. 1988), whereas Lu et al. (1984) found better responses in embryos measuring 0.5 to 2 mm.

Leaf explants were obtained from 3–4-week-old plants growing under aseptic conditions on basal MS medium. A clear gradient in the response from the base to the apex of the leaves was noticed (Linacero and Vázquez 1986; Vázquez et al. 1991). Figure 2 shows that the basal part was more suitable for the induction of embryogenic calli. Moreover, we observed that leaves cultured from poorly developed plants failed to proliferate, thus the selection of the most vigorous plants was important for the success of induction. A similar gradient along the leaves in the response has been noted in other Gramineae in which leaves have been cultured: sorghum (Wernicke and Brettel 1980), rice (Wernicke et al. 1981), wheat (Wernicke and Milkovits 1984), and barley (Ruiz et al. 1992).

The length of the immature inflorescence, which again is related to its developmental stage, is also important. Lengths of 1 to >20 mm in *S. cereale* (Linacero and Vázquez 1990) produced embryogenic calli (Table 2), but not all the explants (inflorescence segments) formed them, and inflorescences larger than 20 mm produced less. However, inflorescence lengths of 5 to 45 mm always produced embryogenic calli in all the explants of *S. vavilovii* (Vázquez et al. 1991).

Table 2. The effect of the inflorescence size (mm) on embryogenic callus formation (Linacero and Vázquez 1990)

Size	No. of embryogenic calli / No. of explants	(%)
1–10	730/1850	39.45
11–20	617/1835	33.62
> 20	279/1085	25.71

Table 3. Percentage of embryogenic callus formation from different immature explants and from different genotypes of *Secale*

Immature explant	*S. cereale*				*S. vavilovii*
	Elbon (%)	D. Enrique (%)	Merced (%)	Ailes (%)	
Embryos	11.45	16.97	21.24	26.40	18.65
Inflorescences	21.92	31.85	32.41	49.64	100
Leaves	18.75	26.98	30.47	38.57	25.64

A comparison of the results obtained from the three types of explants in *Secale* and other gramineae species is difficult because different genotypes were used. Nevertheless, in a few cases, a comparative analysis was carried out. In *S. vavilovii* the immature inflorescence was the better explant to produce somatic embryos and regenerated plants. A similar result was obtained in *Tritordeum* (Barceló et al. 1989). In *Hordeum vulgare* the results varied between genotypes, in some cases the best explant was the immature inflorescence (Thomas and Scott 1985), and in others immature embryos or young leaves were better (Ruiz et al. 1992). In *Secale cereale* immature inflorescences also gave better results than other explants (Table 3). However, in this case, the differences between them were not as obvious as in the case of *S. vavilovii.* On the other hand, each explant has its advantages, so we were able to select one according to our requirements. For instance, young leaves as explant were most rapidly obtained since they could be taken from plantlets growing under controlled conditions. With immature inflorescences as explants it is possible to multiply a particular genotype, because of the great number of explants obtained from the same plant.

3.2 Culture Conditions

3.2.1 Culture Medium

The critical factor regarding an embryogenic response is the inclusion of an appropriate growth factor in the medium and in most cases this should be 2,4-D. In general, the phenoxyacetic acids are more effective than other auxins in initiating embryogenic callus and embryoid formation (Lu et al. 1984).

In immature embryos, Lu et al. (1984) reported that 2.5 mg/l was the optimum concentration for those tested (0–5 mg/l). Higher concentrations (5 mg/l) reduced both germination and adventitious root formation. However, the 2,4-D concentration had little effect on the percentage of embryos forming callus or the ratio between soft and embryogenic calli.

In *S. vavilovii*, 1 mg/l 2,4-D was slightly better (21.18% of embryogenic callus) than 2 mg/l (18.65% of embryogenic callus). However, Rybcznski and Zdunczyk (1986) observed an increase in the embryogenic ability of the immature embryos of several *Secale* species when the 2,4-D concentration was increased.

We usually noticed that a high 2,4-D concentration promoted better callus formation, but was inversely related to embryogenic callus formation in young leaves and immature inflorescence cultures. Thus, in the case of the young leaves, there were two optimal concentrations, 0.5 and 1 mg/l, to initiate callus with embryogenic competence in *S. cereale*. For the immature inflorescences of the same species, 1.5 mg/l 2,4-D was the best concentration to obtain embryogenic calli. The frequency of soft calli was high in the 2 mg/l concentration, while poorly developed calli were formed in the lower (0.5 mg/l) hormone concentration (Linacero and Vázquez 1986, 1990).

In the lower 2,4-D concentration media, the embryogenic calli formed embryoids which often germinated precociously, however, the embryoids on the 2 mg/l 2,4-D supplemented medium did not germinate in the presence of the hormone. So it is convenient to use high levels of 2,4-D (2–5 mg/l) in order to prevent precocious germination and to maintain embryogenic competence for longer periods (Vasil 1987). Although the main requirement for obtaining embryogenic callus is the presence of 2,4-D, there are other factors in the media which could modify the percentage of embryogenic callus obtained.

MS basal medium was mostly employed; a double concentration of the mineral salts had no effect on embryoid formation (Lu et al. 1984). Lu et al. (1984) also found that an increase in sucrose concentration from 3 to 6% resulted in a 15–20% increase in the amount of embryogenic calli. However, more embryoids were formed on 3% sucrose than on 6% sucrose.

3.2.2 Light Conditions

While the best condition to induce embryogenic calli in *S. cereale* from immature embryos is darkness (Lu et al. 1984), the best response of the immature inflorescences was obtained when these explants were cultured under a diffuse light regime of 16 h (see Sect. 5; Linacero and Vázquez 1990), and for the young leaves when they were cultured under a light regime of 16 h (Linacero and Vázquez 1986). In *S. vavilovii* some of the young leaves were cultured in the dark and others under a light regime of 16 h. Notable differences were obtained: the responses in the dark varied between 4.35 and 6.67%, and in the light from 14.25 to 25.64% (Vázquez et al. 1991).

3.3 Genotype

The influence of genotype upon the in vitro response has been very controversial. We have proved that it is one of the major factors implicated in the response of rye.

3.3.1 Differences Between Species

The frequency of embryogenic callus formation from immature embryos was better in *S. africanum* (20%) and *S. segetale* (15.5%) than in *S. ancestrale* and *S. kuprianovii* (9.2%) and *S. vavilovii* (2.7%) (Rybczynski and Zdunczyk 1986). In the last species, Vázquez et al. (1991) reported a maximum embryogenic callus formation from immature embryos of 33.33%, while the optimum result obtained by the first authors was 8.5%. In our case and using the same culture conditions, *S. vavilovii* showed a better response than *S. cereale* when immature inflorescences were used as the initial explant. However, as Table 3 shows for other explants, *S. cereale* appears more efficient in producing embryogenic calli, except in the case of the cultivar Elbon.

3.3.2 Differences Between Cultivars

Great differences between cultivars of the same species have been reported in Gramineae, barley (Lührs and Lörz 1987), maize (Duncan et al. 1985) and rice (Abe and Futsuhara 1985). We also found differences between different cultivars in *Secale cereale* in all the explants used. The cultivar Elbon showed the worst response, while Ailes was the more successful cultivar with which to obtain an embryogenic culture (Table 3). In the same species, Krumbriegel-Schroeren et al. (1984) used five different cultivars, but the frequencies of embryogenic callus formation were too low (1–3%), and it was not possible to determine whether there were differences between these cultivars.

Other authors consider the physiological and developmental stages of the plant or the explant at the time of culture to be the most important factors in the response. In our case, even when these factors affected the response, the genotype seemed to have a fairly strong influence on the culture behavior.

However, we do not deny the existence of a genotype x explant interaction in embryogenic callus formation in rye, which occurs in several Gramineae species such as barley (Ruiz et al. 1992) and wheat (Lazar et al. 1983).

3.3.3 Differences Within Cultivars

Rye has a great variability due to its allogamous condition. Obviously, plants of the same cultivar are genetically more homogeneous than plants from different cultivars. But, even in this case, plants from the same cultivar were not genetically identical, so by having a different genotype they could show a different response to the culture in vitro.

We studied the capacity of inflorescences belonging to the same plant and so having the same genotype to develop embryogenic calli. Table 4 shows the number of explants, plants, and their distribution for each cultivar in relation to their embryogenic callus formation frequencies. The frequency of plants capable of developing embryogenic calli (competent plants) varied between the different cultivars. The cultivar Elbon had the lowest frequency of the competent plants. Also, it was observed that in the competent plants, the frequency of explants which produced embryogenic calli was variable between cultivars. The better plants showed embryogenic calli formation in 30–40% of the Elbon explants, and in 60–70% of the Ailes explants (Linacero and Vázquez 1990).

Through the study of the response of individual plants, we have provided evidence that the genotype effect on the somatic embryogenic response in rye is due to two factors: the frequency of competent plants, and the response level of each plant. This knowledge is important in order to perform genetic selection experiments to improve the ability of in vitro embryogenesis.

Plants regenerated from tissue culture can be used as parents for the next generations, to carry out this type of experiment, since we assume that these plants have a high probability of giving a good genetic combination to produce embryogenic calli. We used regenerated plants from the Elbon cultivar which was the cultivar showing the worst response. The regenerated plants formed the R_0 generation. The next generation, R_1, was obtained by manual crosses between two R_0 regenerated plants, and all the R_1 plants from the same R_0 cross constituted a family. The R_2 generation was obtained through manual crosses between two R_1 plants from the same family.

The embryogenic response of R_2 progenies from ten different R_0 plants was studied by analyzing the immature embryos obtained from different R_1 crosses per regenerated plant. R_2 embryos, recovered on the plant used as female in each cross, were cultured.

The frequency of embryogenic calli obtained from the immature embryos belonging to one individual plant was considered the "embryogenic in vitro response". The embryogenic response per plant in the original Elbon population varied between 0 and 25%. The progenies of normal regenerated plants increased this response up to 50%. Figure 3 shows the plant distribution according to embryogenic response. A one-way ANOVA was performed to ascertain whether the differences between the original Elbon population and the regenerated plants response were significant. The result ($F_{1,58}$ = 7,602) showed they were highly significant. Thus, embryo culture per se could be a method to select genes which increase the response level. Similar results have been reported in other crops (Bingham et al. 1975). We obtained a regenerated plant which segregated polyembryonic kernels in its progeny, and also showed a very high embryogenic response. This response and their distribution were different from that shown by the normal regenerated plants and the original population (Linacero and Vázquez 1992).

Table 4. Number of competent and noncompetent plants. (Linacero and Vázquez 1990)

Cv.		Total	Without response	With response	10–20	20–30	Distribution of the competent plants depending upon the percentage of embryogenic callus formation			
							30–40	40–50	50–60	60–70
E	No. of plants	14	8	6	1	1	4	–	–	–
	(%)		(57.14)	(42.86)	(7.14)	(7.14)	(28.58)	–	–	–
	No. of explants	515	297	218	29	65	124	–	–	–
DE	No. of plants	19	10	9	–	1	–	8	–	–
	(%)		(52.63)	(47.37)	–	(5.26)	–	(42.11)	–	–
	No. of explants	1196	671	525	–	60	–	465	–	–
MD	No. of plants	18	6	12	–	1	2	–	9	–
	(%)		(33.33)	(66.67)	–	(5.56)	(11.11)	–	(50.00)	–
	No. of explants	616	181	435	–	39	110	–	266	–
As	No. of plants	11	4	7	–	–	2	–	–	5
	(%)		(36.36)	(63.64)	–	–	(18.18)	–	–	(45.46)
	No. of explants	687	211	476	–	–	106	–	–	370

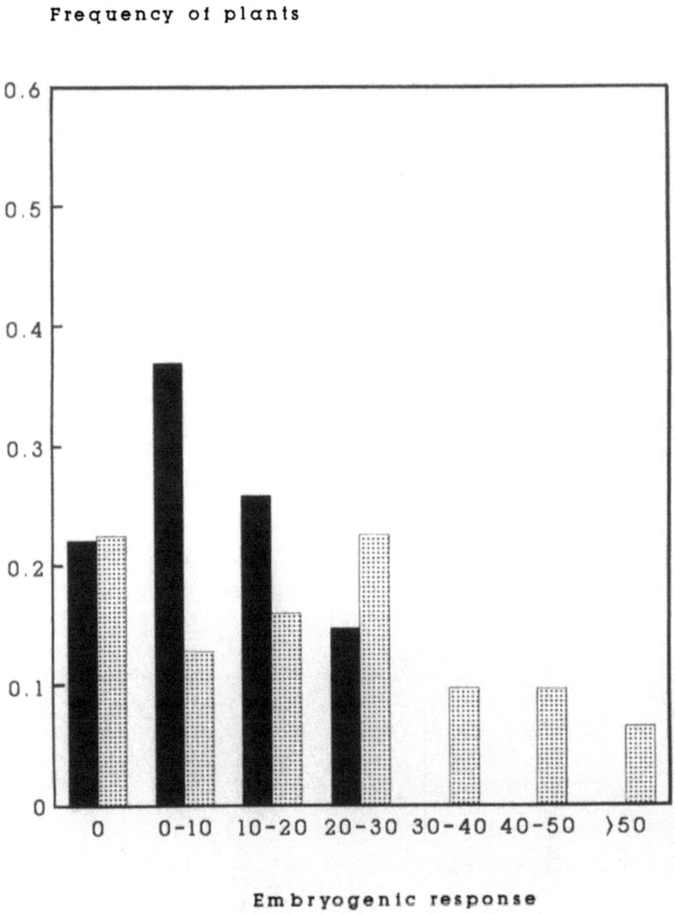

Fig. 3. Distribution of the plants according to their ability to produce embryogenic calli in vitro. ■ Plants from the original population of Elbon; ⊞ R_1 plants

4 Conclusions

Rye has been one of the most recalcitrant cereal species producing embryogenic cultures. However, using immature inflorescences as explants, it is possible to obtain a large number of regenerated plants of the same genotype. This fact is especially important for the propagation of an interesting genotype in allogamous species such as *Secale cereale*, or in hybrids.

On the other hand, it is also possible to improve the capacity to develop embryogenic callus in less responsive genotypes by selection, as the progenies of regenerated plants show an increase in the frequency of embryogenic calli formation.

5 Protocol for the Induction of Embryogenic Callus from Immature Inflorescences

1. Rye plants are raised from seed in the field.
2. Shoots containing young, unemerged inflorescences (3 or 4 leaves) are rinsed for a few seconds with 70% ethnol, surface sterilized in a sodium hypochlorite solution (5%) with a few drops of Tween 80 for 15 min, and washed three times in sterile water.
3. The leaves are removed and immature inflorescences are cut into segments of about 2 mm which are placed in Petri dishes on MS medium with 3% sucrose, 1% Difco agar, and supplemented with 1–2 mg/l 2,4-D. All media are adjusted to pH 5.7 and autoclaved.
4. After sealing with Parafilm, Petri dishes are maintained at 25±1 °C in a 16-h regime of diffuse light (using filter papers) for 28 days.
5. Embryogenic calli are transferred to test tubes with MS basal medium without hormones to promote the maturation and germination of the somatic embryos. The cultures are maintained at 25 ± 1 °C in a 16-h light regime.
6. When the plantlets have developed good root and shoot systems, they can be transferred to soil.

Note: The leaf explants are obtained from 3–4-week-old plants growing under aseptic conditions in test tubes on basal MS medium and incubated at 25±1 °C under a light regime of 16 h.

Immature embryos (1–2 mm length) are obtained from developing seeds.

Acknowledgments. This work has been supported by a grant from the CICYT of Spain (N 0678-86). The authors thank Mr. E. Wiltshire for critical reading of the manuscript.

References

Abe T, Futsuhara Y (1985) Efficient plant regeneration by somatic embryogenesis from root callus tissues of rice. J Plant Physiol 121: 111–118

Barceló P, Vazquez AM, Martin A (1989) Somatic embryogenesis and plant regeneration from *Tritordeum*. Plant Breed 103: 235–240

Bebeli P, Karp A, Kaltsikes PJ (1988) Plant regeneration and somaclonal variation from cultured immature embryos of sister lines of rye and triticale differing in their content of heterochromatin. Theor Appl Genet 75: 929–936

Bingham ET, Hurley LV, Katz DM, Saunders W (1975) Breeding of alfalfa which regenerates from callus tissue in culture. Crop Sci 15: 719–721

Bushuk W (1976) History. World distribution, production, and marketing. In: Rye: production, chemistry and technology vol V. American Association of Cereal Chemists, St Paul pp 1–11

Carew PD, Schwarting AE (1958) Production of rye embryo callus. Bot Gaz 119: 237–239

Duncan DR, Williams ME, Zehr BE, Widholm JM (1985) The production of callus capable of plant regeneration from immature embryos of numerous *Zea mays* genotypes. Planta 165: 322–332

Eapen S, Rao PS (1982) Callus induction and plant regeneration from immature embryos of rye and triticale. Plant Cell Tissue Organ Cult 1: 221–227

Evans LE, Scoles GJ (1976) Cytogenetics. Plant breeding, and agronomy. In: Rye: production, chemistry and technology vol V. American Association of Cereal Chemists, St Paul, pp 13–26

Kumbriegel-Schroeren G, Finger J, Schroeren V, Binding H (1984) Embryoid formation and plant regeneration from callus of *Secale cereale* L. Z. Pflanzenzücht 92: 89–94

Lazar MD, Collins GB, Vian WE (1983) Genetic and environmental effects on the growth and differentiation of wheat somatic cell cultures. J Hered 74: 353–357

Linacero R, Vázquez AM (1986) Somatic embryogenesis and plant regeneration from leaf tissue of *Secale cereale* L. Plant Sci 44: 219–222

Linacero R, Vázquez AM (1990) Somatic embryogenesis from immature inflorescences of rye. Plant Sci 72: 253–258

Linacero R, Vázquez AM (1992) Somatic embryogenesis in polyembryonic *Secale cereale* L. Plant Cell Rep 12: 26–28
Lu C, Chandler SF, Vasil IK (1984) Somatic embryogenesis and plant regeneration from cultured immature embryos of rye (*Secale cereale* L). J Plant Physiol 115: 237–244
Lührs R, Lörz H (1987) Plant regeneration in vitro from embryogenic cultures of spring and winter type barley (*Hordeum vulgare* L) varieties. Theor Appl Genet 75: 16–25
Maddok SE, Lancaster VA, Riscott R, Franklin J (1983) Plant regeneration from cultured immature embryos and inflorescences of 25 cultivars of wheat (*Triticum aestivum*). J Exp Bot 34: 915–926
Murashige T, Skoog F (1962) A revised medium for rapid growth and bioassays with tobacco tissue cultures. Physiol Plant 15: 473–497
Roberts E, Street HE (1955) The continuous culture of excised rye roots. Physiol Plant 8: 238–262
Ruiz ML, Rueda J, Peláez MI, Espino FJ, Candela M, Sendino AM, Vázquez AM (1992) Somatic embryogenesis, plant regeneration and somaclonal variation in barley. Plant Cell Tissue Organ Cult 28: 97–101
Rybczynski JJ, Zdunczyk W (1986) Somatic embryogenesis and plantlet regeneration in the genus *Secale*. I. Somatic embryogenesis and organogenesis from cultured immature embryos of five wild species of rye. Theor Appl Genet 73: 267–271
Rybczynski JJ, Stolarz A, Slusarkiewic-Jarzina A (1980) Differentiation of shoot elements from the rachis of *Secale cereale* L. Acta Soc Bot Polon 49: 161–168
Thomas MR, Scott KJ (1985) Plant regeneration by somatic embryogenesis from callus initiated from immature inflorescences of *Hordeum vulgare*. J. Plant Physiol 121: 159–169
Vasil IK (1987) Developing cell and tissue culture systems for the improvement of cereal and grass crops. J Plant Physiol 128: 193–218
Vázquez AM, Espino FJ, Rueda J, Candela M, Sendino AM (1991) A comparative study of somatic embryogenesis in *Secale vavilovii*. Plant Cell Rep 10: 265–268
Vences FJ, Vaquero F, Vázquez AM, Pérez de la Vega M (1986) Isozymatic changes during in vitro morphogenetic process in self-pollinated species of *Secale*. J Plant Physiol 123: 31–36
Wernicke W, Brettell RIS (1980) Somatic embryogenesis from *Sorghum bicolor* leaves. Nature 287: 138–139
Wernicke W, Milkovits L (1984) Developmental gradients in wheat leaves–response of leaf segments in different genotypes cultured in vitro. J Plant Physiol 115: 49–58
Wernicke W, Brettell R, Wakizuka T, Potrykus I (1981) Adventitious root formation from rice leaves. Z Pflanzenphysiol 103: 361–365

I.5 Somatic Embryogenesis in Oat (*Avena sativa* L.)

P.P. BREGITZER[1], S.K.C. MILACH[2], H.W. RINES[2,3], and D.A. SOMERS[2]

1 Introduction

Tissue cultures of most gramineous species may undergo plant generation via either organogenesis or somatic embryogenesis. Callus cultures capable of somatic embryogenesis are more likely to exhibit rapid growth rates, friability, and high levels of plant regeneration than cultures capable only of organogenesis. Because of these desirable characteristics, much of the effort to produce regenerable tissue cultures of monocotyledoenous plants has been focused on the establishment of friable, embryogenic (FE) callus. FE callus is a valuable source of cells for the establishment of embryogenic suspension cultures and is currently regarded as the preferred source of target cells for genetic transformation of most cereal species. Development of FE callus in oat and the ensuing demonstration of plant regeneration via somatic embryogenesis, and of factors affecting plant regeneration were important contributions to the development of a genetic transformation system in oat (Somers et al. 1992). This chapter reviews the current status of investigations of somatic embryogenesis in oat.

2 Establishment of Callus Culture and Somatic Embryogenesis

2.1 Tissue Cultures

In hexaploid cultivated oats, (*Avena sativa* L.) as in other gramineous species, plant regeneration from tissue culture appears to proceed by either organogenesis or somatic embryogenesis. Initiation of oat tissue cultures was reported initially in 1967 (Carter et al.). Plant regeneration from oat tissue cultures was reported by two groups independently in 1976 (Cummings et al.; Lörz et al.). In these studies, plant regeneration was observed from nonfriable callus that was

[1] USDA-ARS, National Small Grains Germplasm Research Facility, P.O. Box 307, Aberdeen, Idaho 83210, USA
[2] Department of Agronomy and Plant Genetics, University of Minnesota, St. Paul, Minnesota 55108, USA
[3] USDA-ARS, Plant Science Research Unit, St. Paul, Minnesota 55108, USA

Biotechnology in Agriculture and Forestry, Vol. 31
Somatic Embryogenesis and Synthetic Seed II (ed. by Y.P.S. Bajaj)

characterized by the presence of organized, green primordia. Plants were presumed to be regenerated via organogenesis, primarily based on morphological and developmental similarities to maize cultures that had been characterized as organogenic (Green and Phillips 1975; Springer et al. 1979). Similarly organized, nonfriable callus cultures were described later (Heyser and Nabors 1982; Nabors et al. 1982, 1983). In these studies, embryogenic sectors were identified based on similarities to embryogenic callus described for other crops. These sectors had an improved frequency and longevity of plant regeneration (Nabors et al. 1983). These descriptions of embryogenic oat callus, and the development of FE maize callus through friable sector selection (Green 1982; Armstrong and Green 1985), ultimately served as the impetus for the development of FE oat callus (Bregitzer et al. 1989).

2.2 Friable, Embryogenic Callus

Establishment of FE oat callus is dependent on the source of the explant, its physiological stage and genotype, and the composition of the culture initiation medium. The majority of our work has utilized immature zygotic embryos cultured on an MS salts medium (Murashige and Skoog 1962) containing 2mg/l 2,4-dichlorophenoxyacetic acid (2,4-D) (Bregitzer et al. 1989; Rines et al. 1992). Other explants exhibiting callus formation include mesocotyl of germinated seedlings (Heyser and Nabors 1982; Bregitzer et al. 1989), mature seeds (Carter et al. 1967; Lörz et al. 1976; Heyser and Nabors 1982), and basal sections of immature leaves (Y. Matsuda, H.W. Rines, and W.R. Bushnell, unpubl.). In our experience, immature embryos provided the highest and most reproducible callus initiation frequency (Bregitzer et al. 1989). Probably most important in our work is the use of specific genotypes that were bred and selected for culture response (Bregitzer et al. 1989). These oat lines are designated GAF and GAF/Park to signify that they are F_4 derivations of a cross of Garland and the wild oat *Avena fatua* 1221 (GAF) or the subsequent cross of GAF by Park (GAF/Park).

Friable, embryogenic oat callus is established in two stages. Callus initiated from immature embryos consists of a mixture of callus types, most of which are composed of vacuolated nonregenerable cells. FE callus sectors are identified in newly initiated callus as friable, semi-opaque, white callus from which numerous suspensor-borne, globular somatic embryos are produced (Fig. 1a). These FE callus sectors are isolated and continuously selected during subculture for FE appearance at 2- to 3-week intervals. The resulting FE oat callus cultures (Fig. 1b) are strikingly similar to the FE maize callus in which somatic embryogenesis has been documented (Green 1982; Armstrong and Green 1985; McCain and Hodges 1986; McCain et al. 1988).

Scanning electron micrographs (SEM) (Fig. 1c) and light microscopy were used to characterized the stage of development of somatic embryos in FE callus (Bregitzer et al. 1989, 1991; Milach et al. 1992). These studies indicated that the callus consists of globular embryoids supported on multicellular suspensors, which is consistent with descriptions of somatic embryogenesis in other species (Haccius 1978). Histological studies were conducted to further characterize

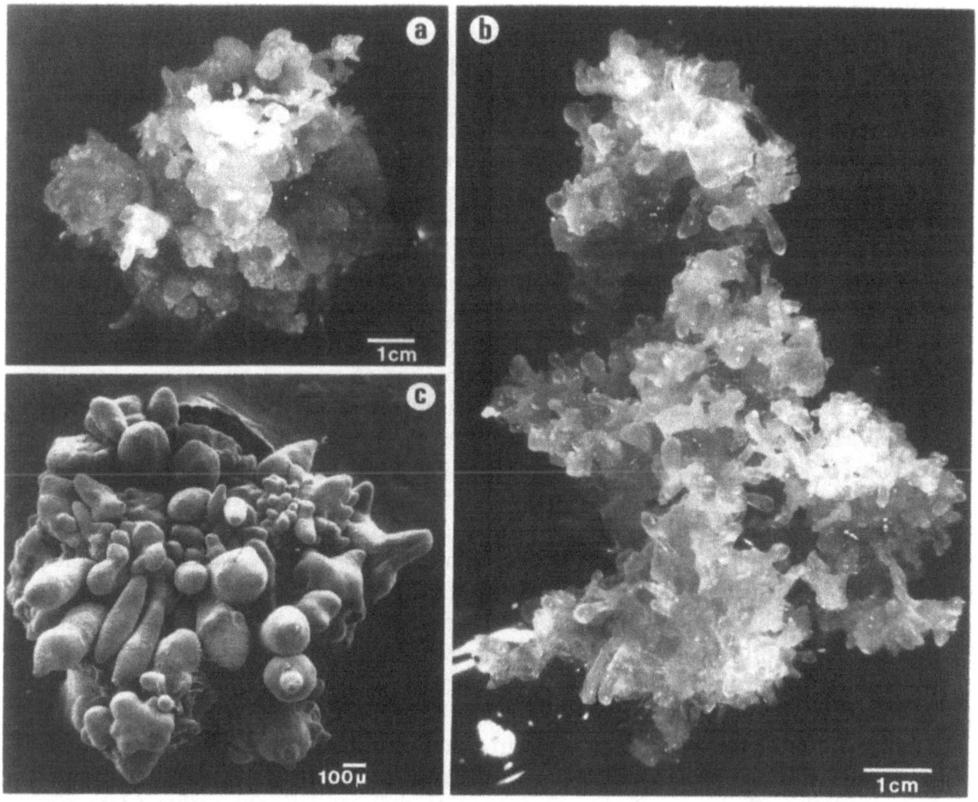

Fig. 1. Development of FE oat callus from nonfriable callus derived from immature FI, embryos of GAF/Park. **a** Embryogenic sectors on nonfriable callus; **b** FE callus 26 weeks after initial selection; **c** scanning electron micrograph of FE callus. (Bregitzer et al. 1989)

somatic embryos in FE callus initiated from the cultivar Starter as described in the next section. Sectors of the Starter FE callus appeared to be similar to GAF and GAF/Park FE callus (data not shown). Sectioned FE callus contained an array of apparent somatic embryoids at various stages of development, ranging from globular to torpedo stages (Milach et al. 1992). Generally, embryoid structures exhibited a well-defined layer of epidermal cells surrounding densely cytoplasmic cells. Vascular system development was observed in the more mature embryoids. These studies also indicated that suspensor-like structures were multicellular. The progenitor cells that give rise to the somatic embryos (as described in maize by McCain et al. 1988) were not resolved by Milach et al. (1992). However, the similarity of oat somatic embryos to observations of somatic embryogenesis in maize (McCain et al. 1988) and other species (Haccius 1978) suggests a unicellular origin of somatic embryos.

The structural integrity of oat embryos was confirmed by observing somatic embryo maturation when FE callus was plated on embryo maturation medium and in studies in which somatic embryos were isolated from the callus and cultured individually on different media. During this maturation process, the identification of somatic embryos in FE callus was made based on well-defined embryogenic structures such as scutella and coleoptilar nodes (Fig. 2a,b). Individual somatic embryos were isolated and cultured to characterize the effects of various media and somatic embryo development. The relatively undifferentiated, globular embryos were induced to mature and germinate by manipulating the concentrations of growth regulators and sucrose in the growth medium (Bregitzer et al. 1989). The choice of growth regulators and sucrose concentration in the culture medium profoundly affected the morphology of regenerating structures developing from isolated somatic embryos. Elevating the media sucrose concentration to 6% in the absence of growth regulators (MS-6S medium) induced embryo maturation (Fig. 2b,c); subsequent reduction of the sucrose concentration allowed the development of normal-appearing, bipolar plants from isolated somatic embryos (Bregitzer et al. 1989). An alternative regeneration process using an MS medium containing 2 mg/l napthyleneacetic acid and 0.2 mg/l benzylaminopurine (MS-N+B medium), a medium developed for plant regeneration from oat callus cultures (Rines and McCoy 1981), did not result in recognizable embryo maturation. Often in this medium, individual embryos enlarged and appeared to proliferate as nonfriable callus and turned green, while supporting the development of multiple shoot meristems (Bregitzer et al. 1991). During this stage of shoot regeneration, the shoots did not exhibit parallel formation of roots, but roots developed after transfer to MS medium containing no growth regulators. This latter regeneration process was perhaps less visually appealing, but roughly double the efficiency of plant regeneration on a callus fresh weight basis. Subsequent research showed that addition of either BAP or kinetin to plant regeneration media at concentrations of 1 mg/l or greater increased the efficiency of shoot regeneration from FE oat callus (data not shown). Similar effects of cytokinins on callus morphology and plant regeneration have been reported in wheat (Armstrong et al. 1987).

2.3 Secondary Somatic Embryogenesis

Isolated somatic embryos were induced to undergo secondary embryogenesis by plating on callus maintenance medium (Bregitzer et al. 1991). Isolation was accomplished by vortexing FE callus followed by fractionation based on size. Somatic embryos were cultured on MS-2D medium in 8 × 10 mm wells in a 96-well microtiter plate. Fifty-five percent of the individually cultured somatic embryos reinitiated FE callus. Approximately one-half of these callus cultures regenerated fertile plants. Secondary initiation of FE callus from somatic embryos also appears to occur in FE callus (data not shown). Therefore, it is possible that somatic embryos comprise the progenitor cells from which FE callus proliferates.

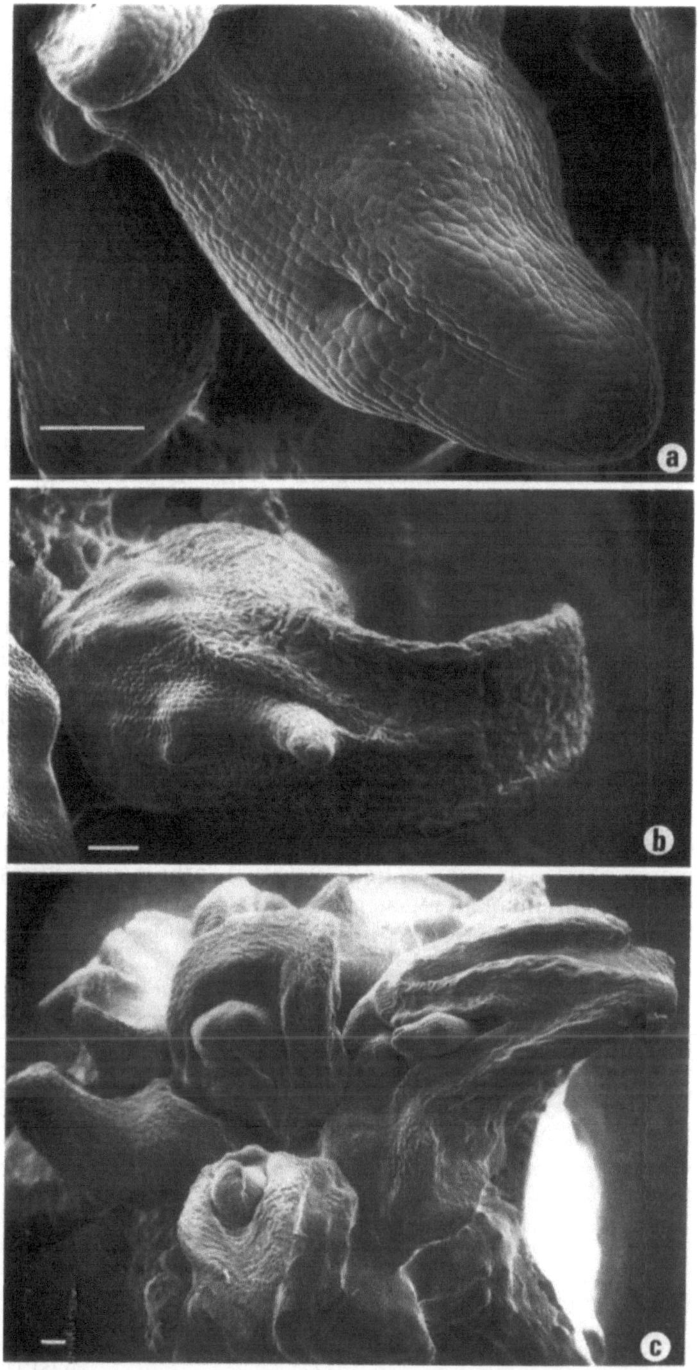

Fig. 2. a Scanning electron micrograph of oat somatic embryos from FE callus of GAF/Park on MS-2D maintenance medium; **b** after 3 days on MS-6S embryo maturation medium; **c** after 9 days on MS-6S medium. (Bregitzer et al. 1989)

2.4 Genotype × Media Interactions

Variation for callus initiation among oat genotypes has been reported (Cummings et al. 1976; Rines and McCoy 1981; Bregitzer et al. 1989). Generally, the frequency of FE callus initiation of some of the elite germplasm lines has been very low. As previously mentioned, FE callus cultures in oat most commonly have been established from immature embryos of the GAF and GAF/Park genotypes plated on MS basal medium supplemented with high levels (2–5 mg/l) of auxin, usually 2,4-D (Rines and Luke 1985; Bregitzer et al. 1989). Although this culture medium has been successful with GAF and GAF/Park, these genotypes may be of limited value to future plant improvement efforts because of their poor agronomic qualities. Therefore, extensive efforts to investigate other genotypes for initiation of FE callus are warranted.

A mixture of embryogenic and nonembryogenic callus has been identified on tissue cultures initiated from the cultivar Starter by placing immature embryos on MS salts medium containing 0.2 mg/l 2,4-D and 0.2 mg/l benzylaminopurine (BAP). By selecting the embryogenic sectors and subculturing them on MS medium containing 2 mg/l 2,4-D, it has been possible to establish FE callus. Somatic embryogenesis in Starter callus appears to be similar to that previously described for the GAF/Park FE callus (Bregitzer et al. 1989). Plant regeneration was obtained from these cultures by transferring them to the MS-N+B regeneration medium. An increase in the frequency of FE callus initiation has been achieved recently for several oat elite lines by culturing immature embryos on MS salts medium containing 0.2 mg/l 2,4-D and 0.2 mg/l BAP (Table 1). However, the frequency of FE callus initiation from GAF/Park immature embryos was not improved by using the modified medium (Table 1). Embryogenic callus sectors from callus initiated from the oat cultivars (Table 1) were isolated at about 21 days and transferred to MS medium supplemented with 2 mg/l 2,4-D for callus increase and maintenance. Following two or three subcultures at 21-day intervals on this medium, plant regeneration from tissue cultures of all genotypes listed in Table 1 was obtained by transfer to the MS-N+B regeneration medium (data not

Table 1. Effect of culture media on initiation of embryogenic callus from immature embryos of four oat genotypes

Genotype	Medium	No. of immature embryos cultured	Number of immature embryos producing embryogenic callus (% embryogenic callus initiation)
Donald	1[a]	164	27 (16.5%)
	2[b]	145	42 (29.6%)
Ogle	1	114	2 (1.8%)
	2	116	5 (4.3%)
Starter	1	189	12 (6.3%)
	2	191	41 (21.5%)
GAF/Park	1	148	96 (64.9%)
	2	137	70 (51.1%)

[a] 1 = MS + 2 mg/l 2,4-D
[b] 2 = MS + 0.2 mg/l 2,4-D + 0.2 mg/l BAP

shown). The variation in establishment of FE callus from the different oat genotypes tested on the two types of callus initiation media suggests that genotype × media interactions may be an important factor in further improving tissue culture methodologies in oat.

2.5 Cell Suspension and Protoplast Cultures

Friable, embryogenic callus can be inoculated into liquid medium and, following continued selection, will produce rapidly growing dedifferentiated suspension cultures. Furthermore, plants were regenerated from GAF suspension cultures for longer than 1 year after the cultures were initiated; however, all plants were male-sterile (Gu and Somers,unpubl.). Protoplasts isolated from embryogenic suspension cultures reformed cell walls and divided to form callus which, in rare instances, produced small shoot-forming structures; however, mature plants could not be recovered.

2.6 Embryogenic Cells as Targets for Transformation

Evidence of somatic embryogenesis and the extremely friable nature of FE oat callus indicated that cells within the FE callus would be suitable targets for the introduction of DNA and regeneration of transgenic plants. Conditions were established to isolate and culture globular somatic embryos (Bregitzer et al. 1991). Callus proliferation seemed to occur at least in part via secondary embryogenesis from the surface of somatic embryos, suggesting that the surface cells of somatic embryos were totipotent. The putative epidermal location of embryogenic cells suggested that DNA delivery via microinjection could be accomplished. Iontophoretic microinjection was used to deliver Lucifer yellow into peripheral cells of isolated somatic embryos immobilized in agarose (Bregitzer 1989). This dye served as a sensitive marker for establishing optimal conditions for dye delivery and cell viability (Fig. 3a–c). However, DNA delivery via iontophoresis proved to be technically difficult, and techniques such as microprojectile bombardment appeared more promising as a means to deliver DNA into embryogenic cells.

Microprojectile bombardment-mediated DNA delivery was applied to both FE callus and suspension cultures as targets (Somers et al. 1992). Tissue cultures expressing the introduced selectable marker gene(s) were analyzed using Southern analysis to confirm integration of the transgene sequences. The results of the initial experiments indicated that there was no difference in selection of transgenic tissue cultures from either bombarded suspension cultures or FE callus. Because of the problems in plant regeneration and fertility associated with long-term suspension cultures (Gu and Somers,unpubl.), suspension cultures were not further employed in transformation studies. Transgenic plants regenerated from FE callus were uniformly transformed as opposed to being chimeric for transgene expression. Furthermore, transgenic phenotypes were inherited in expected Mendelian segregation ratios confirming that they were stably transformed.

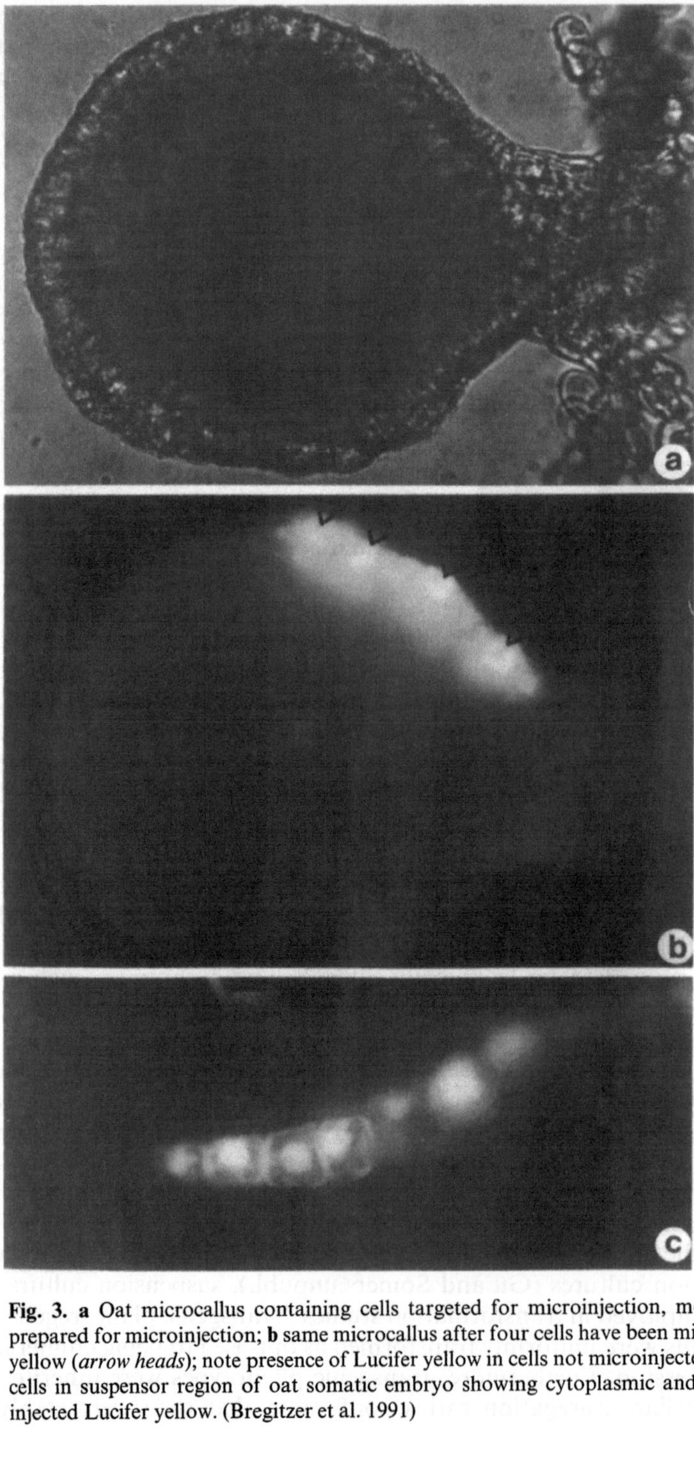

Fig. 3. **a** Oat microcallus containing cells targeted for microinjection, mounted in agarose, and prepared for microinjection; **b** same microcallus after four cells have been microinjected with Lucifer yellow (*arrow heads*); note presence of Lucifer yellow in cells not microinjected; **c** relatively vacuolate cells in suspensor region of oat somatic embryo showing cytoplasmic and nuclear localization of injected Lucifer yellow. (Bregitzer et al. 1991)

Further modification of FE callus has resulted in improved regeneration of fertile transgenic oat plants. In the absence of intense selection during subculturing for well-developed, globular-stage embryos, FE oat callus becomes noticeably less differentiated. These less differentiated callus cultures are more rapidly initiated from immature embryos, easier to maintain, provide a more uniform tissue for bombardment, and retain the capacity of prolific, long-term plant regeneration. Although well-developed somatic embryos are not present in this type of culture at high frequencies, renewed selection for sectors bearing well-developed somatic embryos resulted in FE callus, indicating that these cultures may consist, at least in part, of progenitor cells for FE callus.

3 Summary

Somatic embryogenesis in oat (*Avena sativa* L.) is associated with both non-friable and friable callus. Embryogenic callus is routinely initiated by culturing immature embryos of selected genotypes on Murashige and Skoog medium containing 2,4-D. This callus is a mixture of various callus types. Persistent selection during subculturing eventually yields uniformly friable, embryogenic callus. Embryogenic callus inducton has been achieved from a number of cultivars by manipulation of 2,4-D and benzylaminopurine levels in the culture medium. Friable, embryogenic callus has been successfully used as a source of totipotent target cells for microprojectile bombardment-mediated DNA delivery, resulting in the production of transgenic tissue cultures that regenerated fertile, transgenic plants.

References

Armstrong CL, Green CE (1985) Establishment and maintenance of friable, embryogenic maize callus and the involvement of L-proline. Planta 163: 207–214

Armstrong TA, Metz GS, Mascia PN (1987) Two regeneration systems for the production of haploid plants from wheat anther culture. Plant Sci 51: 231–237

Bregitzer PP (1989) Development and characterization of friable, embryogenic oat callus and cellular microinjection of DNA. PhD Thesis, University of Minnesota, St. Paul, 62 pp

Bregitzer P, Somers DA, Rines HW (1989) Development and characterization of friable, embryogenic oat callus. Crop Sci 29: 798–803

Bregitzer P, Bushnell WR, Rines HW, Somers DA (1991) Callus formation and plant regeneration from somatic embryos of oat (*Avena sativa* L.). Plant Cell Rep 10: 243–246

Carter O, Yamada Y, Takahashi E (1967) Tissue culture of oats. Nature (Lond) 214; 1029–1030

Cummings DP, Green CE, Stuthman DD (1976) Callus induction and plant regeneration in oats. Crop Sci 16: 465–470

Green CE (1982) Somatic embryogenesis and plant regeneration from the friable callus of *Zea mays*. In: Fujiwara A(ed) Plant tissue culture, Maruzen, Tokyo, pp 107–108

Green CE, Phillips RL (1975) Plant regeneration from tissue cultures of maize. Crop Sci 15: 417–421

Haccius B (1978) Question of unicellular origin of non-zygotic embryos in callus cultures. Phytomorphology 28: 74–81

Heyser JW, Nabors MW (1982) Long term plant regeneration, somatic embryogenesis, and green spot formation in secondary oat (*Avena sativa*) callus. Z Pflanzenphysiol 107: 153–160

Lörz H, Harms CT, Potrykus I (1976) Regeneration of plants from callus in *Avena sativa* L.Z Pflanzenzucht 77: 257–259

McCain JW, Hodges TK (1986) Anatomy of somatic embryos from maize embryo cultures. Bot Gaz 146: 453–460

McCain JW, Kamo KK, Hodges TK (1988) Characterization of somatic embryo development and plant regeneration from friable maize callus cultures. Bot Gaz 149: 16–20

Milach SCK, Rines HW, Somers DA, Gu W, Grando M (1992) Improvements in embryogenic callus culture in oat (*Avena sativa* L.) In: Int Crop Science Congr Abstr. Iowa State University, Ames, p 61

Murashige T, Skoog F (1962) A revised medium for rapid growth and bioassays with tobacco tissue cultures. Physiol Plant 15: 473–497

Nabors MW, Krosky CS, McHugh DM (1982) Green spots are predictors of high callus growth rates and shoot formation in normal and salt stressed tissue culture of oat (*Avena sativa* L.). Z Pflanzenphysiol 105: 341–349

Nabors MW, Heyser JW, Dyker TA, DeMott KJ (1983) Long-duration high-frequency plant regeneration from cereal tissue cultures. Planta 157: 385–391

Rines HW, Luke HH (1985) Selection and regeneration of toxin-intensitive plants from tissue cultures of oats (*Avena sativa*) susceptible to *Helminthosporium victoriae*. Theor Appl Genet 71: 16–21

Rines HW, McCoy TJ (1981) Tissue culture initiation and regeneration in hexaploid species of oats. Crop Sci 21: 837–842

Rines HW, Phillips RL, Somers DA (1992) Application of tissue cultures to oat improvement. In: Marshall HG Sorrells ME (eds) Oat science and technology Am Soc Agron, Madison, pp 777–791

Somers DA, Rines HW, Gu W, Kaeppler HF, Bushnell WR (1992) Fertile, transgenic oat plants. Bio/ Technology 10: 1589–1594

Springer WD, Green CE, Kohn KA (1979) A histological examination of tissue culture initiation from immature embryos of maize. Protoplasma 101: 269–281

I.6 Somatic Embryogenesis in Ryegrass (*Lolium multiflorum*) and Tall Fescue (*Festuca arundinacea*)

G. ALIBERT[1], A. SALLANDROUZE[1], and C. GRAND[2]

1 Introduction

Ryegrass (*Lolium multiflorum* Lam.) and tall Fescue (*Festuca arundinacea* Scherb.) are two of the most important grasses cultivated in the world as forage and turf. For example, around 160 000 tonnes of grain seed (forage and turf) are marketed every year in western Europe. This total includes 75 000 t of perennial ryegrass, and 30 000 t of tall fescue.

At present, the cultivars are all synthetic varieties. Partly for this reason, the genetic improvement through conventional breeding of dry matter yield in both tall fescue and ryegrass (as forage) has been small compared to that obtained in grain crops. The major increases in agronomical performances concern characters such as persistency, resistance to certain diseases, and winter hardiness. Unfortunately, they are negatively correlated with each other and/or with herbage yield (Van Den Bogaert 1991).

There are many complementary characters in *Lolium* and *Festuca* species (Table 1) which, if combined together in a single genotype, would have considerable adaptive value in grassland management. As early as 1933 these properties led Peto to try to produce interspecific hybrids between these species. The so-called *Festulolium* are now easily produced through in vitro regeneration of immature embryos (Humphreys 1989; Bonnecuelle 1992). However, the hybrids are generally characterized by low fertility (Bückner et al. 1961; Lewiss 1966; Rybozynski et al. 1983). In order to solve this problem, the introgression into one of the parents of the amphiploid hybrids is viewed (Kleijer 1988; Humphreys 1989).

Alternative approaches deal with cell or protoplast regeneration in order to perform protoplast fusion or gene transfer. Plant regeneration from protoplasts produced from established embryogenic cell suspension cultures was described by several authors. Dalton (1988) was the first to regenerate whole plants from *Festuca arundinacea* and *Lolium perenne* protoplasts, however, most of the regenerated plants were albino. Green shoots were obtained from protoplasts by Creemers-Molenaar et al. (1989) in *L. perenne*, and recently by Takamizo et al. (1990) and Wang et al. (1992) in *Festuca arundinacea*. The plants, when

[1] Laboratoire de Biotechnologie et Amélioration des Plantes (BAP), INP-ENSAT, Unité associée á l'INRA, 145 Avenue de Muret, 31076 Toulouse Cedex, France
[2] RAGT, 18 rue de Séguret-Saincric, 12003 Rodez Cedex, France

Biotechnology in Agriculture and Forestry, Vol. 31
Somatic Embryogenesis and Synthetic Seed II (ed. by Y.P.S. Bajaj)

Table 1. Comparison of ryegrass and tall fescue for their agronomical traits. (Kleijer 1988)

Characters	Italian ryegrass	Tall fescue
Ease of culture	++	–
Yield the first year	++	+/–
Hardiness	–	+
Resistance to cold	–	+
Resistance to draught	–	+
Forage quality	+	–
Earliness	–	+

transferred to soil, gave no indications as to the fertility of the protoclones. We also developed a protocol for plant regeneration from protoplasts evolved from cell suspension cultures established from mature germinating seeds (Radilofe 1991) however, when cultured in soil, none of the plants produced seeds.

Recent works by Sam et al. (1992), on the one hand, and Takamizo et al. (1992), on the other, using protoplasts, showed that gene transfer and somatic hybridization could be developed in ryegrass and fescue. However, until now, no information is available about the normality of the putative transformants or the somatic hybrids regenerated.

Because somatic embryogenesis is the best way to produce normal plants, this procedure was explored in the past by several authors. Starting from the pioneer works of Lowe and Conger (1979) on root and shoot induction on callus of tall fescue, Dale (1980) was the first to produce embryoids in *L. multiflorum* by in vitro culture. Later, Torello and Symington (1984) on *L. perenne*, Torello et al. (1984) in red fescue, and Creemers-Molenaar et al. (1989) in *L. perenne* obtained plant regeneration by somatic embryogenesis induced in cell suspension cultures initiated from immature inflorescences or from seed-derived embryos.

Recently, we (Radilofe-Rajoelina et al. 1990) tested different protocols for regeneration applicable to both ryegrass and fescue, and have devised an improved process of plant production by somatic embryogenesis, which is discussed here.

2 Somatic Embryogenesis

Plant regeneration of grasses by somatic embryogenesis required the production of embryogenic calli and/or embryogenic cell suspension cultures. In monocots, these are generally produced from either mature or immature embryos (Vasil 1987). In the case of ryegrass and fescue, we used mature seeds.

2.1 General Protocol (Radilofe-Rajoelina et al. 1990)

2.1.1 Seed Sterilization and Culture

Seeds were dipped in concentrated hydrochloric acid for 15 min, and then immersed for 10 min in 10% commercial sodium hypochlorite solution. After three washes with sterile water, the seeds were cultured in Petri dishes (55 × 10 mm, ten seeds per dish) containing 5 ml of MS medium (Murashige and Skoog 1962) without glycine. The culture medium contained 30 g/l sucrose, 8 g/l Difco-Agar, and 10 mg/l 2,4-D. The pH was adjusted to 5.7 with M KOH (MS10). The Petri dishes were kept in the dark at 25 °C for 8 weeks.

2.1.2 Callus Production and Culture

Calli appearing on the seedlings (Fig. 1) were recovered and transferred to the same medium, but containing 5 mg/l 2,4-D. The culture was maintained under the same conditions as above until the calli reached 0.5 cm (this procedure takes about 8 weeks).

2.1.3 Initiation of Cell Suspensions

About 5 g of calli was disrupted in 40 ml of MS medium containing 5 mg/l 2,4-D, 0.2 mg/l BAP, 1 g/l casein hydrolysate, and 30 g/l sucrose. The pH was adjusted to 5.7 with M KOH (MS5). The callus suspension, was transferred to 250-ml Erlenmeyer flasks and continuously shaken at 150 rpm in the dark at 25 °C. Seven days later the suspension was allowed to settle, 20 ml of the supernatant was discarded and substituted with 20 ml of the same fresh medium, and shaken for an additional week. Then the same operation as above was again performed, except that the new medium contained 2 mg/l of 2,4-D (MS2). From this stage the suspension was half diluted every week with this medium.

2.1.4 Plant Regeneration

The initiated suspension was composed of isolated elongated cells, small clusters of dense cytoplasmic cells, microcolonies, and calli of about 1–3 mm. Every 2 weeks, these calli were removed from the suspension before dilution and transferred to culture tubes containing 10 ml of a half-strength MS medium, 20 g/l sucrose, 8 g/l Difco-Agar, and 0.1 mg/l BAP, pH 5.7. The cultures were incubated under light (25 μE m^{-2} s^{-1}) at 25 °C.

Several plantlets developed directly on each callus after about 3–5 weeks of culture (Figs. 2 and 3). Before transferring to soil the plants were separated from each other and cultured individually in pots (1/3 soil-1/3 mold-1/3 sand) under a water saturated atmosphere to avoid excessive drying. After 2 weeks the plants could be acclimatized in a greenhouse (Fig. 4) and then transferred to the field (Fig. 5).

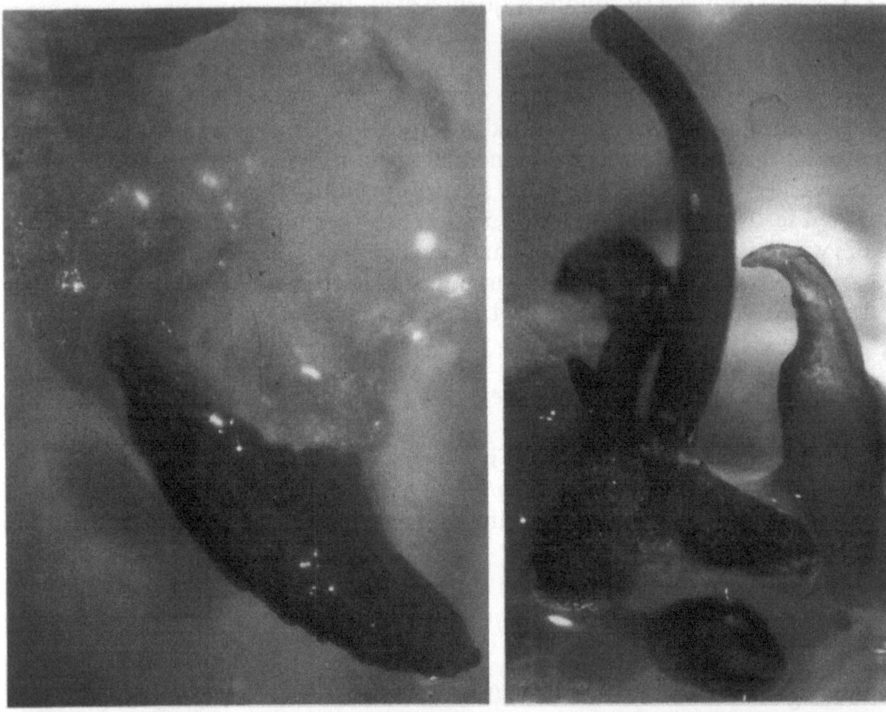

1 2

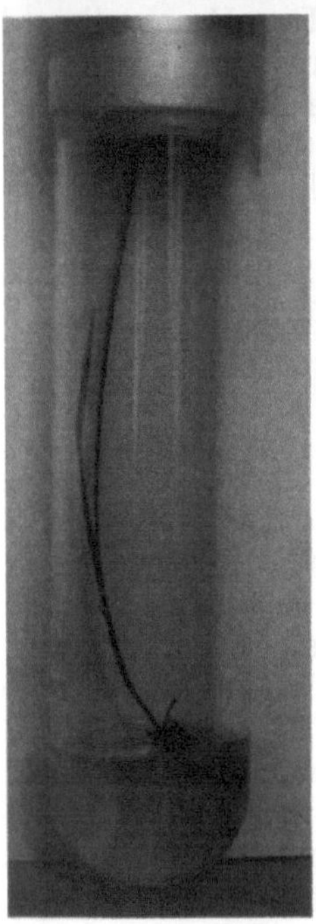

4,5

Figs. 1–5. Plant regeneration of tall fescue (*Festuca arundinacea* Schreb.) by somatic embryogenesis. (Similar patterns of developments are obtained for ryegrass). **1** Callus induction on a mature seed (8 weeks; MS medium + 30 g/l sucrose, 8 g/l Difco-Agar, and 10 mg/l 2,4-D). **2** Plantlets appearing on a callus 4 weeks after transfer to regeneration medium (1/2 MS + 20 g/l sucrose; 8 g/l Difco-Agar and 0.1 mg/l BAP). **3** A plant in a culture tube (8 weeks after transfer). **4** Plant acclimated in a pot in a greenhouse. **5** Cultured plants in the field

3

3 Factors Controlling Plant Regeneration

1. High 2,4-D concentrations are necessary to induce calli on the young seedlings, however, prolonged exposure to the high hormone concentration produced deleterious effects on the calli.
2. A stepwise decrease in the 2,4-D concentration (10-5-2 mg/l) was necessary to produce an embryogenic cell suspension.
3. Keeping the suspension in the dark was a critical point. If illuminated, the suspension rapidly turned brown and the cells died.
4. Transferring the calli to a medium containing 3 mg/l 2,4-D, changing the volume of the suspension from 40 to 100 ml, and culturing the suspension in the light, gave rise to a dispersed homogeneous cell suspension without any embryogenic capacity. However, protoplasts could be isolated from this suspension and plants regenerated following the protocol of Dalton (1988) (Radilofe-Rajoelina et al. unpubl.).
5. Somatic embryogenesis developed only on the calli. Green spots emerged from the mass of calli and developed directly into shooted and rooted plants.
6. While initiated the suspension produced embryogenic calli for more than 6 months. Subsequently, the embryogenic capacity of the suspension decreased and a new suspension must be initiated.

4 Application of the Protocol to Other *Lolium* and *Festuca* Species, and Efficiency of regeneration

The protocol worked well with both *Lolium multiflorum* and *Festuca arundinacea*. It was applied without modifications to several species of these two genera, in contrast to the results described by Jackson and Dale (1989) on *Lolium multiflorum* lines. The main problem encountered concerned the protocol for seed sterilization (concentration of HCl and duration of the treatment), which had to be adapted for each case. Indeed, some seeds were very difficult to sterilize: prolonged incubation in concentrated hydrochloric acid inhibited germination, while insufficient treatment led to contamination.

The protocol developed by our group (protocol 3) was compared to the techniques described by Dale (1980) (protocol 1) and Torello et al. (1984) (protocol 2). In protocol 1, calli were induced on 0.5–1 mm immature embryos and cell suspension cultures were established from the embryogenic calli developed as described in protocol 3. In protocol 2, the calli induced on mature seeds were directly plated on the regeneration medium avoiding the cell suspension step.

Table 2 shows that protocol 2 was the least efficient. Protocol 3 was as efficient as protocol 1 for green plant regeneration in ryegrass as well as in fescue; however, some albino plants were produced. Moreover, it was easier to induce calli on mature seeds than on immature embryos, indicating the need to cultivate plants under rigorously controlled conditions and to sample the embryos at the right stage of development. Lastly, the regeneration capacity of the cell suspension cultures initiated from immature embryos ceased after 15 weeks of culture, whereas when initiated from mature embryos the cultures continued to produce plants after 24 weeks without any loss of regeneration capacity or alterations in the normality of the regenerated plants.

Most of the regenerated plants established by protocol 3 were tested in field. The number of chromosomes was identical to the parent plants and all plants were fertile.

Table 2. Comparison of the efficiency of several protocols of regeneration for somatic embryogenesis and albino plant production of ryegrass and fescue (Radilofe-Rajoelina et al. 1990)

Species	Lines	Protocols[a]	Callus-regenerating plants (%)	Albino plants (%)
Ryegrass				
	1	1	58	0
		2	3	0
		3	nd[b]	nd
	2	1	58	4
		2	12	0
		3	53	7
Fescue				
	12	1	100	8
		2	12	4
		3	nd	nd
	13	1	50	0
		2	15	15
		3	73	10

[a] Protocol 1: Dale (1980); protocol 2: Torello et al. (1984); Protocol 3: Radilofe-Rajoelina et al. (1990).
[b] nd, Not determined.

5 Conclusion

An efficient protocol for plant regeneration by somatic embryogenesis, applicable to several *Festuca* and *Lolium* species, was designed. It involved the establishment of an embryogenic cell suspension culture from calli induced on mature seeds under strictly controlled culture conditions. Regenerated plants were all fertile and identical to the parent plants in morphology and chromosome number. Embryogenic cell suspension cultures could be good starting material for direct gene transfer by projectile bombardment.

References

Bonnecuelle MH (1992) Obtention d'hydrides *Festulolium* par sauvetage d'embryons immature issus de croisements interspécifiques. Le Selectioneur Francais 42: 41–46

Bückner RC, Hill HD, Burrus PD (1961) Some characteristic of annual and perennial ryegrass x tall fescue hybrids and the amphiploid progenies of annual ryegrass × tall fescue. Corp Sci 1: 75–80

Creemers-Molenaar J, Van Der Valk P, Loeffen JPM, Zaal MACM (1989) Plant regeneration from suspension cultures and protoplasts of *Lolium perenne* L. Plant Sci. 63: 167–176

Dale PJ (1980) Embryoids from cultured immature embryos of *Lolium multiflorum*. Z Pflanzenphysiol 100: 573–577

Dalton SJ (1988) Plant regeneration from cell suspension protoplasts of *Festuca arundinaceae* Schreb. (tall fescue) and *Lolium perenne* L. (perennial ryegrass). J Plant Physiol 132: 170–175

Humphreys MW (1989) The controlled introgression of *Festuca arundinacea* genes into *Lolium multiflorum*. Euphytica 42: 105–116

Jackson JA, Dale PJ (1989) Somaclonal variation in *Lolium multiflorum* and *L. temulentum* L. Plant Cell Rep 8: 161–64

Kleijer G (1988) Les croisements interspécifiques en amélioration des plantes. I. Raygrass d'Italie x fétuque élevée. Rev Suisse Agric 20: 165–170

Lewiss FJ (1966) The production and manipulation of new breeding material in *Lolium-Festuca*. In Hill AGG (ed) Proc 10th Int Grassland Congr. Finnish Grassland Association, Helsinki, pp 688–693

Lowe KW, Conger BV (1979) Root and shoot formation from callus culture of tall fescue. Corp Sci 19: 397–400

Murashige T, Skoog F, (1962) A revised medium for rapid growth and bioassays with tobacco tissue cultures. Physiol Plant 15: 473–497

Peto FM (1933) The cytology of certain intergeneric hybrids between *Lolium* and *Festuca*. J Genet 28: 113–157

Radilofe S (1991) Amélioration génétique de la fétuque élevée et du raygrass: approche biotechnologique. Thèse doctorat, Institut National Polytechnique Toulouse, France

Radilofe-Rajoelina S, Alibert G, Planchon C (1990) Continuous plant regeneration from established embryogenic cell suspension cultures of italian ryegrass and tall fescue. Plant Breed 104: 265–271

Rybozynski JJ, Zwierzykowski Z, Slusarkiewicz-Jardina A (1983) Plant regeneration with doubled chromosome number in tissue culture of F1 *Lolium-Festuca* hybrids. Genet Pol 21: 1–7

Sam B, Wu FS, Thorne TK (1992) Transgenic tall fescue (*Festuca arundinacea* Schreb.) plants regenerated from protoplasts. Cell Biochem Sup 16F: 207

Takamizo T, Suginobu KI, Ohsugi R (1990) Plant regeneration from suspension culture derived protoplasts of tall fescue (*Festuca arundinacea* Schreb.) of a single genotype. Plant Sci. 72: 125–131

Takamizo T, Wang Z, Suginobu K, Perez-Vincente R, Potrykus I, Spangenberg G (1992) Biotechnological approaches to forage grass improvement: *Festuca* and *Lolium*. J Cell Biochem Suppl. 16F: 212

Torello WA, Symington AG (1984) Regeneration from perennial ryegrass callus tissues. Hort Science 19: 56–57

Torella WA, Symington AG, Rufner R (1984) Callus initiation, plant regeneration and evidence of somatic embryogenesis in red fescue. Crop Sci 24: 1037–1040

Van Den Bogaert CW (1991) Achievements in fodder crop breeding. In: Den Nijs APM, Elgersma A (eds) Fodder crop breeding: achievements, novel strategies and biotechnology. Pudoc, Wageningen pp 77–78

Vasil IK (1987) Developing cell tissue culture systems for the improvement of cereal and grass crops. J Plant Physiol 128: 193–218

Wang Z, Takamizo T, Iglesia JA, Osusky M, Nagel J, Potrykus I, Spangenberg G (1992) Transgenic plants of tall fescue (*Festuca arundinacea* Schreb.) obtained by direct gene transfer to protoplasts. Bio/Technology 10: 691–696

I.7 Somatic Embryogenesis in Orchardgrass (*Dactylis glomerata* L.)

B.V. Conger[1], R.N. Trigiano[2], D.J. Gray[3], and J.K. McDaniel[1]

1 Introduction

The genus *Dactylis* comprises a small and systematically rather isolated group in the tribe Festuceae of the grass family (Stebbins and Zohary 1959). The best-known and most economically important species is *Dactylis glomerata* L., commonly referred to as "orchardgrass" in the United States and as "cocksfoot" in Europe, especially in the British Isles. It is a cool-season forage grass grown for both pasture and hay. Although it is a native of Europe, it has existed in North America for more than 200 years (Jung and Baker 1985).

In the USA, orchardgrass is found from Maine to the Gulf states and from the Atlantic coast to the eastern Great Plains. It is also found in the western parts of northern California, Oregon, and Washington (Jung and Baker 1985). It is common in Europe, except in the most northern sections, and also found in temperate regions of South America, New Zealand, and Australia.

Orchardgrass is a perennial with a bunch-type growth habit. It reproduces sexually by seed formation and has a high degree of self-incompatibility. Synthetic breeding methods are the most commonly used for development of new cultivars. However, long periods (15–20 years) may be required and the progress in achieving significant gains is often slow. For example, only 2 and 4% increases in dry matter yield were obtained over a 20-year period in France and Italy, respectively (Veronesi 1991). Therefore, grass breeders have long been interested in alternate approaches to synthetic cultivar development procedures. These include the incorporation of various recent biotechnological techniques. However, most, if not all, of these techniques such as micropropagation, anther culture, in vitro selection, gene transfer, etc. require the regeneration of whole plants from cells or tissues cultured *in vitro*. In the grass family, the most reliable method of regeneration is by somatic embryogenesis (Conger and Gray 1984; Morrish et al. 1987; Vasil 1987; Gray 1990). In 1982 we reported somatic embryogenesis in orchardgrass from mature embryo (McDaniel et al. 1982) and

[1] Department of Plant and Soil Science, University of Tennessee, Knoxville, Tennessee 37901-1071, USA
[2] Department of Ornamental Horticulture and Landscape Design, University of Tennessee, Knoxville, Tennessee 37901-1071, USA
[3] Central Florida Research and Education Center, University of Florida, I.F.A.S., 5336 University Avenue, Leesburg, Florida, 34748, USA

Biotechnology in Agriculture and Forestry, Vol. 31
Somatic Embryogenesis and Synthetic Seed II (ed. by Y.P.S. Bajaj)

leaf (Hanning and Conger 1982) explants. The systems developed with this species are among the most efficient and repeatable existing in cereals and grasses and will be described in this chapter.

2 Leaf Culture

Our initial experiments utilized either whole caryopses (Conger and Carabia 1978) or mature embryos (McDaniel et al. 1982) as explants. Regeneration was primarily by organogenesis and was inefficient. Also, since orchardgrass has a high degree of self-incompatibility, and is therefore naturally cross-pollinated, it was difficult to control and maintain specific genotypes. The report by Wernicke and Brettell (1980) on leaf culture of 10-day-old seedlings of *Sorghum bicolor* (L.) Moench provided a stimulus for our leaf culture experiments with orchardgrass.

Orchardgrass leaves are folded with united edges rather than whorled. This makes separation very easy and the identity of each can be easily maintained. Tillers from plants grown in a greenhouse or growth chamber are excised slightly above the soil level. The outer leaves are removed and the basal 2–3 cm of the two innermost are split longitudinally along the midvein. The four leaf halves are then sterilized and cut transversely into six segments of 3–4 mm length. The segments are plated serially, from the base, onto solidified SH medium (Schenk and Hildebrandt 1972) containing 30 μM dicamba (3,6-dichloro-2-methoxybenzoic acid). This medium is hereafter referred to as SH-30. Plated leaf segments are shown in Fig. 1. Specific details of the protocol are given in Section 7.1.

Another advantageous characteristic is that experiments can be conducted with paired half-leaf segments. Segments from one leaf half can be used for a specific treatment while the corresponding (sister) segments from the other half can serve as a control. This provides not only for a positive control but also permits the application of "more powerful" paired statistics. We have taken advantage of the paired-leaf system to study the effects of plant growth regulators (Trigiano et al. 1987; Wenck et al. 1988; Bell et al. 1993) and for histological investigations (Trigiano et al. 1989).

Between 4 and 6 weeks after incubation in the dark at 25 °C, cultured leaf segments exhibit embryogenic callus, direct embryo formation, or a combination of both depending on the original position of the segment within the leaf (Conger et al. 1983). Segments from the most basal portion form highly embryogenic callus. The callus formation response decreases and the incidence of direct embryo formation increases on the next segments taken distally from the base. The embryogenic response decreases on the most distal segments. On the older of the two leaves, direct embryo formation begins closer to the base. A leaf segment with numerous embryos is shown in Fig. 2. A scanning electron micrograph (SEM) of a well-developed embryo arising directly from the leaf surface is shown in Fig. 3.

Reasons for the gradient response of segments along the leaf are unknown. However, since the grass leaf grows from the base, a differentiation gradient exists with cells at the extreme basal portion actively dividing and nondifferen-

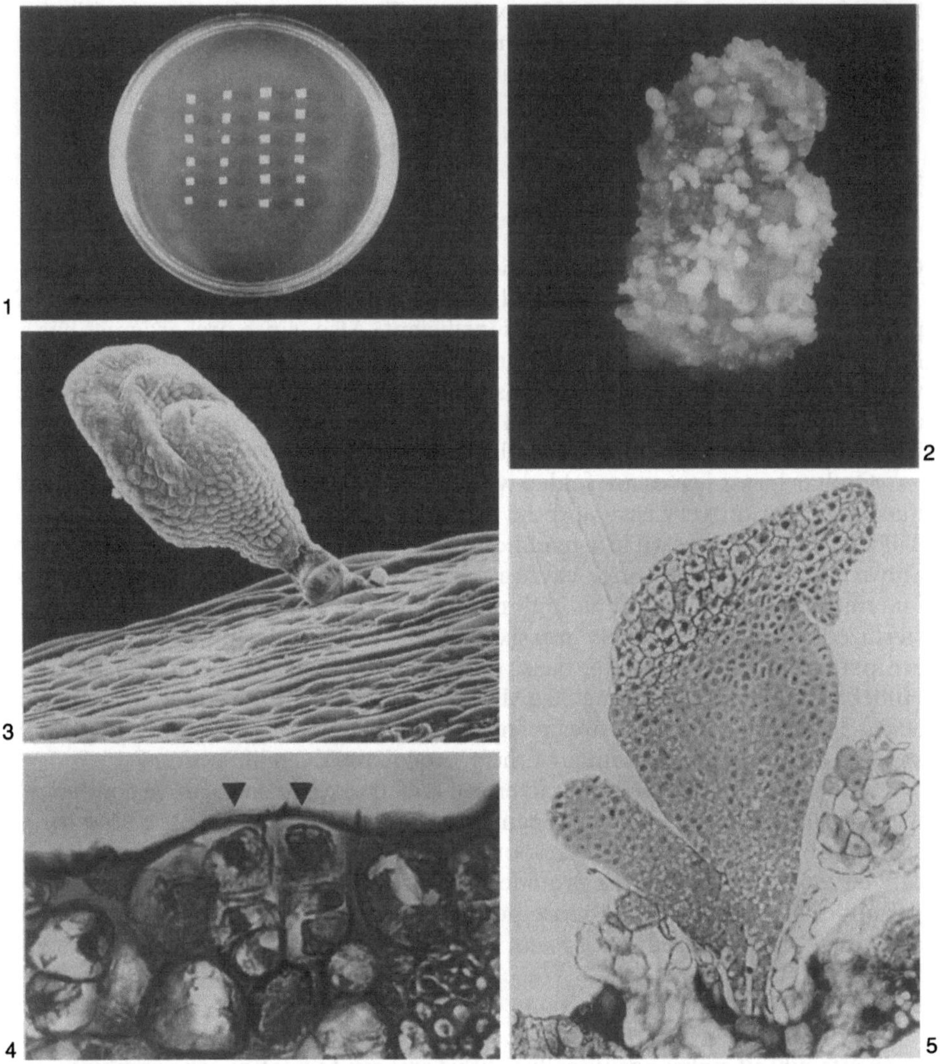

Figs. 1–5. Somatic embryogenesis from *Dactylis glomerata* leaf segments. **Fig. 1.** Basal segments of the two innermost leaves of a tiller explanted onto SH-30 medium; ×.35. **Fig. 2.** Embryogenic callus and somatic embryos formed on a lower leaf segment after 4 weeks in culture; ×84. **Fig. 3.** SEM of a somatic embryo formed directly from an upper leaf segment after 4 weeks in culture; ×50. **Fig. 4.** Early periclinal cell divisions in the mesophyll immediately subjacent to the epidermis; ×500. **Fig. 5.** Two somatic embryos with uniseriate suspensors; ×175. (Fig. 3 from Conger et al. 1983; Figs. 4 and 5 from Trigiano et al. 1989)

tiated, whereas those in more distal segments are less mitotically active and more differentiated. In our system, there seems to be an optimum point on the differentiation gradient where direct embryogenesis is more frequent. This hypothesis also accounts for the shifting of the embryogenic response toward the base of the older compared to the younger of the two leaves, since as leaves become older the meristematic area is restricted further toward the base. In fact, a response is not always obtained from the uppermost segments of older leaves where, presumably, cells have already completed differentiation.

Embryos physically removed from the leaf surface and placed on SH medium without auxin (SH-o) germinate and form individual plantlets. On SH-30 medium, they form embryogenic callus from which many more embryos can be recovered.

Histological studies of direct embryo formation show cells with dense cytoplasm and initial divisions 4 days after culture initiation (Trigiano et al. 1989). Dividing cells are usually separated by several other nondividing cells and located primarily in the first cell layer of the leaf mesophyll subjacent to both the adaxial and abaxial leaf surfaces. Few divisions are observed in either epidermal cells or bundle sheath parenchyma and no divisions are seen in vascular tissue.

Divisions of daughter cells occur 6–8 days after culture initiation. These divisions follow primarily three basic patterns. In the first pattern, one or both of the daughter cells divide periclinally to form a uniseriate column of cells which presumably produce the suspensor and proembryo (Fig. 4). The base of the suspensor is usually completely surrounded by necrotic explant tissue and/or mature, vacuolated cells incapable of division (Fig. 5).

Serial sections through samples harvested between 20 and 28 days showed that 11 of 132 embryos examined had uniseriate suspensors and that 6 of these 11 embryos were formed from divisions of an epidermal cell. After 4 days in culture, approximately 2% of the epidermal cells had divided. Thus, additional divisions must have occured later to produce these embryos (Trigiano et al. 1989).

A second early pattern is an anticlinal division of the upper daughter cell without division of the basal cell. Subsequent anticlinal and periclinal divisions form a multiseriate suspensor and embryo. Proembryos with multiseriate suspensors and a single basal cell are usually observed within leaf tissue 2 weeks after culture initiation. Approximately 20% of the somatic embryos which form individually on leaf segments are attached by multiseriate suspensors that range from two to seven cells in width. Microscope examination of these embryos at a more advanced stage shows that they originate from a single mesophyll cell. The basal cell of the suspensor is typically surrounded by necrotic tissue and/or large vacuolated cells which supports the hypothesis of a unicellular origin .

The third type of early division pattern is the formation of internalized embryogenic cell masses analogous to the proembryonal cell complexes described by Haccius (1978). Sections from leaf explants cultured for 1–2 weeks exhibit variously sized spherical clusters of small, cytoplasmically dense cells that are usually located directly under both epidermises. Since cells capable of division are usually well separated from each other, these proembryonal cell complexes are products of single cells and resultant daughter cells which divide in randomly oriented planes after the initial periclinal division. Therfore, embryos arising from these groups of cells must have been ultimately derived from a single cell (Haccius 1978). Somatic embryos with a common point of attachment to the explant are frequent. Histological examination of similar aggregates shows that these embryos originate from proembryonal cell complexes and is the most common type of development. Embryos that originate from these have either uniseriate of multiseriate suspensors. Presumably, suspensor and embryo development proceed as previously described from single cells of the complex.

3 Suspension Culture

Orchardgrass is unique among the cereals and grasses in that fully developed embryos capable of germination and growth to plants are produced directly in a single liquid medium (Gray et al. 1984; Gray and Conger 1985; Trigiano and Conger 1987). For all other gramineous species, somatic embryo development is limited to globular stages and maturation only proceeds when they are transferred to solidified medium (cf. Lu and Vasil 1981; Ho and Vasil 1983; Kamo and Hodges 1986). Complete embryo development in a single liquid medium affords advantages for investigating the influence of culture medium on somatic embryogenesis (Gray and Conger 1985; Trigiano and Conger 1987; Trigiano et al. 1987, 1992), and somatic embryo ontogeny (Conger et al. 1989).

The protocol for establishing embryogenic suspension cultures has been described in detail (Gray et al. 1984; Trigiano and Conger 1987) and is given at the end of this chapter in Section 7.2. Initiation of embryogenesis is controlled by the addition of various compounds containing reduced nitrogen to the basal medium. The addition of 3–4 g/l of casein hydrolysate, composed of 18 amino acids (Sigma Chemical Co., St. Louis, Missouri, analytical data), stimulated embryogenesis, whereas deletion of the addendum halted the process (Gray et al. 1984; Trigiano and Conger 1987). Individual embryos often accumulated at the air-medium interface along the surface of the flask and also as freely suspended individuals 2–4 weeks after the inclusion of casein hydrolysate in the medium. Several thousand recognizable individual embryos were collected on a 180-μm sieve from 30 ml of medium (Fig. 6). Cultures consisted of very large clumps of cells, smaller aggregates, and individual cells. Microscopic and histological examination of the larger cell clumps revealed that many were proembryonal cell complexes with embryos in various stages of development attached by either uniseriate or multiseriate suspensors (Fig. 7). Although embryo ontogeny was asynchronous and between 30–40% had germinated precociously, completely developed embryos (Fig. 8), capable of growth to plants (Fig. 9) when transferred to SH-O medium, were isolated routinely.

Long-term embryogenic cultures are maintained by replenishing spent medium weekly with SH-30 supplemented with casein hydrolysate. Increase in culture mass occurs both from cells that are sloughed from dedifferentiating and proliferating epidermal cells of developing embryos (Fig. 8) and fragmentation of proembryonal cell complexes. Sloughed cells are capable of dividing and ultimately differentiating into fully developed embryos. Theoretically, such a cyclic system (Fig. 10) would be perennially totipotent (Conger et al. 1989).

As mentioned previously, various compounds containing reduced nitrogen have been used to stimulate embryogenesis in suspension cultures. Specific compounds profoundly influence not only the growth characteristics of the

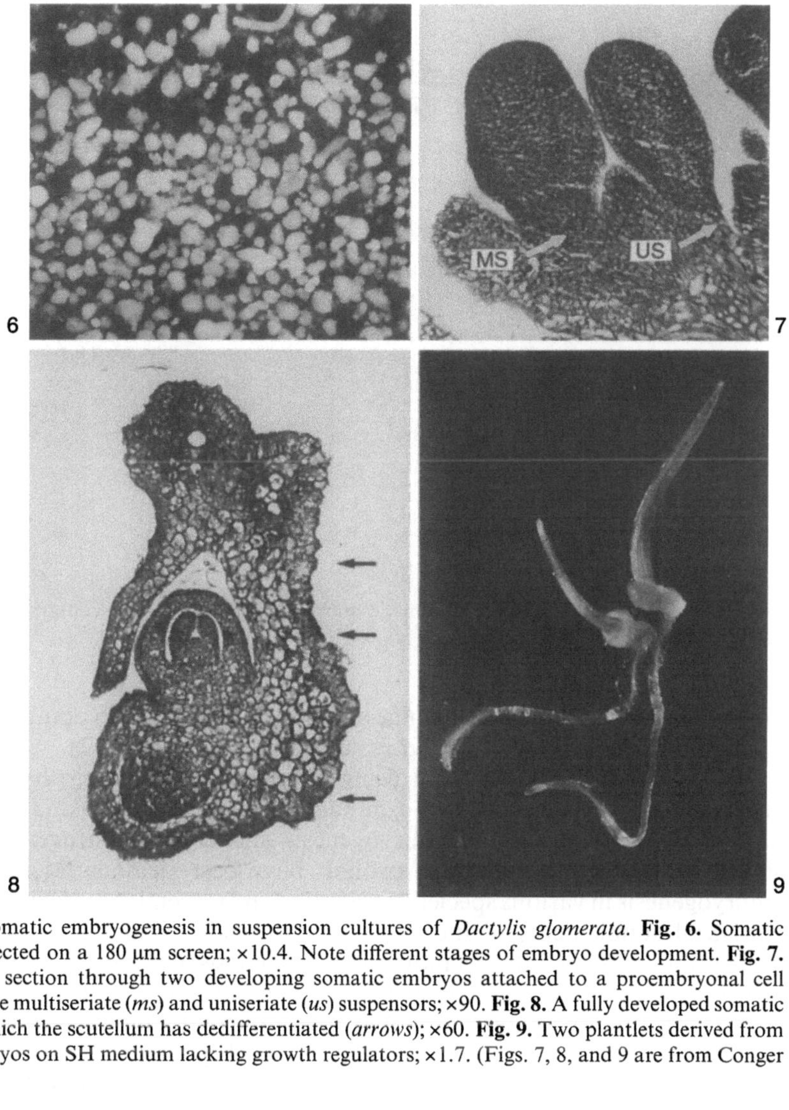

Figs. 6–9. Somatic embryogenesis in suspension cultures of *Dactylis glomerata*. **Fig. 6.** Somatic embryos collected on a 180 μm screen; ×10.4. Note different stages of embryo development. **Fig. 7.** Longitudinal section through two developing somatic embryos attached to a proembryonal cell complex. Note multiseriate (*ms*) and uniseriate (*us*) suspensors; ×90. **Fig. 8.** A fully developed somatic embryo in which the scutellum has dedifferentiated (*arrows*); ×60. **Fig. 9.** Two plantlets derived from somatic embryos on SH medium lacking growth regulators; ×1.7. (Figs. 7, 8, and 9 are from Conger et al. 1989)

cultures, i.e., dry mass and cell types, but also mediate the rate of embryogenesis and quality of somatic embryos. For example, when cells were grown in SH-30 medium supplemented with either 12.5 or 25.0 mM ammonium ion as $(NH_4)_2 \cdot SO_4$, cultures were very clumpy and embryogenesis and embryo development were unregulated. Embryos callused more extensively than observed with casein hydrolysate and produced secondary and tertiary embryos from epidermal cells similar to repetitive embryogenesis. The majority of the embryos formed in medium containing the aforementioned concentrations of ammonium ion failed to develop into plantlets when transferred to medium lacking growth regulators.

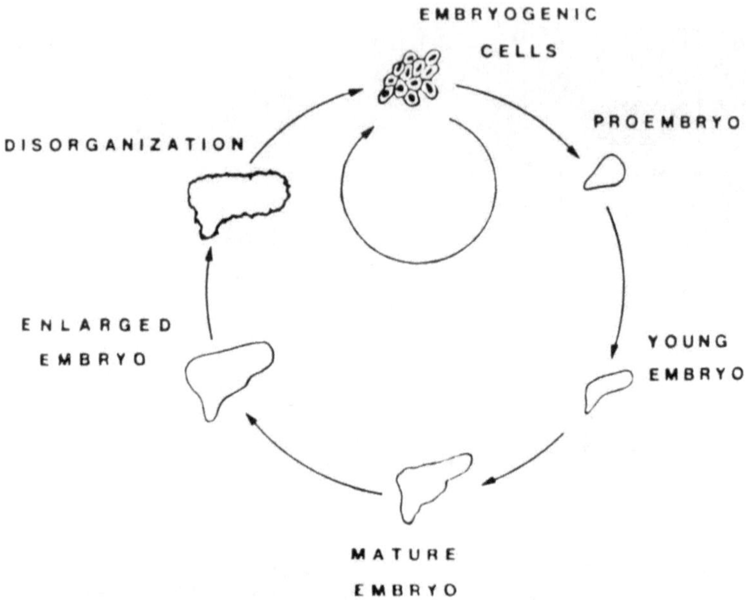

Fig. 10. Diagram of cyclic patterns of somatic embryogenesis in suspension cultures of orchardgrass. (Conger et al. 1989)

Growth in MS medium (Murashige and Skoog 1962), which contains 19.6 mM ammonium ion, produced similar results (Trigiano et al. 1992).

A logical progression from using amino acid-rich casein hydrolysate to regulate embryogenesis was to examine the effects of individual and combinations of amino acids on embryogenesis and embryo ontogeny. Although several amino acids, commonly proline, have been demonstrated to enhance embryogenesis in various species (Armstrong and Green 1985; Meijer and Brown 1987), all supplements of individual amino acids evaluated in orchardgrass suspension cultures failed to induce embryogenesis and most inhibited growth. However, equimolar (12.5 or 25.0 mM) combinations of proline with either serine (Trigiano and Conger 1987) or threonine (Trigiano et al. 1992) restored growth of a fine suspension and stimulated embryogenesis. These amino acid combinations apparently also exerted control of the developmental process since somatic embryos exhibited very little epidermal dedifferentiation and precocious germination was reduced to below 20%. Germination rates were significantly higher than for embryos grown in casein hydrolysate-supplemented medium.

Synchronization of embryo development and inhibition of precocious germination were attempted using abscisic acid (ABA) (Conger et al. 1988). Treatments with either 15 or 30 μM ABA dramatically lowered the rate of embryo formation and failed to provide a developmentally homogeneous population of embryos. However, precocious germination was sharply reduced and there was little proliferation of epidermal cells. In these respects, ABA exerted similar regulatory effects on embryo development observed for combinations of amino acids.

4 Genotype

As with most plant species, plant regeneration from *in vitro* cultures of orchardgrass is highly dependent on genotype and appears to be under genetic control. Orchardgrass is self-incompatible and naturally cross-pollinated. Therefore, each seed embryo, even from the same plant, must be considered to be a different genotype. We discovered somatic embryogenesis in a preliminary experiment testing sterilization procedures on leaves of seed-grown plants of the cultivar Potomac. Two somatic embryos were found on a callus piece. These embryos were germinated and the plants established in a greenhouse. When leaves of these somatic embryo-derived plants were cultured, a highly embryogenic response was obtained from both plants including direct embryogenesis (Hanning and Conger 1982). In a latter experiment, again with Potomac, we found that less than 5% of 200 seed-grown plants produced embryogenic calli (Hanning and Conger 1986).

The best plant from the initial screening experiment has been used in almost all of our experiments since 1982 and was recently released as Embryogen-P (Conger and Hanning 1991). Orchardgrass is a perennial and, therefore, specific genotypes are easily maintained by vegetative propagation. Plants are divided and one to several tillers are placed in new pots with a fresh soilless mix.

When Embryogen-P was crossed reciprocally to two nonembryogenic genotypes, the embryogenic trait was transmitted through both male and female gametes and expressed in approximately 50% of the F_1 plants (Gavin et al. 1989). This result indicates a dominant nuclear gene(s). Since orchardgrass is a purported autotetraploid, the 1:1 ratio fits the model of a simplex (Aaaa) embryogenic genotype crossed to a nulliplex (aaaa) nonembryogenic genotype. Additional data from F_2 and backcross progeny are needed to confirm this model.

The genetic and physiological bases for the embryogenic trait are unknown. However, endogenous cytokinin levels were found to be approximately three times higher in two nonembryogenic genotypes tested than in Embryogen-P (Wenck et al. 1988). Also, low levels of zeatin (10^{-9} M) added exogenously to the medium greatly reduced somatic embryogenesis in the embryogenic genotype. Abscisic acid at 0.01 to 10.0 μM increased embryo production between 1.4- and 1.8-fold in Embryogen-P (Bell et al. 1993). Experiments with senescence inhibitors and promoters may provide additional information on hormonal mechanisms involved in the embryogenic response.

5 Protoplast Culture and Gene Transfer

Orchardgrass was the first forage grass to be regenerated from protoplasts (Horn et al. 1988a) and, until recently, the only one reported to be genetically transformed (Horn et al. 1988b). Transgenic plants have subsequently been obtained in tall fescue (*Festuca arundinacea* Schreb.) by Wang et al. (1992). In orchardgrass, suspension cultures were enriched for small clusters of embryogenic cells. These were used for protoplast isolation. The isolated protoplasts divided at a frequency of 0.5–10% when plated in an agarose-solidified culture medium. Conditioned medium in which embryogenic orchardgrass suspension cultures had been grown increased the rate of colony formation. Protoplast-derived colonies grew rapidly in a bead-type culture system of floating agarose slabs in liquid medium. Embryogenic colonies grew out of the agarose and they regenerated green plants when plated onto a solid medium without auxin. The plants were established in soil and grown to maturity (Horn et al. 1988a).

Using this system, a chimeric hygromycin-resistance gene was introduced into protoplasts using heat shock followed by electroporation or polyethylene glycol treatment. Cell colonies resistant to 20 µg/ml hygromycin were selected. Transformed calli were identified by Southern hybridization and induced to regenerate plants. The transformed plants were shown to contain the hygromycin-resistance gene by Southern blot hybridization (Horn et al. 1988b).

6 Conclusions

In orchardgrass, the origin of somatic embryos from mesophyll cells in cultured leaf segments and the full development of embryos in a single liquid medium remain unique among the cereals and grasses. The highly efficient and repeatable systems in this species can serve as a model to study genetic and hormonal factors influencing somatic embryogenesis in other gramineous species. The systems also offer unique opportunities for gene transfer using microprojectiles or other methods.

7 Protocols to Induce Somatic Embryogenesis

7.1 Leaf Culture

Prepare SH medium (Schenk and Hildebrandt 1972) with 5 mg/l thiamine, 1 g/l inositol, 30 g/l sucrose, 8 g/l agar, and 6.6 mg/l (30 µM) dicamba (3,6-dichloro-2-methoxybenzoic acid). This medium is hereafter referred to as SH-30. Sterilize medium by autoclaving at 1.06 kg/cm^2 for 15–20 min. Allow to cool and dispense approximately 25 ml per 10-cm-diameter Petri dish. Harvest tillers from known embryogenic plants, e.g., Embryogen-P (Conger and Hanning 1991). Remove and discard all but the innermost two leaves from each tiller. Excise the upper portion of each leaf and retain the basal 3 cm. Split this portion along the midvein so that there are two halves of each leaf. Maintain the identity of each of the two leaves.

Sterilize the four leaf halves by stirring in 50% NaOCl (commercial bleach) with a few drops of surfactant for 2 min. This is usually done in a 100-ml beaker and using a magnetic stirrer. Next, rinse the leaf portions three times in sterile distilled water contained in sterile 100-ml beakers. After the final rinse, place the leaf bases in a sterile empty Petri dish. Beginning at the base, cut the leaves transversely into six segments of approximately 3 mm length with a sterile scapel. Plate the leaf segments serially from the basal to the distal segment into Petri plates containing the medium described above. Again, maintain the identity of each leaf half as well as the basal and distal segments. Seal the dishes with Parafilm and place in an incubator at 25 °C in the dark.

Fully developed (mature) embryos will appear on leaf surfaces 4–5 weeks after explanting. Embryos may be germinated individually by placing them onto SH medium without dicamba (SH-o) and incubating in the light at 25 °C. They may also be germinated en masse by transferring whole leaf segments to SH-o medium and incubating as above. In either case, plants can be recovered in 3–4 weeks.

7.2 Suspension Culture

Remove embryos from leaf cultures and place them onto solid SH-30 medium (20 per Petri plate) and incubate in the dark at 25 °C for 10 days. Place 80 callused embryos into 15 ml of filter sterilized (0.22 µm mesh) SH-30 medium amended with 3 g/l casein hydrolysate (SH-30C). We use 125-ml Erlenmeyer flasks with screw-on-tops and a shaker speed of 60 rpm in a lighted incubator.

The cultures are allowed to grow for 2–3 weeks (depending on rate of development) until the suspension becomes viscous. At that time 15 ml of fresh SH-30C is added. After another 2 weeks, the cultures are sieved with a 710-µm mesh screen to remove the large clumps. The embryogenic cells are collected on a 200-µm mesh screen and returned to culture in SH-30C medium. Small developing embryos are observed within 10 days. These cultures can be maintained indefinitely with repeated subculturing or the embryos may be allowed to mature (3 weeks). After maturation, the embryos are collected and germinated as described above.

References

Armstrong CL, Green CE (1985) Establishment and maintenance of friable, embryogenic maize callus and the involvement of L-proline. Planta 164: 207–214

Bell LM, Trigiano RN, Conger BV (1993) Relationship of abscisic acid to somatic embryogenesis in *Dactylis glomerata.* Environ Exp Bot 33: 495–499

Conger BV, Carabia JV (1978) Callus induction and plantlet regeneration in orchardgrass. Crop Sci 18: 157–159

Conger BV, Gray DJ (1984) In vitro culture in forage grass improvement. In: Proc 28th grass breeders work planning conf. USDA/ARS Mandan, North Dakota, pp 50–64

Conger BV, Hanning GE (1991) Registration of Embryogen-P orchardgrass germplasm with a high capacity for somatic embryogenesis from in vitro cultures. Crop Sci 31: 855

Conger BV, Hanning GE, Gray DJ, McDaniel JK (1983) Direct embryogenesis from mesophyll cells of orchardgrass. Science 221: 850–851

Conger BV, Trigiano RN, Gray DJ (1988) Cell culture of the Poaceae (Gramineae). In: Pais MSS, Mavituna F, Novais JM (eds) Plant cell biotechnology. Springer, Berlin Heidelberg New York 49–61

Conger BV, Hovanesian JC, Trigiano RN, Gray DJ (1989) Somatic embryo ontogeny in suspension cultures of orchardgrass. Crop Sci 29: 448–452

Gavin AL, Conger BV, Trigiano RN (1989) Sexual transmission of somatic embryogenesis in *Dactylis glomerata.* Plant Breed 103: 251–254

Gray DJ (1990) Somatic embryogenesis and cell culture in the Poaceae. In: Kasperbauer MJ (ed) Biotechnology in tall fescue improvement. CRC, Boca Raton, pp 25–57

Gray DJ, Conger BV (1985) Influence of dicamba and casein hydrolysate on somatic embryo number and culture quality in cell suspensions of *Dactylis glomerata* (Gramineae). Plant Cell Tissue Organ Cult 4: 123–133

Gray DJ, Conger BV, Hanning GE (1984) Somatic embryogenesis in suspension and suspension-derived callus cultures of *Dactylis glomerata.* Protoplasma 122: 196–202

Haccius B (1978) Question on the unicellular origin of non-zygotic embryos in callus cultures. Phytomorphology 28: 74–81

Hanning GE, Conger BV (1982) Embryoid and plantlet formation from leaf segments of *Dactylis glomerata* L. Theor Appl Genet 63: 155–159

Hanning GE, Conger BV (1986) Factors influencing somatic embryogenesis from cultured leaf segments of *Dactylis glomerata*. J Plant Physiol 123: 23–29

Ho WJ, Vasil IK (1983) Somatic embryogenesis in sugarcane (*Saccharum officinarum* L.). II. Growth and plant regeneration from embryogenic cell suspension cultures. Ann Bot 51: 719–726

Horn ME, Conger BV, Harms CT (1988a) Plant regeneration from protoplasts of embryogenic suspension cultures of orchardgrass. Plant Cell Rep 7: 371–374

Horn ME, Shillito RD Conger BV, Harms CT (1988b) Transgenic plants of orchardgrass (*Dactylis glomerata* L.) from protoplasts. Plant Cell Rep 7: 469–472

Jung GA, Baker RF (1985) Orchardgrass. In: Heath ME, Barnes RF, Metcalfe DS (eds) Forages: the science of grassland agriculture, 4th edn. Iowa State Univ Press, Ames, pp 224–232

Kamo KK, Hodges TK (1986) Establishment and characterization of long-term embryogenic maize callus and cell suspension cultures. Plant Sci 45: 111–117

Lu CY, Vasil IK (1981) Somatic embryogenesis and plant regeneration from freely suspended cells and cell groups of *Panicum maximum* Jacq. Ann Bot 47: 543–548

McDaniel JK, Conger BV, Graham ET (1982) A histological study of tissue proliferation, embryogenesis and organogenesis from tissue cultures of *Dactylis glomerata* L. Protoplasma 110: 121–128

Meijer EGM, Brown DCW (1987) Role of exogenous reduced nitrogen and sucrose in rapid high frequency somatic embryogenesis in *Medicago sativa*. Plant Cell Tissue Organ Cult 10: 11–19

Morrish F, Vasil V, Vasil IK (1987) Developmental morphogenesis and genetic manipulation in tissue and cell cultures of the Gramineae. Adv Genet 24: 431–499

Murashige T, Skoog F (1962) A revised medium for rapid growth and bioassays with tobacco tissue cultures. Physiol Plant 15: 473–497

Schenk RU, Hildebrandt AC (1972) Medium and techniques for induction and growth of monocotyledonous and dicotyledonous plant cell cultures. Can J Bot 50: 199–204

Stebbins GL, Zohary D (1959) Cytogenetic and evolutionary studies in the genus *Dactylis* I. Morphology, distribution and interrelationships of the diploid subspecies. Univ Calf Berkeley Publ Bot 31: 1–39

Trigiano RN, Conger BV (1987) Regulation of growth and somatic embryogenesis by proline and serine in suspension cultures of *Dactylis glomerata*. J Plant Physiol 130: 49–55

Trigiano RN, Songstad DD, Conger BV (1987) Effects of mefluidide and dicamba on in vitro growth and embryogenesis of *Dactylis glomerata* (orchardgrass). J Plant Growth Regul 6: 133–146

Trigiano RN, Gray DJ, Conger BV, McDaniel JK (1989) Origin of direct somatic embryos from cultured leaf segments of *Dactylis glomerata*. Bot Gaz 150: 72–77

Trigiano RN, May RA, Conger BV (1992) Reduced nitrogen influences embryo quality and plant regeneration from suspension cultures of orchardgrass. In Vitro Cell Dev Biol Plant 28P: 187–191

Vasil IK (1987) Developing cell and tissue culture systems for the improvement of cereal and grass crops. J Plant Physiol 128: 193–218

Veronesi F (1991) Achievements in fodder crops breeding in maritime Europe. In: den Nijs APM Elgersma A (eds) Fodder crops breeding: achievements, novel strategies and biotechnology. Pudoc, Wageningen, pp 25–30

Wang Z-y, Takamizo T, Iglesias VA, Osusky M, Nagel J, Potrykus I, Spangenburg G (1992) Transgenic plants of tall fescue (*Festuca arundinacea* Schreb.) obtained by direct gene transfer to protoplasts. Bio/Technology 10: 691–696

Wenck AR, Conger BV, Trigiano RN, Sams CE (1988) Inhibition of somatic embryogenesis in orchardgrass by endogenous cytokinins. Plant Physiol 88: 990–992

Wernicke W, Brettel R (1980) Somatic embryogenesis from *Sorghum bicolor* leaves. Nature 287: 138–139

I.8 Somatic Embryogenesis in Bluestem Grasses (*Bothriochloa* Species)

CHANDRA I. FRANKLIN[1]

1 Introduction

A group of grasses commonly known as Old World bluestems, represented by three genera (*Bothriochloa*, *Dichanthium* and *Capillipedium*), belong to the generic section Bothriochloininae of the tribe Andropogoneae (Chheda and Harlan 1962). The genus *Bothriochloa*, native to Europe and Asia (Harlan 1963) was first introduced into the United States in 1914 (C. Dewald, pers. comm.). Three species of the genus *Bothriochloa*, namely, *B. caucasica*, *B. intermedia*, and *B. ischaemum*, have been widely cultivated for forage in the southern Great Plains of the USA, as they are well adapted to the warm season and different soil conditions in this region (Sims and Dewald 1982). These grasses exhibit an apomictic mode of reproduction (Taliaferro et al. 1989). Seeds of an apomict contain embryos that are produced asexually from the maternal tissue (Nogler 1984). Improvement of an apomictic species through conventional breeding is very difficult, because it cannot be used as a female parent to produce a hybrid. The development of efficient in vitro regeneration protocols may allow the use of biotechnological techniques for the improvement of these crop species. In our laboratory, a tissue culture method to regenerate plants via somatic embryogenesis was developed as an initial step in attempts to improve nutritional qualities and biomass production in caucasian bluestem [*Bothriochloa caucasica* (*Trin.*) C.E. Hubbard].

2 Somatic Embryogenesis and Plant Regeneration

Plant regeneration through somatic embryogenesis in Old World bluestems, like in most other grasses, is achieved through two stages; first, the induction and maintenance of embryogenic callus and, second, the development of somatic embryos and their "germination" to plantlets. Immature inflorescences as well as mature seeds have been used as explants (see Table 1.). Plantlets have been regenerated via the embryogenic callus stage approximately 8–12 weeks after culture initiation.

[1] Plant Biology Division, The Samuel Roberts Nobel Foundation, P.O. Box 2180, Ardmore, Oklahoma 73401, USA

Biotechnology in Agriculture and Forestry, Vol. 31
Somatic Embryogenesis and Synthetic Seed II (ed. by Y.P.S. Bajaj)

Table 1. Summary of published protocols for somatic embryogenesis in *Bothriochoa* spp

Species	Explant	Basal Medium[a]	Medium supplements	Growth regulators[b]	Results	References
B. ischaemum	Young Inflorescence	LS	Sucrose 3%, agar 0.7%	23 μM 2,4-D	E-callus[b] induction and growth	Metzinger et al. (1987)
		1/2 MS	Sucrose 1.5%, agar 0.5%	2 μM 2,4-D 18 μM IAA 5 μM Kinetin	Plant regeneration	
		1/2 MS	Sucrose 1.5%, agar 0.7%	14 μM 2,4-D 6 μM IAA	E-callus induction and growth	Artunduaga et al. (1988)
		1/2 MS	Sucrose 1.5%, agar 0.7%	2 μM 2,4-D 5 μM Zeatin	Plant regeneration	
	Seeds	MS salts and B5 vitamins	12 mM Proline, 2% sucrose	14 μM 2,4-D	E-callus induction and growth	Johnson and Worthington (1987)
		MS salts and B5 vitamins	12 mM Proline, 2% sucrose	5 μM Zeatin	Plant regenaration	
B. intemedia	Young inflorescence	LS	Sucrose 3%, agar 0.7%	23 μM 2,4-D	E-callus induction and growth	Metzinger et al. (1987)
		LS	Sucrose 1.5%, agar 0.5%	2 μM 2,4-D 18 μM IAA 5 μM Kinetin	Plant regeneration	
B. caucasica	Seeds	MS salts and B5 vitamins	Sucrose 2%, agar 0.8%, 12 mM Proline	5 μM 2,4-D	E-Callus induction, growth and plant regeneration	Franklin et al. (1990)

[a] LS = Linsmaier and Skoog (1965) medium; MS = Murashige and Skoog (1962) medium, 1/2 MS = half-strength MS medium; B5 Gamborg et al. (1968).
[b] 2,4-D = 2,4-Dichlorophenoxy acetic acid; IAA = Indole acetic acid; E-callus = embryogenic callus

2.1 Callus Induction, Growth, and Somatic Embryogenesis

Two types of calli are produced when the explants are cultured on defined medium containing 2,4-dichlorophenoxy acetic acid (2,4-D). One type, referred to as non-embryogenic callus, is translucent and friable (Fig. 1C, ne), and is composed of cells of various shapes and sizes with large vacuoles. The other callus type, referred to as embryogenic callus, is white and has a compact appearance (Fig. 1C, e); it is composed of cells uniform in size and spherical or oblong in shape with dense cytoplasm.

In Caucasian bluestem, callus induction from seed explants occurs approximately 8–10 days after culture initiation. The first sign of callus initiation is indicated by the characteristic swelling (Fig. 1A) at the base of the coleoptile above the cotyledon (mesocotyl region). Callus initiation from the immature inflorescences of *B. ischaemum* and *B. intermedia* occurred from the floral primordia, rachis, and rachilla (Metzinger et al. 1987). In Caucasian bluestem, 2 to 3 weeks after culture initiation, the swelling at the mesocotyl region further increased, and it was accompanied by the production of callus from the coleoptile and radical tissues (Fig. 1B). The embryogenic callus appears to be initiated from the tissues inside the swollen region, presumably from the apical meristem or the meristematic tissue at the bases of young leaves.

In Caucasian bluestem, 4–6 weeks after culture initiation, proliferation of both callus types was observed (Fig. 1C). Somatic embryos were produced only from the embryogenic callus. The induction and the maturation of somatic embryos occur randomly and, as a result, somatic embryos of various developmental stages can be observed in an embryogenic callus. In addition to somatic embryos, several modular structures can also be seen in embryogenic callus. The somatic embryos can be easily distinguished from the modulated structures by the characteristic appearance and the relative positions of the coleoptile and scutellum (see Fig. 1D). The nodular structures generally proliferate further to produce more embryogenic callus. "Germinating" somatic embryos show elongated coleoptiles and withered scutellar tissue (Fig. 1E). The development of a root system normally follows further elongation of the coleoptile or in some cases the development of leaves (Fig. 1F).

In Caucasian bluestem, callus clumps from 8–12-week-old cultures were used for initiating cell suspension cultures. However, in *B. ischaemum* it has been reported (Johnson and Worthington 1987) that seed explants were directly used for initiating cell suspension cultures. Cell suspensions of Caucasian bluestem were cultured in a medium identical to the CBS-5 medium except that the 2,4-D concentration was increased to 20 μM and agar was omitted. The cell cultures were maintained at 25 ± 2 °C in the dark or at low light (100 lx. fluorescent light) with a 16-h photoperiod on rotary shakers at 140 rpm. The cell suspension cultures were subcultured at weekly intervals.

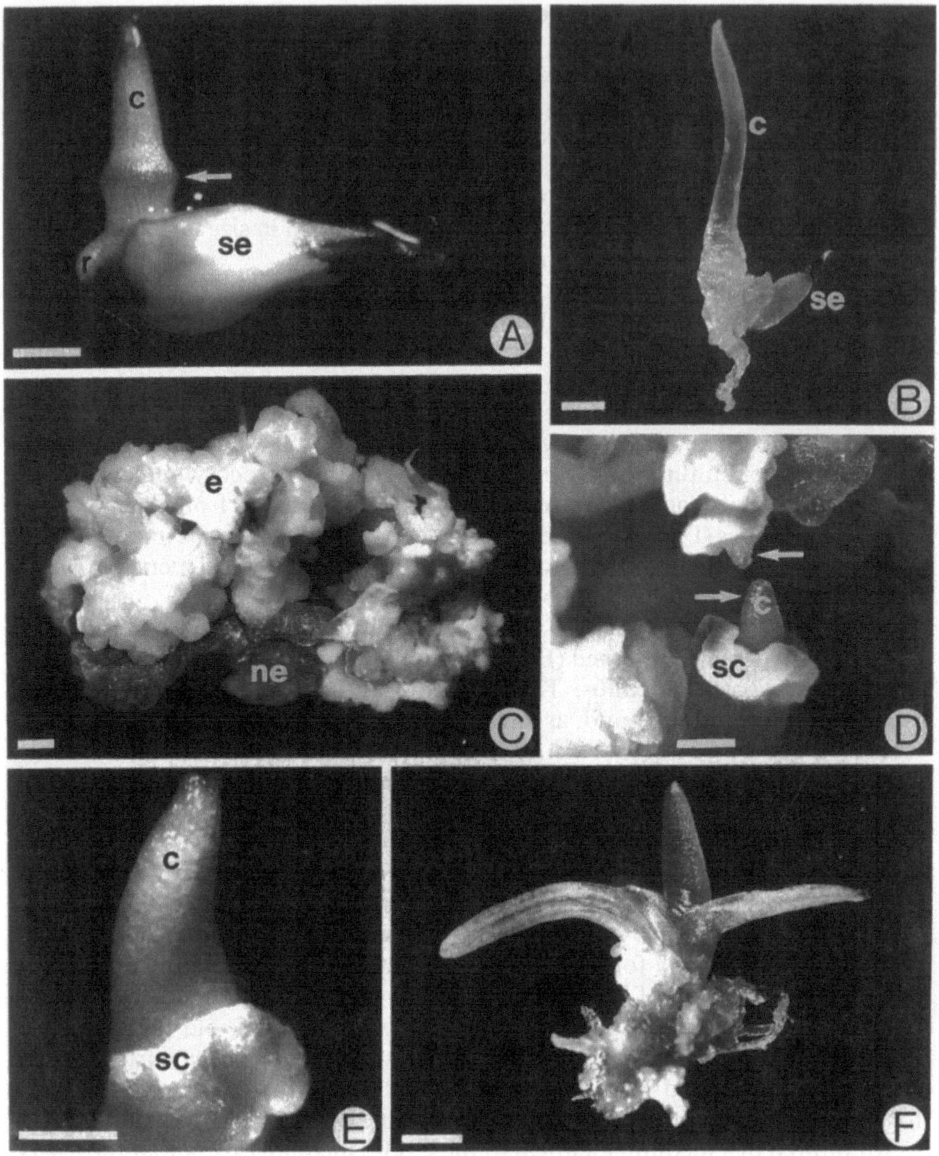

Fig. 1A–F. Different stages of callus induction, growth, and somatic embryogenesis from the seed explants of Caucasian bluestem. *Bars* indicate 0.5 mm in **A**, **D** and **E**; 1 mm in **B**, **C**, and **F**. *c* coleoptile; *e* embryogenic callus; *ne* nonembryogenic callus; *sc* scutellum; *se* seed; *r* radicle. **A** A seed explant 10 days after culture initiation showing the characteristic swelling (*arrow*) at the base of the coleoptile. **B** Response of a seed explant 3 weeks after culture initiation. Note the initiation of callus from coleoptile and radicle. **C** A callus clump, 6 weeks after culture initiation showing both embryogenic (*e*) and nonembryogenic (*ne*) callus types. **D** Young somatic embryos (*arrows*). **E** "Germinating" somatic embryo. **F** Young plantlet from a "germinated" somatic embryo

2.2 Factors Influencing Induction and Growth of Embryogenic Callus and Somatic Embryogenesis

2.2.1 Explants

Mature seeds and immature inflorescences have been used as explants in bluestem grasses for plant regeneration via somatic embryogenesis. When immature inflorescences of *B. ischaemum* and *B. intermedia* were used as explants, age of the explant as determined by the size (6–9 mm) was critical for obtaining optimum results (Metzinger et al. 1987). Mature seeds provide a convenient source of explant material when compared to immature organs such as immature embryos, inflorescence, etc. Hairy outgrowths from the glumes of mature seeds make the surface sterilization procedure ineffective, therefore, it is essential to remove the glumes carefully without damaging the seed. Surface sterilization of seeds after the removal of glumes is described in Section 3. In bluestem grasses, many of the caryopses produced contain partially filled seeds, and they do not germinate. Therefore, it is necessary to choose only the fully matured seeds as explants. Using a dissecting microscope, such seeds can be identified by the presence of completely filled endosperm without wrinkles on the outer surface (Fig. 2A). Among the fully matured seeds of Caucasian bluestem, responding to callus induction treatment (described in Sect. 3), there is a tremendous variation with respect to the time taken for callus induction and the amount of callus produced per explant. Approximately 80–90% of fully matured seeds germinate 3–5 days after culture initiation, and 60–70% of these seeds produce embryogenic callus 3–4 weeks after culture initiation. About 5–10% of seeds germinate 7–12 days after culture initiation and produce callus 4–6 weeks after culture initiation, and roughly 5–10% of fully matured seeds fail to germinate.

2.2.2 Media Requirements

In bluestem grasses, callus induction and growth have been achieved in MS (Murashige and Skoog 1962) or LS (Linsmaier and Skoog 1965) based medium with supplements. Ammonium to nitrate ratios different from those present in MS or LS media did not significantly influence the embryogenic callus production, growth, or somatic embryogenesis from immature inflorescences of *B. ischaemum* or *B. intermedia* (Metzinger et al. 1987). Addition of organic nitrogen supplement, namely, proline (12 mM) to the culture medium promotes embryogenic callus production and somatic embryogenesis. Other nitrogen supplements such as ammonium phosphate or ammonium citrate as well as different media formulations such as SH, B5, or GD (see Table 2) did not have a significant influence on embryogenic callus production or somatic embryogenesis from Caucasian bluestem seed explants. Use of B5 vitamins instead of MS vitamins in the culture medium (CBS medium, see Sect. 3 for medium components) showed a slight increase in callus production and embryo differentiation in Caucasian bluestem.

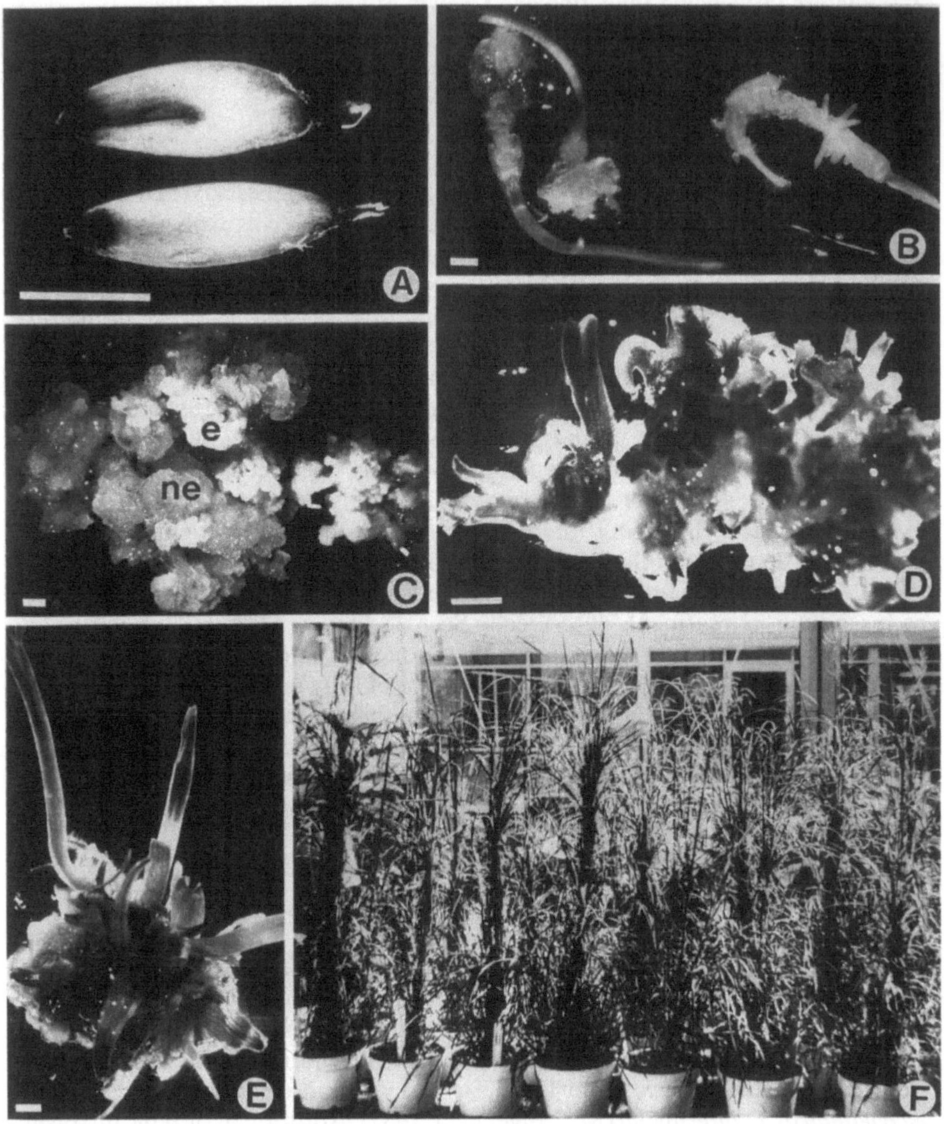

Fig. 2A–F. Various stages of plant regeneration from seed explants of Caucasian bluestem; *bars* 1 mm **A** Seeds at the time of culture initiation (after removing glumes and other floral tissue). **B** Callus initiation from germinated seeds, 20 days after culture initiation. **C** Callus produced from the explants 4 weeks after culture initiation; *e* embryogenic callus; *ne* nonembryogenic callus. **D** Regeneration of plantlets from the callus 8 weeks after culture initiation. **E** Regenerated plantlets 3 weeks after transfer to 2,4-D-free medium, showing root and shoot systems. **F** Regenerated plants grown in the greenhouse for 16 weeks. (Franklin et al. 1990)

Table 2. Influence of various factors on the growth of Caucasian bluestem embryogenic callus. Unless otherwise stated, the following medium components were supplemented to the CBS-10 medium

Media components/culture conditions	Promoters of embryogenic callus growth	Inhibits or retards embryogenic callus growth	No significant influence
Benzyladenine (2 μM)	–	–	√
Cocunut water (5–10% v/v)		–	√
Ammonium citrate (12 mM)	–	–	√
Amonium phosphate (12 mM)	–	–	√
L-Glutamine (12 mM)	–	√	–
SH-Medium[a]	–	–	√
B5- Medium[a]	–	–	√
GD- Medium[a]	–	√	–
SI-Medium[b]	–	√	–
Total darkness or low light intensity (≃ 100 lx)	√	–	–
High light intensity (≃ 3000 lx)	–	√	–

[a] Basal salt and vitamin composition of these media were supplemented with 10 μm 2,4-D, 2% sucrose, and 0.8% agar. SH= Schenk and Hildebrandt (1972); B5= Gamborg et al. (1968); and GD-Gresshoff and Doy (1974).
[b] SI medium (Franklin et al. 1991) consists of all the components of MS (Murashige and Skooge 1962) salts except the inorganic nitrogen source, MS vitamins, 10 μM 2,4-D, 2% sucrose, and 0.8% agar. In this medium, the inorganic nitrogen is replaced by 10 mM L-glutamine.

In Caucasian bluestem, plant regeneration from the embryogenic callus occurred rapidly (8–10 weeks after culture initiation) when 2,4-D at a concentration of 5 μM was used (see Sect. 2.3.1). For prolonged maintenance (4–5 months) of cultures in the callus stage (without plant regeneration), it was necessary to increase the 2,4-D concentration in the medium to 10 μM (CBS-10 medium). At this concentration, the frequency of plant regeneration was rapidly reduced, and after two to three subcultures, only callus proliferation occurred without any plantlet regeneration.

2.2.3 Culture Conditions

In bluestem grasses, calli were cultured at 25 ± 2 °C in the dark or with a 16-h photoperiod under low light conditions (100 lx). When Caucasian bluestem calli were cultured under high light intensity (3000 lx), embryogenic callus growth was retarded, production of nonembryogenic callus increased, and anthocyanin pigmentation of embryogenic callus occurred.

2.3 Plant Regeneration

2.3.1 From Callus Cultures

In Caucasian bluestem, two regeneration systems were established. These systems differ from each other with respect to the amount of embryogenic callus production and the duration of culture period. In the regeneration system, referred to as rapid regeneration system (described in Sect. 3), plantlets were produced 8–10 weeks after culture initiation, and the following developmental stages were observed. Seeds (Fig. 2A) germinated 3–5 days after culture initiation. Nonembryogenic callus first developed from the lower portion of the coleoptile by 7–10 days after culture initiation (Fig. 2B). Small pockets of embryogenic callus with the characteristic white color and compact appearance appeared 4 weeks after culture initiation (Fig. 2C, e). Between 6–8 weeks after culture initiation, the embryogenic callus produced several somatic embryos and plantlets with small roots (Fig. 2D). Both root and shoot systems were well established by 3 weeks after the plantlets were transferred to the 2,4-D-free medium (Fig. 2E). About 95% of the plantlets survived greenhouse acclimatization and produced healthy plants (Fig. 2F) and viable seeds.

In the rapid regeneration system, the callus induction and plant regeneration were dependent on the 2,4-D concentration in the medium. Five μM 2,4-D was optimal for callus production and plantlet regeneration (Fig. 3). The amount of callus produced and the number of plantlets regenerated per explant increased as a function of 2,4-D concentration up to 5 μM. Above this concentration, a rapid decline in the number of plantlets and a gradual decrease in callus production were observed. At 5 μM 2,4-D, 40–50% of the callus produced was of the embryogenic type, and an average of 60 plants was regenerated per seed explant over a 9-week culture period.

In this system it was not necessary to alter the medium components or growth regulator levels to regenerate plants. An advantage of this rapid regeneration system is that plantlets can be regenerated in a short time period. Also, the short duration of the culture period minimizes the potential for the induction of

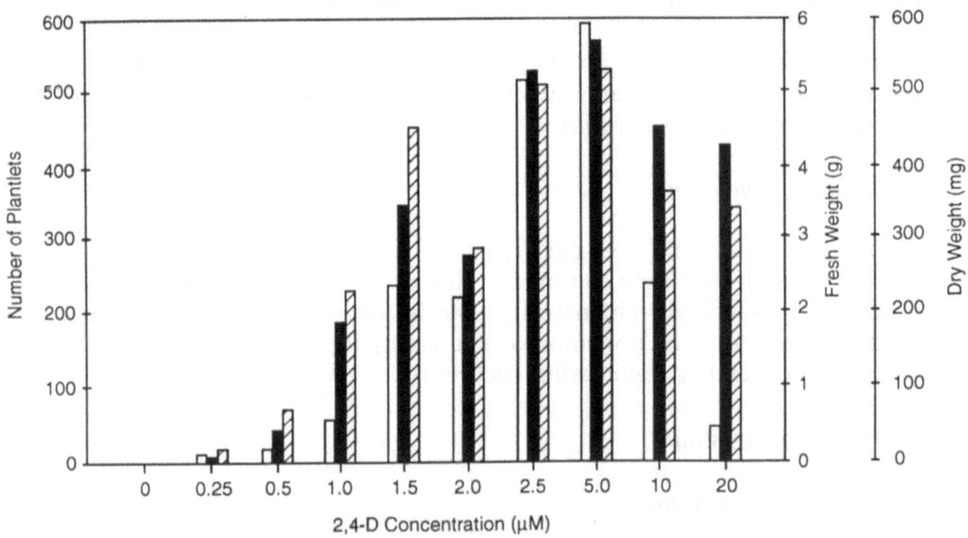

Fig. 3. Influence of 2,4-D concentration on callus growth and plantlet regeneration from seed explants of Caucasian bluestem. Pooled values from 10 explants per 2,4-D concentration are given. ▭, Number of plantlets; ■, fresh weight; ▨, dry weight. (Franklin et al. 1990)

somaclonal variation. This system is attractive for the genetic transformation of Caucasian bluestem, because it is desirable to avoid somaclonal variation among transgenic plants.

The other regeneration system in Caucasian bluestem was specifically developed for inducing mutation and variation in culture. In this system, the cultures could be maintained in the callus stage (without plant regeneration) for a long period (4–5 months), and the callus cultures could be treated with various mutagens to induce desired variations. Factors influencing the prolonged maintenance of embryogenic calli are discussed in Section 2.2.2. Plant regeneration occurred 4–8 weeks after these callus clumps were transferred to CBS-5 medium and cultured under conditions described in Section 3.

2.3.2 From Cell Suspension Cultures

For plant regeneration, cell clumps from Caucasian bluestem suspension cultures were first transferred to CBS-10 medium. The embryogenic callus proliferated and somatic embryos were produced from these cell clumps 4–8 weeks after the transfer. The embryogenic callus was then transferred to CBS-5 medium and cultured as described in Section 3 for regenerating plants. The age of cell suspension culture is critical for regenerating plants. Cultures older than 6 months lost their ability to regenerate plants.

2.3.3 Growth and Establishment of Plantlets

Plantlets 10 mm or taller produced from embryogenic callus on CBS-5 medium were transferred to CBS-0 medium (CBS-5 medium without 2,4-D) contained in Magenta boxes (Magenta Corporation, Chicago, Illinois) and cultured under high light intensity (3000 lx). This medium enhanced the further growth of root and shoot systems of Caucasian bluestem plantlets. To enhance the further growth of *B. ischaemum* and *B. intermedia* plantlets, it was necessary to use half-strength MS medium supplemented with 5.4 μM NAA and 1.5% sucrose (Metzinger et al. 1987). Recovery of albino plants from tissue cultures among grasses is not uncommon. However, in Caucasian bluestem, none of the plants regenerated from calli or cell suspension cultures were albinos. Also, reports on other bluestem grasses do not mention the recovery of albino plants from tissue culture (Metzinger et al. 1987; Johnson and Worthington 1987 Artunduaga et al. 1988.

3 A Protocol for Regenerating Plants from Caucasian Bluestem Seed Explants

1. After removing the glumes and other parts, wash the seeds in tap water containing a few drops of commercial liquid detergent and rinse them thoroughly in tap water.
2. Treat seeds with 70% ethanol for 30 s to 1 min and rinse again with tap water.

3. Soak seeds in 20% commercial bleach (Clorox) solution (plus a few drops of liquid detergent) with constant agitation for 20 min, and subsequently rinse 4 times in sterile distilled water.
4. Transfer seeds to medium (referred to as CBS-5 medium) contained in Petri dishes (5 seeds per dish). CBS-5 medium consists of MS salts (Murashige and Skoog 1962), B5 vitamins (Gamborg et al 1968), 12 mM proline, 2% sucrose, 0.8% agar, and 5 μM 2,4-D.
5. Incubate the seeds in a growth room at 25 ± 2 °C under low light conditions (100 lx, fluorescent light) with a 16-h photoperiod.
6. After 4 weeks, selectively remove the callus produced from the explant and transfer it to fresh CBS-5 medium and incubate the cultures under similar light and temperature regimes as described in (5).
7. Subculture the callus along with the plantlets (if present) on fresh CBS-5 medium at 4-week intervals.
8. Four weeks after the first subculture (i.e., 8 weeks after culture initiation), and subsequently during each subculture, remove plantlets 10 mm or taller and transfer them to CBS-0 medium (CBS-5 medium without 2,4-D) contained in Magenta boxes, and grow them under higher light intensity (3000 lx, fluorescent light) with a 16-h photoperiod at 23 ± 2 °C.
9. Transfer plantlets 5 cm or taller to the greenhouse. Follow standard procedures for greenhouse acclimatization of these plantlets.

4 Applications

The rapid regeneration system described in Section 3 for Caucasian bluestem is ideal for regenerating large numbers of plantlets in a short period of time, and it may be suitable for obtaining transgenic plants. In our laboratory, attempts are now in progress to develop a genetic transformation protocol in which transgenic Caucasian bluestem plants can be regenerated using this rapid regeneration system. The genetic transformation protocol will be useful for future genetic manipulations of Caucasian bluestem to improve its nutritional quality. An alternate approach for the improvement of crop plants, especially the apomicts (Taliaferro et al. 1989) is to select tissue culture-induced mutants or somaclonal variants showing useful characters. Somaclonal variation can be beneficial if it is heritable for useful traits. For example, a somaclonal variant that can undergo normal sexual reproduction could be of great value in a breeding program for apomicts (C. Taliaferro, pers. comm.). The regeneration system modified for the prolonged maintenance of embryogenic callus and the cell culture system is now being used in our laboratory to induce variation and to select variants for useful characters such as higher biomass production and a non-apomictic mode of reproduction.

Acknowledgments. The author thanks Drs. P.L. Sims, C.L. Dewald, C.M. Taliaferro, and B.B. Johnson for helpful suggestions and information, R.L. Dalrymple for providing Caucasian bluestem seeds, Drs. A. Oommen and J. Ponsamuel for a critical review of this article, J. Brightwell and T. Myers for graphics and photography, and A. Wilkins for typing the manuscript.

References

Artunduaga IR, Taliaferro CM, Johnson BL (1988) Effects of auxin concentration on induction and growth of embryogenic callus from young inflorescence explants of Old World bluestem (*Bothriochloa* spp.) and bermuda (*Cynodon* spp.) grasses. Plant Cell Tissue Organ Cult 12: 13–19

Chheda HR, Harlan JR (1962) Mode of chromosome association in *Bothriochloa* hybrids. Caryologia 15: 461–476

Franklin CI, Trieu TN, Gonzales RA (1990) Plant regeneration through somatic embryogenesis in the forage grass Caucasian bluestem (*Bothriochloa caucasica*). Plant Cell Rep 9: 443–446

Franklin CI, Trieu TN, Gonzales RA, Dixon RA (1991) Plant regeneration from seedling explants of green bean (*Phaseolus vulgaris* L.) via organogenesis. Plant Cell Tissue Organ Cult 24: 199–206

Gamborg OL, Miller R, Ojima K (1968) Nutrient requirements of suspension cultures of soybean root cells. Exp Cell Res 50: 151–158

Gresshoff PM, Doy CH (1972) Development and differentiation of haploid *Lycopersicon esculentum*. Planta (Berl) 107: 161–170

Harlan JR (1963) Natural introgression between *Bothriochloa ischaemum* and *B. intermedia* in West Pakistan. Bot Gaz 124: 294–300

Johnson BB, Worthington M (1987) Establishment of suspension cultures from seeds of plains bluestem (*Bothriochloa ischaemum* (L.) Keng.) and regeneration of plants via somatic embryogenesis. In Vitro Cell Dev Biol 23: 783–788

Linsmaier EM, Skoog F (1965) Organic growth factor requirements of tobacco tissue cultures. Physiol Plant 18: 100–127

Metzinger BD, Taliaferro CM, Johnson BB, Mitchell ED Jr (1987) In vitro regeneration of apomictic bluestem grasses. Plant Cell Tissue Organ Cult 10: 31–38

Murashige T, Skoog F (1962) A revised medium for rapid growth and bioassays with tobacco tissue cultures. Physiol Plant 15: 473–497

Nogler GA (1984) Gametophytic apomixis. In: Johri BM (eds) Embryology of angiosperms. Springer, Berlin Heidelberg New York, pp 475–518

Schenk RU, Hildebrandt AC (1972) Medium and techniques for induction and growth of monocotyledonous and dicotyledonous plant cell cultures. Can J Bot 50: 161–167

Sims PL, Dewald CL (1982) Old World bluestems and their forage potential for the southern Great Plains. USDA Agric. Rev Manual 28: 1–15

Taliaferro CM, Dabo SM, Mitchell ED, Johnson BB, Metzinger BD (1989) Morphologic, cytogenetic, and enzymatic variation in tissue culture regenerated plants of apomictic old-world bluestem grasses (*Bothriochloa* sp.). Plant Cell Tissue Organ Cult 19: 257–266

I.9 Somatic Embryogenesis in Sugarcane (*Saccharum* Species)

E. Guiderdoni[1], B. Mérot[2], T. Eksomtramage[1], F. Paulet[1], P. Feldmann[3], and J.C. Glaszmann[1]

1 Introduction

1.1 Taxonomy and Origin of Sugarcane

Sugarcane is the name given to sacchariferous, cultivated species and descendants of interspecific hybrids in the genus *Saccharum,* family Graminae, tribe Andropogonae. *Saccharum* species are highly polyploid with no known diploid form. A basic chromosome number has not yet been firmly established, but is estimated in the range of 5–10 (Stevenson 1965). Until the late 1800s, sugarcane varieties belonged mostly to *S officinarum* (2n = 8x = 80), the "noble" cane originating from New Guinea, and also to *S sinense* (2n = 106–120) and *S. barberi* (2n = 80–120). Interspecific crosses between *S officinarum* and *S. spontaneum*, followed by two or three backcrossings to the noble cane – a process called nobilization – yielded the first modern commercial varieties (*Saccharum* ssp.) in the early 1900s in Java and later in India (Daniels and Roach 1987). The original hybrids were, in turn, crossed with each other to produce new commercial varieties. The limited genetic basis of the original hybridizations, which according to Arceneaux (1967) involved no more than 20 noble clones and fewer than 10 *S. spontaneum* derivatives, led various countries to reinitiate nobilization work in the early 1960s in order to transfer specific, desirable characters, such as disease resistance, from wild germplasm to commercial hybrids (Berding and Roach 1987). Current breeding goals are to develop cultivars with increased sugar yield and resistance to diseases (smut, mosaic, rust, leaf scald) and insects.

Breeding sugarcane has proven difficult and empirical. Complex cytogenetics, prevalence of abnormal meiosis resulting in male and female sterility, and high levels of heterozygosity in hybrid clones make it difficult to interpret the patterns of inheritance. Modern cultivars have a variable chromosome number – sometimes displaying chromosome number variation from cell to cell in the same clone – and often rarely flower.

[1]CIRAD-CA, BP 5035, 34032 Montpellier Cedex, France
[2]Laboratoire d'Amélioration des Plantes, Bât 360, Orsay-Paris XI University, 91405 Orsay Cedex O5, France
[3]CIRAD-CA, Station de Roujol, 97170 Petit Bourg, Guadeloupe, French West Indies

Biotechnology in Agriculture and Forestry, Vol. 31
Somatic Embryogenesis and Synthetic Seed II (ed. by Y.P.S. Bajaj)

1.2 Distribution and Economics

Sugarcane is grown over 18 million ha (FAO 1993) and accounts for 65% of the world sugar production, which reached 111 million tons in 1992/93 (Ahfield 1993). Seventy percent of the canesugar is produced in developing countries. It is used for domestic consumption, and often exported, or, as in Brazil, converted to alcohol fuel. It is therefore of primary economic importance for these countries. Average yields range from 4 to 115 tons of harvested cane per hectare (FAO 1993). India, Brazil, Cuba, and China are the four major canesugar producers. During the last 20 years, per capita consumption of sucrose has been decreasing in developed countries due to competition with other products (isoglucose, fructose, and sweeteners), while it has been increasing in developing countries. Sugarcane also has to face competition with sugar beet grown in temperate areas, the relative yield improvement of which has been much greater during the last decade. There is therefore an urgent need to improve the competitiveness of sugarcane cropping methods and varieties. Varietal improvement will result from use of modern biotechnological tools.

1.3 Significance of Somatic Embryogenesis in Sugarcane Improvement

Sugarcane is propagated from vegetative stalk cuttings called setts, which bear one to three axillary buds. Planting sugarcane fields with new setts, generally every 2 to 5 years – often once a year in Asia – is a tedious, time consuming, labor-intensive and costly task. Moreover, vegetative propagation by stem cuttings favors systemic disease transmission even if cuttings can be treated by heat therapy or chemicals before planting.

An alternative to the hardly profitable clonal propagation through meristem and axillarly bud in vitro cultures is the mass production in cell suspensions cultures of somatic embryos (SE) that can be further encapsulated and used as synthetic seeds for direct planting. Setting up such a large-scale delivery system would be of invaluable interest in sugarcane and could be facilitated by the fact that most sugarcane hybrids tested in the literature proved capable of producing embryogenic calluses (see Sect. 1. 4). However, the establishment of long-term, maintainable, finely dispersed cell suspensions from these calli is not yet a routine procedure (see Sect. 1. 4), and plating of the cell suspension onto solid medium is still a prerequisite for differentiation of somatic embryos. Nevertheless, the possibility to convert calcium alginate-encapsulated sugarcane embryogenic nodules to plants has already been demonstrated (Mérot and Guiderdoni 1987).

In sugarcane, microprojectile bombardment of leaf tissues (Gallo-Meagher and Irvine 1993) and cell suspensions (Frank and Birch 1991) has been used for transient expresion studies. The later technique has permitted production of stably transformed, herbicide-resistant callus (Chowdhury and Vasil 1992) and recently, transgenic plants of cv. Pindar have been obtained through bombardment of embryogenic calli with an efficiency of one to three transgenic plants per treated plate (Bower and Birch 1992). As SE may originate from single cells of the epidermal layer of the embryogenic callus (Ho and Vasil 1983a; see Sect. 2.4), this method should yield nonchimeric transgenics, contrasting with bombardment of meristems, which has so far led to the formation of sectorial transformants (Gambley et al. 1993). As most modern cultivars can produce embryogenic cultures, particle bombardment of calli appears to be the most promising and universal technique amongst the currently available gene transfer technologies.

Protoplasts preparations isolated from cell suspension cultures established from embryogenic callus have been successfully used for (1) regenerating plants (Chen et al. 1988b; Srinivasan and Vasil 1986); (2) for producing stably transformed cell cultures through electroporation (Chen et al. 1987; Chowdhury and Vasil 1992; Rathus and Birch 1992b), and (3) producing somatic hybrid tissues through fusion with *Pennisetum purpureum* protoplasts (Tabaiezadeh et al. 1986).

The compact structure and high regeneration potential of somatic embryos and embryogenic cultures make them attractive structures for cryopreservation of genetic resources. various tissues and callus cultures of sugarcane have been cryopreserved (Bajaj et al. 1987), and high frequency plant regeneration from cryopreserved calli (Jian et al. 1987) and cell suspensions (Gnanapragasam and Vasil 1990) have been reported. Eksomtramage et al. (1992) have recently developed an efficient method for cryopreservation of embryogenic cultures which appears applicable to a wide range of genotypes.

2 Review of Sugarcane Somatic Embryogenesis

Sugarcane tissue culture was initiated in Hawaii by HSPA scientists in 1961 as a tool for physiological and biochemical studies. Subsequent research focused on producing somaclones exhibiting improved agronomical value and disease resistance from existing commercial varieties. The main developments have been reviewed (Heinz et al. 1977; Liu 1981, 1984, Maretzki 1987).

Explants. Although every part of the sugarcane plant is capable of callus induction (Liu 1981), only immature inflorescence and young rolled leaves (from mature plants or in vitro plantlets) are able to initiate embryogenic tissue cultures. Young rolled leaves were most often used due to their availability throughout the year; 3–5-mm sections of the spindle of the four to five innermost furled leaves aseptically dissected 1 to 60 mm above the apical meristem were used in most studies. Although scutellum cells of cereal embryos are well known for their embryogenic potential, sugarcane caryopse is not used due to genetic heterogeneity and limited size.

Genotypes. Leaf tissues of modern hybrid clones are usually more responsive than those of traditional varieties or wild relatives (Chagvardieff et al. 1981; Liu 1981; Chen et al. 1988a). Although quantitative differences in response are usually found among commercial hybrid clones, nearly all of the 18 genotypes tested in a recent study were found capable of initiating regenerable white compact embryogenic callus cultures (Taylor et al. 1992).

Culture Medium. Murashige and Skoog's (1962) salts, vitamins, and iron source or modifications have been most often used for callus production, maintenance, and regeneration as well as for cell suspension establishment and regeneration in sugarcane. N6 (Chu et al. 1975) basal medium formulation was found 40% less effective than MS medium in inducing embryogenic calli from leaf explants but in contrast it permitted better maintenance of their totipotency (Fitch and Moore 1993) (Ho and Vasil 1983a). Larkin (1982) developed an original CS medium consisting of modified MS salts, half-strength iron EDTA, and increased vitamins for leaf tissue, cell suspension, and protoplast cultures.

2,4-D, used in a range of 1 to 3 mg/l, appears to be the most efficient auxinic substance for initiating embryogenic calli. In a test survey of about 79 chemicals, 25 were found efficient for callus induction from leaf explants, of which 80% permitted plant regeneration in medium devoid of growth regulator (Irvine et al. 1983). Among these chemicals, picloram is used in Hawaii for formation and maintenance of organogenic green calli over long culture periods (Fitch and Moore 1990), although Ho and Vasil (1983a) found that the chemical was not effective in inducing white embryogenic callus. Although the use of cytokinin in combination with 2,4-D has generally been found unfavorable for callus formation in enhancing the release of phenolics from the explant, Larkin (1982) supplemented callus initiation and maintenance medium with 3 mg/l 2,4-D and 0.25 mg/l kinetin. A combination of 5 mg/l NAA and 2 mg/l kinetin has also been used to promote direct plant regeneration from leaf tissue sections, without callus formation, 10 days after inoculation (Irvine and Benda 1985). However, the morphogenetic pathways for plant regeneration were not detailed in the two latter studies.

Callus induction and cell suspension culture media have also been supplemented by either 50 mg/l arginine (Nickell and Maretzki 1969; Ho and Vasil 1983a;.Chen et al. 1988a; Taylor et al. 1992), 0.5–1 g/l yeast extract, 5–10% coconut water, or 500 mg/l casein hydrolysate. The addition of activated charcoal (10 g/l) has been found beneficial in preventing detrimental action of phenolics, both in embryogenic callus formation (Ho and Vasil 1983a; Srinivasan and Vasil 1986) and cell suspension plating (Ho and Vasil 1983b).

Light is generally considered unfavorable for embryogenic callus formation and maintenance, leading to precocious germination of the SE (Chen et al. 1988a). However, exposure to diffuse, continuous light is necessary for identifying and maintaining long-term, regenerable green organogenic calluses on either 2,4-D or picloram-containing media (Fitch and Moore 1990).

Transfer of the embryogenic callus to MS medium lacking growth regulators is usually sufficient for promoting plant regeneration. Although transfer of young shoots to SH (Schenk and Hildebrandt 1972) modified medium or MS medium containing 5 mg/l NAA (Nadar and Heinz 1977) was originally found necessary for triggering root formation (Liu and Chen 1974), subculture on full or half-strength MS medium containing 6 to 8% sucrose was later found sufficient for obtaining intense root formation (Maretzki and Hiraki 1980).

Types of Callus Formation Produced by Leaf Explants. Taylor and coworkers (1992) have recently provided a precise description, based on SEM observations,

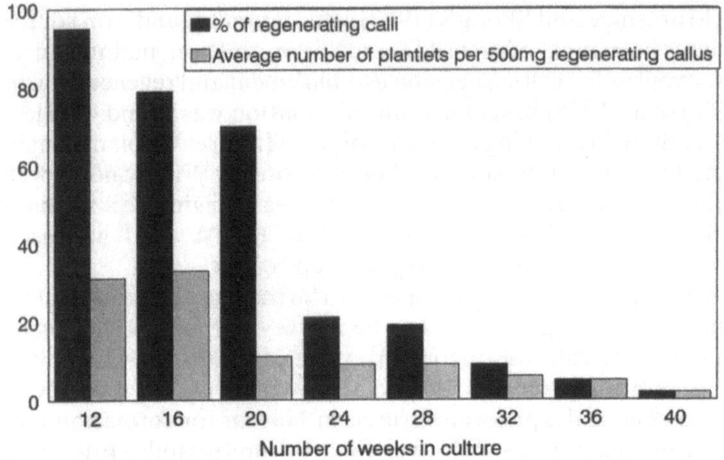

Fig. 1. Green plant-regenerating frequency (*black bar*) and average number of plantlets recovered per 500 mg of callus transferred to regeneration medium (*gray bar*) in CO6415 leaf-derived embryogenic cultures as a function of 4-week passages on callus maintenance medium (60 calli were tested at each step). (Guiderdoni and Demarly 1988)

and have identified four types of formations: type 1, observed first at the cut end of the leaf explant after 5 days of culture is semitranslucent and consists of loose, large, elongated cells. Type II develops among the type-1 formations after 13 days of culture and is soft, mucilaginous, gray-yellow, and contains round, vacuolated, highly dissociated cells. Type III, appearing after 13 to 21 days at the cut end of the leaf explants and around the vascular tissue, consists of compact, hard, smooth-surfaced globular nodules, yellow at first and later becoming white, and made up of small, round, densely cytoplasmic cells. Type IV is yellow, friable, presents clusters of semiorganized structures, and consists of small, round, densely cytoplasmic cells. Types I and II were found to be nonmorphogenic, whereas types III and IV can regenerate plants with a high frequency through embryogenesis and with a low frequency through organogenesis, respectively. The type-III nodular callus is referred to as embryogenic callus (Fig. 1). This heterogeneity is due to the fact that the explant is a transversal section of several leaves, differing in degree of maturity and diversity of callusing tissue (see Sect. 2. 1).

Decline of Morphogenic Potential in Embryogenic Cultures. A dramatic decrease in the regeneration ability of the embryogenic callus is usually observed after five to six 4-week passages on MS medium containing 1–3 mg/l 2,4-D (Fig. 2; Chen et al. 1988a; Guiderdoni and Demarly 1988; Taylor et al. 1992; Fitch and Moore 1993). Chen and coworkers were able to maintain a 70% regeneration potential over 30 months by alternating the medium monthly by supplementation with either 1 or 3 mg/l 2,4-D (Chen et al. 1988a). Long-term regeneration was also achieved by Larkin (1982) and Fitch and Moore (1993) by replacing MS medium with CS medium supplemented with 3 mg/l 2,4-D and 0.25 mg/l kinetin and

N6 medium supplemented with 1mg/l 2,4-D. Fitch and Moore (1990) have reported plant regeneration from 1–4-year-old callus cultures of various commercial hybrids and *S. spontaneum* clones by substituting 2,4-D with picloram and selecting an organogenic green callus type, which was most likely a type-IV callus.

Plant Regeneration from Embryogenic Cell-Suspensions. The establishment of finely dispersed embryogenic cell suspensions from the compact nodular type-III callus has proven difficult since it hardly dissociates and abundantly produces phenolics in liquid medium. Frequent replacement of the culture medium in the early phases of the establishment of the suspension, followed by selective subculturing, has proven necessary. Larkin (1982) and Ahloowalia and Maretzki (1983) were the first to claim the establishment of embryogenic cell suspensions containing isolated globular embryos. However, these formations appeared sporadic or occurred in quite heterogeneous cell suspensions still containing large aggregates. Ho and Vasil (1983b) first reported clear evidence of the establishment of an embryogenic cell suspension of the clone 68-1067, which was able to form SE and regenerate plants through somatic embryogenesis upon plating onto solid medium. Using Ho and Vasil's procedures, Taylor et al. (1992) tested embryogenic calluses of 18 commercial clones for their cell suspension-forming ability. Eight out of the 18 genotypes were capable of producing maintainable and regenerable heterogeneous embryogenic cell suspensions, among which 6 later formed homogeneous cell suspensions, cytologically identical to the 68-1067 cell suspension, but in contrast were unable to regenerate plants.

Plant Regeneration form Protoplast Cultures. Establishment of cell suspensions from friable, nonembryogenic calli, isolation of protoplasts from these cell suspensions, and obtainment of protocolonies, first reported in the pioneering work of Maretzki and Nickell (1973), have proven repeatable in various genotypes of sugarcane. In contrast, enzymatic isolation of a large amount of protoplasts from compact embryogenic callus is impossible (Mérot 1988). Heterogenous cell suspensions consisting of aggregates of various sizes, generally release a low yield of protoplasts of poor viability which do not sustain divisions, whereas protoplasts prepared from homogeneous cell suspensions efficiently produce protocolonies which failed to regenerate (Taylor et al. 1992). Srinivasan and Vasil (1986) first regenerated five green plants from protoplast preparations isolated from a 4–6-month-old, fast-growing, finely dispersed cell suspension of the genotype B4362. Chen and coworkers (1988b) similarly reported plant regeneration from protoplasts of genotype F164. However, these latter two groups were unable to repeat their successes. Sugarcane protoplasts are generally embedded in an agarose media based on Kao and Michayluk (1975) or Kao (1977) complex formulations and cultured. Nurse cultures have also been used to improve plating efficiency in transformation experiments (Rathus and Birch 1992a).

Table 1. Nodular embryogenic callus-forming abilities of the two cultured basal leaf segments of in vitro plantlets of various commercial hybrids. (E. Guiderdoni and B. Mérot, unpubl.)

Hybrids	Origin	No. of leaf explants tested	Explants forming NEC (%)	Explants forming EC (%)	Frequency of EC Induction (%)
CO6415	India	150	40	26	40
CO421	India	120	36	3	7
FR80-236	France	72	47	13	26
FR80-412	France	287	53	34	70
POJ2878	Indonesia	90	41	6	16
Q75	Australia	100	52	32	62

3 Structural and Biochemical Aspects of Sugarcane Somatic Embryogenesis

The following sections focus on somatic embryogenesis in the two 3–4-mm-long basal leaf segments of in vitro cultured plantlets which can derive from either meristem, axillary bud, seed, or callus cultures. The use of such material presents several practical advantages (aseptic conditions, synchronism in plantlet development, possibility of monitoring and modifying the plantlet environment). Variation in the embryogenic callus-forming abilities of the cultured, two basal leaf segments of plantlets of commercial hybrids is illustrated in Table 1. Leaf segments of clones CO6415, Q75 and Q96 and clone FR 80412 have been used as materials for studying the histology and biochemistry of somatic embryogenesis, respectively.

3.1 Ontogeny of the Embryogenic Callus

The two 3–4-mm-long basal segments of the actively growing blade of the innermost leaf and the first basal segment of the second innermost leaf are capable of producing EC. The more mature parts of those leaves or the basal segments of older leaves generally give rise to only nonmorphogenic friable or mucilaginous formations. These quantitative and qualitative gradients, which

Fig 2. a Nodular (*nod*) and friable (*fr*) callus formations arising from a cultured basal leaf segment of an in vitro grown plantlet, clone CO6415, 4 weeks after inoculation onto induction medium (*bar* = 1 mm). **b–d** Ontogeny of the nodular embryogenic callus in basal segments of the cultured, two innermost leaves. **b** Detail of a cross section performed 3mm above the shoot meristem in the basal leaf segment, 6 days after inoculation, showing cell reactivation in all the tissues of the innermost leaves, clone CO6415. *le* Lower epidermis; *ue* upper epidermis; *bs* bundle sheath cells; *bar* = 25 μm (Guiderdoni and Demarly 1988). **c** Section across a 4-week-old nodule showing persistent sectorial organization. *me* Meristematic cells; *dc* densely cytoplasmic cells; *ep* epidermal cells; *lb* former position of the leaf bundles; *bar* = 100 μm (Guiderdoni and Demarly 1988). **d** Cross section of an established 3-month-old embryogenic callus, clone Q96; *bar* = 100 μm, (**b** and **c**: paraffin-embedded tissue; **d**: resin-embedded tissue)

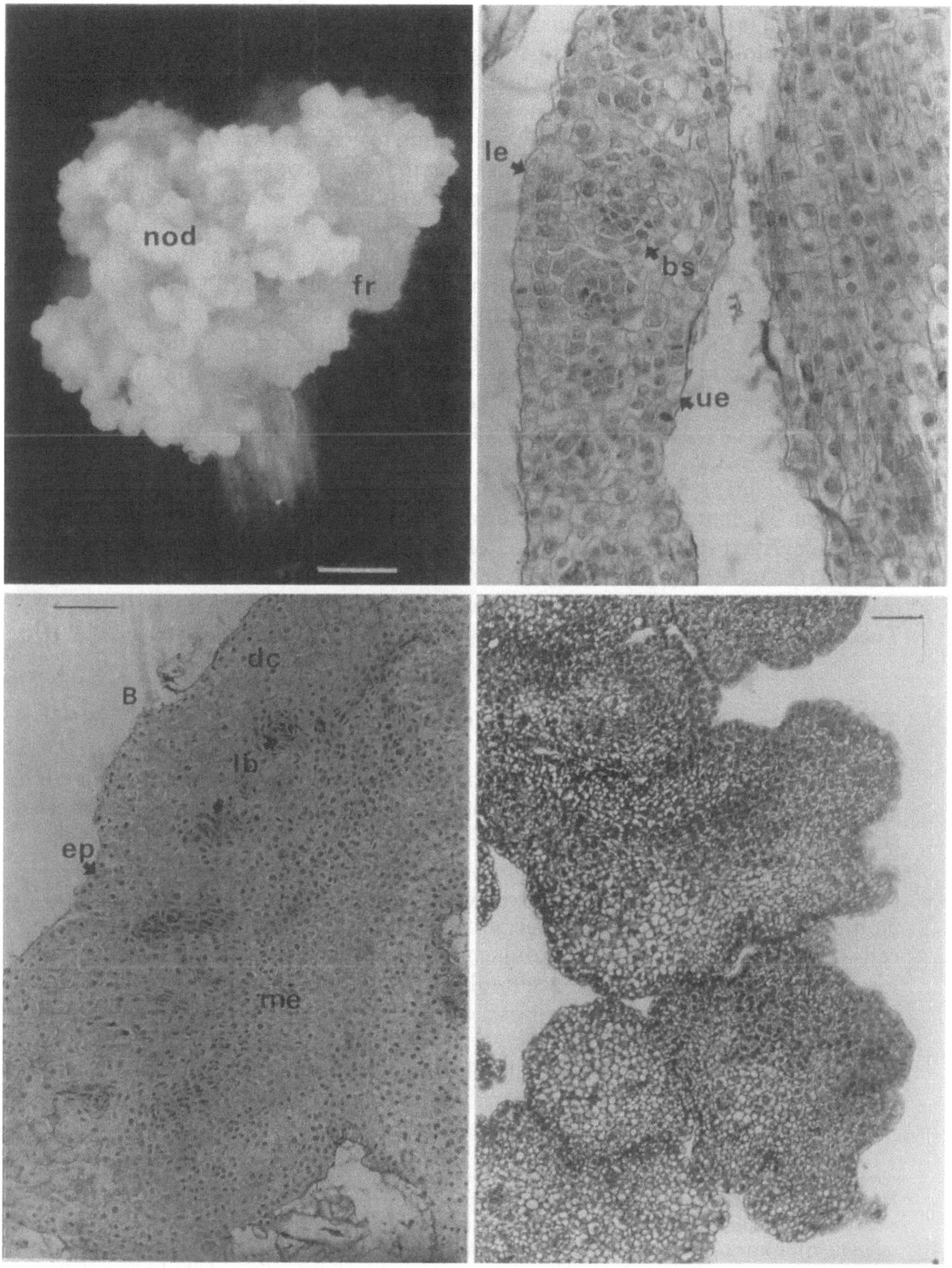

a b c d

are also found in cultured leaves of mature plants (Chagvardieff et al. 1981; Ho and Vasil 1983a), are a consequence of the basipetal mode of leaf development in Graminae. Along the same line, histology of callus induction has shown that the range of tissues which participate in callus formation in a given explant varies with the rank of the leaf and the position of the excised segment along the leaf (Guiderdoni and Demarly 1988). EC was found to arise from the immature parts of the leaf blade or the very young leaf sheath, which contain elongating cells. Divisions, at first taking place in subepidermal fiber cells, spread to all tissues of the leaf explant (i.e., epidermal, nonchlorophyllous parenchyma, mesophyll, bundle sheath, phloem and xylem parenchyma cell) 6 days after inoculation of the leaf segment (Fig. 2b). Proliferations are more intense at the phloem/lower epidermis pole. Within 30 days, a highly organized callus slowly develops from regions close to the cut end of the leaf segment through divisions of meristematic cells first organized in a pseudocambium layer (Fig. 2c). Intensive folding of that tissue leads to nodular formations of the embryogenic callus.

In cultured segments of furled leaves of mature plants, the pluritissular origin of the morphogenic callus has also been shown (Bonnel et al. 1983; Ho and Vasil 1983a). In contrast, Liu and coworkers (1982) described morphogenic callus formation originating only from the phloem parenchyma cells. Bonnel et al. (1983) identified the segments of the elongating blade of the three innermost leaves as the most responsive explants, whereas Ho and Vasil (1983a) conversely claimed that EC only originates from segments of the sheath of leaves 4 and 5. These discrepancies may be due to genotypic and/or physiological reasons.

3.2 Organization of the 2-Month-Old Embryogenic Callus

Two-month-old EC consists of lobed nodular units delimited by an invaginated epidermal layer of isodiametric cells which give the nodule a smooth appearance (Fig. 2d). The nodules exhibit a sectorial organization and consist of: (1) actively dividing, small meristematic cells displaying a high nucleoplasmic ratio, a conspicuous nucleus and nucleolus, small vacuoles, and a few small starch grains (Fig 3a,b); (2) looser subepidermal areas, most likely deriving from the meristematic cells of large, oval, and richly cytoplasmic cells exhibiting large vacuoles and a lower nucleoplasmic ratio and accumulating larger starch granules (Fig. 3c,d). The latter cells may evolve in a loose, undividing storage tissue accumulating large amounts of protein and starch (Fig. 4a), and, as shown by TEM observations, lipids (Fig. 3d). These cells closely resemble those of the subepithelial and adaxial tissues of the scutellum of gramineous embryos. (3) Epidermal cells,

Fig 3a–d. Light microscopy and TEM observations of the structural organization of a 2-month-old embryogenic callus. **a** Detail of meristematic cell area exhibiting intense mitotic activity (*arrows* = mitoses; cytoplasmic *dark dots* = starch grains; *bar* = 50 μ. **b** TEM observation of epidermal and subepidermal meristematic cells showing tight arrangement of the cells and thickening of the midlamellae. *v* Vacuole; *n* nucleus: *sg* starch grain: *p* proteins *bar* = 10 μ. **c** Detail of the loose structure of the densely cytoplasmic cell area (*bar* = 50 μ). **d** TEM observation of the loose areas of densely cytoplasmic cells showing peripheral accumulation of lipids. *sg* Starch grain: *l* lipids; *bar* = 20 μ

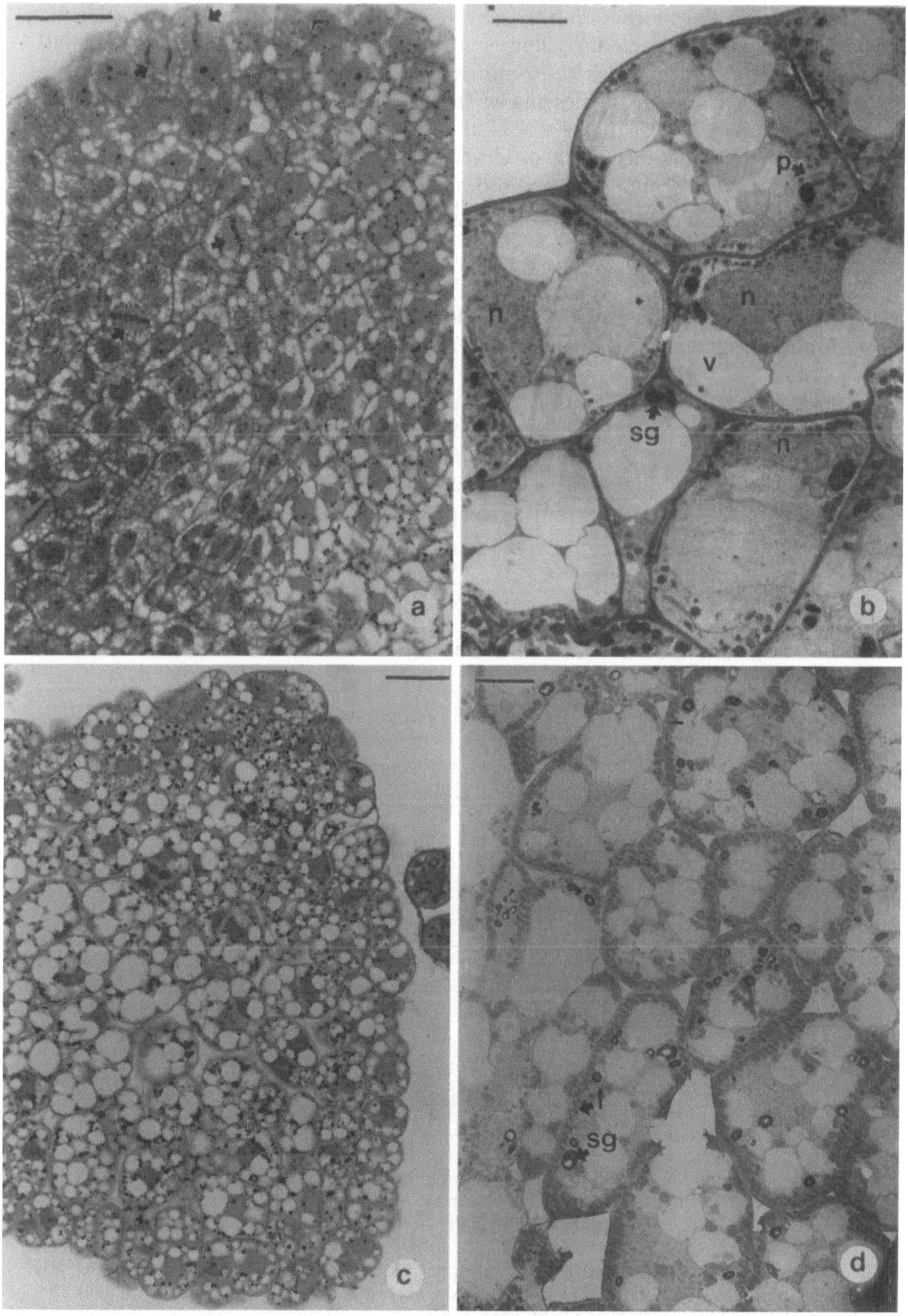
a
p
n
n
v
sg
n
b
c
sg
d

which exhibit the characteristics of embryogenic cells (very high nucleoplasmic ratio, conspicuous nucleolus, dense cytoplasm with a few small vacuoles, small starch and protein granules, thickening of the midlamella) and which can actively divide (Fig. 4b); Such structuring of the embryogenic callus has been recently confirmed in cv. NCo310 (Brisibe et al. 1993).

Shoot meristems are rarely observed as long as the cultures are maintained in the dark. However, amalgamated shoot meristems or embryos may differentiate in regions on top of the callus, probably due to a sectorial decrease in 2,4-D concentration.

3.3 Comparative Biochemistry and Endogenous Growth Regulators of Embryogenic and Nonembryogenic Cultures

Endogenous Sugars and Polysaccharides. Amylolysis of dehydrated tissues of 2-month-old EC and NEC (nonembryonic callus) derived from the same leaf segments shows that EC contains half the amount of neutral sugars, in both soluble and insoluble fractions, and twice the amount of starch compared to NEC (Table 2). The latter result is in agreement with the abundance of storage tissue in the EC.

Table 2. Relative importance of sugars, polysaccharides, and nonpolysaccharide compounds in the soluble and insoluble fractions of EC and NEC extracts (clone FR80 412). (B. Mérot, M.T. Tollier, and X. Rouau, unpubl.)

Callus type	Fraction studied	Sample dry weight (%)	Neutral sugars (%)	Uronic acid (%)	Nonpolysaccharide compounds (%)
Embryogenic	Insoluble fraction	57.6	*29.9*	*2.7*	*67.4*
	Soluble fraction	8.8	*23.8*	*1.6*	*74.6*
	Extraction losses (starch)	33.6	–	–	–
Non-embryogenic	Insoluble fraction	62.1	*55.3*	*3.0*	*41.7*
	Soluble fraction	20.2	*56.7*	*2.7*	*40.6*
	Extraction losses (starch)	17.7	–	–	–

Fig. 4a, b. Structural organization of a 2-month-old embryogenic callus, clone Q96 (PAS-Naphtol Blue Black double stain). **a** Detail of the epidermal (*ep*) cell layer and the subepidermal, densely cytoplasmic cell area exhibiting the accumulation of clumps of storage starch globules (*bold face arrows*) and lipoprotein bodies (*boldface double arrows*); (*bar* = 50 μ). **b** Outgrowth of potentially embryogenic cell clumps deriving from the epidermal cell layer of the embryogenic nodules; *bar* = 50 μm. **c** Green plant regeneration in a 3-month-old embryogenic callus, clone CO6415, 10 days after transfer to 2,4-D-free medium and light; *bar* = 1 mm

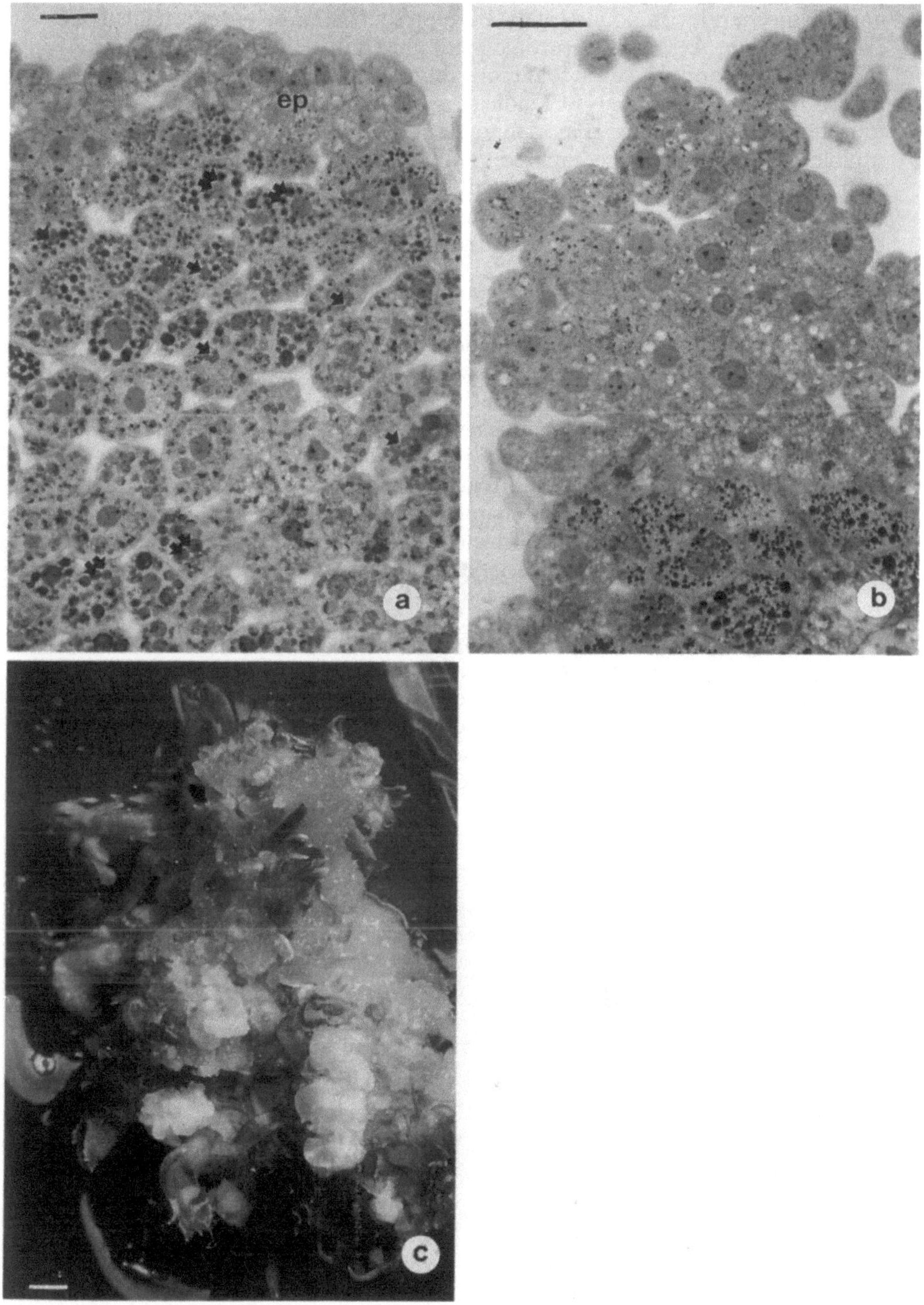
ep
a
b
c

The soluble fraction of the NEC extract is twice that of the EC. The nonpolysaccharide compounds of the EC represent 68.3% of its dry weight, if one excludes starch. Acid sugars (uronic acid) are equivalent in the two types of culture. Ribose appears seven fold more abundantly in the insoluble fraction of the EC than in that of the NEC, indicating an active metabolism. A respective four and three fold increase in mannose and glucose contents is found in the soluble fraction of the EC compared to that of NEC. On the other hand, NEC contains three fold more fucose (B. Mérot, M.T. Tollier, and X. Rouau, unpubl.). This latter result shows that meristems in NEC resemble root meristems, the caps of which are known to be particularly fucose-rich.

Endogenous Growth Regulators. Analysis of plant hormones in EC and NEC calli was performed following the extraction, HPLC purification, and ELISA measurement procedures detailed elsewhere (Maldiney et al. 1986). Relative contents of IAA, ABA, iPA, ZR, and their precursors/conjugates in EC and NEC are given in Table 3. The endogenous level of IAA in EC is twice that of NEC. Although the overall cytokinin content appears equal in the two callus types, the relative abundance of these growth regulators varies: zeatin is four fold more abundant in NEC, whereas EC is characterized by a higher level of iP and the presence of iPA. Zeatin riboside is detected at a similar level in EC and NEC. EC is also characterized by the presence of ABA which has not been detected in NEC.

Amino Acids. EC is characterized by a higher overall amino acid content (representing 20.4% dry weight vs. 14% in NEC), comparable to that of seeds (22.3%). This agrees with the abundance of nonpolysaccharide compounds in EC reported above. EC also exhibit to a higher endogenous level of arginine compared to NEC. Glutamic acid appears to be the most abundant amino acid in EC (16% of the overall amino acid content) and NEC (17%) as well as in seeds (25.2%).

Table 3. Endogenous levels of growth regulators in EC and NEC (clone FR80412) (results are given in picomole per gram dry weight). B. Mérot, P. Le Maréchal, B. Sotta, M. Bonnet, and E. Miginiac, unpubl.)

Immunoglobulin tested	Hormone identified	Embryogenic callus	Nonembryogenic callus
Anti-IAA	IAA	221	108
	precursor/conjugate	11	5.7
Anti-ABA	ABA	87.1	0.5
	precursor/conjugate	21.4	1.7
Anti-iPA	iP	82.7	48.5
	precursor/conjugate	2.7	4.4
	iPA	25.3	0.5
	precursor/conjugate	4.9	1.2
Anti-ZR	Zeatin	13.8	57.2
	Zeatin riboside	3.2	2.7

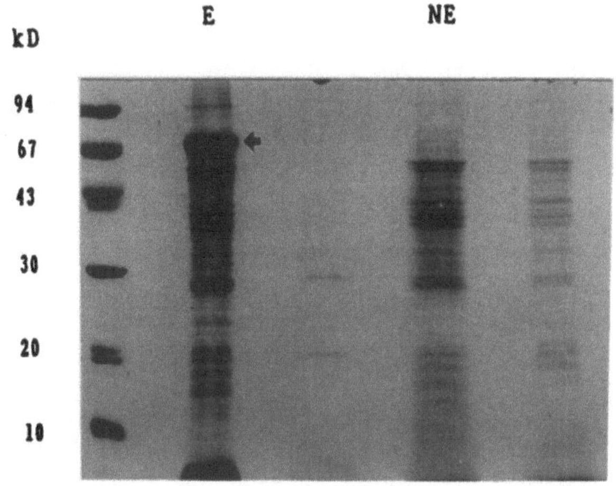

Fig. 5. SDS-PAGE analysis of EC and NEC proteins showing existence of a 70-KD EC-specific protein (*arrow*). (B. Mérot, F. Ambard Bretteville, and R. Rémy, unpubl.)

Proteins. Monodimensional SDS-PAGE analysis of EC and NEC crude extracts permits the identification of an abundant, high molecular weight (65/70 KDa) protein which appears specific to EC (Fig. 5). This protein accounts for more than one-third of the total protein content of the EC and is specifically found in its insoluble fraction. These results are confirmed by SDS-PAGE two-dimensional electrophoresis (Fig. 6). Further analyses have shown that the EC-specific protein consists of two polypeptides (Fig. 7), exhibiting similar molecular weights (65 and 67 kDa) and an identical isoelectric point (6.5), and appears to be glycosylated. Differential centrifugations have shown that the protein is specific to a heavy fraction where nucleus fragments, organelles, and starch grains are also found. When the EC is transferred to regeneration medium under light, the level of the EC-specific protein decreases at day 10, being undetectable at day 13. Further experiments have shown that culture under light is not responsible for that loss. An increase in the 2,4-D concentration in the maintenance medium leads to a decrease in the content of the EC-specific protein in EC. The protein was also found to be specific to EC in another commercial clone, R570 (Mérot 1988).

Abundance and glycosylation of the EC-specific protein, as well as its disappearance during "germination" of SE, are very similar in seed storage proteins, although the polypeptide is not found in that form in sugarcane seeds (Mérot 1988). Along the same line, Chen and Luthe (1987) have identified an EC-specific 57-KDa polypeptide in rice, which is also found in seeds and is most likely a storage protein.

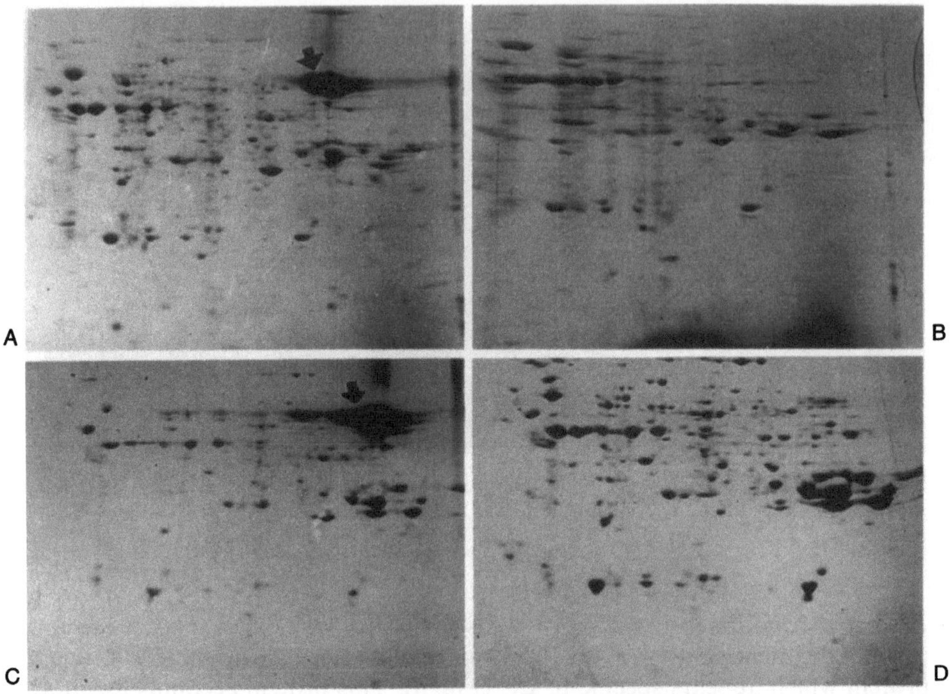

Fig. 6a–d. SDS-PAGE two-dimensional analysis of the total protein of EC (**a**) and NEC (**b**) and the insoluble (**c**) and soluble (**d**) fraction of EC demonstrating the specific presence of the protein in the insoluble fraction of the EC. (Mérot, Ambard Bretteville, and Rémy, unpubl.)

3.4 Plant Regeneration from Embryogenic Cultures

Two- to 3-month-old EC usually produces green shoots with a 90% frequency, 15 days after transfer to MS medium devoid of 2,4-D and under light (Fig. 4c). In contrast to that observed for embryogenic callus formation, little variation is observed in plant regeneration frequency among genotypes. Albina, striata, and xantha plantlets are sometimes simultaneously regenerated. Sugarcane embryogenic callus possesses an outstanding regeneration potential since one gram of fresh weight of EC yields an average of 130 to 200 plantlets (Guiderdoni and Demarly 1988).

SEM examination of the EC 3 days after its transfer to regeneration medium shows the presence of lobed, globular structures delimited by isodiametric cells and development of a scutelar notch (Fig. 8a). Cross sections show well-organized bipolar embryos possessing a shoot meristem, a shoot-root axis, and a distinct scutellum. At this early stage, the SE consist of a meristematic and storage cell area and are delimited by a nonproliferating epidermis, i.e., the same cell types found in the embryogenic nodules (Fig. 8c). Further stages of SE development are observed 10 days after the transfer: SE exhibit at this stage a shoot meristem producing young leaf primordia surrounded by a coleoptile and

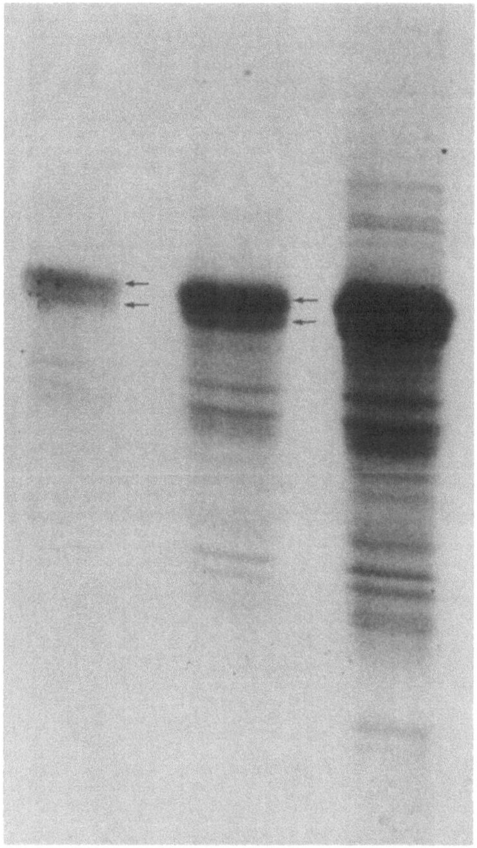

Fig. 7. Electrophoresis in a 12–18% polyacrylamide gradient showing that the EC-specific protein consists of two polypeptides (arrows) of 65 and 67 kDa molecular weight (B. Mérot, F. Ambard Bretteville and R. Rémy, unpubl.)

a distinct root meristem (Fig. 8b,d). Early development of embryo axillary buds, resulting in multiple shoot formation, was sometimes observed. The coleoptile is sometimes absent and root development is often delayed in embryonic structures. The most complete and individualized SE were found to differentiate on top of the nodular callus, whereas atypical, multiembryonic configurations and structures with multiple shoots are observed in callus sectors in contact with the medium.

In analyzing the early phase of EC formation, 2–4 weeks after inoculation of the leaf segment on the induction medium, Ho and Vasil (1983a) showed that SE arise directly from single cells of the epidermal layer of the young EC, in the presence of 2,4-D. However, in 2–3-month-old cultures, the origin of the somatic embryos appears less clear since meristematic and storage tissues of the callus have actively proliferated and folded probably similar to the tissues of the early, induced proembryos. The shoot meristem and the scutellar epidermal cells of the SE are very responsive and may, respectively, develop axillary shoots and rapidly produce adventitious SE, resulting in complex, fused structures. It is therefore

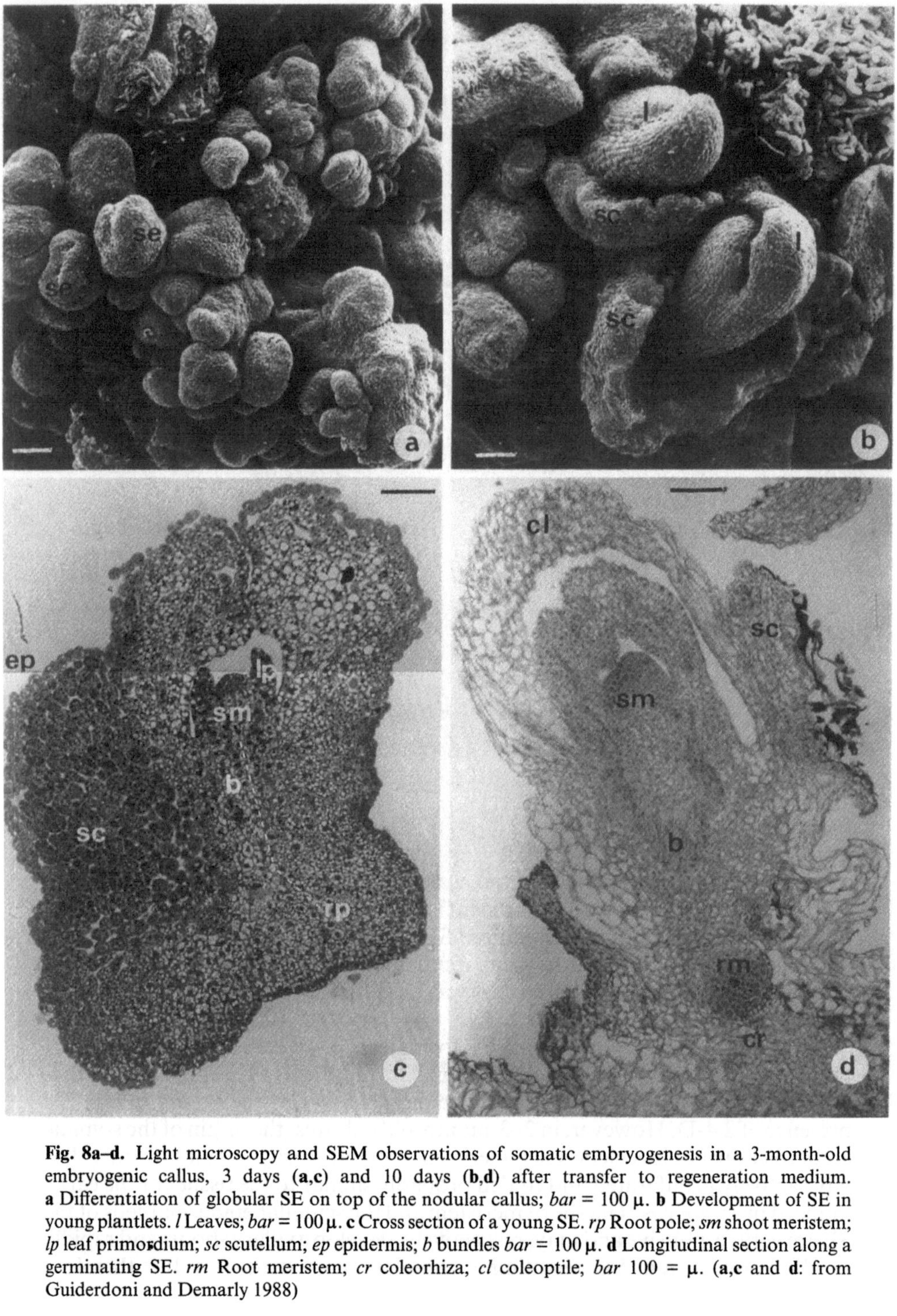

Fig. 8a–d. Light microscopy and SEM observations of somatic embryogenesis in a 3-month-old embryogenic callus, 3 days (**a,c**) and 10 days (**b,d**) after transfer to regeneration medium. **a** Differentiation of globular SE on top of the nodular callus; *bar* = 100 μ. **b** Development of SE in young plantlets. *l* Leaves; *bar* = 100 μ. **c** Cross section of a young SE. *rp* Root pole; *sm* shoot meristem; *lp* leaf primosdium; *sc* scutellum; *ep* epidermis; *b* bundles *bar* = 100 μ. **d** Longitudinal section along a germinating SE. *rm* Root meristem; *cr* coleorhiza; *cl* coleoptile; *bar* 100 = μ. (**a,c** and **d**: from Guiderdoni and Demarly 1988)

difficult to determine whether SE derive from derepression of the cleaved embryogenic tissue, from single cells, or both.

Therefore, the morphogenetic pathway used for plant regeneration in sugarcane tissue cultures has long been debated. Organogenesis (Liu and Chen 1974; Bonnel et al. 1983), embryogenesis (Nadar et al. 1978; Ho and Vasil 1983a), or both pathways (Chen et al. 1988a) have been described. Differences in the genotype, the callus type, the proliferating stage, and/or the age of the cultures under study may explain these contrasting reports.

4 Summary and Conclusions

The white, compact nodular callus, called EC, can be induced from leaf explants of most commercial sugarcane hybrids, despite a wide variation in frequency, and can readily regenerate plants, insofar as the transfer to regeneration medium is achieved early. Somatic embryogenesis in leaf explants is indirect through the formation of a nodular, meristematic EC. The EC has a pluritissular origin and consequently exhibits a sectorial organization. It consists of meristematic and storage cell areas delimited by an epidermal layer of potential embryogenic cells. SE may arise either directly from the latter cells or through cleaving of the proembryogenic callus tissues. Young SE were found to be made up of the same types of cells as those found in the EC.

Comparative biochemical analyses of EC and NEC have shown that EC is characterized by a high starch, protein, and amino acid content, the presence of endogenous iPA and ABA, and a higher level of endogenous IAA compared to NEC. Mono- and two-dimensional SDS-PAGE analyses of proteins have permitted the identification of an EC-specific glycoprotein consisting of two polypeptides of 65 and 67 kDa molecular weight respectively. This protein disappears during plant regeneration and is most likely a storage protein.

Embryogenic callus appears to be a suitable material for cryopreservation and transformation work. Procedures for routine establishment of synchronized embryogenic cell suspensions in a wide range of genotypes must still be developed for profitable clonal propagation through SE encapsulation.

Better knowledge of somatic embryo environmental requirements will also lead to reduction in somaclonal variation. Such a reduction is an important goal for clonal propagation, transformation, and preservation of genetic resources. Ongoing development of molecular markers in sugarcane (D'Hont et al. 1994) will further the knowledge of the genetic basis of such variation, making it possible to limit its occurrence in tissue cultures.

5 Protocol for the Induction and Expression of Somatic Embryogenesis

Plant Material. 10- to 15-cm-tall plantlets derived from either meristem, axillary bud, germinating seed, or callus cultures are cultured on solid or liquid MS basal medium without growth regulators in test tubes or glass jars.

Under Laminar Flow:

1. Remove the roots, the outer mature leaves, and expanded leaf blades of the plantlet using forceps.
2. With a scalpel cut the shoot 1 mm above the shoot apex and slice sequentially the cylinder of the 2–3 innermost immature leaves into 2–4-mm-long segments.
3. Inoculate 6 segments onto 25 ml of MS medium supplemented with 3 mg/l 2,4-D, 6 g/l agarose, and 30 g/l sucrose in 100-mm-diameter Petri dishes. Incubate the cultures in the dark for 4 weeks. Swelling and callusing of the cultured tissue can be seen after 4–5 and 10–15 days of culture, respectively.
4. Remove the unresponsive browning parts of the leaves and transfer the whole callus formation to MS medium; the 2,4-D concentration is decreased to 1 mg/l (4 calluses per 100-mm-diameter Petri dish). Incubate a further 4 weeks.
5. Carefully separate the white, compact, nodular sectors having a dry appearance from the rest of the callus and either subculture them onto maintenance medium for a further 4 weeks of growth or transfer them to MS basal medium under light for plant regeneration.
6. Plant regeneration occurs within 2–3 weeks with fragmentation and intense growth of the callus. Transfer 5-cm-long individual shoots or stool of shoots that can be easily separated from one another with foreceps to test tubes containing MS basal medium supplemented with 60 g/l sucrose for rooting, vigorous plantlet development, and tillering.

Note. This protocol can be applied to 5-mm sections of a 6-cm-long cylinder of the 3–4 innermost furled leaves taken 2 cm above the shoot apex and aseptically extracted from the surface sterilized outer leaf sheath of greenhouse or field-grown mature plants (Chagvardieff et al. 1981).

Acknowledgments. Part of this work is the result of a CIRAD-ADAR research project carried out at the Laboratoire d' Amélioration des Plantes, URA CNRS 115, Bât 360, Orsay University, F-91405 Orsay France. Analysis of endogenous growth regulators was carried out by P. Le Maréchal, B. Sotta, M. Bonnet, and Professor E. Miginiac at the Laboratoire de Physiologie du Développement des Plantes, URA CNRS 1180, Pierre and Marie Curie University, F-75252 Paris. Biochemical analyses were carried out by M.T. Tollier and X. Rouau (sugars) and A. Risterrucci (amino acids) at CIRAD, Montpellier. Identification of the EC-specific protein is the result of collaboration with F. Ambard Bretteville and R. Rémy, Laboratoire de Génétique Moléculaire Végétale, Bât 430, Orsay University, F-91405 Orsay. EG and BM wish to thank the Electron Microscopy Laboratories of the University of the Sciences and Techniques of Languedoc (USTL) Montpellier, for SEM observations and of Orsay University, for TEM observations.

References

Ahfield H (ed) (1993) FO LICHT's world sugar and sweetener yearbook: world statistics 1992/93. FO LICHT, Ratze burg

Ahloowalia BS, Maretzki A (1993) Plant regeneration via somatic embryogenesis in sugarcane. Plant Cell Rep 2: 21–25

Arceneaux G (1967) Cultivated sugarcanes of the world and their botanical derivations. Proc 12th ISSCT Congr, 28 March- to April 1965, Puerto Rico. Elsevier, Amsterdam, pp 844–854

Bajaj YPS, Korneva SB, Gutierrez R, Maribona RH (1987) Freeze preservation of plantlets, excised meristems and cell cultures of sugarcane (*Saccharum officinarum* L.) at –196 °C. In: Bose A, Sengupta P (eds) Proc Int Conf Cryogenics. McGraw-Hill, New Delhi, pp 222–226

Berding N, Roach BT (1987) Germplasm, collection, maintenance and use. In: Heinz DJ (ed) Sugarcane improvement through breeding. Developments in crop science, vol 11. Elsevier, Amsterdam, pp 143–210

Bower R, Birch RG (1992) Transgenic sugarcane plants via microprojectile bombardment. Plant J 2: 409–416

Brisibe EA, Nishioka D, Miyake H, Tanigushi T, Maeda E (1993) Developmental electron microscopy and histochémistry of somatic embryo differentiation in sugarcane. Plant Sci 89: 85–92

Chagvardieff P, Bonnel E, Demarly Y (1981) La culture in vitro de tissues somatiques de canne à sucre (*Saccharum* sp.). L'Agron Trop 36: 266–278

Chen LJ, Luthe DS (1987) Analysis of proteins from embryogenic and nonembryogenic rice (*Oryza sativa* L.) calli. Plant Sci 48: 181–188

Chen WH, Gartland KMA, Davey MR, Sotak R, Gartland JS, Mulligan BJ, Power JB, Cocking EC (1987) Transformation of sugarcane protoplasts by direct uptake of a selectable chimaeric gene. Plant Cell Rep 6: 297–301

Chen WH, Davey MR, Power JB, Cocking EC (1988a) Control and maintenance of plant regeneration in sugarcane callus cultures. J Exp Bot 39: 251–261

Chen WH, Davey MR, Power JB, Cocking EC (1988b) Sugarcane protoplasts: factors affecting division and plant regeneration. Plant Cell Rep 7: 344–347

Chowdhury MKV, Vasil IK (1992) Stably transformed herbicide resistant callus of sugarcane via microprojectile bombardment of cell suspension cultures and electroporation of protoplasts. Plant Cell Rep 11: 494–498

Chu CC, Wang CC, Sun CS, Hsu C, Yin KC Chu YC, Bi FY (1975) Establishment of an efficient medium for anther culture of rice through comparative experiments on the nitrogen source. Sci Sin 18: 659–668

Daniels J, Roach BT (1987) Taxonomy and evolution. In: Heinz DJ (ed) Sugarcane improvement through breeding. Developments in crop science vol. 11. Elsevier, Amsterdam, pp 7–84

D'Hont A, Lu YH, Gonzalez de Leon D, Grivet L, Feldmann P, Lanaud C, Glaszmann JC (1994) A molecular approach to unravelling the genetics of sugarcane, a complex polyploid of the Andropogonae tribe. Genome 37: 222–230

Eksomtramage T, Paulet F, Guiderdoni E, Glaszman JC, Engelman F (1992) Development of a cryopreservation process for embryogenic calluses of a commercial hybrid of sugarcane (*Succharum* sp) and application to different varieties. Cryo Lett 13: 239–252

FAO (1993) Production, yearbook: 1992, vol 46. FAO statistic series 112, FAO, Rome

Fitch MMM, Moore PH (1990) Comparison of 2,4-D and picloram of longer-term totipotent green callus culture of sugarcane. Plant Cell Tissue Organ Cult 20: 157–163

Fitch MMM, Moore PH (1993) Long term culture of embryogenic sugarcane callus. Plant Cell Tissue Organ Cult 32: 335–343

Franks T, Birch RG (1991) Gene transfer into intact sugarcane cells using microprojectile bombardment. Aust J Plant Physiol 18: 471–480

Gallo-Meagher M, Irvine JE (1993) Effects of tissue type and promoter strength on transient GUS expression in sugarcane following particle bombardment. Plant Cell Rep 12: 666–670

Gambley RL, Frod R, Smith GR (1993) Microprojectile transformation of sugarcane meristems and regeneration of shoots expressing β-glucoronidase. Plant Cell Rep 12: 343–346

Gnanapragasam S, Vasil IK (1990) Plant regeneration from a cryopreserved embryogenic cell suspension of a commercial sugarcane hybrid (*Saccharum* sp.) Plant Cell Rep 9: 419–423

Guiderdoni E, Demarly Y (1988) Histology of somatic embryogenesis in cultured leaf segments of sugarcane plantlets. Plant Cell Tissue Organ Cult 14: 71–88

Heinz DJ, Krishnamurthi M, Nickell LG, Maretzki A (1977) Cell, tissue and organ culture in sugarcane improvement. In: Reinert J, YPS Bajaj (eds) Applied and fundamental aspects of plant cell, tissue, and organ culture. Springer, Berlin Heidelberg New York, pp 3–17

Ho WJ, Vasil IK (1983a) Somatic embryogenesis in sugarcane (*Saccharum Officianarum* L.) 1. The morphology and physiology of callus formation and the ontogeny of somatic embryos. Protopalsma 118: 169–180

Ho WJ, Vasil IK (1983b) Somatic embryogenesis in sugarcane (*Saccharum officinarum* L.): growth and plant regeneration from embryogenic cell suspension cultures. Ann Bot 51: 719–726
Irvine JE, Benda GTA (1985) Sugarcane mosaic virus in plantlets regenerated from diseased leaf tissue. Plant Cell Tissue Organ Cult 5: 101–106
Irvine JE, Fitch M, Moore PH (1983) The induction of callus in sugarcane tissue cultures by selected chemicals. Plant Cell Tissue Organ Cult 2: 141–149
Jian LC, Sun DL, Sun LH (1987) Sugarcane callus cryopreservation. In: Li PH (ed) Plant biology, vol 5. Plant cold hardiness. A R Liss, New York, pp 323–337
Kao KN (1977) Chromosomal behaviour in somatic hybrids of soybean – *Nicotiana glauca*. Mol Gen Genet 150: 225–230
Kao KN, Michayluk MR (1975) Nutritional requirements for growth of *Vicia hajastana* cells and protoplasts at a very low population density in liquid media. Planta 126: 105–110
Larkin PJ (1982) Sugarcane tissue and protoplast culture. Plant Cell Tissue Organ Cult 1: 149–163
Liu MC (1981) In vitro methods applied to sugarcane improvement In: Thorpe TA (ed) Plant tissue culture; methods and applications in agriculture. Academic Press, New York, pp 299–323
Liu MC (1984) Sugarcane. In: Sharp WR, Evans DA, Ammirato PV, Yamada Y (eds) handbook of plant cell culture, 2. Crop species. MacMillan, New York, pp 573–605
Liu MC, Chen WH (1974) Histological studies on the origin and process of plantlet differentiation in sugarcane callus mass. Proc 15th ISSct Congr, 13–29 June 1974, Durban, South Africa, pp 1–12
Liu MC, Chen WH, SC Shih (1982) Histogenesis of sugarcane callus originating from young leaf and stem tip explants. Can J Bot 60: 2889–2895
Maldiney R, Leroux B, Sabbagh I, Sotta B, Sossountzov L, Miginiac E (1986) A biotin-avidine based enzyme immunassay to quantify three phytohormones: auxin, abscisic acid and zeatin-riboside. J Immunol Methods 90: 151–158
Maretzki A (1987) Tissue culture: its prospects and problems. In: Heinz DJ (ed) Sugarcane improvement through breeding. Developments in crop science vol. 11. Elsevier, Amsterdam, pp 343–384
Maretzki A, Hiraki P (1980) Sucrose promotion of root formation in plantlets regenerated from callus of *Saccharum* spp. Phyton 38: 85–88
Maretzki A, Nickell LG (1973) Formation of protoplasts from sugarcane cell suspensions and the regeneration of cell cultures from protoplasts. In: Protoplastes et fusion de cellules somatiques végétales, colloques Int CNRS No 212. CNRS Editions, Gifs/Yvette, France, pp 51–63
Mérot B (1988) Etudes de l'embryogenèse somatique de la canne à sucre (*Saccharum* sp.): optimisation de la production, application aux cultures de cellules et de protoplastes, analyses cytologiques, biochimiques et hormonales. Plant Development and Improvement. Doctoral Thesis, Orsay University, Paris XI, France, 295 pp
Mérot B, Guiderdoni E (1987) Somatic embryogenesis in cultured leaf segments of sugarcane (*Saccharum* sp.) plantlets: first report of in vitro conversion of encapsulated somatic embryos to plants. In: Abstr 1st Int Cong Plant Tissue Cult, Tropical Species, 21–25 Sept 1987. Bogota, Colombia
Murashige T, Skoog F (1962) A revised medium for rapid growth and bioassays with tobacco tissue culture. Physiol Plant 15: 473–497
Nadar HM, Heinz DJ (1977) Root and shoot development from sugarcane callus tissue. Crop Sci 17: 814–816
Nadar HM, Soepraptopo S, Heinz DJ, Ladd SL (1978) Fine structure of sugarcane (*Saccharum* sp.) callus and the role of auxin in embryogenesis. Crop Sci 17: 210–216
Nickell LG, Maretzki A (1969) Growth of suspension cultures of sugarcane in a chemically defined medium. Physiol Plant 22: 117–125
Rathus C, Birch RG (1992a) Optimization of conditions for electroporation and transient expression of foreign genes in sugarcane protoplasts. Plant Sci 81: 65–74
Rathus C, Birch RG (1992b) Stable transformation of callus from electroporated sugarcane protoplasts. Plant Sci 82: 81–89
Schenk RU, Hildebrandt AC (1972) Medium and techniques for induction and growth of monocotyledonous and dicotyledonous plant cell cultures. Can J Bot 50: 199–204

Srinivasan C, Vasil IK (1986) Plant regeneration from protoplasts of sugarcane (*Saccharum officinarum* L.). J Plant Physiol 126: 41–48

Stevenson GC (1965) Genetics and breeding of sugarcane. Tropical Science Series. Longman, London, 284 pp

Tabaiezadeh Z, Feri RJ, Vasil IK (1986) Somatic hybridization in the Graminae: *Saccharum officinarum* L. (sugarcane) and *Pennisetum americanum* (L.) K. Schum. (pearl millet). Proc Natl Acad Sci USA 83: 5616–5619

Taylor PWJ, Ko HL, Adkins SW, Rathus C, Birch RG (1992) Establishment of embryogenic callus and high protoplast yielding suspension cultures of sugarcane (*Saccharum* spp. hybrids). Plant Cell Tissue Organ Cult 28: 69–78

Section II
Somatic Embryogenesis in Vegetables and Fruits

II.1 Somatic Embryogenesis in Asparagus

A. Levi[1] and K.C. Sink[2]

1 Introduction

The genus *Asparagus* belongs to the family Liliaceae, and is comprised of about 150–300 herbaceous and woody perennial species spread throughout the temperate and tropical regions of the world (Lawrence 1982). Among these species *Asparagus officinalis* L. is the most well-known edible, and is considered one of the oldest garden vegetable plants (Luzny 1979). It is an European-Siberian continental plant which belongs to the northeast Moditerranean vegetation gene center (Luzny 1979), where it is well adapted to sandy soils. Asparagus is widely distributed and grown as a high-value crop in different parts of the world.

Asparagus crowns establish a rhizome which bears a few lateral buds that give rise to spears that develop into 4- to 6-ft-tall ferns. Elongation of spears occurs first at the lowest internode and then in successive internodes, resulting in vigorous spear growth (Peirce 1987).

Asparagus fficinalis is a dioecious species with male and female plants (sex ratio 1 : 1). The male flowers have well-developed anthers and an aborted ovary. The female flowers are smaller than those of male plants and have nonfunctional anthers, a well-developed ovary and style, and three stigma. Occasionally, plants will produce perfect flowers. This is due to a mutation which reverts male plants to the andromonecious mode (Wricke 1979). The asparagus fruit is a red berry at maturity that contains up to six small globular seeds.

Asparagus faces breeding difficulties due to being a perennial species and having a dioecious reproductive mode. The plants begin to yield after 3 to 4 years in the field, and the average harvesting period is about 10 to 15 years. Therefore, the development of new cultivars requires many years (Ellison 1986).

As an open-pollinated species with seperated male and female plants, asparagus is considered highly heterozygous due to random mating (Ellison 1986). Homozygous asparagus lines cannot be readily obtained through traditional inbreeding procedures, due to the dioecious biology. Consequently, considerable variation still exists in yield of individual plants of the same variety. Thus, a primary breeding goal is to increase genetic uniformity between plants of a cultivar. It is desirable to combine uniformity with asparagus crowns that have a high yield spears 3/8 to 3/4 in. in diameter.

[1] USDA-ARS, PSI, Fruit Laboratory, BARC-West, Beltsville, Maryland 20705, USA
[2] Department of Horticulture, Michigan State University, East Lansing, Michigan 48824, USA

Biotechnology in Agriculture and Forestry, Vol. 31
Somatic Embryogenesis and Synthetic Seed II (ed. by Y.P.S. Bajaj)

The production of double haploid plants through polyembryonic seeds, use of andromonoecious plants, or anther culture are currently the only avenues to obtain pure male plant lines that can be used for breeding and production of uniform hybrid lines (Doré 1990).

There are major differences in the growth pattern of asparagus female versus male plants. Females produce fewer spears that are larger in diameter than those of male plants. On the other hand, male plants produce a larger number of spears and have greater longevity than females. Moreover, male plants do not manifest seedling weeds that may foster diseases and compete with the crowns that are already established in the field. For these reasons vital attempts are made to produce all male varieties (Ellison 1986).

It is established that the sexual mode in asparagus is controlled by one locus; female plants are homozygous (mm), while male plants are heterozygous (Mm) (Loptien 1979). Therefore, double haploids generated through polyembryonic seeds or anther culture are homozygous female (mm) or supermale (MM) genotypes. Such plants can be used for further breeding and production of genetically uniform all-male hybrid (Mm) plants (Doré 1990). In addition to the above-mentioned aspects of in vitro culture, a somatic embryogenic system can be a useful tool for rapid propagation of elite asparagus crowns, and for future transformation of asparagus with genes having valuable agronomic traits. Our work (Levi and Sink 1990, 1991a,b, 1992) on somatic embryogenesis is summerized here.

2 Somatic Embryogenesis

Asparagus has a low multiplication rate using conventional propagation methods. Only a limited number of propagules can be obtained by division of a mature crown. Experiments in establishing rooted cuttings were of limited success. Yang and Clore (1973, 1975) offered an alternative method of propagating aerial crowns of asparagus stems in potted plants kept under high moisture. Following enlargement, shoots and roots are formed and are separated artificially to form new plants. However, this technique is slow and not feasible in producing a large number of plants. In contrast, tissue culture techniques have alleviated the constraints of crown division and led to greater efficiency in propagation of asparagus lines (Hasegawa et al. 1973; Yang and Clore 1973; Chin 1982; Reuther 1984; Volokita et al. 1987). Moreover, the estimated number of clonal plants that can be produced through somatic embryogenesis in asparagus can be at least five-to tenfold higher than micropropagation techniques.

Somatic embryogenesis in asparagus can be used as a rapid means for production of elite asparagus crowns, selected hybrid varieties, and potential parents of such hybrids. Additionally, it can be a valuable tool for clonal propagation of double haploid plants that may be used as parents of hybrids (Doré 1990).

Several studies have been devoted to the development of somatic embryogenic systems in asparagus. Wilmar and Hellendoorn (1968) reported that hypocotyl-derived callus formed a large number of dense, globular, embryo-like bodies when placed on Linsmaier and Skoog (1965) basal medium (LS) + 1 mg/l 2,4-D and 0.3 mg/l kinetin. Globular embryos continuously differentiated in liquid suspension of the same composition and developed to mature plantlets. Steward and Mapes (1971) reported the formation of somatic embryos from asparagus stem segments through a sequence of callus induction in White's liquid medium supplemented with coconut milk and NAA, followed by replacement of NAA with 2,4-D, induction of embryos in high salt medium supplemented with NAA, and formation of root and shoot primordia by transferring embryos to medium supplemented with coconut milk and IAA. Reuther (1977) reported that shoot tip- and shoot segment-derived callus formed on LS + 1 mg/l NAA and 1 mg/l kinetin gave rise to somatic embryos when transferred to LS + 1 mg/l IAA and 0.1 mg/l BA, and subsequently to plantlets when transferred to LS or LS + 0.5 mg/l IAA.

In our study (Levi and Sink 1991) crowns established in vitro and lateral bud explants were more suitable for initiation of embryogenic callus than spear sections. When cultured on MS (Murashige and Skoog 1962) medium + 0.1 mg/l NAA and 0.01 mg/l kinetin, crowns established in vitro and lateral buds occasionally formed new shoots and roots, and, simultaneously, a yellowish, friable embryogenic callus. Following two subcultures, at 4-week intervals, such calli were transferred to embryo induction medium of MS + 0–10 mg/l 2,4-D or NAA and kinetin. After 4 weeks calli were transferred to embryo development MS medium + 0.1 mg/l NAA and 0.01 mg/l kinetin. The results of this study indicated that NAA (0.3–10 mg/l) was sufficient for the induction of asparagus somatic embryos in callus derived from lateral buds, or crowns established in vitro (Fig. 1) and it was more suitable than 2,4-D of comparable levels for the induction of normal embryos. Embryos developed and converted to plants within 4 weeks on MS + 0.01 mg/l NAA and 0.1 mg/l kinetin (Levi and Sink 1991).

Carbohydrate source and concentration also had a marked effect on the frequency and development of asparagus somatic embryos (Levi and Sink 1990). Sucrose, fructose, or glucose levels of 5–6% were optimal for the development and conversion of asparagus somatic embryos into plants.

In an additional study (Levi and Sink 1992), callus derived from stems of an elite asparagus crown was used for the establishment of embryo suspension cultures in liquid MS + NAA (10–20 mg/l). The development and conversion of embryos to plants were enhanced two- to four-fold by transfer of embryos from suspension to solid MS with high carbohydrate levels (4–10%), for 2 weeks, followed by transfer of the embryos to a solid MS with a low carbohydrate level (2%) (Fig. 2). These results indicated that an efficient somatic embryogenic system in asparagus may require high carbohydrate levels during embryo formation and maturation, followed by transfer of embryos to a lower carbohydrate level (2%) for conversion. Four hundred plants of an asparagus male line (M.S.U. 88-10) were obtained through suspension culture and planted during the spring of 1989 at the horticulture research stations in Michigan (Fig. 3F). The performance of these asparagus crowns is under evaluation.

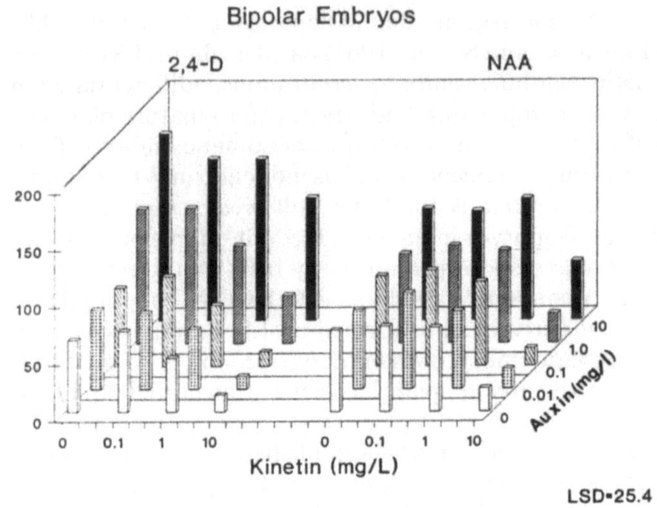

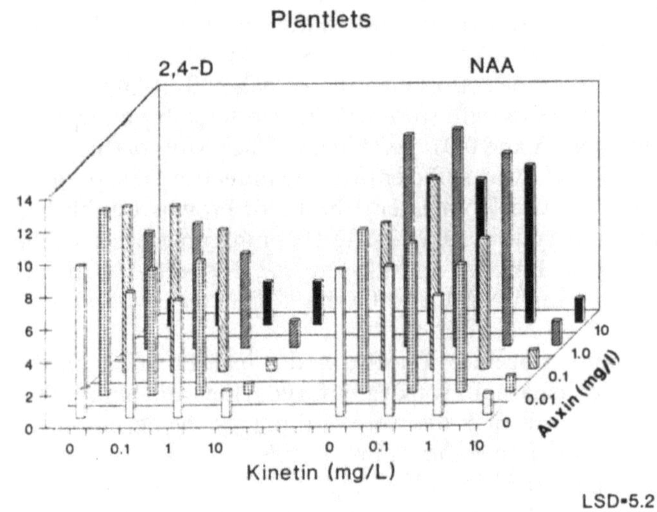

Fig. 1. The effect of auxins (2,4-D and NAA) and kinetin in induction medium on the number of embryos and plantlets (per g callus) formed within 6 weeks after transfer of callus to embryo development medium. (Levi and Sink 1991a)

3 Conclusions

For initiation of embryogenic callus, lateral buds and in vitro established crowns were the most suitable explants. Asparagus genotypes may have different requirements for auxin type and concentration for embryo induction and development.

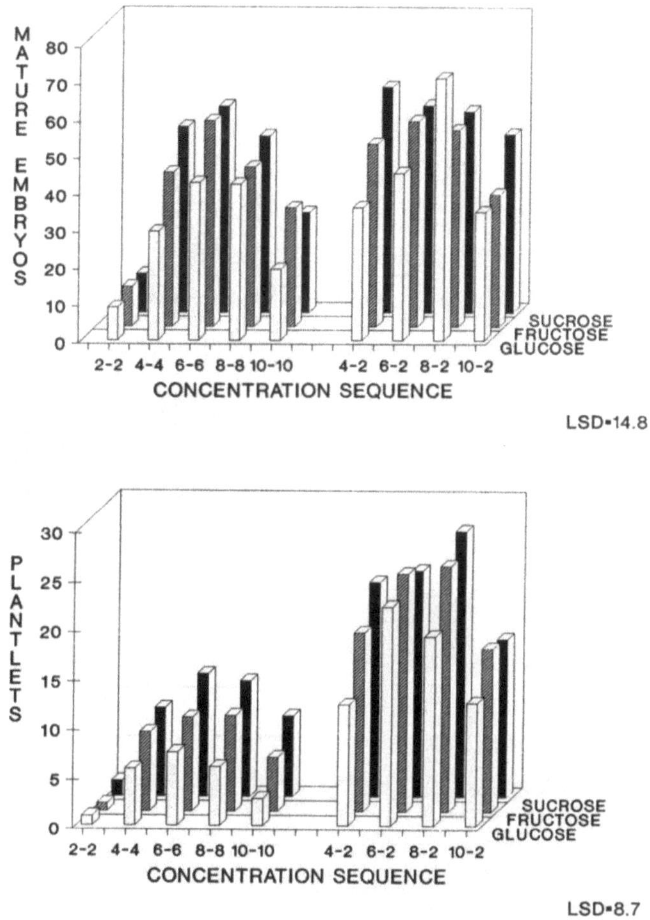

Fig. 2. Number of mature embryos and plantlets (per g callus) following subculture or transfer of embryos from MS medium with a high (4–10%) to low carbohydrate level (2%). (Levi and Sink 1992)

Like many other somatic embryogenesis systems, frequency of conversion of embryos to plantlets is considered a limiting factor. High conversion rates of embryos can be obtained by using high carbohydrate levels (5–6%) during embryo induction and development, and a low carbohydrate level (2%) for conversion of embryos to plantlets.

4 Protocol

Collect spears (3–7 days old) from mature greenhouse-grown plants. Surface sterilize for 30 min in aqueous sodium hypochlorite (1.5% v/v). Rinse thoroughly three times with sterile distilled water. Dissect lateral buds from spear and remove leaflets.

Place 3–5 lateral buds in a Petri dish containing MS + 0.5 mg/l NAA and 0.1 mg/l kinetin, for 4 weeks. Calli are formed at the periphery of the explants. Transfer callus to MS + 0.1 mg/l NAA and

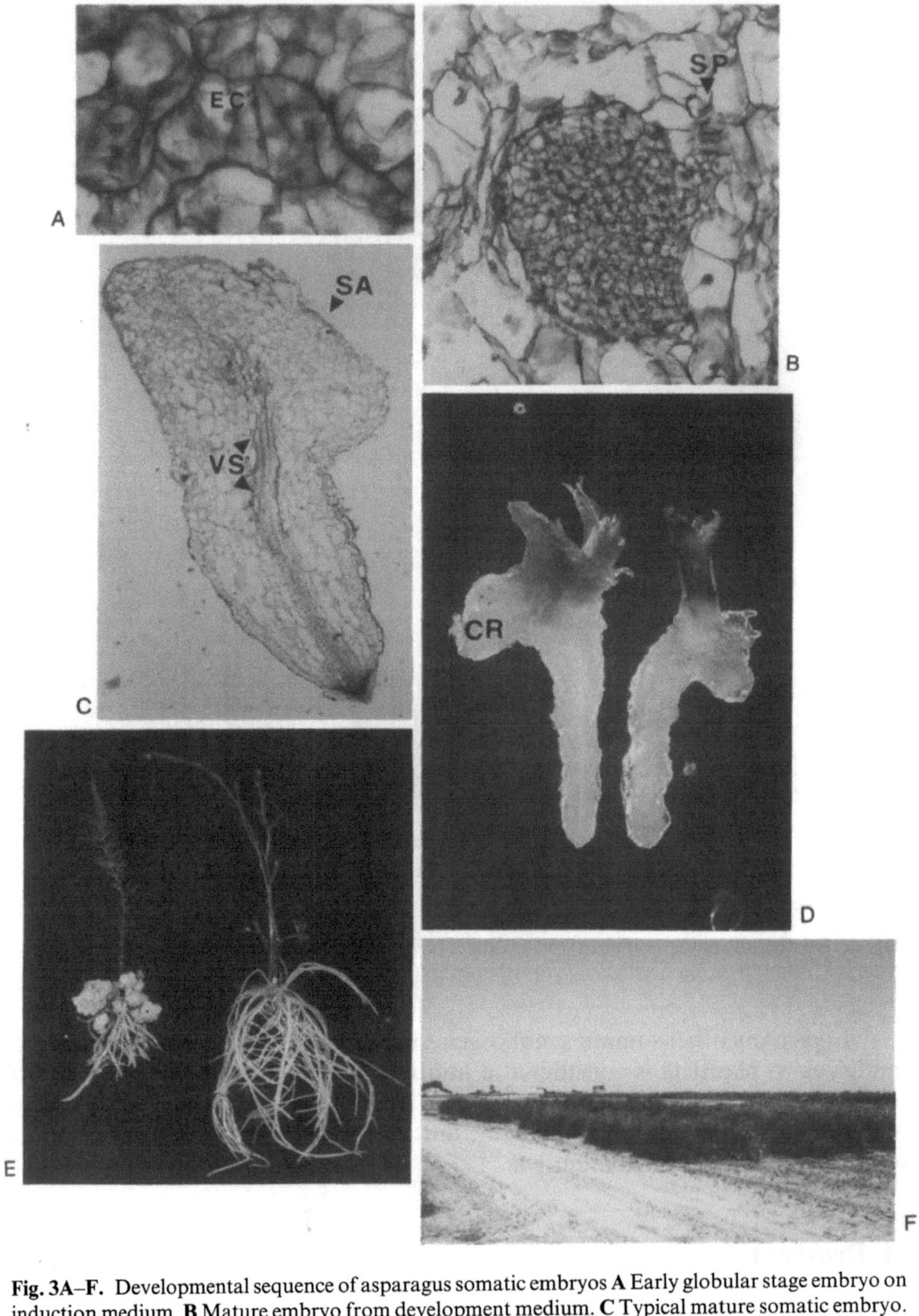

Fig. 3A–F. Developmental sequence of asparagus somatic embryos **A** Early globular stage embryo on induction medium. **B** Mature embryo from development medium. **C** Typical mature somatic embryo. **D** Young plantlets converted from mature somatic embryos. **E** Crown formed in 5 weeks from young plantlet. **F** Somatic embryo-derived plants in a field trial. (Levi and Sink 1991b)

0.01 mg/l kinetin for further proliferation and subculture. After two subcultures at 4-week intervals, transfer callus to embryo induction medium of MS containing 6% sucrose + 1–10 mg/l NAA. After 2 to 4 weeks of culture, transfer embryos to MS + 0.01 mg/l NAA and 0.01 mg/l kinetin for development and conversion of embryos to plantlets.

To enhance conversion embryos may be transferred to MS medium containing 2% sucrose. To establish a well-developed root system, culture plantlets on MS medium containing 2% glucose + 0.1 mg/l NAA and 0.01 mg/l kinetin.

References

Chin CK (1982) Promotion of shoot and root formation in asparagus in vitro by ancymidol. HortScience 17: 590–591

Doré C (1990) Asparagus anther culture and field trials of dihaploids and F1 hybrids. In: Bajaj YPS (ed) Biotechnology in agriculture and forestry, vol 12. Haploids in crop improvement I. Springer, Berlin Heidelberg, New York, pp 322–345

Ellison JH (1986) Asparagus breeding. In: Bassett MJ (ed) Breeding vegetable crops. Avi, Westport, pp 523–569

Hasegawa PM, Murashige T, Takatori FH (1973) Propagation of asparagus through shoot apex culture. II. Light and temperature requirements, transplantability of plants, and cytohistological characteristics. J Am Soc Hortic Sci 98: 143–148

Lawrence GHM (1982) Taxonomy of vascular plants. Macmillan, New York, 823 pp

Levi A, Sink KC (1990) Differential effects of sucrose, glucose and fructose during somatic embryogenesis in asparagus. J Plant Physiol 137: 184–189

Levi A, Sink KC (1991a) Somatic embryogenesis in asparagus: the role of explants and growth regulators. Plant Cell Rep 10: 71–75

Levi A, Sink KC (1991b) Histology and morphology of asparagus somatic embryos. HortScience 10: 1322–1324

Levi A, Sink KC (1992) Asparagus somatic embryos: Production in suspension culture and conversion to plantlets on solidified medium as influenced by carbohydrate regime. Plant Cell Tissue Organ Cult 31: 115–122

Linsamier EM, Skoog F (1965) Organic growth factor requirements of tobacco tissue cultures. Physiol Plant 18: 100–127

Loptien H (1979) Identification of the sex chromosome pair in asparagus (*Asparagus officinalis* L.). Z Pflanzenzücht 82: 162–173

Luzny J (1979) The history of asparagus as a vegetable, the tradition of its growing in Czechoslovakia (CSSR) and prospect of its further propagation and breeding. In: Reuther G (ed) Proc 5th Int Asparagus Symp, Eucarpia Section Vegetables. Geisenheim Forschungsanstalt, Germany

Murashige T, Skoog F (1962) A revised medium for rapid growth and bioassays with tobacco tissue cultures. Physiol Plant 15: 473–497

Peirce LC (1987) Vegetables: characteristics, production and marketing. John Wiley, New York

Reuther G (1977) Adventitious organ formation and somatic embryogenesis in callus of asparagus and iris and its possible application. Acta Hortic 78: 217–224

Reuther G (1984) Asparagus. In: Sharp WR, Evans DA, Ammirato PV, Yamada Y (eds) Handbook of plant cell culture, vol 2. Macmillan, New York, pp 211–242

Steward FC, Mapes MO (1971) Morphogenesis and plant propagation in aseptic cultures of asparagus. Bot Gaz 132: 70–79

Volokita M, Levi A, Sink KC (1987) A revised protocol for micropropogation of asparagus. Asparagus Res Newslett 4: 8–17

Wilmar C, Hellendoorn M (1968) Growth and morphogenesis of Asparagus cells cultured in vitro. Nature 217: 369–370

Wricke G (1979) Breeding research in *Asparagus officinalis*: introductory remarks. In: Reuther G (ed) Proc 5th Int Asparagus Symp, Ecarpia Section Vegetables. Geisenheim Forschungsanstalt, Germany, pp 1–7

Yang HJ (1977) Tissue culture technique developed for asparagus propagation. HortScience 12: 140–141
Yang HJ, Clore WJ (1973) Aerial crown in *Asparagus officinalis* L. HortScience 8: 33
Yang J, Clore WJ (1975) In vitro reproductiveness of asparagus stem segments with branch-shoots at a node. HortScience 10: 411–412

II.2 Somatic Embryogenesis in Chicory (*Cichorium* Species)

J. Vasseur, J. Dubois, J.L. Hilbert, and J.P. Couillerot[1]

1 Introduction

1.1 Importance, Morphology, and Distribution of *Cichorium*

Cichorium belongs to the family Asteraceae and is subdivided into three European species: *C. endivia*, *C. intybus*, and *C. spinosum*. The first two are particularly interesting for their agronomical characteristics. *C. endivia* L. is represented by "scarole" (cv. latifolia) and "curly" (cv. crispa) chicories which are used as salads, while *C. intybus* L. is long known for its use in medicine and as food. The latter species is composed of several varieties (Longly and Louant 1987; Leteinturier et al. 1991):

1. var. *sativum* Biskoff: industrial chicory cultivated for its roots and used as a coffee additive after roasting. Known as "chicorée à café", it comprises the Brunswick chicory (serrated leaves and red veins) and also the Magdebourg chicory (unindented leaves and white veins). The roots of these plants are also used for their inulin content, a polyfructosan which is a potential source of fructose.
2. var. *foliosum* Biskoff: Witloof or Bruxelles chicory whose roots are able to produce etiolated buds under artificial culture conditions (forcing). These plants are the result of the improvement of the Magdebourg-type chicory.
3. var. *silvestre* Biskoff: wild, improved chicory with red leaves (Trévise, Vérone), green leaves (Pain de sucre, Barbe de Capucin), or variegated leaves (Castelfranco, Chioggia).

Despite being a perennial plant in the wild form, chicory (*C. intybus* L.) is cultivated according to a biennal cycle. For flowering, this long-day plant requires a cold treatment period. The first-year plant, which possesses a rosette of leaves, produces a floral stem the second year. The vegetative phase therefore includes the germination stage of seeds in spring and continues until the formation of a tuberous root in autumn, the latter progressively accumulating sugars, mainly as inulin in the time course. The plant also contains various phenolics and sesquiterpene lactones involved in the characteristic bitterness of the species. Chicory is a self-incompatible (the floral morphology is unfavorable for

[1] Université des Sciences et Technologies de Lille, Laboratoire de Physiologie Cellulaire et Morphogenèse Végétale, Bt SN 2, 59655 Villeneuve d' Ascq Cedex, France

Biotechnology in Agriculture and Forestry, Vol. 31
Somatic Embryogenesis and Synthetic Seed II (ed. by Y.P.S. Bajaj)

self-fertilization) and entomogamous plant. Furthermore, it possesses a physiological auto-incompatibility system which hinders the germination of the auto-pollen or inhibits pollen tube growth.

Cichorium intybus may have originated from Eurasia or Europe and could have been then introduced in North America primarily from Mediterranean regions (Kohls and Rather 1941; Martin and Warren 1949; Cichan and Palser 1982). Chicory is actually common in the temperate regions. It is found everywhere in Europe, except the further septentrional zones, in North Africa and part of Asia.

Witloof chicory (Leteinturier et al. 1991) is mainly cultivated in France (16 000 ha), Belgium (8600 ha), and The Netherlands (5860 ha). It is also produced in some countries such as Germany, Switzerland, Spain, and the United Kingdom (about 1000 ha in all). Comparatively, industrial chicory cultures represent 7000 to 8000 ha mostly in Europe (France and Belgium, in particular, as well as in Austria) and also in Poland, Hungary, and Yugoslavia.

1.2 Review of Work and Significance of Somatic Embryos in *Cichorium*

Vasil et al. (1964) and Vasil and Hildebrandt (1966) showed for the first time the presence of somatic embryos in callus cultures several years old derived from mature embryos of *Cichorium endivia* L. In liquid medium the tissues dissociated into a thick suspension of cells or cell groups and many small, round structures designated as embryoids, first developing roots and then shoots. On an agar medium shoots arose first from marginal meristematic areas, while roots developed later from deeper callus regions. These structures did not survive after transfer to soil. Twenty years later, by using cubic pieces of mature roots, Heirwegh et al. (1985) obtained embryoid formation in *C. intybus* L., and this resulted in their further development into plantlets. In both cases, however, indirect embryogenesis occurred with an intermediate callus phase. These studies, together with our team's first results published in 1988, are to our knowledge the first works on chicory somatic embryogenesis.

Somatic embryogenesis is first of all a cloning technique of interest for the rapid and exponential multiplication of particular genotypes (plant hybrids or male sterile plants), provided the genetic conformity of the obtained plants is verified. It is also a suitable method to maintain genetic variability created by cellular or genetic manipulations which often generate plants whose sexual reproduction can be disrupted.

2 Somatic Embryogenesis

2.1 Origin

Somatic embryos were first obtained from anthers and styles taken from young inflorescences of hybrid *Cichorium intybus* × *C. endivia* cv. 474 (Dubois et al. 1988; Guedira et al. 1989). After 10 days of culture, 60 to 90% of anthers and

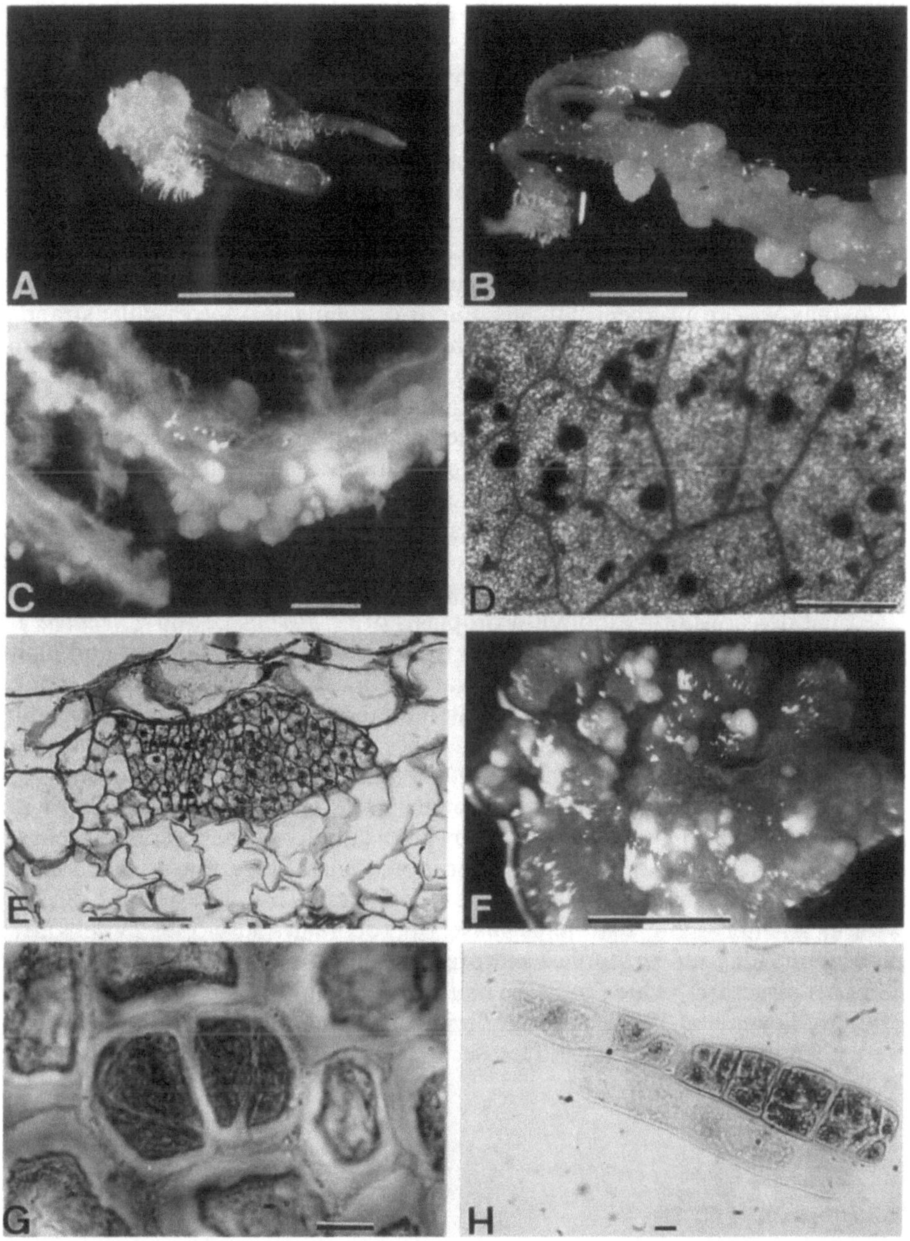

Fig. 1A–H. Somatic embryogenesis in reproductive and vegetative tissues of *Cichorium*. Embryos on an anther (**A**) and on a style (**B**) of *C. intybus*, cv. 1106 cultured for 2 weeks on a solid induction medium. Numerous embryos in roots (**C**) and leaves (**D**,**E**) of *C. intybus* × *C. endivia*, cv. 474 cultured for 10 days in a liquid-agitated induction medium. (**F**) Globular and heart-shaped embryos in a 4-week-old protoplast-derived microcallus of *Cichorium* hybrid, cv.474. Early stages of embryos originated from the segmentation of an epidermal cell of a style (**G**) and from a long cortical cell of a root (**H**), of the *Cichorium* hybrid, cv. 474. In **A, B, C, D, F**, *bar* = 1 mm; **E**, bar = 50 μm; **G, H**, *bar* = 10 μm. (**G** from Dubois et al. 1988; **H** from Dubois et al. 1990)

100% of styles produced embryos. Styles have greater embryogenic potentialities than anthers which require more precise culture conditions. Anthers and styles of *C. intybus* var. Magdebourg (cv. 1106) produce embryos under the same conditions (Fig. 1A, B). Roots (Fig. 1C) and leaves (Fig. 1D,E) of 2- to 3-month-old plantlets regenerated from the first batches of embryos are better embryogenic material, because of their permanent availability all year-round and high yield of embryos (Guedira et al. 1990; Dubois et al. 1991).

Somatic proembryos develop in vitro from several sources of diploid vegetative cells, without any intervening callus development. Histological observations clearly demonstrate that embryos originate from single cells. In styles, proembryos may arise from epidermal cells (Fig. 1G), mainly near the vascular strands. Sections and squashes of roots during the first 6 days of culture do not show any visible modification. After the 6th day, some epidermal and cortical parenchyma cells undergo linear segmentation without any increase in volume; globular segmentation occurs in each daughter cell, without apparent polarity (Fig. 1H). In the leaves, epidermal embryogenic cells are rare; mesophyll cells are more frequently embryogenic, but we did not note any difference between adaxial and abaxial sides or between the apical or basal part of the leaf. In contrast to *Dactylis glomerata* (Conger et al. 1983; Trigiano et al. 1989), parenchyma cells near the vascular bundles are very responsive in *Cichorium*.

First divisions of mesophyll protoplasts of *C. intybus* L. and *C. endivia* L. were reported by Binding et al. (1981), but protoplast development and plant regeneration were problematic for the various lines and hybrids tested (Crepy et al. 1982). Fertile plants were obtained in our laboratory from *C. intybus* Zoom, a witloof hybrid chicory (Saksi et al. 1986) and from *C. intybus* Pévèle, an industrial chicory (Rambaud et al.1990). Plant regeneration from protoplast-derived callus was always achieved through shoot formation. Experiments with the embryogenic *Cichorium* cv. 474 showed the ability of the mesophyll cells to retain full embryogenic properties after a protoplast stage. Embryos occurred indirectly via callus proliferation and differentiation (Fig. 1F) or quite directly without an intervening, real callus phase. Microcolonies developed 2 weeks after plating of 7-day-old protoplast cultures on an induction medium, and some compact structures, which must be considered as embryoids, arose from them (Sidikou-Seyni et al.1992). After acclimatization, all protoplast-derived plants were phenotypically normal and fertile.

2.2 Induction

2.2.1 *Induction Factors*

The most important physical factor involved in the induction of somatic embryos is the temperature; 35 °C in darkness is the best condition. Anthers and styles are particularly sensitive to this temperature: at 35 °C, 98% of the styles are embryogenic, each giving 10 to 50 proembryos on the 10th day. At 30 °C, the production is not reliable, and the mean percentage of embryogenic styles is 39 %; at other temperatures, no embryos are formed (Dubois et al. 1988). Roots and leaves are not so sensitive. Nevertheless, at 35 °C in darkness, embryogenesis is the only morphogenetic event, while leaves cultured in light, particularly at 20 or 25 °C gave rise to callus formation leading to the appearance of buds.

Several chemical factors are involved in the induction of embryos; the most important being the content of exogenous growth regulators in the medium and the composition of glucidic and nitrogen compounds added as nutrients.

The role of nitrogen in embryogenesis is somewhat obscure because of conflicting results. A series of preliminary experiments showed that the most favorable quantity of nitrogen for embryogenesis in anthers was about 337mg/l. For comparative tests between anthers and styles, we used this amount as KNO_3, NH_4NO_3, NH_4Cl, or glutamine. If nitric-nitrogen seems to be more suitable for anther embryogenesis, style embryogenesis is enhanced by the corresponding nitrogen-reduced form. From the different compounds, glutamine alone is the least effective, inducing embryogenesis on only 4.2% of the anthers; with the styles which are more reactive we obtained few proembryos. In contrast, best results were obtained when NH_4NO_3 was associated to glutamine in the proportions 850 to 250 mg/l.

After several screening experiments regarding sugar requirements, sucrose was found to be the best carbon source for the induction of somatic embryos. In root explants, increasing the sucrose concentration from 15 to 360 mM enhances somatic embryogenesis, but inhibits differentiation and development of the embryos (Guedira et al. 1990). The data presented in Table 1 show that glucose, fructose, maltose, sucrose, and raffinose are good carbon sources for embryo induction in styles, while the embryo yield from anthers is increased in the presence of maltose or sucrose (Couillerot et al. 1992). The incorporation of glycerol in induction medium as osmoticum was followed by its partial utilization (Robatche-Claive et al. 1992).

Table 1. Effects of different sugars (120 mM) on the percentage of embryogenic explants (%) of the 474 hybrid cultured for 10 days in induction medium (darkness, 35 °C) and on the average number of somatic embryos (AN ± SE; 0.05)

Carbohydrate source	Root fragments, 2.5 mm		Anthers		Styles	
	(%)	AN ± SE (n=20)	(%)	AN ± SE (n=105)	(%)	AN ± SE (n=60)
Fructose	95.2	5.1 ± 2.8	6.0	1.1 ± 0.1	100	25.9 ± 4.3
Glucose	80.9	3.9 ± 1.6	13.3	1.6 ± 0.1	98.1	18.9 ± 3.7
Galactose	0	0	0	0	36.6	2.7 ± 0.4
Sucrose	95.0	11.2 ± 4.0	18.1	2.3 ± 2.1	100	25.8 ± 3.5
Maltose	84.2	6.0 ± 2.8	21.9	1.5 ± 0.2	100	25.6 ± 3.9
Cellobiose	33.3	0.3 ± 0.2	8.5	1.1 ± 0.1	63.3	2.5 ± 0.5
Trehalose	0	0	0	0	0	0
Melibiose	0	0	0	0	0	0
Lactose	0	0	0	0	61.6	3.1 ± 0.5
Raffinose	95.2	2.6 ± 0.8	10.5	1.4 ± 0.2	100	17.1 ± 2.3

2.2.2 Cytochemical Markers

For histological and SEM studies of early stages of embryogenesis, leaves provide the best material. They are more easily available than anthers and styles and produce better histological sections than roots. In *Cichorium* leaves, most embryos arise from fully differentiated cells. No sign of dedifferentiation could be seen until the 3rd day, when the first deposits of callose became visible in the walls of the embryogenic cells. These have very faint callose patches, presumably at the site of plasmodesmata, as in embryogenic root cells of *Cichorium* (Dubois et al. 1990). Such callosic walls were more obvious on the 5th day (Fig. 2A,B), and disappeared as the embryos grew (Fig. 2C). Callose was also observed in embryogenic cells of *Trifolium* (Maheswaran and Williams 1985). One of its probable roles is to isolate the embryogenic cells from the influence of surrounding tissue, while they follow a new metabolic pathway.

Extracellular glycoproteins have been related to somatic embryogenesis in carrot (Satoh et al. 1986; De Vries et al. 1988), and some bind with DBA (*Dolichos biflorus* agglutinin) and UEA (*Ulex europaeus* agglutinin). Out of a few FITC-labeled lectins tested on embryogenic *Cichorium* leaves, only DBA showed specific binding with embryogenic cells and small globular proembryos (Fig. 2D, E,F); (Dubois et al. 1991).

After 7–10 days of culture, a fibrillar network could be seen on the surface cells of proembryos. First observations were made in roots cultured in an agitated medium (Fig. 2G,H), then a similar network was seen on the surface cells of proembryos just emerging through the leaf epidermis (Fig. 2I,J,K). Sondahl et al. (1979) previously described analogous phenomena on embryoids of *Coffea*, and, more recently, an extracellular network was observed on the surface cells of leaf embryoids of *Camellia* sp. (S. Pais, pers. Comm.). While many plants have already produced embryoids, one could wonder why only *Coffea*, *Cichorium*, and *Camellia* show strands linking the surface cells of proembryos. This may be due to the fact that early stages in somatic embryogenesis have not often been observed. Big globular embryos have a protoderm, and no longer display a network (Fig. 2L,M; Dubois et al. 1991). Effects of solvents and enzymatic digestions on the network were consistent with a proteinaceous nature of this structure, and not with a pectic one (Dubois et al. 1992). Specific glycoproteins with an N-acetylgalactosamine binding site were previously revealed with an FITC-labeled lectin in the cells of young globular proembryos, but it

Fig. 2A–M. Cytological and cytochemical markers of early stages of somatic embryogenesis in *Cichorium* hybrid, cv. 474 **A, B** Callose deposits in walls surrounding the embryogenic cells on the 5th day. **C** On the 7th day, a multicellular embryo shows no callose except in the plates of dividing cells. FITC-labeled DBA bound with an embryogenic cell (**D**) and young proembryos (**E, F**) showing the presence of a specific glycoprotein with galactosamine binding sites. SEM of embryogenic roots (**G, H**) and leaves (**I, J, K**) after 10 days of induction in a liquid-agitated medium: a reticulated proteinaceous network links the surface cells of the globular proembryos. After 12 days, the surface of the torpedo-shaped embryos (**L**) is covered with a smooth protoderm, without a network (**M**). *Bar* = 20 μm, except **G, I, L**, *bar* = 100 μm. **A, B, C, I, J, L, M** from Dubois et al. 1991; **H** from Dubois et al. 1990

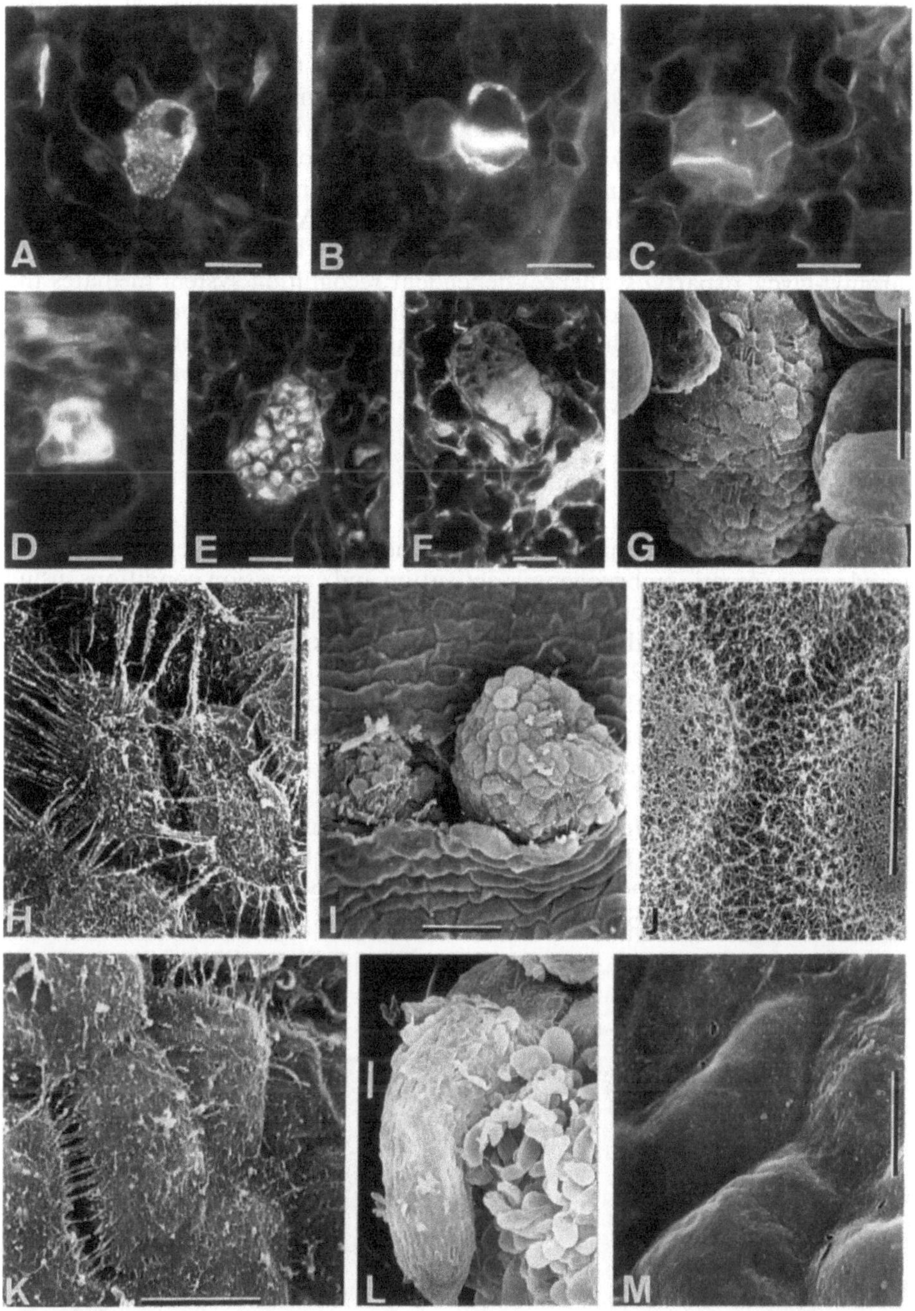
A
B
C
D
E
F
G
H
I
J
K
L
M

was not possible to determine whether the fluorescence originated from the extracellular matrix, the cell wall, or the cytoplasm. If further biochemical analysis can prove the glycoproteinaceous nature of the network, it could be a vizualization of the embryogenic extracellular glycans reported by Satoh et al. (1986) and De Vries et al. (1988) during carrot embryogenesis.

The extracellular network could be a point of comparison in animal cells, where surface glycoproteins of the extracellular matrix play a role in adhesion and recognition, and are linked to the cytoskeleton. Adhesion proteins between cell wall and plasmalemma have already been shown in many plants. The RDG (arginyl-glycyl-aspartyl sequence)-dependent recognition system mediates the coordination and control of plant cell division and proper cell wall biosynthesis (Schindler et al. 1989). The network is perhaps a part of an organized extracellular proteoglycan complex that may promote coordination of early developmental stages of proembryos.

2.2.3 Molecular Markers

In studies aiming to examine changes in gene expression, proteins were monitored under somatic embryogenesis activation and expression conditions, using two-dimensional polyacrylamide gel electrophoresis. Proteins were extracted from induced or noninduced leaves, and during expression of embryogenesis (Hilbert et al. 1991, 1992). Most of the polypeptides observed in induced tissues (92%) were already synthesized in control leaves. Among the 360 polypeptides analyzed on 2-D gels, only two acidic polypeptides (37.7 kDa, pI 4 and 4.5), named $E_{37.7a}$ and $E_{37.7b}$, appeared to be specifically accumulated in the 5-day-induced tissues (Fig.3B), whereas they were not observed in control tissues (Fig. 3A). They were detected only after induction and remained present during expression (Fig. 3C,D). The association of $E_{37.7a}$ and $E_{37.7b}$ polypeptides with two embryogenic lines and their absence from a nonembryogenic line identify them as good candidates for embryogenesis-specific markers studies.

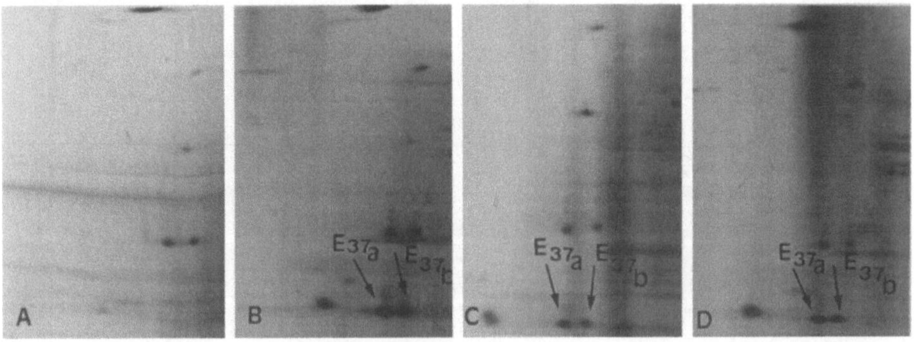

Fig. 3A–D. Silver-stained two-dimentional polyacrylamide gels of total cellular protein extracts. **A** Non induced leaves ; **B** 5-day-old activated leaves; **C** 5-day-old activated leaves after 3 days on expression medium **D** 5-day-old activated leaves after 7 days on expression medium of line 474. *E* indicates embryogenesis-related polypeptides and the *numbers* denote molecular weight . (Hilbert et al. 1991, unpubl.)

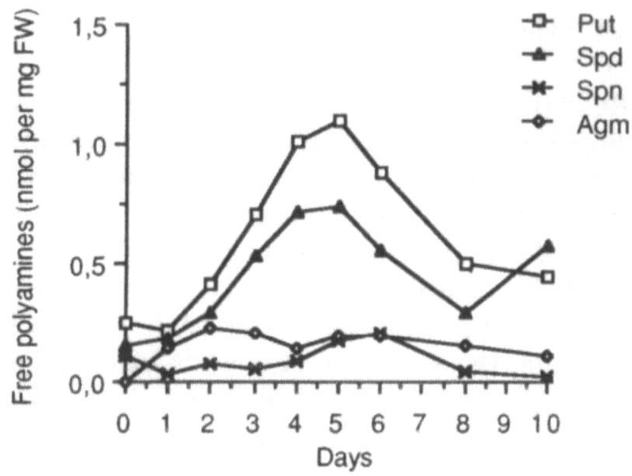

Fig. 4. Changes in the content of free putrescine (*Put*), spermidine (*Spd*), spermine (*Spn*), and agmatine (*Agm*) during the culture of root fragments (2.5 mm) of the hybrid 474 (35 °C, darkness) (Couillerot et al. 1992, unpubl. results)

Polyamine content has been studied during somatic embryogenesis. In *cv.* 474 root explants, free putrescine and spermidine increase up to day 5 and then decrease, while agmatine and spermine levels remain low (Fig. 4). This peak of polyamine prior to the first linear segmentation of competent cells (Fig. 1H) is not detected in roots of a nonembryogenic line, suggesting a correlation between somatic embryogenesis and polyamine content. Such an increase in polyamines and especially putrescine has been observed during the early stage of the carrot somatic embryogenesis system (Evans and Malmberg 1989). Actually, in our system, we are studying relations between polyamine metabolism and somatic embryogenesis.

2.3 Development and Plant Regeneration

Somatic embryos of *Cichorium* can develop directly in the induction medium and whole fertile plants are produced after transfer of mature embryos onto a solid Heller medium (Guedira et al. 1990). However, factors increasing the yield of proembryos (high sucrose concentration, long contact with the induction medium) inhibit maturation and/or germination. We define "maturation" as the process of conversion from a globular proembryo to a mature embryo with a protoderm and two meristematic areas, and "germination" as the development of the cotyledons and the emergence of the root, resulting in vegetative growth towards a plantlet. Better development of embryos is obtained when roots and leaves are transferred after 1 week to an agitated maturation liquid medium containing mineral salts of Heller solution, some vitamins, 15 mM sucrose, and 0.15 μM GA_3. After a few days, the embryos (most of them with heart and torpedo shapes) settle at the bottom of the flask. They are then transferred to a Petri dish containing a static liquid medium of the same composition to achieve maturation in the light. Mature embryos with two cotyledons and a root are planted in test tubes on a solid Heller medium without a growth regulator.

2.4 Synchronization Assays

Some progress has been made in inducing synchronization of somatic embryos, since the early experiments of Fujimura and Komanine (1975). Nevertheless, synchronization is still a confusing problem because studies focused sometimes only on the induction or the maturation of the embryos and at times only on the yield of somatic embryogenesis.

It is possible to induce synchronization in roots of *Cichorium* with high sucrose concentrations. After 14 days of incubation in medium containing 360 mM sucrose, we observed 66.5 % globular embryos, 32. 5 % oblong embryos, and only 1% elongated embryos, indicating an obvious synchronization of developmental steps. Unfortunately, only the embryos obtained in a low sucrose concentration grew normally into plants. By adding mannitol and thus increasing the osmotic pressure, we obtained a better synchronization and a normal development, except for the highest tested concentration. Recent experiments showed that addition of 330 mM glycerol in the induction medium stopped the embryogenic differentiation of proembryos in *Cichorium* leaves at an early stage. The transfer of the explants onto an expression medium with 30 mM sucrose as carbon source synchronized the first division of the embryogenic cells: 48 h after the transfer more than 95% of the competent cells divided. When cultured under suitable conditions, most proembryos developed cotyledons and germinated into normal plantlets. Mannitol or sucrose used in place of glycerol in the induction medium suppressed the synchronization (Robatche-Claive et al. 1992).

3 Protocol for the Induction of Somatic Embryos

The protocol for induction of somatic embryos and development into plants is summarized in Fig. 5. Inflorescences at tetrad and microspore stages were removed from the field-grown plants. They were surface-sterilized and rinsed with sterile water. Anthers and styles were cultured on the induction medium, I (see Table 2) solidified with agar Biokar (6 g/l) in 90-mm Petri dishes sealed with Parafilm and incubated at 35 °C in the dark for 10–12 days. Maturation and germination of the embryos were achieved in a liquid medium, G (see Table 2) at 20–24 °C under a 16-h photoperiod. Germinated embryos were transferred to a solid medium, D (see Table 2). Within 8 weeks rooted plantlets could be transferred to the soil (acclimatization in the green house caused no major problems).

Roots of 2- to 3-week-old plantlets regenerated as indicated in Fig. 5A are better embryogenic material than floral explants because of the availability and yield of embryos. Ramified roots were cut and placed in the agitated liquid induction medium, I, in darkness at 35 °C for 7 days (Fig. 5B). Maturation, germination, and development of embryos were carried out as indicated above. Under these conditions more than 50% of the proembryos produced normal plantlets.

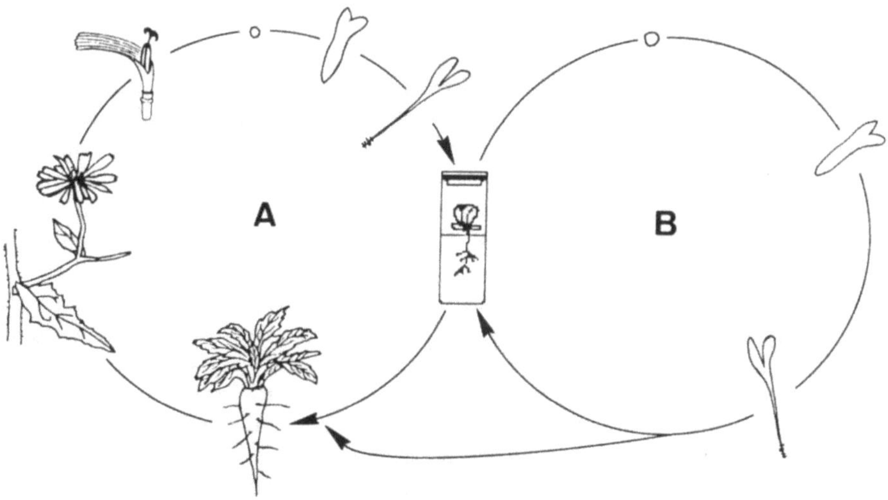

Fig. 5 A,B. Various steps for somatic embryogenesis in *Cichorium* from **A** reproductive and **B** vegetative explants

Table 2. Composition (mg/l) of the culture medium for the induction, maturation, germination, and development of somatic embryos

Components	Induction (I)	Maturation (M) and Germination (G)	Development (D)
$CaCl_2 \cdot 2H_2O$	220	75	75
KCl	75	750	750
KH_2PO_4	85	–	–
$MgSO_4 \cdot 7H_2O$	185	250	250
$NaH_2PO_4 \cdot H_2O$	–	125	125
$NaNO_3$	–	600	600
NH_4NO_3	825	–	–
Glutamine	250	–	–
Sucrose	10 000	5 000	5 000
Oligoelements	Heller (1953)	Heller (1953)	Heller (1953)
Vitamins	Morel and Wetmore (1951)	Morel and Wetmore (1951)	Morel and Wetmore (1951)
ANA	0.02	–	–
2 iP	0.5	–	–
AG_3	–	0.05	–
Agar	–	–	5 000
pH	5.5	5.5	5.5
$FeSO_4 \cdot 7H_2O$	27.8	27.8	27.8
$EDTA \cdot Na_2$	37.3	37.3	37.3

4 Summary and Conclusion

Chicory species provide good experimental plant material for studies at both basic and applied levels. In vitro regeneration is developed in our laboratory from cells or protoplasts with the aim of furthering cellular and molecular knowledge about the earliest events of bud or embryo differentiation. The challenge, in *Cichorium* as in other genera, is possibility of inducing genotypes that are still resistant to somatic embryogenesis. Among various tested chicory genotypes, only two of them produce somatic embryos. All experiments described here have been conducted with the chicory hybrid 474. It is not an agronomically interesting plant, but offers us possibilities to set up an embryogenic system in which embryos develop directly without an intervening callus phase and which might serve for the analysis of somatic embryos ab initio. In this context, our aim is to determine the nature and the regulation of early cytochemical and molecular events and to find a relation between these events. This approach is compatible with the general concept of morphogenesis in plants occurring as a consequence of spatial and temporal gene expression (Racusen and Shiavone 1990) and with the working hypothesis that the formation of embryos in vitro could be a stress response (wounding, hormone concentration, artificial growth conditions, etc.) which could play a role in the activation of an embryogenic program.

Acknowledgments. Our thanks are due to the other members of our team who helped us to compile results and also for their interest in writing this review. This work is supported in part by a grant: Contrat Plan Etat-Région.

References

Binding H, Nehls R, Kock R, Finger J, Mordhorst G (1981) Comparative studies on protoplast regeneration in herbaceous species of the Dicotyledoneae class. Z Pflanzenphysiol 101: 119–130

Cichan MA, Palser BF (1982) Development of normal and seedless achenes in *Cichorium intybus* (Compositae). Am J Bot 69: 885–895

Conger BV, Hanning GE, Gray DJ, Mc Daniel JK (1983) Direct embryogenesis from mesophyll cells of orchardgrass. Science 221: 850–851

Couillerot JP, Warnot F, Decout E, Dubois J, Vasseur J (1992) Utilization and metabolism of carbohydrates during direct somatic embryogenesis in roots of *Cichorium*. Présentation, Congr Pays de Galles Aberystwyth, 3 juillet 1992

Crepy L, Chupeau MC, Chupeau Y (1982) The isolation and culture of leaf protoplasts of *Cichorium intybus* and their regeneration into plants. Z Pflanzenphysiol 107: 123–131

De Vries SC, Booij H, Janssens R, Vogels R, Saris L, Lo Schiavo F, Terzi M, Van Kammen A (1988) Carrot somatic embryogenesis depends on the phytohormone-controlled presence of correctly glycosylated extracellular proteins. Genes Dev 2: 462–476

Dubois T, Dubois J, Guedira M, Vasseur J (1988) Embryogenèse somatique directe sur les styles de *Cichorium*: effets de la température et origine des embryoïdes. C R Acad Sci Paris 307: 669–675

Dubois T, Guedira M, Dubois J, Vasseur J (1990) Direct somatic embryogenesis in roots of *Cichorium*: is callose an early marker ? Ann Bot 65: 539–545

Dubois T, Guedira M, Dubois J, Vasseur J (1991) Direct somatic embryogenesis in leaves of *Cichorium*. A histological and SEM study of early stages. Protoplasma 162: 120–127

Dubois T, Dubois J, Guedira M, Diop A, Vasseur J (1992) SEM characterization of an extracellular matrix around somatic pro-embryos in roots of *Cichorium*. Ann Bot 70: 119–124

Evans PT, Malmberg RL (1989) Do polyamines have roles in plant development? Annu Rev Plant Physiol Plant Mol Biol 40: 235–269

Fujimura T, Komamine A (1975) Effect of various growth regulators on the embryogenesis in a carrot cell suspension culture. Plant Sci Lett 5: 359–364

Guedira M, Dubois-Tylski T, Vasseur J, Dubois J (1989) Embryogenèse somatique directe à partir de cultures d'anthères du *Cichorium* (Asteraceae). Can J Bot 67: 970–976

Guedira M, Dubois T, Dubois J, Vasseur J (1990) Influence du saccharose sur le rendement, la morphologie et le développement des embryons somatiques chez *Cichorium*. C R Acad Sci Paris 310: 453–459

Heirwegh KMG, Banerjee N, Van Nerum K, De Langhe E (1985) Somatic embryogenesis and plant regeneration in *Cichorium intybus* L. (Witloof, Compositae). Plant Cell Rep 4: 108–111

Heller R (1953) Recherches sur la nutrition minérale des tissus végétaux cultivés in vitro. Ann Sci Nat Bot Biol Vég 14: 1–223

Hilbert JL, Dubois T, Vasseur J (1991) Protein patterns associated with somatic embryogenesis in *Cichorium*. Presentation, Congr Plant sciences today, La Colle sur Loup, 16–18 octobre 1991

Hilbert JL, Dubois T, Vasseur J (1992) Detection of embryogenesis-related proteins during somatic embryo formation in *Cichorium*. Plant Physiol Biochem 30: 733–741

Kohls HL, Rather HC (1941) Growing chicory seed. Mich Q Bull 23 : 243–250

Leteinturier J, Cochet JP, Marle M, Benigni M (1991) L' endive: guide pratique. Editions CTIFL, Paris 271 pp

Longly B, Louant BP (1987) Mécanismes de la reproduction chez la chicorée de Bruxelles: fondements et applications à la sélection. IRSIA Bruxelles, 108 pp

Maheswaran G, Williams EG (1985) Origin and development of somatic embryoids formed directly on immature embryos of *Trifolium repens* in vitro. Ann Bot 56: 619–630

Martin John H, Leonard Warren H (1949) Principles of fields crop production. Macmillan, New York, 1185 pp

Morel G, Wetmore RH (1951) Fern callus tissue culture. Am J Bot 38: 141–143

Racusen RH, Schiavone FM (1990) Positional cues and differential gene expression in somatic embryos of higher plants. Cell Differ Dev 30: 159–169

Rambaud C, Dubois J, Vasseur J (1990) Some factors related to protoplast culture and plant regeneration from leaf mesophyll protoplasts of Magdeburg chicory (*Cichorium intybus* L. var. Magdeburg). Agronomie 10: 767–772

Robatche-Claive AS, Couillerot JP, Dubois J, Dubois T, Vasseur J (1992) Embryogenèse somatique directe dans les feuilles de *Cichorium* hybrid 474: synchronisation de I'induction. C R Acad Sci Paris 314: 371–377

Saksi N, Dubois J, Millecamps JL, Vasseur J (1986) Régénération de plantes de Chicorée Witloof cv. "Zoom" à partir de protoplastes: influence de la nutrition glucidique et azotée. C R Acad Sci Paris 302: 165–170

Satoh S, Kamada H, Harada H, Fujii T (1986) Auxin-controlled glycoprotein release into the medium of embryogenic carrot cells. Plant Physiol 81: 931–933

Schindler M, Meiners S, Cheresh DA (1989) RGD-dependant linkage between plant cell wall and plasma membrane: consequences for growth. J Cell Biol 108: 1955–1965

Sidikou-Seyni R, Rambaud C, Dubois J, Vasseur J (1992) Somatic embryogenesis and plant regeneration from protoplasts of *Cichorium intybus* L. × *Cichorium endivia* L. Plant Cell Tissue Organ Cult 29: 83–91

Sondahl MR, Salisbury JL, Sharp WR (1979) SEM characterization of embryogenic tissue and globular embryos during high frequency somatic embryogenesis in coffee callus cells. Z Pflanzenphysiol 94: 185–188

Trigiano RN, Gray DJ, Conger BV, Mc Daniel JK (1989) Origin of direct somatic embryos from cultured leaf segments of *Dactylis glomerata* Bot Gaz 150: 72–77

Vasil IK, Hildebrandt AC (1966) Variations of morphogenetic behaviour in plant tissue cultures. I. *Cichorium endivia*. Am J Bot 53: 860–869

Vasil IK, Hildebrandt AC, Riker AJ (1964) Endive plantlets from freely suspended cells and cell groups grown in vitro. Science 146: 76–77

II.3 Somatic Embryogenesis in Cucurbits

P.P: CHEE[1]

1 Introduction

1.1 Economic Importance of Cucurbits

The family Cucurbitaceae consists of about 90 genera and 750 species, almost equally divided between the New and Old World tropics (Whitaker and Davis 1962). The cultivated species of this family (Table 1; Jelaska 1986) are important crops in the tropics, subtropics, and milder portions of the temperate zones of both hemispheres. They are used as sources of carbohydrates when cooked (squash, pumpkin, marrow, chayote), as dessert or breakfast fruits (watermelon, muskmelon), as ingredients of salads (cucumber, gherkin), or as pickles (cucumber, gherkin). In 1989, commercial plantings of cultivated cucurbits occupied 1.47 million ha in more than 70 countries; of these, 60% were cucumbers and gherkins; 40% were pumpkins, squash. Table 2 shows the total production of cucurbits in various parts of the world and the distribution of harvested areas.

1.2 Significance of Somatic Embryogenesis in Cucurbits

Numerous diseases attack the cultivated species and cause tremendous annual losses. The main objective of a breeding program is thus not only to increase yield and quality, but also to select cultivars resistant to diseases. One method of obtaining these traits is to hybridize the wild species with the cultivated species (Deakin et al. 1971). Because of sexual incompatibility, the transfer of many desirable traits are precluded. The problem is especially acute in the family of Cucurbitaceae. The recent advances in genetic engineering seem to be a promising alternative to sexual propagation techniques for improvement of this economically important crop (Chee 1990b, 1991; Fang and Grumet 1990; Slightom et al. 1990; Chee and Slightom 1991, 1992). The success of any gene transfer technique depends on the availability of high frequency regeneration methods to recover plantlets from cells and/or tissues cultured in vitro. This can be achieved by two pathways. In somatic embryogenesis, both meristems

[1] Molecular Biology Division, The Upjohn Company, 301 Henrietta Street, Kalamazoo, Michigan 49007, USA

Biotechnology in Agriculture and Forestry, Vol. 31
Somatic Embryogenesis and Synthetic Seed II (ed. by Y.P.S. Bajaj)

Table 1. The botanical and common names of some cultivated species of the Cucurbitaceae. (Jelaska 1986)

Latin name	Common name
Citrullus vulgaris Schrad	Watermelon
Cucumis anguria L.	West India gherkin
Cucumis melo L.	Muskmelon
Cucumis sativus L.	Cucumber
Cucurbita ficifolia Bouche	Malabar gourd, fig-leaf gourd
Cucurbita maxima Duch.	Winter squash, turban squash, marrow squash, pumpkin
Cucurbita mixta Pang.	Winter squash, pumpkin
Cucurbita moschata Poir	Winter squash, pumpkin
Cucurbita pepo L.	Summer squash, winter squash, marrow squash, pumpkin
Lagenaria siceraria (Mol) standl	White flower gourd
Luffa cylindrica Roem.	Dish-rag gourd, loofah gourd
Sechium edule SW.	Chayote

Table 2. Distribution and production of cucurbits in different parts of the world. (FAO 1989)

World region	Harvest area (10^3 ha)	(% of total)	Yield (kg/ha)	Total production (10^3 MT[a])	(% of total)
Cucumber and gherkins					
Africa	22	2.5	16 421	356	2.8
Asia	463	52.6	15 488	7175	56.2
Europe	131	14.9	20 994	2744	21.5
North and Central America	82	9.3	12 543	1029	8.1
South America	3	0.3	16 315	49	0.4
All developed countries	390	44.3	15 162	5913	46.3
All developing countries	490	55.7	13 994	6861	53.7
World total	880	100.0	14 512	12 774	100.0
Pumpkins, squash, gourds					
Africa	85	14.5	11 745	1002	15.3
Asia	229	39.1	13 167	3010	45.8
Europe	131	22.4	20 962	1339	20.4
North and Central America	61	10.4	4 962	304	4.6
South America	69	11.8	11 210	773	11.8
All developed countries	174	29.7	11 190	1942	29.6
All developing countries	412	70.3	11 214	4626	70.4
World total	586	100.0	11 207	6568	100.0

[a] Metric tonnes.

necessary for complete plant development are initiated in one step. Thus, regeneration of plants through somatic embryogenesis is often more advantageous than organogenesis for plant breeding programs. In some cases, plants regenerated from somatic embryogenesis may develop from a single cell and may exhibit less variability. In the process, the frequency of plantlets per unit time and space is enhanced, making genetic manipulation, particularly genetic transformation, more efficient. Somatic embryos may prove useful for long-term storage, such as in germplasm banks, mass production of plant varieties, micropropagation of sterile plants such as triploids, male-sterile genotypes, and interspecific hybrids whose zygotic development is recalcitrant (Fassuliotis and Nelson 1988). For cucurbit seeds of added value, mass production of somatic embryos and their encapsulation may lead to commercialization of synthetic seeds.

1.3 Brief Review of Somatic Embryogeneis in Cucurbits

Table 3 lists the members of the Cucurbitaceae that have generated somatic embryos in culture. This report describes the limited studies on somatic embryogenesis and successful plant regeneration in cucurbits. These systems can be used for cucurbit improvement via genetic transformation.

2 Somatic Embryogenesis in *Cucumis sativus*

2.1 Somatic Embryogeneis in Cucumber

There are a few reports describing the successful regeneration of cucumber. Somatic embryogenesis in cucumber was first reported by Malepszy and Nadolska-Orczyk (1983) in leaf explants from 16-day-old seedlings. The explants were cultured on MS medium supplemented with edamin and 1.2 mg/l 2,4,5-T in the dark for 6–8 weeks. Proembyonic masses and structures resembling somatic embryos were formed in a secondary medium containing 0.02 mg/l 2,4-D after 2–4 weeks of incubation in the light. Histological studies (Malepszy et al. 1982) showed that the callus tissues obtained from leaf explants have different stages of embryoids with developmental anomalies and meristem structures. Different patterns of development were observed on media containing 2,4-D vs. those containing 2,4,5-T during the first 20 days of culture. On media supplemented with 2,4-D, meristematic cells developed and proliferated around vascular bundles, and further forming centers of meristematic cells. Proembryoids and embryoids were developed simultaneously on different parts of the explants within 20 to 35 days of culturing and secondary embryoids developed from many of these embryoids. Rajasekaran et al. (1983) reported that the mode of regeneration was dependent on the concentration of 2,4-D present in the medium. A high concentration of 2,4-D promoted embryogenesis and a low concentration promoted organogenesis. However, because of a lack of histological analysis, the possibility that these plants formed by organogenesis due to abnormalities in embryoid conversion could not be excluded.

A cucurbit plant regenerated from cell suspension cultures was reported by Chee (1988). Embryogenic callus was initiated from the primary leaves of 14–17-day-old plants. Suspension cultures were composed of a population of cells that were densely cytoplasmic and potentially embryogenic. These cells were grown in liquid MS medium containing 5 μM 2,4,5-T and 4 μM BA. Differentiation of embryos was enhanced by washing the suspension-cultured cells with MS basal medium. Sixty to 70% of the embryos prewashed with activated charcoal germinated into plantlets with normal morphology. Embryos obtained from suspension-cultured cells without prewashing with activated charcoal organized into plantlets with abnormal primary leaves. However, morphologically normal plantlets were obtained by excising the shoot tips and transferring them to fresh medium. Protoplast culture of six cultivars, three gynoecious lines, and two F_1 hybrids of cucumber was reported by Jia et al. (1986). Sustained divisions of protoplast cultures were obtained when protoplasts were cultured in liquid medium supplemented with 0.5 mg/l 2,4-D and 1 mg/l kinetin. The frequency of division reached 46% after 8 days in culture. After transfer of cell colonies onto MS medium supplemented with 0.01 mg/l 2,4-D and 1 mg/l BA or 0.2 mg/l IAA and 0.5 mg/l BA, all 11 lines formed callus. However, only the callus of the F_1 hybrid 37-1Gx78-50 produced a large number of somatic embryos on medium with the hormone combination of 0.2 mg/l IAA and 5 mg/l kinetin, or 2 mg/l BA. Plantlets developed when embryos were transferred onto a medium with half-strength MS medium without any growth regulators. Somatic embryogenesis and plantlet recovery from root explants were reported by Trulson and Shahin (1986), however, the frequency of regeneration in this procedure was extremely low. More recently, Chee reported a simple procedure for regeneration of cucumber plants from cotyledon and hypocotyl explants. Somatic embryogenesis was induced on MS medium supplemented with 2,4-D at 2 mg/l. Development of embryos was accomplished on MS medium with 1 mg/l NAA and 0.5 mg/l kinetin. Furthermore, Chee (1990) reported that transgenic cucumber plants can be obtained by the application of this efficient regeneration procedure in conjunction with the *Agrobacterium*-mediated transfer or microprojectile bombardment system (Chee and Slightom 1992). Transformed cucumber plants that express the CMV-C coat protein gene appeared morphologically normal and fertile (Chee and Slightom 1991; Slightom et al. 1990). In addition, the CMV-C coat protein gene was stably inherited in the progeny plants without detrimental effects on the host plants. Interestingly, the field test results for the transformed progeny plants containing the CMV-C coat protein gene suggest that CMV-C coat protein gene expression does offer protection from CMV-C infection.

2.2 Mode of Regeneration of Cucumber

Chee (1988) reported that the frequency of embryogenesis depends on the explant sources. The highest overall frequency of regeneration was obtained from cotyledon explants (Table 4). The frequencies of regeneration from cotyledon, hypocotyl, and leaf explants, under the best hormonal combination tested, were 50, 25, 17%, respectively. Chee (1989) also reported that regeneration from the cotyledon explants of cucumber cv. Poinsett 76 took place in all growth regulator combinations tested with the highest response occurring at 2,4-D +

Table 3. Somatic embryogenesis in the Cucurbitaceae

Species	Explants	Medium	Growth Regulators		Reference
			Primary Medium (µM)	Secondary Medium (µM)	
Cucumis sativus L. (cucumber)	Cotyledon	MS	2,4-D (10)	None	Cade et al. (1990)
	Cotyledon, leaf	MS	2,4,5-T(5), BA(4)	None	Chee (1988) Chee (1990 a–c)
	Cotyledon, hypocotyl	MS	2,4-D(10), Kinetin(2.5)	None	Chee (1990 a–c)
	Coty, protoplast	MS	2,4-D(2.5), Kinetin(5)	None	Jia et al. (1986)
	Leaf	MS	2,4,5-T(5), BA(4)	BA(2)	Malepszy and Nadoska-Orczyk (1993)
	Petiole	MS	2,4-D(5), BA(5)	None	Punja et al. (1990)
	Cotyledon, root	MS	2,4-D(5), NAA(5), BA(2.5)	NAA(5), BA(2.5)	Truslon and Shahin (1986)
	Hypocotyl	MS	2,4-D(5), BA(1)	None	Ziv and Gadasi (1986)

Cucumis melo L. (muskmelon)	Cotyledon	MS	2,4-D(5)	None	Oridate and Oosawa (1986)
	Mature seed	MS	2,4-D(5), NAA (11), BA(5)	None	Homma et al. (1991)
	Stem, leaf, shoot apex	MS	2,4-D(5), NAA(10)	BA(0.5)	Kageyama et al. (1990)
	Mature seed	MS	2,4-D(5), NAA(5), BA(0.5)	None	Kageyama et al. (1991)
	Cotyledon, hypocotyl, stem, petiole	MS	2,4-D(10), BA(0.5)	GA_4(0.6)	Tabei et al. (1991)
	Cotyledon	MS	2,4-D(5), NAA(5), BA(2.5)	NAA(5), BA(2.5)	Truslon and Shahin (1986)
Cucurbita pepo L. (pumpkin)	Cotyledon, hypocotyl	MS	IBA(4.9)	2,4-D(1.4), watermelon sap (10%)	Jelaska (1974)
Cucurbita pepo L. (summer squash)	Cotyledon shoot apex	MS	2,4,5-T(5), BA(4), kinetin(0.5)	NAA (3), kinetin (0.25)	Chee (1991)
	Cotyledon	MS	2,4-D(22.7)	NAA (0.5), kinetin(0.25)	Chee (1992)

Table 4. Effects of growth regulator composition on induction of somatic embryogenesis from cotyledon, hypocotyl, and leaf explants of cucumber

Growth regulators[a] (mg/l)		Explants producing embryogenic callus (%)		
		Cotyledon	Hypocotyl	Leaf
2,4-D	*Kinetin*			
1	0.5	9	0	0
1	1.0	9	0	0
2	0.5	50	25	0
2	1.0	1	0	0
2	2.0	1	0	0
2,4,5-T	*BA*			
1.2	0.8	5	0	17

[a] BA = 6-benzylaminopurine; 2,4-D = 2,4-dichlorophenoxyacetic acid; 2,4,5-T = 2,4,5-trichlorophenoxyacetic acid.

kinetin (Table 4). Different callus types developed 4 weeks after the explants were plated onto culture media. A particular type of tissue with a translucent, smooth, and gelatinous appearance (Fig. 1) was developed on the surface of the explants in contact with culture media. The upper part of the gelatinous tissue contained small putative embryogenic tissues. Microscope examination showed that these tissues consisted of cytoplasmically dense, multicellular aggregates, resembling proembryonic masses. These aggregates later organized into embryos. The embryogenic potential of the gelatinous callus was not lost with successive subcultures on induction medium for at least 6 months. In contrast to the gelatinous tissue, callus that was hard and nodular or white and friable was not found morphologically competent.

After 4 weeks, the dense sectors were selected for transfer to conversion medium (MS + 1 mg/l NAA + 0.5 mg/l kinetin). Upon transfer to light, the putative proembryonic structures developed into yellowish, embryo-like structures (Fig. 2). A histological section of this embryonic callus tissue is shown in Fig 3. Somatic embryos closely resemble their zygotic counterparts with the appropriate root, shoot, and cotyledonary organs and have no extraneous proliferations. Formation of these structures was not synchronized across explants. The proembryonic tissue also proliferated, eventually forming a homogeneous mass

→

Fig. 1. Embryogenic callus derived from a cotyledon explant after culture for 1 month on induction medium; *bar* = 1 mm. (Chee 1990a)

Fig. 2. Clusters of yellowish somatic embryos formed on the surface of gel-like callus after culture for 2 weeks on MS medium + 1 mg/l NAA + 0.5 mg/l kinetin; *bar* = 1 mm. (Chee 1990a)

Fig. 3. **A** Section through an embryogenic callus tissue showing early multicellular stages of embryo differentiation. **B** Longitudinal section of a somatic embryo with a bipolar structure; *bar* = 375 mm. (Chee 1990a)

Fig. 4. A cluster of developing somatic embryos with well-developed cotyledons cultured on conversion medium; *bar* = 1 mm. (Chee 1990a)

Fig. 5. A converted somatic embryo; *bar* = 1 mm. (Chee 1990a)

Fig. 6. Regenerated plants from somatic embryos 1 week after transplanting to pots containing planting medium. (Chee 1990a)

1

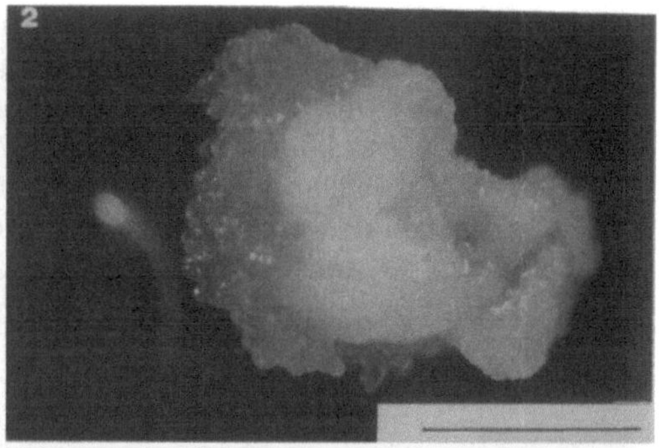
2

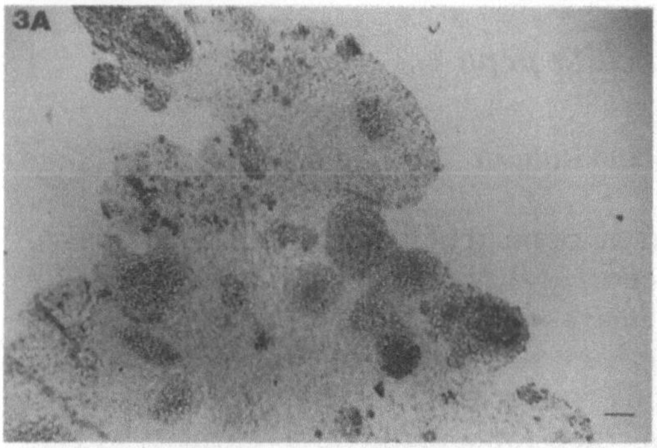
3A

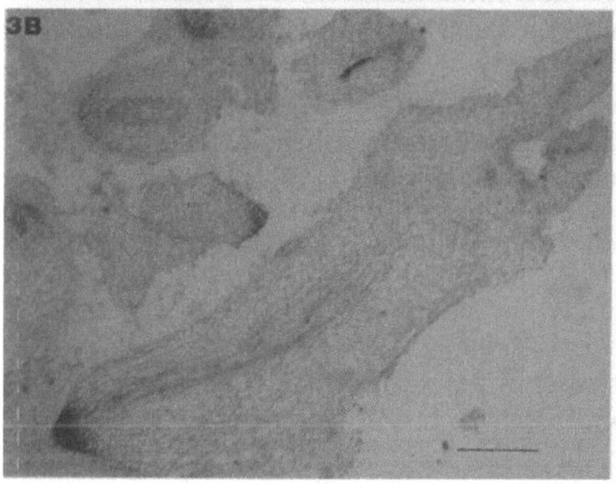
3B

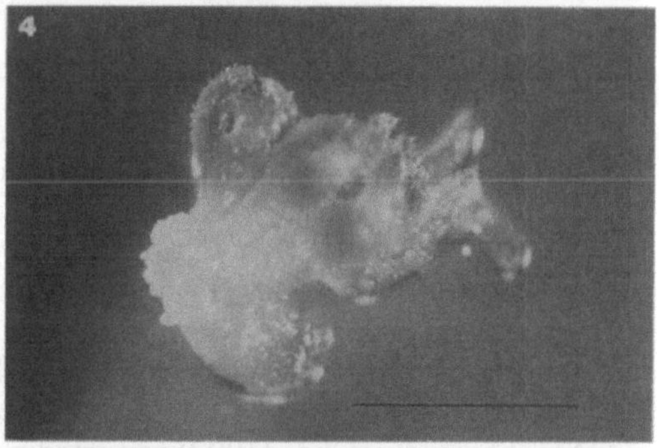
4

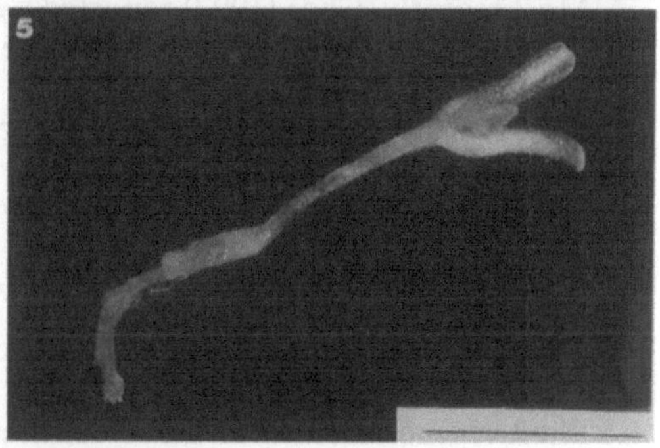
5

6

of granular, yellow tissue. The putative somatic embryos had a tendency to recallus when they were maintained on the conversion medium for more than 2 weeks. Therefore, the embryos were transferred onto growth regulator-free MS medium 2 weeks later when plantlet development followed (Figs. 4 and 5). About 80 to 85% of the embryos developed into phenotypically normal plantlets (Fig. 6). Whole plants were able to flower and set seeds normally. The combination of 2 mg/l 2,4-D and 0.5 mg/l kinetin also had an induction effect on mature cotyledons from dry seeds. However, the frequency of explants forming embryogenic callus was only 5%.

3 Somatic Embryogenesis in *Cucurbita pepo* L.

3.1 Somatic Embryogenesis in Pumpkin and Squash

There have been very few studies on the development of tissue culture techniques of *Cucurbita pepo* L. Schroder (1968) reported the production of embryogenic tissue from pericarp of zucchini squash. Jelaska (1974) reported somatc embryogenesis in hypocotyl- and cotyledon-derived callus of pumpkins and demonstrated that embryos could develop into normal plants at a very low frequency. Most recently, Chee (1992) reported a method for somatic embryogenesis with plant regeneration from shoot apex-derived callus (1991) and another method from cotyledons of mature seeds. The frequency of regeneration according to these two methods has ranged from 8 to 29%. Transformed squash plants could be obtained by applying these regeneration procedures and the gene transfer approach *via* microprojectile bombardment (Chee, unpubl.).

3.2 Mode of Regeneration of Squash

Plant regeneration from tissue cultures of summer squash has been observed (Chee 1991, 1992). Explants of *Cucurbita pepo* cv. YC60 callused less rapidly than those of *C. sativus.* Embryogenic callus tissues were initiated when cotyledons of mature seeds were excised and cultured on MS medium supplemented with 5 mg/l 2,4-D or 1.2 mg/l 2,4,5-T. Callus with different morphologies formed 6 weeks after the explants cultured on MS containing 2,4-D. Among these, the embryonic type was translucent, smooth, and gelatinous in appearance. Most of this callus developed in the area of the embryonic axis and at the site of explant contact with the medium. After another 8 to 9 weeks of incubation, approximately 10 to 20 putative somatic embryos were observed on the surface of the gelatinous callus (Fig. 7). A similar callus morphology was observed in embryogenic callus tissue dervied from cucumber explants. A longitudinal cross section of a bipolar somatic embryo before transfer to maturation medium is shown in Fig. 8. The development of the somatic embryos on maturation medium (MS + 0.05 mg/l NAA + 0.05 mg/l kinetin) was initially quite abnormal. The embryos

Fig. 7. Cluster of somatic embryos formed on an embryonic axis of a squash explant. (Chee 1992)
Fig. 8. Longitudinal section of a squash somatic embryo indicating bipolar structure. (Chee 1992)
Fig. 9. A maturing squash somatic embryo 2 weeks after culture on maturation medium. (Chee 1992)
Fig. 10. Plantlet derived from a squash somatic embryo

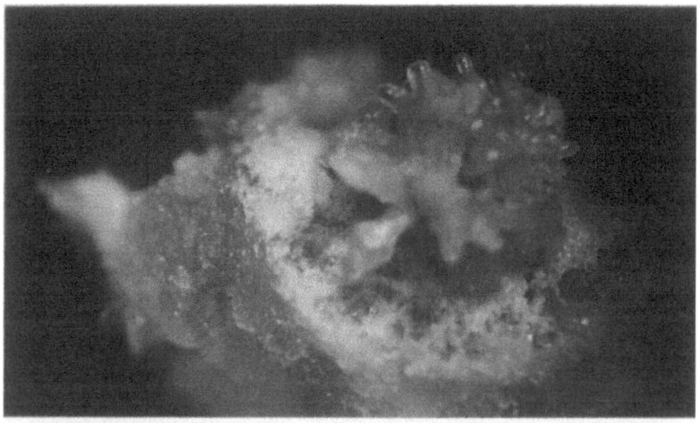

7

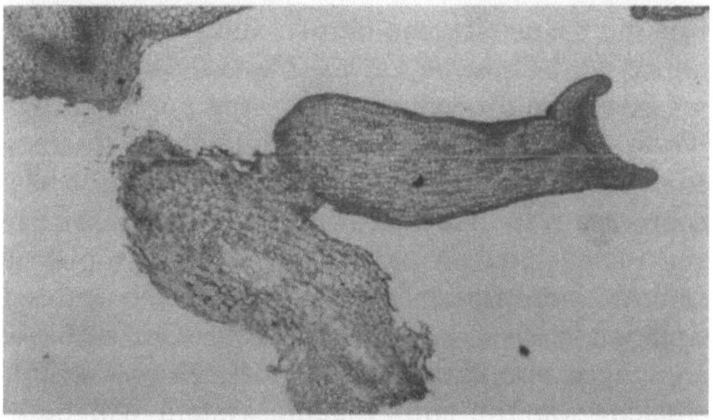

8

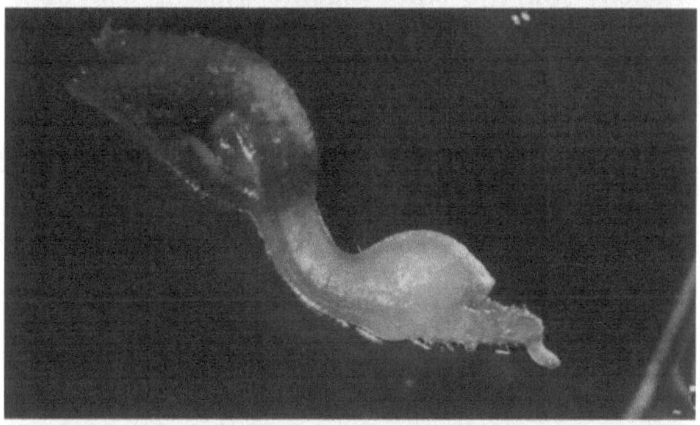

9

10

were elongated with a distorted morphology (Fig. 9); however, about 75% of these embryos developed into normal plantlets (Fig. 10). Root formation was obtained when the mature embryos were grown on half-strength MS basal medium without growth regulators. The success of transfer of the plantlets from agar to planting medium was 100%. None of the regenerated plants displayed gross phenotypic abnormalities; they flowered and set viable seeds.

4 Conclusion

For the Cucurbitaceae family, successful somatic embryogenesis has been obtained with *Cucumis sativus*, *Cucumis melo* L., and *Cucurbita pepo*, although at present these procedures have been applied only on a limited number of genotypes. Certainly, the greatest challenge in the future is the induction of somatic embryogenesis in genotypes where it has not been observed or where it occurs sporadically or at low frequencies. The source of initial explants, mode of culture, and the physical properties of the medium play important roles in determining somatic embryogenesis in tissue and cell cultures. Clearly, the quality of somatic embryos needs to be improved to provide embryos with a structure, physiology, and vigor comparable to zygotic embryos. Finally, the development of an efficient somatic embryogenic system will be extremely useful in obtaining genetically transformed plants and in the development of synthetic seeds.

5 Protocols for the Induction of Somatic Embryos Used in Our Laboratory

5.1 Somatic Embryogenesis and Regeneration of Cucumber Plants

1. Soak seeds of cucumber (Poinsett 76) in tap water for approximately 15 min and remove the seed coats manually. Surface sterilize the decoated seeds with 70% alcohol for 1 min, followed by treatment with 25% (v/v) aqueous solution of Clorox for 25 min, and finally rinse four times with sterile distilled water.
2. Germinate the sterilized seeds at 28 °C on 0.8% water agar (Difco) in darkness.
3. Excise the cotyledons from 3–5-day-old in vitro grown seedlings and cut into 5-mm^2 sections.
4. Place cotyledon sections in the Petri dishes with induction medium (MS salts and vitamins, 2 mg/l 2,4-D, 0.5 mg/l kinetin, 3% sucrose, 0.6% Phytagar (Gibco), pH 5.8). Incubate culture for 4 weeks at 26 °C in darkness. Every 4 weeks transfer the cultures to fresh medium.
5. After 6–8 weeks select embryoid callus and transfer to maturation medium (MS salts and vitamins, 1 mg/l NAA, 0.5 mg/l kinetin, 3% sucrose, 0.6% Phytagar). Incubate these cultures at 26 °C under diffuse cool-white fluorescent lamps (4 klx) with a 16-h photoperiod for 2 weeks.
6. After 2 weeks, transfer embryoid structure to MS medium containing no growth regulators.
7. Transfer plantlets with an extensive root system to pots containing planting medium and cover with clear storage bags for 1 week for acclimatization.
8. Subsequently transfer the regenerated plants to pots containing a soil mixture (Metro Mix 300R; A.H. Hummert Seed Co., St. Louis).

5.2 Somatic Embryogenesis and Plant Regeneration of Squash Plants

1. Surface sterilize the seeds (YC60) with 15% Clorox for 10 min and rinse three times with sterile water.
2. Remove the seed coat, treat the seeds again with 25% Chlorox for 25 min, and rinse three times with sterile water, 1 min with 70% alcohol, and finally three times with sterile water.
3. Cut the embryos transversely into two unequal sections; one section consists of the embryonic axis and one-third of the cotyledons; the other, the remainding cotyledons.
4. Discard the section consisting only of the cotyledons.
5. Place the sections containing the embryonic axes in Petri dishes containing induction medium (MS salts and vitamins, 5 mg/l 2,4-D, 3% sucrose, 0.8% Phytagar, pH 5.8). Incubate the cultures in the dark at 26 °C.
6. After 9 weeks, with one subculture at 4.5 weeks, select the embryoid callus and transfer to maturation medium (MS salts and vitamins, 0.05 mg/l NAA, 0.05 kinetin, 3% sucrose, 0.8% Phytagar). Incubate the cultures at 28 °C under diffuse cool-white fluorescent lamps (4 klx) with a 16-h photoperiod.
7. After another 3–4 weeks, transfer the embryoid structures to half-strength hormone-free MS medium.
8. Transplant the plantlets to pots containing planting medium and cover with clear plastic storage bags for 1 week for acclimatization.
9. Transfer plants to pots containing soil mixture (Metro Mix 300R; A.H. Hummert Seed Co., St. Louis) and grow in a greenhouse.

References

Cade RM, Wehner TC, Blazich FA (1990) Somatic embryos derived from cotyledons of cucumber. J AM Soc Hortic Sci 115: 691–696

Chee PP (1988) Somatic embryogenesis and plant regeneration from cell suspension cultures of *Cucumis sativus* L. Plant Cell Rep 7: 274–277

Chee PP (1990a) High frequency of somatic embryogenesis and recovery of fertile cucumber plants. Hortsci 25: 792–793

Chee PP (1990b) Transformation of cucumber via *Agrobacterium tumefaciens*. In: Nijkamp HJJ, Van der Plas LHW, Van Aartrijk J (eds) Progress in plant cellular and molecular biology. Kluwer, Dordrecht, 85–90

Chee PP (1990c) Transformation of *Cucumis sativus* tissue by *Agrobacterium tumefaciens* and the regeneration of transformed plants. Plant Cell Rep 9: 245–248

Chee PP (1991) Somatic embryogenesis and plant regeneration of squash *Cucurbita pepo* L cv. YC 60. Plant Cell Rep 9: 620–622

Chee PP (1992) Initiation and maturation of somatic embryos of summer squash *Cucurbita pepo* L. Hort Science 27: 59–60

Chee PP, Slightom JL (1991) Transfer and expression of cucumber mosaic virus coat protein gene in the genome of *Cucumis sativus* L. Am Soc Hortic Sci 116: 1098–1102

Chee PP, Slightom JL (1992) Transformation of cucumber tissues by microprojectile bombardment: identification of plants containing functional and non-functional genes. Gene 118: 255–260

Deakin JR, Bohn GW, Whitaker TW (1971) Interspecific hybridization in *Cucumis*. Econ Bot 25: 195–211

Fang G, Grumet R (1990) *Agrobacterium tumefaciens* mediated transformation and regeneration of muskmelon plants. Plant Cell Rep 9: 160–164

FAO (1989) FAO production yearbook, Rome, 43: 183–186

Fassuliotis G, Nelson BV (1988) Interspecific hybrids of *Cucumis metuliferus* x *C. anguria* obtained through embryo culture and somatic embryogenesis. Euphytica 37: 53–60

Homma Y, Sugiyama K, Oosawa K (1991) Improvement in production and regeneration of somatic embryos from mature seed of melon (*Cucumis melo*) on solid media. Jpn J Breed 41: 543–551

Jelaska S (1974) Embryogenesis and organogenesis in pumpkin explants. Physiol Plant 31: 257–261
Jelaska S (1986) Cucurbits. In Bajaj YPS (ed) Biotechnology in agriculture and forestry, vol 2. Crop I. Springer, Berlin Heidelberg New York, pp 371–386
Jia SR, Fu YY, Lin Y (1986) Embryogenesis and plant regeneration from cotyledon protoplast culture of cucumber (*Cucumis sativus* L.). Plant Physiol 124: 393–398
Kageyama K, Kazunori Y, Miyajimas S (1990) Somatic embryogenesis and plant regeneration from stem, leaf and shoot apex of melon (*Cucumis melo* L.). Plant Tissue Cult Lett 7: 193–198
Kageyama K, Yabe K, Miyajima S (1991) Somatic embryogenesis in suspension culture of mature seed of *Cucumis melo* L. Jpn J Breed 41: 273–278
Malepszy S, Nadoska-Orczyk A (1983) In vitro culture of *Cucumis sativus*. 1. Regeneration of plantlets from callus formed by leaf explants. Z Pflanzenphysiol 111: 273–276
Malepszy S, Niemirowicz-Szczytt K, Wiszniewska J (1982) Cucumber (*Cucumis sativus* L.) somatic embryogenesis in vitro. Acta Biol 10: 218–220
Oridate T, Oosawa K (1986) Embryogenesis and organogenesis from suspension callus culture in melon (*Cucumis melo* L.). Jpn J Breed 36: 424–428
Punja ZK, Abbas N, Sarmento GG, Tang FA (1990) Regeneration of *Cucumis sativus* and *C. sativus* var. *hardwickii*, *C. melo* and *C. meruliferus* from explants through somatic embryogenesis and organogenesis. Plant Cell Tissue Organ Cult 21: 93–102
Rajasekaran K, Mullins MG, Nair Y (1983) Flower formation in vitro by hypocotyl explants of cucumber (*Cucumis sativus* L.). Ann Bot (Lond) 52: 417–420
Schroder CA (1968) Adventive embryogenesis in fruit pericarp tissue in vitro. Bot Gaz 129: 374–376
Slightom JL, Chee PP, Gonsalves D (1990) Field testing of cucumber plants which express the CMV coat protein gene: field plot design to test natural infection pressures. In: Nijkamp HJJ, van der Plas LHW, Van Aartrijk J (eds) Progress in plant cellular and molecular biology. Kluwer, Dordrecht, pp 201–206
Tabeiy, Kanno T, Nishino T (1991) Regulation of organogenesis and somatic embryogenesis by auxin in melon, *Cucumis melo* L. Plant Cell Rep 10: 225–229
Tisserat E, Esan EB, Murashige T (1979) Somatic embryogenesis in angiosperms. Hortic Rev 1: 1–78
Trulson AJ, Shahin EA (1986) In vitro plant regeneration in the genus *Cucumis*. Plant Sci 47: 35–43
Whitaker TW, Davis GN (1962) Cucurbits (botany, cultivation and utilization). Leonhard Hill, London Interscience, New York (World Crops Book)
Ziv MA, Gadasi G (1986) Enhanced embryogenesis and plant regeneration from cucumber (*Cucumis sativus* L.) callus by activated charcoal in solidified double-layer cultures. Plant Sci 47: 115–122

II.4 Somatic Embryogenesis in Cucumber (*Cucumis sativus* L.)

J.A.R. Ladyman[1]

1 Introduction

Cucumbers are an important commercial crop in many parts of the world, and they are grown both for fresh market consumption and for pickling. Cucumbers grown in the field typically require pollination to set fruit, while those grown in greenhouses, particularly as practiced in The Netherlands and other European countries, reproduce parthenocarpically (de Rougemont 1989). The plant has a trailing or climbing growth habit and can be trained up trellises, strings, and stakes to conserve growing space (Russo et al. 1991). It is a subtropical species that grows well in temperate climates, however, no varieties have been identified as being frost-resistant.

Both hybrid and open-pollinated cucumber seed are produced by seed growers. Hybrid seed are frequently preferred by growers because of superior yields, improved disease-resistance, and good flavor. Hybrid vigor has long since been documented (Winnick and Vetusnjak 1952; Lower et al. 1982). However, hybrid seed production relies on a complex sex expression of the mature plant that is environmentally sensitive and often requires that parents be treated with gibberellin or silver nitrate to ensure that the correct sex is expressed. This process can be laborious when hand pollinations are required. These factors lead to a relatively high cost of producing hybrid seed. These factors supported our interest in developing artificial seed of cucumber. Once the initial hybrid seed were made it should be possible to clone the cross by generating somatic embryos.

Improvement of cucumber by exploiting the genes for disease and insect resistance carried by cultivated and wild relatives is limited by sterility barriers which exist between cucumber and its relatives (Deakin et al. 1971). Successful crosses can only be made between cucumber and *Cucumis hardwicki* which is a variety of *C. sativus* rather than distinct subspecies (Deakin et al. 1971). These barriers preclude transfer of many desirable traits. Genetic engineering provides an opportunity to introduce useful genes into the *Cucumis sativus* gene pool.

Cucumber somatic embryos have been derived from a variety of explants, e.g., cotyledons (Malepsky et al. 1982; Trulson and Shahin 1986; Ziv and Gadasi

[1] Plant Cell Research Institute, Dublin, California, USA. Present address: JnJ Associates, 1701 Father Sky Ct N E, Albuquerque, New Mexico 87112, USA
The use of a Trademark name for a product provides no endorsement of the products. They are used solely for complete explanation of protocols.

Biotechnology in Agriculture and Forestry, Vol. 31
Somatic Embryogenesis and Synthetic Seed II (ed. by Y.P.S. Bajaj)

1986; Chee 1990a), primary leaves (Chee and Tricoli 1988), immature zygotic embryos (Kim and Janick 1989), and hypocotyl segments (Rajasekaran et al. 1983; Chee 1990a). Cucumbers have proved relatively difficult to routinely regenerate. Somatic embryos produced from immature zygotic embryos had malformed cotyledons and did not regenerate into seedlings (Kim and Janick 1989). The procedures developed by Trulson and Shahin (1986) and Ziv and Gadasi (1986), which used cotyledonary explants, led to well-formed embryos which germinated successfully, but did not include a multiplication phase of the primary embryos in the protocol. Chee and Tricoli (1988) and Chee (1990a) developed a procedure whereby suspension cultures were established from embryogenic callus derived from the primary leaves of young plants. In the regeneration method described in this Chapter a "conditioning" step in liquid suspension in the presence of abscisic acid, in addition to the induction and multiplication stages on semisolid media, has been found to be beneficial to the ultimate regeneration of the embryos.

2 Somatic Embryogenesis

In order to clarify the following discussion a scheme of the protocol for the successful induction, multiplication, and regeneration of somatic embryos is described in Fig. 1. A detailed, stepwise, protocol is described later in Section 5.

Somatic embryos form on the cut edges of cotyledonary explants, and can be then regenerated immediately or multiplied, and then regenerated. Chee (1990a) and Malepszy (1988) described a type of embryogenic tissue similar to that which we observed during the multiplication phase, although the conditions and hormones were significantly different in all three protocols. Tissue in the multiplication phase of somatic embryogenesis was frequently a gelatinous matrix in

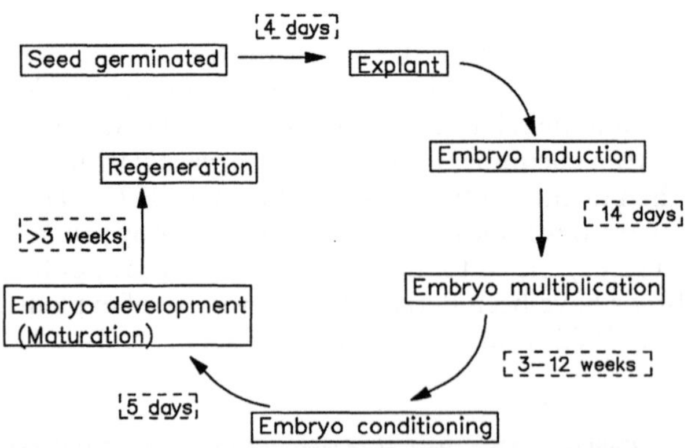

Fig. 1. Scheme of the protocol for the induction, multiplication, and regeneration of somatic embryos of cucumber

which early (globular) embryos were apparently suspended. Under some conditions, which will be described later, the total tissue during the multiplication phase was embryogenic but in other cases more than 60% was undifferentiated callus. After tissue multiplication, embryogenic tissue was placed in a suspension culture in the presence of abscisic acid (ABA). After this "conditioning step" the embryos are transferred to semisolid medium for regeneration and finally transplanted into soil.

In all experiments reported upon here the pH was adjusted prior to autoclaving which may be an important factor when interpreting results (Owens et al. 1991).

2.1 Induction

Induction of somatic embryos on the cut edges of cotyledonary explants proceeded on all cultivars tested. The extent to which the embryos formed differed, but this phenomenon was not tested statistically. Gelling agent type and concentration had a significant impact on the induction of somatic embryogenic tissue (Ladyman and Girard 1992a). Agar (0.35, 0.7, and 1.4%; Baltimore Biological Laboratories, Cockeysville, MD) was compared to Gelrite (gellan gum; 0.15, 0.3, and 0.6%; Kelco Division, San Diego, CA) for suitability as the gelling agent for the induction phase. Significant differences were observed between treatments. Induction was found to be best on 0.7% agar. In this case the putative somatic embryos which formed around the cut edge were smooth, yellowish in color and very little nonembryogenic tissue (callus) developed. Gelrite at 0.15% caused excessive expansion of the explants and the tissue became vitrified. In addition, no somatic embryos and very little callus were formed. Less expansion and less vitrification occurred on 0.3 and 0.6% Gelrite, but the tissue produced more callus and fewer embryos than that on 0.7% agar. Tissue quality from cultures on 0.35% agar resembled that on 0.6% Gelrite, i.e., well-formed glossy embryos were numerous but they were surrounded by more callus than tissue on 0.7% agar. A greater amount of material was harvested from the Gelrite plates (Table 1) but the embryogenic tissue was less discrete and wetter, and more nonembryogenic callus was inadvertently harvested. The presence of undifferentiated callus tissue was deterimental to somatic embryo multiplication. Explants

Table 1. The fresh weight of the total tissue induced on 9 cotyledonary explants after 14 days on an induction medium. (Ladyman and Girard 1992a)[a]

Gelling agent	Conc. (%)	Weight (g)	Standard error
Agar	0.35	0.388	0.056
	0.70	0.436	0.064
	1.40	0.084	0.01
Gelrite	0.15	0.528	0.073
	0.30	0.567	0.081

[a] Data are means of 7 plates; 9 cotyledonary explants/plate. Each explant was 0.196 cm^2.

on 1.4% agar did not expand, vitrify, or produce callus, but the production of embryos was also very limited (Table 1).

2.2 Multiplication

In the tissue multiplication phase the cultivar does influence the extent to which somatic embryogenic tissue and undifferentiated callus multiply. For example, the fresh market cucumber Corona always produced less tissue than did the pickling cucumber Cinton. Different concentrations of 2,4-dichlorophenoxyacetic acid (2,4-D) and one concentration of 2,4,5-trichlorophenoxyacetic acid (2,4,5-T) were tested for their effect on the multiplication of tissue of the variety Corona (Fig. 2). Although the concentration of 2,4-D did not give statistically significant differences in quantity of tissue produced, the numbers of regenerates differed. Those somatic embryos multiplied on 4 μM 2,4-D regenerated two to three times more frequently into morphologically normal plants than those on 5 μM 2,4-D Levels of 2,4-D higher than 8 μM gave rise to morphologically abnormal (very small, dark green, and pubescent) seedlings. No regenerates were obtained from 2,4,5-T. Therefore, for routine regeneration, e.g., in transformation experiments, although 4 μM 2,4-D was used in media for Clinton, 5 μM 2,4-D was used for Corona.

Frequency of transfer of somatic embryogenic tissue during multiplication was also important (Ladyman and Girard 1992b). More frequent transfers, at least every 4 weeks, was necessary to maintain the embryogenic potential of the tissue. After 12 weeks the putative embryogenic tissue multiplied from 100 mg to 7.02 g(± 1.69 g) when biweekly transfers were made, 5.12 g (± 1.39 g) after

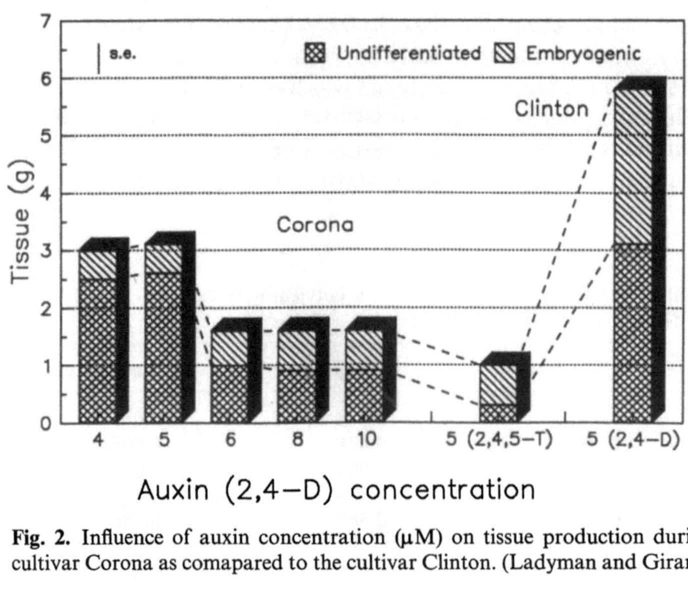

Fig. 2. Influence of auxin concentration (μM) on tissue production during multiplication of the cultivar Corona as comapared to the cultivar Clinton. (Ladyman and Girard, unpubl. results)

4-weekly transfers but only 1.221 g (± 0.49) and 1 g (± 0.17 g) after 6- or 12-weekly transfers, respectively.

In the multiplication phase there were also significant effects of the gelling agent on somatic embryo production. Preliminary experiments (Ladyman and Girard, unpubl.) on the effects of Gelrite and agar on tissue production during the multiplication phase indicated that between 0.1 and 0.4% Gelrite reduced the total quantity of tissue produced during multiplication, but the quality (i.e., a subjective estimate based on color and texture of the tissue) appeared to be better (i.e., more likely to give rise to embryos). This conclusion was supported in further experiments (Ladyman and Girard 1992a). During the multiplication phase, the carbohydrate source was also found to be important. Kim and Janick (1989) suggested that the carbohydrate source may be a critical factor in cucurbit morphogenesis. Two similar, but seperate experiments clearly showed that the gelling agent and carbohydrate source influenced tissue multiplication. Respectively, the following modifications were made to the medium: (1) solidified with 0.7% agar or 0.15% Gelrite, which produces media of comparable firmness (Pasqualetto et al. 1988) in the presence of 3% sucrose, or (2) either fructose (3%), glucose (3%), or maltose (3%) was substituted for sucrose (3%) in the presence of agar (0.7%). About 100 mg of embryogenic tissue was placed onto each Petri plate. The tissue was kept under continuous, cool-white fluorescent light (approximately 36 $\mu E/m^2/s$) at 26 °C. The tissue was transferred after every 4 weeks to fresh multiplication media. At the time of transfer the cream-yellow, smooth embryogenic tissue was separated from hard, slightly pubescent callus tissue which was then discarded. In both experiments the fresh weight of the putative embryogenic tissue was used to quantify multiplication of somatic embryos. The effect of treatment on tissue water retention was not estimated. There were seven Petri plates for each treatment and results were supported in at least one other experiment. During the multiplication phase, in the first experiment, Gelrite (0.15%) gave about the same fresh weight of embryogenic tissue as agar (0.7%). After 12 weeks on multiplication media, the initial 100 mg of tissue yielded 10.9 g (± 2.3 g) tissue on Gelrite and 11.5 g (± 1.9 g) on agar. Tissue on both agar and Gelrite was translucent and smooth, but on Gelrite the embryogenic gelatinous material appeared to be denser. Unfortunately, this parameter was difficult to quantify. Maltose inhibited cucumber tissue multiplication (Table 2). It has been shown that maltose significantly increased the number of somatic embryos produced from alfalfa callus culture (Strickland et al. 1987) and from petunia anthers (Raquin 1983), whereas potato anthers cultured on a medium containing maltose produced significantly fewer embryos than those cultured on sucrose (Batty and Dunwell 1989). The ammoniacal nitrogen (NH_4) concentration (15 to 20 mM) was determined to be important for optimum alfalfa embryo production (Strickland et al. 1987) and a similar concentration (20 mM) was used in this multiplication medium.

The photoperiod significantly influenced the multiplication of the tissue (Ladyman and Girard 1990); this effect is described in Fig. 3. The gelling agent in these experiments was agar (0.7%) and the carbohydrate source was sucrose (3%). The yield of both total undifferentiated and embryogenic tissue benefitted from a dark period. There were statistically significant differences in

Table 2. The effect of carbohydrate source on the multiplication of putative somatic embryogenic tissue on a medium gelled with 0.7% agar. (Ladyman and Girard 1992a)[a]

Time on media (weeks)	Weight of embryogenic tissue (g)							
	Sucrose		Maltose		Fructose		Glucose	
	Mean[a]	SE[b]	Mean	SE	Mean	SE	Mean	SE
0	0.103	0.003	0.104	0.001	0.102	0.002	0.102	0.003
8	5.716	0.495	0.075	0.014	5.652	0.385	5.176	0.620

[a] Data are means of 7 Petri plates. In each Petri plate about 100 mg (time in culture = 0) of embryogenic tissue was placed.
[b] Standard error.

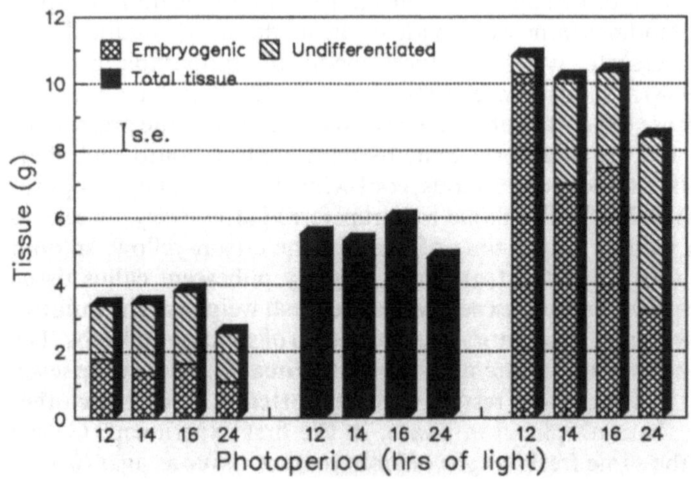

Fig. 3. Graph demonstrating the effect of photoperiod on tissue (cultivar Clinton) production during the multiplication phase

multiplication between the tissue that was exposed to a dark period and that under continuous light. Although no statistically rigorous tests were made, it has been noted in several experiments that light intensities lower than 60 $\mu E/m^2/s$ were preferable to higher intensities.

Activated charcoal (1 g/l, w/v) was added to the multiplication medium in an effort to reduce the effect of any toxic compounds which may be excreted by undifferentiated callus tissue and thus, possibly enhance the production of embryogenic tissue (Ladyman and Girard, unpubl.). It was appreciated that many of the hormones and nutrients in the agar medium may also become less available. The presence of activated charcoal resulted in far less total tissue forming than in its absence (0.6 vs. 2.3 g) but, interestingly, the fraction of undifferentiated callus was also significantly less than in its absence (5 compared to 23%). Premature chlorophyll development and germination of embryos in the presence of activated charcoal were also observed. After multiplication and conditioning in the absence of activated charcoal, the presence of activated charcoal in the medium on which the embryos germinated apparently did not have any effect. Ziv and Gadasi (1986) tested activated charcoal to improve somatic embryo production from hypocotyl segment explants and also found that dedifferentiation and secondary embryogenesis were inhibited. However, meaningful comparisons cannot be made because their system was significantly different from that described here.

2.3 Conditioning and Regeneration

A "conditioning" treatment with abscisic acid (ABA) after multiplication proved to be an important step for good regeneration. An example of the response to ABA is shown in Fig. 4. The regeneration of somatic embryos is successfully accomplished on a medium containing Murashige and Skoog (MS) salts gelled with 1% agar, under a 16-h photoperiod. Conditions that produced the greatest quantity of tissue during the multiplication phase did not necessarily produce the highest number of competent (readily regenerable) embryos. From previous experiments it had been determined that a photoperiod with a dark interlude produced more tissue than under continuous light. Sucrose was indicated to be equivalent to fructose as the carbohydrate source. The finding, that in the multiplication phase where gelrite yielded less tissue but that produced was of subjectively "better quality" than that on agar, was more difficult to interpret. Thus, experiments were performed to compare these factors on the number of regenerates obtained. The initial amount of callus that had been induced on 0.7% agar was weighed carefully to provide an estimate of the number of regenerates per unit of starting material. After multiplication under the conditions described in Table 3, the embryos were treated for 5 days in the dark at 22 °C with 0.2 µM

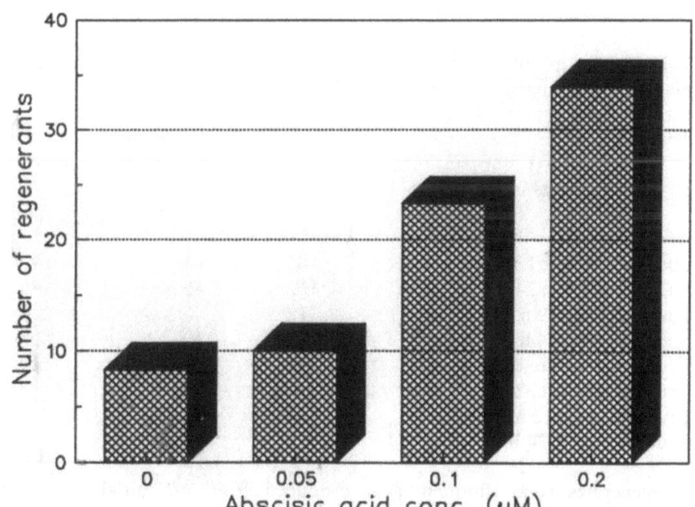

Fig. 4. Example of the influence of abscisic acid on the regeneration of cucumber cultivar Clinton. (Ladyman and Girard, unpubl. results)

ABA. The suspension was passed through filters of 1910 μm followed by 520 μm. The filtrate was centrifuged for 10 min, the supernatant discarded, and the embryos resuspended in twice the bed volume. The embryos started to germinate 14–24 days after plating onto hormone-free medium, however, the ones (small, globular-shaped, muticellular structures) passing through a mesh of 180 μm did not germinate.

The numbers of regenerates per 100 mg of primary somatic embryo tissue were recorded 4 weeks after the embryos had begun to germinate. These results are reported in Table 3. When the multiplication medium contained gelrite and sucrose, the majority of embryos produced both shoots and roots, over half of which were very vigorous and were transferred to soil. When agar was substituted for gelrite, the number of regenerates decreased substantially. When fructose was substituted for sucrose, few embryos germinated from either gelling medium. Those that had been multiplied under a 12-h photoperiod were in general more robust than those multiplied under continuous light.

Embryos that germinated in the Petri dishes were allowed to grow out of the Petri dish within a sterile box (31 × 23 × 10 cm; six Petri plates per box) that was open to ambient air. The seedlings were able to dislodge the lid during growth (Fig. 5). Seedlings treated this way harden off sufficiently well that little shock was experienced after their transplantation into soil.

Germinated embryos that were transplanted into soil were transferred to the greenhouse and developed into healthy robust plants, which flowered and produced fruit (Ladyman and Girard 1990). The fruits from some of the regenerates from a sucrose/Gelrite/12-h photoperiod multiplication (57) were compared to those from sucrose/Gelrite/24-h photoperiod (28) and to plants from zygotic seeds (10). The plants that were compared were all developmentally similar and vigorously growing at the three-leaf stage. They were kept, in randomized positions, in the same greenhouse. Uniformity of the plants derived

Table 3. The number of regenerates after tissue multiplication on either an agar- or Gelrite-gelled medium. (Ladyman and Girard 1992a,b)

Treatment		Regenerants		
Sugar, gelling agent, photoperoid	Total	With no[a] roots	Not into[a,b] soil	Into soil[a,c]
Sucrose, Gelrite, 12 h	184	24	49	111
Sucrose, Gelrite, 24 h	162	45	46	71
Sucrose, agar, 12 h	12	6	6	0
Sucrose, agar, 24 h	24	10	7	7
Fructose, Gelrite, 12 h	<3	1	<1	<1
Fructose, Gelrite, 24 h	<5	3	<1	<1
Fructose, agar 12 h	28	13	9	6
Fructose, agar, 24 h	0	0	0	0

[a] Regenerates (one shoot/embryo) produced from an initial weight of 100 mg of putative embryogenic tissue.

[b] Regenerates having a root system of 2 or fewer roots less than 3 cm long were not transferred to soil. Some seedlings with 2 or fewer roots can survive transplantation into soil with intensive care.

[c] Regenerates having a vigorous root system and shoot.

Fig. 5. Photograph illustrating the hardening off procedure for seedlings. The force produced from the hydrostatic pressure during growth raises the lids and slowly exposes the seedlings to ambient air

from zygotic seed is very high. Figure 6 describes the variation amongst individuals for total weight of fruits, total number of fruits, and the relationship between these parameters. The shape of the fruit derived from somatic embryos was in most cases typical of that obtained from plants from zygotic seed. If the outliner in Fig. 6 (circled), which produced a high number of small fruits,

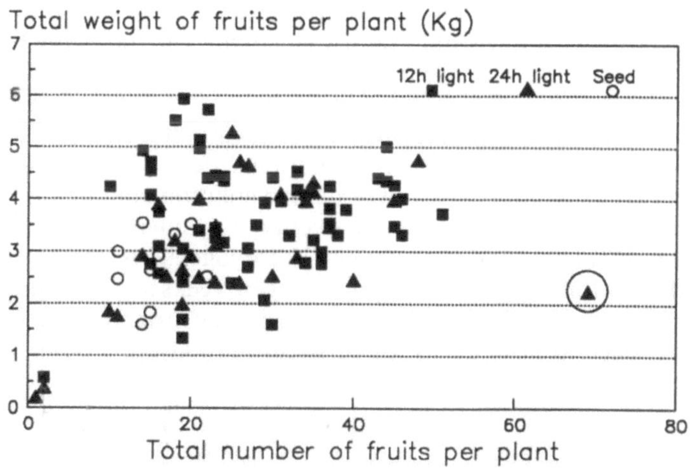

Fig. 6. Graph describing the variation in the total number and total weight of fruit amongst plants derived from somatic embryos that developed under a 12 h photoperiod (■) compared to those developed under continuous light (▲) or from (zygotic) seeds (o). (Ladyman and Girard 1990)

exhibited stable somaclonal variation, that plant may prove useful to the pickling industry.

Several experiments have been carried out to determine the effects of sodium chloride (NaCl) on tissue multiplication and somatic embryo regeneration. Galiba and Yamada (!988) reported the beneficial effects of NaCl on wheat somatic embryogenesis. The most notable result of our experiments was the observation that NaCl(10 mM) consistently enhanced the quantity of tissue produced during multiplication (22.8 g/100 mg primary tissue compared to 15 g/ 100 mg primary tissue) and significantly enhanced the number of embryos that regenerated and grew well in soil (46 seedlings/100 mg primary tissue compared to 7 seedlings/100 mg primary tissue). In all experiments with NaCl multiplication was carried out under continuous light and on agar (0.7%), thererfore, the next step must be to determine the effects of NaCl in conjuction with Gelrite and a 12–16-h photoperiod.

3 Transformation

Transformation of cucumber somatic embryos has been reported by Trulson et al. (1986), Chee (1990b), and Forsman et al. (1991). We found that the step that promotes multiplication of somatic embryos also magnifies the initial transformation events (Forsman et al. 1991). Overnight liquid cultures of *Agrobacterium tumafaciens* (LBA 4404::pBI121 or pZBin) were grown in ABS plus acetosyringone and 25 mg/l kanamycin. Regeneration of cucumber was accomplished following the protocol in Section 6 with modifications to accommodate manipulations with *Agrobacterium*. Prior to inoculation, the bacteria were centrifuged at 7000 × g and resuspended in fresh media plus acetosyringone but without antibiotics. A suspension of tobacco cells was plated onto embryo induction media and covered with filter paper to act as a nurse culture. Cotyledonary disks from 4-day-old seedlings were preincubated on the feeder layer (nurse) layer for 24 h. The explants were inoculated by immersion into bacterial suspensions ranging from 108 to 109 CFU/ml ($OD_{550\,nm}$ = 0.1–0.6). After 10 min the explants were blotted dry and returned to the nurse cultures. Cocultivation took place for 48 h at 26 °C under near-dark conditions. Following cocultivation, explants were rinsed in 20 mM phosphate buffered solution containing 150 mg/l cefotaxime, blotted dry, and returned to induction medium minus feeder layers but containing cefotaxime and vancomycin. Embryogenic tissue was excised and placed on the multiplication medium for about 10 weeks and regenerated according to the protocol described in Section 6. No transformed, regenerable embryos were obtained without the use of nurse cultures. Using this procedure, about 2.8% of embryos regenerated into transformed plants.

4 Summary and Conclusions

Somatic embryogenesis is now a straightforward process in *Cucumis sativus*. Agar (0.7%) was the most convenient and reliable gelling agent for embryo induction on cotyledonary explants. Transfer of cultured tissue during the multiplication phase should be carried out at least every 4 weeks and preferably at shorter intervals. Gelrite, although allowing less tissue to form during multiplication, did increase the yield of robust regenerates per unit weight of starting material. Multiplication is best carried out under a photoperiod of 12 to 16 h and sucrose was a good carbohydrate source. Maltose was deleterious to cucumber tissue multiplication. Further examination of the beneficial influence of sodium chloride on cucumber somatic embryogenesis would be worthwhile.

The ease with which initial embryos can be multiplied will facilitate the development of artificial seeds in this crop. The process also allows magnification of successful transformation events.

5 A Protocol for the Induction, Multiplication and Regeneration of Somatic Embryos

1. *Explant Source and Preparation*. Cucumber seeds (*Cucumis sativus* L.) are sterilized for 10 min in 10 % Clorox bleach, to which Tween 80 (3 drops/100 ml; Sigma Chemical Co., St. Louis, MO) is added. The seeds are rinsed three times in sterile water and left on moist germination paper for 24 h in the dark. The seeds are then planted on MS medium (Murashige and Skoog 1962) solidified with 1% agar in Magenta boxes (Magenta Corp., Chicago, IL) and left 3 days (16-h photoperiod, 26 °C day/ 18 °C night).

2. *Induction*. Explants made with a cork borer (0.5 cm i.d.) are taken from the cotyledons and placed on the induction medium. The induction medium constituents are similar to CTM2 (Trulson and Shahin 1986) except that the concentrations of plant growth regulators are as follows; 1.1 mg/l 2,4-D; 0.2796 mg/l; naphthalene acetic acid (NAA) and 0.226 mg/l 6-benzylaminopurine (BAP). Agar (0.7%; Baltimore Biological Laboratories, Cockeysville, MD) is the gelling agent. The explants are left in continuous light (approximately 36 $\mu E/m^2/s$) at 26 °C in an incubator.

3. *Multiplication*. Embryogenic tissue is sliced off explants that have been incubated for 14 days on induction medium, and placed on multiplication medium that is gelled with Gelrite (gellan gum; 0.15%; Kelco Division, San Diego, CA). Other constitutes of the multiplication medium are the same as those in the induction medium except that NAA is omitted. Embryogenic tissue (50–100 mg) is placed onto each Petri plate (diameter 9 cm) in 4 or 5 equal piles. The tissue is kept under cool-white fluorescent light (approximately 36 $\mu E/m^2/s$, 16-h photoperiod) at 26 °C. The tissue is transferred after every 2–4 weeks to fresh multiplication media. At the time of transfer best results are obtained if the cream-yellow, smooth embryogenic tissue is separated from hard, slightly pubescent callus tissue which is then discarded.

4. *Conditioning*. After 2 to 10 weeks on multiplication medium, the embryogenic tissue is suspended in liquid "conditioning" media in total darkness (28 °C) on a reciprocal shaker (66 cycles/min). This conditioning medium is similar to the multiplication medium except that it contains 0.2 μM abscisic acid and no gelling agent.

5. *Preparation Before Regeneration.* After 7 days in conditioning media the suspension is sieved sequentially through steel mesh of 860 and 520 μm. The filtrate is centrifuged (approx. 150 × g) for 10 min and the supernatant discarded. The embroys are resuspended in a volume of MS solution twice that of the bed volume and are plated on to MS medium solidified with 1% agar.

6. *Regeneration.* Embryos start to germinate approximately 2–3 weeks after transfer onto the hormone-free media. Seedlings that develop a healthy root and at least three leaves can be transferred to soil and, after slowly acclimating to greenhouse conditions (16-h photoperiod, 26 °C days/18 °C night), grow to be morphologically indistinguishable from conventionally derived plants.

Acknowledgements. The author thanks Dr. J.A. Sanders for reading the manuscript.

References

Batty N, Dunwell J (1989) Effect of maltose on the response of potato anthers in culture. Plant cell Tissue Organ Cult 18: 221–226
Chee PP (1990a) High frequency of somatic embryogenesis and recovery of fertile cucumber plants. HortScience 25: 792–793
Chee PP (1990b) Transformation of cucumber by *Agrobaterium tumafaciens* and the regeneration of transformed plants . Plant Cell Rep 9: 245–248
Chee PP, Tricoli DM (1988) Somatic embryogenesis and plant regeneration from cell suspension cultures of *Cucumis sativus* L. Plant Cell Rep 7: 274–277
Deakin JR, Bohn GW, Whitaker TW (1971) Interspecific hybridization in *Cucumis.* Econ Bot 25: 195–211
de Rougemont GM (1989) A field guide to the crops of Britain and Europe. Collins, London, pp 82–83
Forsman J, Balint R, Steele C, Ladyman JAR (1991) Agrobacterium-mediated transformation of cucumber [*Cucumis sativus* L. (cv. Clinton)] with the coat protein of ZYMV. 3rd Int Cong Plant molecular biology, Tucson, Arizona. Proc No. 889 (Abstr)
Galiba G, Yamada Y (1988) A novel method for increasing the frequency of somatic embryogenesis in wheat tissue culture by NaCl and KCI supplementation. Plant Cell Rep 7 (1): 55–58
Kim YH, Janick J (1989) Somatic embryogenesis and organogenesis in cucumber. HortScience 24 (4): 702
Ladyman JAR, Girard B (1990) Somaclonal variation in *Cucumis sativus.* L. HortScience 25 (9): 1063
Ladyman JAR, Girard B (1992a) Cucumber somatic embryo development on various gelling agents and carbohydrates sources. HortScience 27 (2): 164–165
Ladyman JAR, Girard B (1992b) Non-hormonal factors that improve the multiplication and development of somatic embryos of cucumber *Cucumis sativus.* Acta Hortic 300: 233–236
Lower RA, Nienhuis J, Miller CH (1982) Gene action and heterosis for yield and vegetative characteristic in a cross between a gynoecious pickling cucumber and *Cucumis sativus var. hardwickii.* J Am Soc Hortic Sci 107: 75–78
Malepszy S (1988) Cucumber (*Cucumis sativus* L.) In: Bajaj YPS (ed) Biotechnology in agriculture and forestry, vol. 6. Crops II Springer, Berlin Heidelberg New York, pp 277–292
Malepszy S, Niemirowiczz-Szczytt K, Wiszniewska J (1982) Cucumber (*Cucumis sativus* L.) somatic embryogenesis in vitro. Acta Biol 10 (505): 218–220
Murashige T, Skoog F (1962) A revised medium for rapid growth and bioassays with tobacco tissue cultures. Physiol Plant 15: 473–497
Owens HR, Wengerd D, Miller ER (1991) Culture medium pH is influenced by basal medium, carbohydrate source, gelling agent, activated charcoal and medium storage method. Plant Cell Rep 10: 583–586
Pasqualetto PL, Zimmerman RH, Fordham I (1988) The influence of cation and gelling agent concentration on vitrification of apple cultivars in vitro. Plant Cell Tissue Organ Cult 14: 31–40
Rajasekaran K, Mullins MG, Nair Y (1983) Flower formation in vitro by hypocotyl explants of cucumber (*Cucumis sativus* L.) Ann Bot 52: 417–420

Raquin C (1983) Utilization of different sugars as carbon sources for in vitro cultures of *Petunia*. Z Pflanzenphysiol 111: 453–457
Russo VM, Roberts BW, Schatzer RJ (1991) Feasibility of trellised cucumber production. HortScience 26 (9): 1156–1158
Strickland SG, Nichol JW, McCall CM, Stuart DA (1987) Effect of carbohydrate source on alfalfa somatic embryogenesis. Plant Sci 48: 113–121
Trulson AJ, Shahin EA (1986) In vitro plant regeneration in the genus *Cucumis*. Plant Sci 47: 35–43
Trulson AJ, Simpson RB, Shahin EA (1986) Transformation of cucumber (*Cucumis sativus* L.) plants with *Agrobacterium rhizogenes*. Theor Appl Genet 73 (1): 11–15
Winnik AG, Vetusnjak LF (1952) Intervarietal hybridization of cucumber. Agrobiologia 4: 125–131
Ziv M, Gadasi G (1986) Enhanced embryogenesis and plant regeneration from cucumber (*Cucumis sativus* L.) callus by activated charcoal in solid/liquid double-layer cultures. Plant Sci 47: 115–122

II.5 Somatic Embryogenesis and Organogenesis in Okra (*Abelmoschus esculentus* L. Moench.)

D.A. Cook and A. Brown III[1]

1 Introduction

Okra, *Abelmoschus esculentus* (L.) Moench. is an economically important member of the Malvaceae as a vegetable, and for fiber and oil. In Africa, it is harvested not only for the immature pods, but for the young leaves as well (Olasantan 1988). Okra is a warm weather crop of tropical origin (Kimbrough et al. 1977) that is not drought- or insect-tolerant (Eamus 1986; Olasantan 1986, 1988). Although okara can tolerate simulated insect damage (apical debudding) in the early stages of vegetative growth without significantly reducing crop fruit yield, older plants exhibit reduced fruit yields as a result of this treatment (Olasantan 1986).

In general, the members of the Malvaceae are rather recalcitrant to regeneration. Several cultivars of cotton are regenerable from various explant sources (Davidonis and Hamilton 1983; Price and Smith 1979; Rangan et al. 1984; Gawel et al. 1986; Trolinder and Goodin 1988a,b; Bajaj and Gill 1992) through somatic embryogenesis. Furthermore, *Hibiscus acetosella* (roselle) is also regenerable by somatic embryo production (Reynolds and Blackmon 1983). *Kosteletzkya virginica* (L.) Presl., a halophytic member of the Malvaceae, can be regenerated through organogenesis (Cook et al. 1989).

In okra the induction of callus from ovules (Bajaj 1965) and seeds (Tanda et al. 1980), and somatic embryogenesis has been reported in callused hypocotyl explants (Reynolds et al. 1981), but has only recently been regenerated through the organogenic pathway (Mangat and Roy 1986). Recently, Roy and Mangat (1989) characterized the explant and media conditions necessary for maintaining and regenerating okra via organogenesis.

A consistant regeneration system for okra is required for the exploitation of somaclonal variation for biotic and abiotic stress resistance, crop yield, and asexual propagation. Additionally, such a system is necessary for the possible improvement of okra through plant transformation techniques. Therefore, the conditions for somatic embryogenesis in okra explants are reported herein.

[1] Department of Biological Sciences, Clark Atlanta University, Atlanta, Georgia 30314, USA

Biotechnology in Agriculture and Forestry, Vol. 31
Somatic Embryogenesis and Synthetic Seed II (ed. by Y.P.S. Bajaj)

2. Somatic Embryogenesis

2.1 Explant Preparation

Seeds of okra [*Abelmoschus esculentus* (L.) Moench. cv. Clemson Spineless] were surface sterilized in 1% sodium hypochlorite for 20 min and rinsed three times with sterile distilled water. Cleaned seeds were placed on Murashige and Skoog minimal organics medium (MSMO) (Murashige and Skoog 1962) containing 2g/l Phytagel (Sigma Chemical Co., St. Louis, Missouri) without sucrose or hormones for germination. The pH of all the media was adjusted to 5.7 with 1 N KOH or 1 N HCl prior to the addition of the gelling agent. All media were dispensed in 25 × 150 mm tubes (25 ml) or 250-ml Erlenmeyer flasks (50 ml) and sterilized at 121 °C for 15 min at 15 psi.

2.2 Callus Induction

Hypocotyls, roots, shoot apices with small leaves, cotyledonary nodes, and cotyledons were aseptically excised from seedlings and placed on the media summarized in Table 1 for callus initiation. The concentration of all carbohydrates was 30 g/l. Cultured explants were placed in a growth chamber at 27 °C under a photoperiod of 16-h day/8-h night for callus initiation for up to 100 days.

Callus production occurred between 15 and 30 days in explants using any of the media in Table 1. In general, hypocotyl or cotyledonary explants, especially those cotyledons with attached shoot apices, were best for callus production and embryo development. Greater callus production also occurred in cotyledons oriented with the upper leaf surface down. Roots were the poorest explants for callus production.

Cytokinins, such as benzyladenine, are critical in the formation of callus in okra (Roy and Mangat 1989) and important in the production of embryogenic callus in cotton (Trolinder and Goodin 1988a,b). Therefore, there is a precedent for the use of kinetin as a cytokinin in the regeneration protocol. The carbohydrate source was not especially critical for callus production, but glucose is known to be a more effective carbon source for cotton (Gawel et al. 1986; Shoemaker et al. 1986) and seashore mallow (Cook et al. 1989).

Table 1. Summary of media used for callus initiation in okra

Salt base	Carbohydrate (g/l)	Auxin (mg/l)	Cytokinin (mg/l)
1. MSMO	Glucose, 30	IAA, 2	Kinetin, 1
2. MSMO	Sucrose, 30	IAA, 2	Kinetin, 1
3. MSMO	Sucrose, 30	2,4-D, 1	Kinetin, 1

2.3 Somatic Embryogenesis and Organogenesis

Okra callus was excised aseptically and transferred to a liquid medium consisting of MSMO, 30 g/l sucrose and 1 mg/l 2,4-D for suspension culture development. Callused explants were also transferred to this medium. Liquid cultures were maintained in the laboratory at ambient temperature on a gyratory shaker at 100 rpm.

Somatic embryos were observed to be attached to what remained of the callus, or callused explant, and floating free in the liquid medium within 30 days of transfer of callus material to this liquid medium containing 1 mg/l 2,4-D. Somatic embryos appeared to originate from hard, cream-colored to grayish callus that covered the explant. The grayish color of the embryogenic callus in okra is consistant with that of other members of the Malvaceae (Reynolds and Blackmon 1983; Trolinder and Goodin 1988a,b) and with that of organogenic callus as well (Cook et al. 1989; Roy and Mangat 1989). Some friable callus was also observed floating free in the liquid medium. As a rule, somatic embryos were produced at a low inoculum density in liquid medium.

2.4 2,4-D Experiments

Okra callus and callused explants from the media in Table 1 were transferred to liquid medium containing various levels of 2,4-D (0, 0.01, 0.1, 1 mg/l) for somatic embryo production in order to determine the most effective concentration. Somatic embryos were found floating free in the medium after 7 days in all the treatments. Generally, more embryos were produced in the 0, 0.01 and 0.1 mg/l 2,4-D treatments than in the 1 mg/l concentration (Table 2).

Attempts were made to induce plantlet regeneration by transferring the embryoids to MSMO without hormones or to a half-strength MSMO medium. This was followed by a transfer to a medium containing 0.1 mg/l kinetin. Somatic embryos and embryogenic callus readily developed roots on auxin-free media. In older cultures anthocyanin production also occurred. Since shoots were not produced, it is likely that the embryos are in an arrested state of development because of the relatively high auxin concentrations.

Table 2. The effects of 2,4-D concentration on somatic embryogenesis in okra in 50 ml of growth medium

2,4-D (mg/l)	Number of embryos
0	+++++
0.01	++++++
0.1	++++++
1.0	+++

+ indicates approximately 50 embryos by visual estimation.

The placement and timing of the 2,4-D treatment can be critical in the somatic embryogenesis process. In a second experiment, okra explants were placed on solid media containing either 0 or 1 mg/l 2,4-D concentration for 2, 3, 7, 10, and 14 days for callus or somatic embryo induction prior to being transferred to liquid hormone-free medium or to a medium containing 0.1 mg/l kinetin for 30 days. Organized, cream-colored to gray embryogenic callus was observed on the explants after 7 or 10 days of 2,4-D treatment after transfer to the hormone-free or kinetin medium within a 30-day period. Few somatic embryos were produced under this scheme. Time periods of less than 7 days of 2,4-D treatment were ineffective in the induction of embryogenic callus upon transfer to hormone-free or kinetin medium. Explants gave rise to numerous adventitious roots and very little callus the longer they remained in the presence of 2,4-D.

In a third experiment, explants were left on the solid 2,4-D medium (0, 0.01, 0.1, and 1 mg/l) for 7 and 21 days respectively, before transfer to the hormone-free medium for 30 days. The callused explants were then transferred to media containing 0, 0.1, and 1 mg/l kinetin. The 0.1 and 1 mg/l 2,4-D treatments in this experiment induced more callus production in the presence or absence of kinetin than the 0.01 mg/l concentration. The 2,4-D-free treatment was ineffective in callus production. Embryogenic callus was also observed on the explants after 7 days of 2,4-D treatment after transfer to 2,4-D-free or kinetin medium within a 30-day period. Also, few embryos were observed in either the hormone-free medium or the kinetin-containing media.

2.5 Microscopy

Suspected somatic embryos were fixed in 3% glutaraldehyde in 0.1 M pipes buffer for 1 h at room temperature and rinsed in the same buffer for 30 min. Samples were postfixed in 2% OsO_4 for 1 h and rinsed again with buffer prior to dehydration in an ethonol series (30, 50, 70, 95, and 100% for 10 min). They were then critical-point dried and coated with a mixture of gold and palladium for SEM examination. Specimens for light microscope examination were fixed in 10% formalin or Carnoy's fixative overnight. Fixed tissues were dehydrated in an ethanol series, embedded in Tissue Prep (Fisher Chemical, Pittsburgh, PA), and sectioned at 10 m. Sections were stained with either safranin-fast green or periodic acid-Schiff's reagent (PAS).

Scanning electron microscopy of suspected somatic embryos taken from the liquid 2,4-D concentration study revealed distinct globular, early and late heart-shaped embryos (Fig. 1A,B,C). No torpedo-shaped embryos were observed at this time, although abnormal embryos and polyembryos were present. Specimens analyzed by light microscopy showed globular and heart-shaped structures (data not shown). Little lignin formation occurred in safranin-fast green stained sections. Sections stained with PAS showed carbohydrate deposition. All light microscope specimens revealed either tissue organization, distinct epidermal layers, or embryo development from the callus mass.

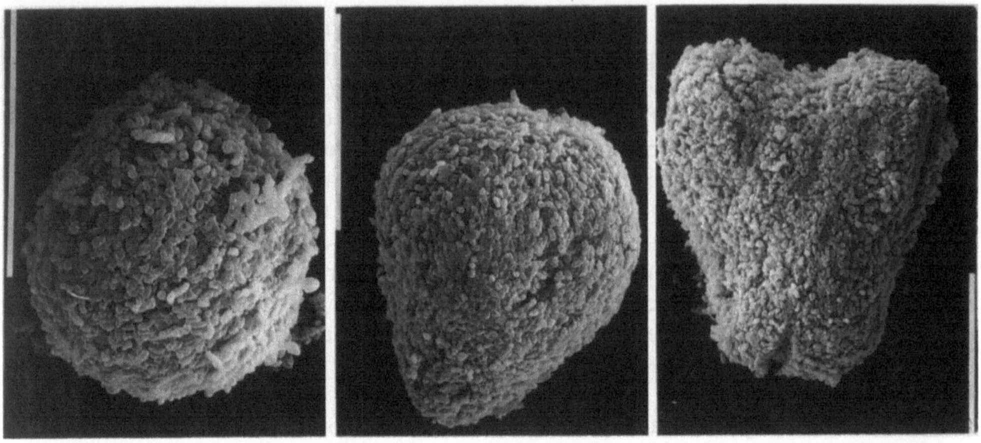
A B C

Fig. 1A–C. SEM analysis of okra somatic embryos in liquid culture. **A** A globular somatic embryo. **B** An early heart-shaped somatic embryo from liquid culture. **C** A late heart-shaped somatic embryo also from liquid culture. *Bars* = 1 mm

3 Summary and Conclusions

Cotyledons were the most effective explants for callus production, roots produced the least. Callus and callused explants produced somatic embryos upon transfer to liquid medium containing 1 mg/l 2,4-D. In additional time course and concentration experiments embryogenic callus was induced by a 1 mg/l 2,4-D treatment of explants for 7 to 14 days; however, large numbers of embryos were not produced in liquid culture from either time course study upon transfer to hormone-free or kinetin-containing media. SEM and light microscope analysis confirmed the presence of somatic embryos in liquid culture. Somatic embryos and associated embryogenic callus produced roots upon transfer to hormone-free, reduced salt, or 0.1 mg/l kinetin-containing media.

4 Protocol

Use explants from aseptically germinated seeds to reduce contamination of cultures. Place cotyledon on solid Murashige and Skoog minimal organics medium with 30 g/l sucrose and 1 mg/l 2,4-D for 7–14 days. Transfer explants to hormone-free medium. Remove explants from hormone-free medium after 30 days or after callus production to 0.1 or 1 mg/l kinetin medium for further development. For greater embryo production, place explants on any of the 3 callus induction media (Table 1) and transfer to liquid media with various 2,4-D concentrations after callus production has occurred. Liquid versions of these media may be equally effective for the somatic embryogenesis process in okra.

Acknowledgements. This research was supported by NSF grants: #DCB-9010198 and #RII-89-278, and a Clark Atlanta University Faculty Development Grant from the Pew Charitable Trust to D.A.

Cook. The author thanks the Center for Advanced Ultrastructural Research of the University of Georgia for SEM analysis, Lawrence Brako for excellent technical assistance, and Jung H. Choi for critical review.

References

Bajaj YPS (1965) Development of ovules of *Abelmoschus esculentus* L. var. Pusa sawani in vitro. Proc Natl Inst Sci (India) 30: 175–185

Bajaj YPS, Gill MS (1992) Micropropagation of cotton. In: Bajaj YPS (ed) Biotechnology in agriculture and forestry, vol 19. High-tech and micropropagation III. Springer, Berlin Heidelberg New York, pp 483–504

Cook DA, Decker DM, Gallagher JL (1989) Regeneration of *Kosteletzkya virginica* (L.) Presl. from callus cultures. Plant Cell Tissue Organ Cult 17: 111–119

Davidonis GH, Hamilton RH (1983) Plant regeneration from callus tissue of *Gossypium hirsutum* L. Plant. Sci Lett 32: 89–93

Eamus D (1986) The responses of leaf water potential and leaf diffusive resistance to abscisic acid, water stress and low temperature in *Hibiscus esculentus*: the effect of water stress and ABA pretreatments. J Exp Bot 37: 1851–1862

Finer JJ, Smith RH (1984) Initiation of callus and somatic embryos from explants of mature cotton (*Gossypium klotzschianum* Andress). Plant Cell Rep 3: 41–43

Gawel NJ, Rao AP, Robacker CD (1986) Somatic embryogenesis from leaf and petiole callus cultures of *Gossypium hirsutum* L. Plant Cell Rep 5: 457–459

Kimbrough WD, Jones LG, Fontenot JF (1977) Okra is produced primarily in the south as a main dish vegetable, and for gumbos. In: Hayes J (ed) Yearbook of agriculture. US Department of Agriculture, Washington DC, pp 224–227

Mangat BS, Roy MK (1986) Tissue culture and plant regeneration of okra (*Abelmoschus esculentus*). Plant Sci 47: 57–61

Murashige T, Skoog F (1962) A revised medium for rapid growth and bioassays with tobacco tissue cultures. Physiol Plant 15: 473–497

Olasantan FO (1986) Effect of apical debudding on growth and yield of okra (*Abelmoschus esculentus*). Exp Agric 23: 307–312

Olasantan FO (1988) Effect of leaf removal on the growth of okra (*Abelmoschus esculentus*) and its relevance to leaf harvesting patterns and pest damage. Exp Agric 24: 449–455

Price HJ, Smith RH (1979) Somatic embryogenesis in suspension cultures of *Gossypium blolzischianum* Andress. Planta 145: 305–307

Rangan TS, Zavala T, Ip A (1984) Somatic embryogenesis in tissue cultures of *Gossypium hirsutum* L. In Vitro 20: 256

Reynolds BD, Blackmon WJ (1983) Embryogenesis and plantlet regeneration from callus of *Hibiscus acetosella*. J Am Soc Hortic Sci 108: 307–310

Reynolds BD, Blackmon WJ, Postek CE (1981) Induction of somatic embryos and proembryos in three mallows: roselle, hollyhock and okra. Environ Exp Bot 21: 438

Roy MK, Mangat BS (1989) Regeneration of plants from callus tissue of okra (*Abelmoschus esculentus*). Plant Sci 60: 77–81

Shoemaker RC, Couche LJ, Galbraith DW (1986) Characterization of somatic embrogenesis and plant regeneration in cotton (*Gossypium hirsutum* L.). Plant Cell Rep 3: 178–181

Tanda AS, Atwal AS, Bajaj YPS (1980) Reproduction and maintenance of root-knot nematodes (*Meloidogyne incognita*) in excised roots and callus cultures of Okra (*Abelmoschus esculentus*). Indian J Exp Biol 18: 1340–1342

Trolinder NL, Goodin JR (1988a) Somatic embryogenesis in cotton (*Gossypium*). I. Effects of source of explant and hormone regime. Plant Cell Tissue Organ Cult 12: 31–42

Trolinder NL, Goodin JR (1988b) Somatic embryogenesis in cotton (*Gossypium*). II. Requirements for embryo development and plant regeneration. Plant Cell Tissue Organ Cult 12: 43–54

II.6 Somatic Embryogenesis and Artificial Seed in Carrot (*Daucus carota* L.)

H. Wake[1], H. Umetsu[2], and T. Matsunaga[3]

1 General Account

Carrot (*Daucus carota* L.) belongs to the Umbelliferae which is composed of over 350 genera and 3225 species, including many medicinal plants. This family is distributed in Eurasia and Africa, with many horticultural varieties, the roots of which are widely used as food throughout the world. The phenomenon of somatic embryogenesis has been extensively investigated using carrot and other members of the family Umbelliferae.

Carrot somatic embryogenesis is used as a model system for early plant development (Wake et al. 1991, 1992a). Physiological and biochemical studies of the process of somatic embryogenesis from cells or cell clusters were performed using this plant system (Sung et al. 1984). A major advantage of this plant system is the relatively simple procedure with which large amounts of somatic embryos can be produced under controlled conditions in liquid culture. Any part of the plant such as the root, peduncle, leaf, stem, or zygotic embryo can produce embryogenic callus.

1.1 Brief Review of Work Done by Others

Since Reinert (1958) and Stewad et al. (1958) independently reported in vitro indirect somatic embryogenesis in carrot, a number of workers have extensively investigated methods to increase carrot embryo formation and to obtain high quality embryos with a high germination frequency potential. There have been several reports describing somatic embryogenesis in carrot (Table 1). It is well known that carrot embryogenesis occurs readily in suspension cultures when embryogenic cells are allowed to grow in a medium containing 2,4-dichlorophenoxyacetic acid (2,4-D) and then transferred to an auxin-free medium (Halperin 1966). The effect of growth regulators such as auxins, cytokinins, gibberellin, and abscisic acid (ABA) on embryo formation has been investigated in detail (Fujimura and Komamine 1975; Kamada and Harada 1979a). Reduced nitrogen sources such as NH_4Cl, amino acids (glutamine, alanine, glutamic acid)

[1] Biotechnology Section, Central Research Laboratory, Pentel Co. Ltd., Soka, Saitama 340, Japan
[2] Department of Bioscience and Biotechnology, Aomori Univeristy, Kobata, Aomori 030, Japan
[3] Department of Biotechnology, Tokyo University of Agriculture and Technology, Koganei, Tokyo 184, Japan

Biotechnology in Agriculture and Forestry, Vol. 31
Somatic Embryogenesis and Synthetic Seed II (ed. by Y.P.S. Bajaj)

Table 1. Published conditions for embryogenesis of carrot

Cultivar	Explants	Basal medium	Method	Remarks	Reference
Kurodagosun	Petiole	Lin and Staba (1961)	Hormone condition	Zeatin 10^{-7} M	Fujimura and Komamine (1975)
US-Harumakigosum	Hypocotyl	Salt of MS	Hormone condition	Addition of 2,4-D during subculture	Kamada and Harada (1979a)
Nantes	Petiole	MS	ABA tretament	Normalized embryo development	Ammirato (1983)
Wild carrot	–	Wetherell (1969)	Reduced nitrogen source	NH_4Cl, glutamine, alanine, glutamic acid	Wetherell and Dougall (1976)
US-Harumakigosun	Hypocotyl	Salt of MS	Reduced nitrogen source	NH_4NO_3, NH_4Cl, alanine, asparagine, aspartic acid	Kamada and Harada (1979b)
S. Valery	Hypocotyl	B5 (Gamborg et al. 1968)	Amino acids	Proline, serine	Nuti-Ronchi et al. (1984)
–	Stem	B5	Conditioned medium	Culture filtrate	Hari (1980)
'Flakkese' SG 766 Trophy	–	B5	Extracellular glyco-protein	Essential for embryogenesis	De Vries et al. (1988)
–	Seedling	B5	Folate analog	Aminopterin 10 to 100 μM in the presence of 5 mg/l 2,4-D	Smith et al. (1987)
Kurodagosun	Hopocotyl	MS	Cyanobacterial extract	Embryos from nonembryogenic cell line	Wake et al. (1991)
Kurodagosun	Hypocotyl	MS	Cyanobacterial extract	Enhanced plantlet formation	Wake et al. (1992)
Kurodagosun	Hypocotyl	Lin and Staba (1961)	Synchronization	Sieving, density gradient. Embryogenesis frequency: >90%	Fujimura and Komamine (1979)
Kurodagosun	Hypocotyl	Lin and Staba (1961)	Embryogenesis from single cell	Sieving, density gradient. 2, 4-D 5×10^{-8} M, zeatin 10^{-6} M Embryogenesis frequency: >85%	Nomura and Komamine (1985)
S. Valery	–	B5	Separation of different stages of embryo	Sieving and differential sedimentation	Giuliano et al. (1983)
US-Harumakigosum	Apical meristem	MS	Stress-induced embryogenesis	0.7 M sucrose, 0.25-1 M cadmium	Kamada et al. (1989)
Scarlet Nantes	Seed	MS	Hormone-free	Mericarp tissues	Smith and Krikorian (1988)
Royal Chantenay	Cultured cell	B5	Protoplast culture	Cell division and embryogenesis	Grambow et al. (1972)
Kurodagosun	Somatic embryo	Lin and Staba (1961)	Protoplast culture	Cell division and embryogenesis	Nomura et al. (1983)

and urea are required as a supplement to nitrate for efficient embryogenesis (Wetherell and Dougall 1976; Kamada and Harada 1979b). Serine and proline, which are the two major constituents of cell wall glycoprotein, extended the time and number of mitotic divisions and stimulated the number of embryos regenerated (Nuti-Ronchi et al. 1984). De Vries et al. (1988) demonstrated that several auxin-controlled extracellular glycoproteins are essential for carrot somatic embryogenesis. Somatic embryos were obtained from a carrot cell line that had no capability of somatic embryogenesis by low and high molecular weight factors in marine cyanobacterial extracts (Wake et al. 1991). High concentrations of aminopterin, a folate analog, trigger somatic embryogenesis in the presence of auxin (Smith et al. 1987). Several studies with carrot cell cultures have shown that inhibition of polyamine biosynthesis causes complete inhibition of somatic embryogenesis (Fienberg et al. 1984; Robie and Minocha 1989). DL-α-difluoromethylornithine, which promoted polyamine biosynthesis, also promoted the development of somatic embryos. Ethylene is a strong inhibitor of carrot somatic embryogenesis. This effect is similar to that of polyamine biosynthesis inhibitors. Roustan et al. (1989) reported stimulation of *Daucus carota* somatic embryogenesis by inhibitors of ethylene synthesis. Furthermore, Khan and Minocha (1991) reported that competition between the polyamine and ethylene biosynthetic pathways for S-adenosylmethionine could be important for the regulation of somatic embryogenesis in carrot. Fujimura and Komamine (1979) reported synchronization of somatic embryogenesis for physiological and biochemical investigation. Cell clusters were prepared by sieving and density gradient centrifugation. High frequency somatic embryogenesis from single cells isolated from cell suspension cultures was performed to clarify a correlation between the potency for embryogenesis and cell morphology (Nomura and Komamine 1985). Methods reported for mass preparation of somatic embryos include synchronization and sieving (Giuliano et al. 1983); abscisic acid treatment (Ammirato 1983); and direct embryogenesis by physiological strees (osmosis, heavy metal) without hormone treatment (Kamada et al. 1989). Carrot "seeds", after the germination of their zygotic embryos, contain cells which can develop into somatic embryos in aseptic culture in a reliable and predictable manner without the use of exogenously added growth regulators (Smith and Krikorian 1988). Carrot cells cultured in vitro as well as protoplasts isolated enzymatically from these cells form embryos and grow into mature plants. Grambow et al. (1972) described conditions for cell regeneration of protoplasts from carrot cell cultures and the production of plants by embryogenesis. Nomura et al. (1983) reported some culture conditions and morphological observations on the cultured protoplasts obtained from carrot somatic embryos. If these protoplasts have a high potential to redifferentiate into embryos and/or to proliferate rapidly, as seen in original embryos, they will be very useful in investigating the nature of embryogenesis.

Methods for using these embryos as a propagation system similar to seeds have recently been developed (Kitto and Janick 1985; Redenbaugh et al. 1991). Artificial seeds encapsulating somatic embryos may have practical applications in agriculture. However, despite extensive trials with various concentrations of plant growth regulators, including carbohydrates and mineral salts, it remains difficult to develop somatic embryos into plantlets at a high frequency with most plant species (Stuart and Redenbaugh 1987). The frequency of artificial seeds which grow into normal seedlings was low when somatic embryos were used without further improvement (Stuart et al. 1985). Kamada et al. (1989) reported stress-induced somatic embryogenesis in carrot and applied this method to synthetic seed production. Recently, we reported the effects of a cyanobacterial extract, and a nondialysate prepared from the extract, on germination of artificial seeds encapsulating carrot somatic embryos and discussed the implications of these results on the improvement of artificial seed systems (Wake et al. 1992b).

1.2 Plant Growth Regulators in Marine Cyanobacteria

Microalgae (green algae and cyanobacteria) require light energy, carbon dioxide, and minerals for their growth, and are now given much attention due to their high potential for producing various biologically active compounds which include antibiotics, algicides, toxins, pharmaceuticals, and plant growth regulators (Metting and Pyne 1986). Recently, we demonstrated the existence of low and high molecular weight factors, which strongly stimulated somatic embryogenesis in embryogenic and nonembryogenic cell lines of *Daucus carota* L. in marine cyanobacterial extracts (Wake et al. 1991). We also reported the effect of nondialysate prepared from an extract of marine cyanobacterium, *Synechococcus* sp. NKBG 042902 on chlorophyll accumulation in and plantlet development from somatic embryos of carrot (Wake et al. 1992a). Marine cyanobacteria are the oldest oxygenic photosynthetic organisms known and are considered to be the evolutionary ancestors of chloroplasts in higher plant cells. Some species of cyanobacteria show cell differentiation into akinetes, hormogonia, and heterocysts, which are essential for the propagation and the survival of the species in their natural habitat. Thus, each cyanobacterial cell has all the functions necessary for propagation and survival. We, therefore, considered whether factors which promote plant development, i.e., embryogenesis and maturation, might exist in marine cyanobacteria as chemical messages for growth, regulation of photosynthesis, and/or cell differentiation.

2 Somatic Embryogenesis

2.1 Plant Regeneration Via Somatic Embryogenesis

Carrot suspension cells are capable of developing into plantlets when they are transferred from MS medium containing 2,4-D to that without 2,4-D. Twenty-five strains of marine cyanobacteria were screened for their ability to promote carrot somatic embryogenesis. Extracts from 21 of these strains of marine cyanobacteria promoted carrot plantlet formation and the relative increases in plantlet numbers, caused by the addition of extracts from *Synechococcus* sp.

Table 2. Effect of extracts from marine cyanobacteria on plantlets formation in carrot suspension cells. (Wake et al. 1991)

Marine cyanobacteria[a]		Plantlet numbers[b]	Relative increase
Synechococcus sp.	N	182 ± 30	5.2
NKBG 042902	L	184 ± 35	5.3
	H	240 ± 35	6.9
Xenococcus sp.	N	153 ± 35	4.4
NKBG 0109	L	83 ± 20	2.3
	H	161 ± 20	4.6
Nostoc sp.	N	132 ± 40	3.8
NKBG 0019	L	148 ± 35	4.7
	H	79 ± 20	2.6
Anabaena sp.	N	72 ± 20	2.1
NKBG 060701	N	201 ± 35	5.7
(–N)[c]	L	194 ± 41	5.5
	H	211 ± 34	6.0
(Another 17 strains)	N	50 to 100	1.4 to 2.8 [d]
Control		35 ± 15	1.0

Cultured with extract (100 mg/l) in hormone-free MS medium for 28 days.
[a] N, Crude extract; H, nondialysate; L, dialysate.
[b] ± indicates standard error in three experiments (n=3)
[c] (–N), cultured in BG-11 medium without nitrogen sources.
[d] The extracts from another 17 strains tested increased plantlet formation about 1.4- to 2.8-fold.

NKBG 042902, *Xenococcus* sp. NKBG 0109, *Nostoc* sp. NKBG 0019, and *Anabaena* sp. NKBG 060701 (-N) (Table 2). Another 17 extracts from marine cyanobacteria slightly promoted plantlet formation in comparison to the control (no addition). Dialysates and nondialysates obtained from extracts of these four strains were tested for their ability to promote plant regeneration (Table 2). Both the dialysates and nondialysates were found to possess the ability to promote plant regeneration in all cases.

In general, a reduced nitrogen source is required for rapid growth and for in vitro somatic embryogenesis of cultured carrot tissue (Wetherell and Dougall 1976; Kamada and Harada 1979b). The effects of amino acids such as α-alanine and glutamine, casein hydrolysate, and yeast extract on plantlet formation were also tested. These were added to a final concentration of 100 mg/l, but were found to promote somatic embryogenesis only slightly (α-alanine, 1.2-fold increase; glutamine, 1.4-fold increase; casein hydrolysate, 1.5-fold increase; yeast extract, 1.2-fold increase) when compared with the marine cyanobacterial extracts also at 100 mg/l.

2.2 Germination of Somatic Embryos

Upon removal of 2,4-D, carrot cells revert to organized division to form embryos of globular, heart, and torpedo stages. Torpedo somatic embryos are mature somatic embryos that developed into plantlets. Somatic embryogenesis may be

Table 3. Effects of various concentration of nondialysate from marine cyanobacteria on plantlet formation in carrot somatic embryos. (Wake et al. 1992a)

Concentration (mg/l)	Number of embryos examined	Plantlets	
		Number	Frequency (%)
1	251	35 ± 6	14
10	311	36 ± 10	12
20	315	49 ± 12	16
50	232	48 ± 11	21
100	350	141 ± 19	40
200	329	192 ± 30	58
0 (Control)	386	34 ± 5	9

Carrot somatic embryos (425–800 μm) were cultured for 6 days; the mean value of triplicate ± SD is shown

divided into three stages: induction, maturation and germination. During induction, carrot cells differentiate and develop into somatic embryos which are globular to heart-shaped. A heart-shaped somatic embryo changes metabolically and morphologically to develop into a plantlet (maturation stage). During the germination stage, a somatic embryo forms a shoot and a root.

The development of somatic embryos into plantlets was demonstrated. Carrot somatic embryos were divided into two categories (>800 and 425–800 μm). The category with the larger embryos (>800 μm) consisted almost completely of torpedo somatic embryos. In the second category, about 60% of the somatic embryos were at the torpedo stage of development, while the remainder (about 40%) were at the heart stage. The frequency of plantlet formation of the large somatic embryos was higher than that of small somatic embryos in all cases.

The effect of a nondialyzing fraction (nondialysate) from an extract of marine cyanobacterium, *Synechococcus* sp. NKBG 042902, on plantlet formation of small somatic embryos (425–800 μm) was also examined. Table 3 shows plantlet numbers and their formation frequency after addition of a nondialysate to the culture medium at various concentrations. The nondialysate promoted plantlet formation of the somatic embryos at concentrations of over 100 mg/l. The nondialysate was employed at a concentration of 100 mg/l for further experiments.

Plantlet formation was stimulated by the addition of a nondialysate (Fig. 1), and 41% of the small somatic embryos (425–800 μm) developed into plantlets after 6 days. This compares to 9% without the addition of the nondialysate. The frequency of plantlet formation of the somatic embryos treated with the nondialysate for 10 days was 61%, whereas that of somatic embryos without the nondialysate was 28%. Initially, 63% of the small somatic embryos (425–800 μm) were at the torpedo stage of development, while the remainder (27%) were at the heart stage. In this case, 97% of torpedo somatic embryos (mature somatic embryos) developed into plantlets. On the other hand, this result also indicated that maturation of somatic embryos was promoted by nondialysate treatment.

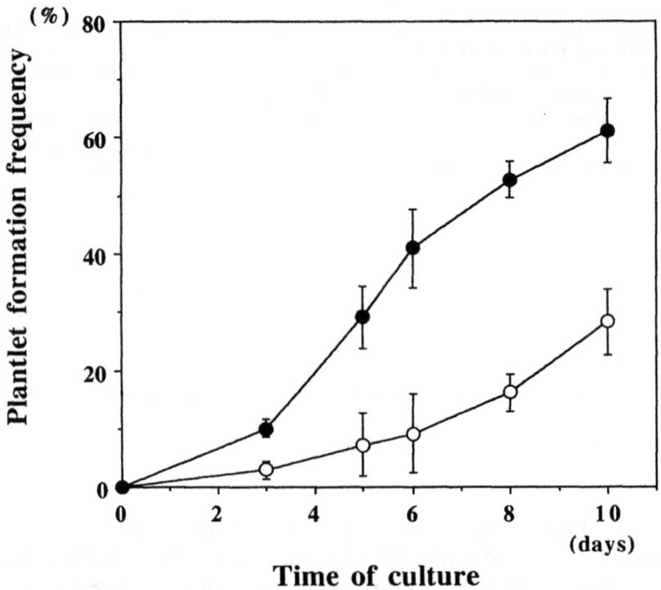

Fig. 1. Time course of plantlet formation from somatic embryos with (100 mg/l ●) or without nondialysate (○). *Bars* indicate SD. (Wake et al. 1992a)

Rapid promotion of greening of buds by germination-promoting factor(s) in the nondialysate was observed. Accumulation of the chlorophyll (Chl) during plantlet formation of a somatic embryo (425–800 μm) was determined (Fig. 2). After 6 days of culture, a somatic embryo treated with nondialysate contained 107 ng Chl compared to 21 ng/embryo without addition.

The Chl a/b ratio during plantlet formation in somatic embryo cultures with or without nondialysate was measured (Fig. 3). In addition, the Chl a/b ratio of seedlings of zygotic embryos was measured 4 and 6 days after germination. The Chl a/b ratio of the seedlings of zygotic embryos was 2.6 ± 0.09 after 4 days of culture and 2.9 ± 0.05 after 6 days of culture. After 4 days of treatment with the nondialysate, these ratios were in the range 2.6 to 3.1. However, the Chl a/b ratio of plantlets induced from somatic embryos without addition was variable, i.e., in the range of 2.1 to 3.5 during the culture period. These results indicate that enhancement of Chl accumulation in somatic embryos is induced by treatment with nondialysate and correlates with plantlet formation.

In general, enhancement of greening has previously been induced by treatment with cytokinins such as kinetin and zeatin. We investigated the effect of various concentrations of zeatin on Chl accumulation and plantlet formation. Optimal concentrations of Chl accumulation in plantlets were over 10^{-7} zeatin (Fig. 4). Zeatin did not promote plantlet formation. Greening occurred only in the buds of plantlets treated with the nondialysate. However, the distribution of cells containing Chl was variable in plantlets treated with zeatin.

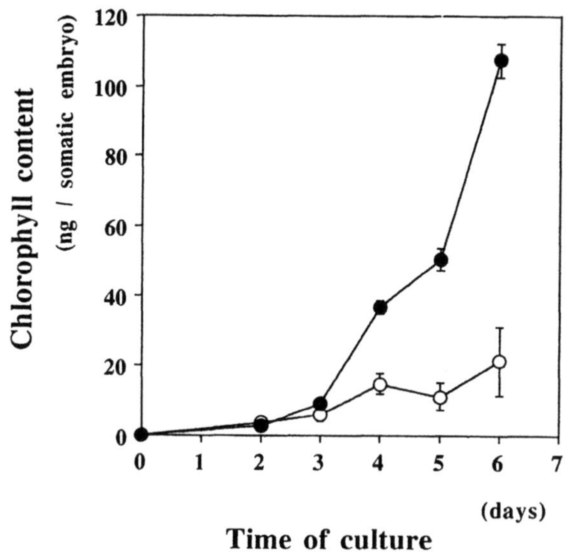

Fig. 2. Accumulation of chlorophyll during plantlet formation of a somatic embryo. ●, Nondialysate (100 mg/l) was added; ○, control (no addition of extract). *Bars* indicate SD. (Wake et al. 1992a)

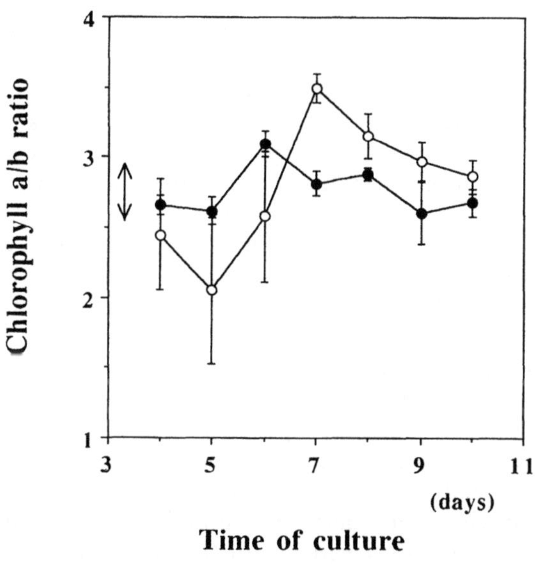

Fig. 3. Change in chlorophyll a/b ratio during plantlet formation. ●, Nondialysate (100 mg/l) was added; ○, control (no addition of extract). Chl a/b ratio of carrot seedlings from zygotic embryos after 4 to 6 days of germination was in the range of 2.6 to 2.9 as shown by the *arrow*. *Bars* indicate SD. (Wake et al. 1992a)

It appears that germination-promoting factor(s) in the extract may have a cytokinin-like activity as well as organogenesis-stimualting activity. Active substances may stimulate conversion from heterotrophic to autotrophic growth. Purification of the active substance is currently in progress.

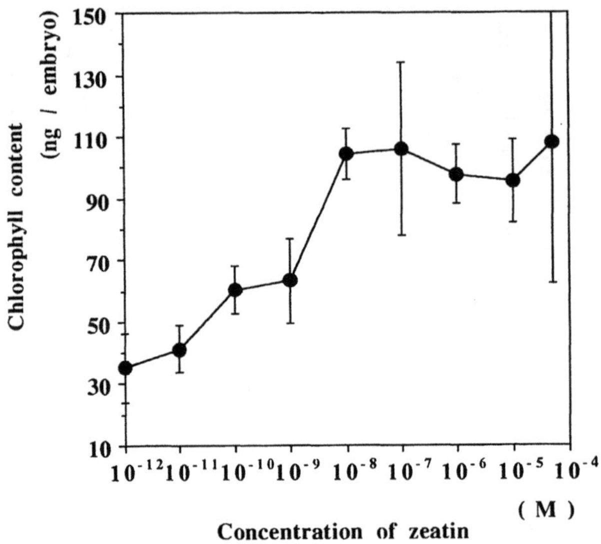

Fig. 4. Accumulation of chlorophyll during plantlet formation of a somatic embryo treated with various concentrations of zeatin. *Bars* Indicate SD

3 Artificial Seed

Recently, we reported the effects of a cyanobacterial extract and a nondialysate prepared from the extract on the germination of artificial seeds encapsulating carrot somatic embryos and discussed the implications of these results for the improvement of artificial seed systems (Wake et al. 1992b).

For high frequency germination, the effects of various concentrations of the extract on germination of artificial seeds encapsulating large somatic embryos (>800 μm) were examined. Germination frequency of the artificial seeds containing the extract at a concentration of 400 mg/l was 91%, whereas that of artificial seeds without the nondialysate was 35%. Kamada et al. (1989) previously reported a germination frequency of 51% for artificial seeds encapsulating the stress-induced somatic embryos of the same size range as our studies, whereas the germination frequency of the control (2,4-D treatment only) was 14%.

The effects of various concentrations of a nondialysate fraction and a dialysate fraction, separated from the extract, on the germination of artificial seeds encapsulating large somatic embryos were examined (Table 4). Both the nondialysate and dialysate were found to possess germination-promoting activity in all cases. The nondialysate was more effective than the dialysate. Germination frequency of the artificial seeds containing the nondialysate at a concentration of 100 mg/l was 94%, whereas that of artificial seeds without the nondialysate was 35%. Even after the addition of the extract at 10 mg/l, a germination frequency of 58% was obtained.

We also achieved high frequency germination of artificial seeds encapsulating small somatic embryos (425–800 μm). The germination frequency of these seeds containing 100 mg/l of nondialysate was 60% and that of artificial seeds without the nondialysate was 9%. When using stress-induced somatic embryos in

Table 4. Effect of various concentration of nondialysate from *Synechococcus* sp. NKBG 042902 extract on germination in artificial seeds. (Wake et al. 1992b)

Concentration (mg/l)	Number of germinated seeds per total artificial seeds	Average frequency (%)
10	42 / 65	
	32 / 58	
	31 / 57	58 ± 6
50	53 / 63	
	45 / 59	
	37 / 52	77 ± 7
100	53 / 65	
	61 / 61	
	57 / 57	94 ± 11
200	55 / 68	
	62 / 62	
	59 / 59	94 ± 11
Control		
0	21 / 55	
	18 / 58	
	16 / 52	
	24 / 63	35 ± 4

Carrot somatic embryos (> 800 μm) were encapsulated in alginate beads; artificial seeds were cultured for 14 days; ± indicates SD

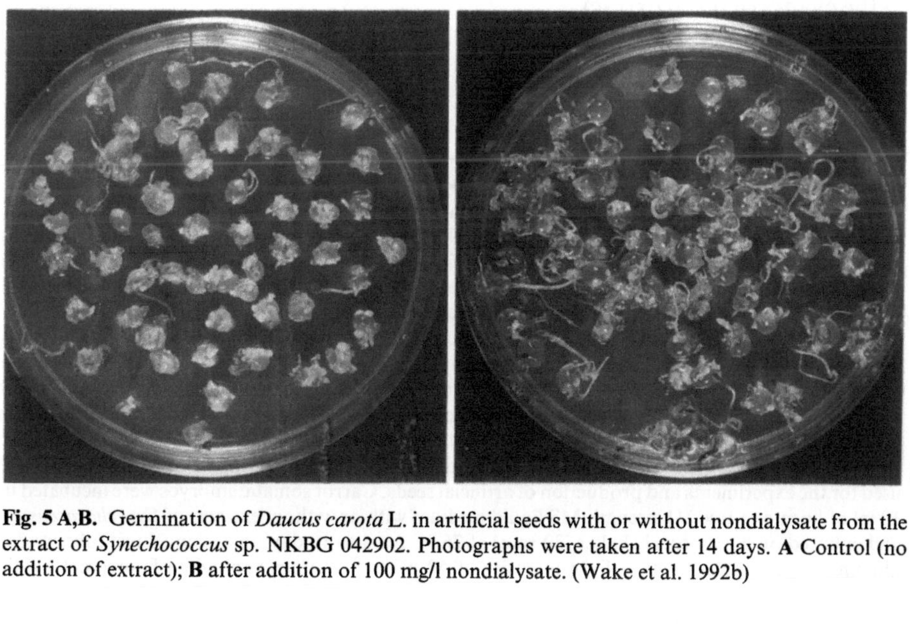

A B

Fig. 5 A,B. Germination of *Daucus carota* L. in artificial seeds with or without nondialysate from the extract of *Synechococcus* sp. NKBG 042902. Photographs were taken after 14 days. **A** Control (no addition of extract); **B** after addition of 100 mg/l nondialysate. (Wake et al. 1992b)

the same size range (0.5–1 mm) as our studies, Kamada et al. (1989) obtained a germination frequency of 47%. Furthermore, the somatic embryos induced by stress formed normal seedlings, but the embryos induced by 2,4-D did not (Kamada et al. 1989). However, even though the small somatic embryos (425-800 μm) showed a lower germination frequency, the plantlet development process was far more vigorous than that of artificial seed encapsulating large somatic embryos (>800 μm) in a manner analogous to the true zygotic seed (Fig. 5).

4 Summary

Somatic embryogenesis and artificial seed in carrot have been described. In addition, the stimulation of somatic embryogenesis using extracts prepared from marine cyanobacteria is also described.

The development of somatic embryos into plantlets was promoted using a nondialysate prepared from marine cyanobacterium, *Syenechococcus* sp. NKBG 042902. Promotion of plantlet formation, shoot greening, and a stable Chl a/b ratio to this extent have not been previously obtained using alternative methods. It is suggested that the physiological investigation of the late phase of somatic embryogenesis may be carried out using the marine cyanobacterial extract. Purification of active substance(s) is currently in progress.

The promoting effects of the cyanobacterial extract, and the nondialysate prepared from the extract, on germination of artificial seeds encapsulating carrot somatic embryos were also observed. The germination-promoting factor(s) in the nondialysate was a high molecular weight compound(s). It is suggested that improvement of artificial seed systems is possible by using the active substance(s) in the cyanobacterial extract.

5 Protocols

Culture of Carrot Cells. Cell suspensions were initiated from cells derived from hypocotyls of domestic carrot, *Daucus carota* L. cv. Kurodagosun. The cells were subcultured every 7 days for 4 to 6 months in MS medium (Murashige and Skoog 1962) containing 0.1 mg/l of 2,4-D. A 7-day-old cell suspension was passed through an 82-μm pore nylon screen, then through a 32-μm pore screen. A 1.5 ml packed cell volume (PCV) of cells > 32 μm in size was inoculated into 100 ml of fresh medium containing 0.1 mg/l 2,4-D. Subcultures were grown at 25 °C in 500-ml Erlenmeyer flasks (100 ml MS liquid medium) on a reciprocal shaker (80 strokes/min and 15 cm amplitude) in the dark. Cells (0.1 ml PCV) were then inoculated into 100 ml hormone-free MS medium and cultured on a reciprocal shaker at 26 °C for 14 days in the dark. The density of the cell suspension was 2×10^4 cells/ml. A 14-day-old culture was passed through an 800-μm pore stainless mesh, then through a 425-μm pore screen. The somatic embryos retained on the 800-μm (>800 μm in size) and on the 425-μm screen (425–800 μm in size) were used for the experiments and production of artificial seeds. Carrot somatic embryos were incubated in 10 ml of hormone-free 1/4 strength MS liquid medium with or without extracts of *Synechococcus* sp. and cultured on a reciprocal shaker (30 rpm) at 26 °C for 4 to 10 days in L-shaped test tubes under illumination (16 h/day) with a light intensity of 50 μE m^{-2} s^{-1} from cool-white fluorescent lights.

Nondialysate from the extracts of *Synechococcus* sp. was added to 10 ml hormone-free 1/4 strength MS liquid medium to a final concentration in the range 1 to 200 mg/l. Media used for the experiments were adjusted to pH 5.7 before autoclaving at 121 °C for 20 min.

Artificial Seeds. Artificial seeds (~ 6 mm Ø) were formed by dropping sodium alginate into a calcium chloride solution. Sodium alginate was dissolved at a concentration of 3% (w/v) in 1/4 strength MS medium with or without the nondialysate. The carrot somatic embryos were mixed with 3% sodium alginate, dropped individually into 75 mM solution of calcium chloride, and kept for 20 to 30 min to produce solid calcium alginate beads. Calcium alginate beads were washed with sterilized water three times. Artificial seeds were cultured on 1/4 strength MS solid medium at 25 °C for 14 days in 16 h light (50 $\mu E\ m^{-2} s^{-1}$) with cool-white fluorescent lights under sterile conditions.

References

Ammirato PV (1983) The regulation of somatic embryo development in plant cell cultures: suspension culture techniques and hormone requirements. Bio/Technology 1: 68–74

De Vries SC, Booij H, Meyerink P, Huiman G, Wilde HD, Thomas TL, van Kammen A (1988) Acquisition of embryogenic potential in carrot cell-suspension cultures. Planta 176: 196–204

Fienberg AA, Choi JH, Lubich LP, Sung ZR (1984) Developmental regulation of polymine metabolism in growth and differentiation of carrot culture. Plant a 162: 532–539

Fujimura T, Komamine A (1975) Effects of various growth regulators on the embryogenesis in a carrot cell suspension culture. Plant Sci Lett 5: 359–364

Fujimura T, Komamine A (1979) Synchronization of somatic embryogenesis in a carrot cell suspension culture. Plant Physiol 64: 162–164

Gamborg OL, Miller RA, Ojima K (1968). Nutrient requirements of suspension cultures of soybean root cells. Exp Cell Res 50: 151–158

Giuliano G, Rosellini D, Terzi M (1983) A new method for purification of the different stages of carrot embryoids. Plant Cell Rep 2: 216–218

Grambow HJ, Kao KN, Miller RA, Gamborg OL (1972) Cell division and plant development from protoplasts of carrot cell suspension cultures. Planta 103: 348–355

Halperin W (1966) Alternative morphogenetic events in cell suspension. Am J Bot 53: 443–453

Hari V (1980) Effects of a cell density change and conditioned media on carrot cell embryogenesis. Z Pflanzenphysiol 96: 227–231

Kamada H, Harada H (1979a) Studies on the organogenesis in carrot tissue cultures. I. Effects of growth regulators on somatic embryogenesis and root formation. Z Pflanzenphysiol 91: 255–266

Kamada H, Harada H (1979b) Studies on the organogenesis in carrot tissue cultures. II. Effects of amino acids and inorganic nitrogenous compounds on somatic embryogenesis. Z Pflanzenphysiol 91: 453–463

Kamada H, Kobayashi K, Kiyosue T, Harada H (1989) Stress induced somatic embryogenesis in carrot and its application to synthetic seed production. In Vitro Cell Dev Biol 25: 1163–1166

Khan AJ, Minocha SC (1991) Polyamines and somatic embryogenesis in carrot. II. The effect of cyclohexylammonium phosphate. J Plant Physiol 137: 446–452

Kitto SL, Janick J (1985) Hardening treatments increase survival of synthetically-coated asexual embryos of carrot. J Am Soc Hortic Sci 110: 285–286

Lin M, Staba J (1961) Peppermint and spearmint tissue cultures. I. Callus formation and submerged culture. Lloydia 24: 139–145

Metting B, Pyne J (1986) Biologically active compounds from microalgae. Enzyme Microb Technol 8: 386–394

Murashige T, Skoog F (1962) A revised medium for rapid growth and bioassays with tobacco tissue cultures. Physiol Plant 15: 473–497

Nomura K, Komamine A (1985) Identification and isolation of single cells that produce somatic embryos at high frequency in a carrot suspension culture. Plant Physiol 79: 988–991

Nomura K, Fukuei K, Nitta T (1983) Culture of protoplasts isolated from somatic embryos of carrot. Plant Sci Lett 29: 1–7

Nuti-Ronchi V, Caligo MA, Nozzolini M, Luccarini G (1984) Stimulation of carrot embryogenesis by proline and serine. Plant Cell Rep 3: 210–214

Redenbaugh K, Fujii J, Slade D, Viss P, Kossler M (1991) Artificial seeds – encapsulated somatic embryos. In: Bajaj YPS (ed) Biotechnology in agriculture and forestry vol 17. High-tech and micropropagation I. Springer, Berlin Heidelberg New York, pp 395–416

Reinert J (1958) Untersuchungen über die Morphogenese an Gewebekulturen. Ber Dtsch Bot Ges 71: 15

Robie CA, Minocha Sc (1989) Polyamines and somatic embryogenesis in carrot. I. The effects of difluoromethylornithine and difluoromethylarginine. Plant Sci 65: 45–54

Roustan JP, Latche A, Fallot J (1989) Stimulation of *Daucus carota* somatic embryogenesis by inhibitors of ethylene synthesis: cobalt and nickel. Plant Cell Rep 8: 182–185

Smith DL, Krikorian AD (1988) Production of somatic embryos from carrot tissues in hormone-free medium. Plant Sci 58: 103–110

Smith RH, Kerns HA, Anthony JL, Wild JR (1987) Methotrexate and aminopterin effects on growth and regenearation in *Daucus carota*. Plant Cell Rep 6: 60–62

Steward FC, Mapes MO, Mears K (1958) Growth and organized development of cultured cells. II. Organization in cultures grown from freely suspended cells. Am J Bot 45: 705–708

Stuart DA, Redenbaugh K (1987) Use of somatic embryogenesis for the regeneration of plants. In: LeBaron HM, Mumma RO, Honeycutt RC, Duesing JH (eds) Biotechnology in agricultural chemistry. Am Chem Soc, Washington, DC. Am Chem Soc Symp Ser 334: 87–96

Stuart DA, Nelsen J, Strickland SG (1985) Factors affecting developmental processes in alfalfa cell cultures. In: Henke RR, Hughes KW, Costantin MP (eds) Tissue culture in forestry and agriculture. Plenum Press, New York, pp 59–73

Sung ZR, Feinberg A, Chorneau R, Borkird C, Turner I, Smith J, Terzi M, LoSchiavo F, Giuliano G, Pitto L, Nuti-Ronchi V (1984) Developmental biology of embryogenesis from carrot cultures. Plant Mol Biol Rep 2: 3–14

Wake H, Umetsu H, Ozeki Y, Shimomura K, Matsunaga T (1991) Extracts of marine cyanobacteria stimulated somatic embryogenesis of *Daucus carota* L. Plant Cell Rep 9: 655–658

Wake H, Akasaka A, Umetsu H, Ozeki Y, Shimomura K, Matsunaga T (1992a) Promotion of plantlet formation from somatic embryos of carrot treated with a high molecular weight extract from a marine cyanobacterium. Plant Cell Rep 11: 62–65

Wake H, Akasaka A, Umetsu H, Ozeki Y, Shimomura K, Matsunaga T (1992b) Enhanced germination of artificial seeds by marine cyanobacterial extract. Appl Microbiol Biotechnol 36: 684–688

Wetherell DF (1969) Phytochrome in cultured wild carrot tissue. I. Synthesis. Plant Physiol 44: 1734–1737

Wetherell DF, Dougall DK (1976) Sources of nitrogen supporting growth and embryogenesis in cultured wild carrot tissue. Physiol Plant 37: 97–103

II.7 Somatic Embryogenesis in Bananas and Plantains (*Musa* Clones and Species)

A.D. Krikorian and M.E. Scott[1]

1 Introduction

1.1 Distribution, Importance, and Morphology

The genus *Musa* (bananas and plantains, family Musaceae) comprises about 25 species. The main center of diversity of these tree-like herbaceous monocotyledons is southeast Asia, but edible fruit and ornamental varieties have long been grown widely in the tropics and subtropics of both hemispheres.

The morphology and growth habit of these giant herbs are highly distinctive and characteristic. While in the vegetative condition, the shoot apical growing point is more or less at soil level. Leaves are produced on the flanks of the meristem with very little vertical displacement of the growing tip, and leaves wrap around themselves forming a false trunk or pseudostem (Barker and Steward 1962a). The lateral (leaf-opposed) buds in *Musa* are adventitious (cf. Barker and Steward 1962a; Fisher 1978; Devos 1985). When the vegetative growing point undergoes transition to the flowering mode, the true stem tip moves upwards vertically through the hollow of the pseudostem (Barker and Steward 1962b; Mohan Ram et al. 1962).

The group as a whole has been inadequately studied taxonomically. The best method available for dealing with the cultivated clones involves the use of a short-hand technique that takes into account the amount of the contribution of "A" genome (from *M. acuminata*) and "B" genome (from *M. balbisiana*) and some 15 characteristics (Simmonds and Shepherd 1955). A generalization is that the more of the A genome, the sweeter the fruit and, hence, it is usable as a table fruit; the more B, the more starchy it is, hence, it needs cooking. The better known clones are generally parthenocarpic triploids (3n = 33), although diploids and tetraploids exist as well and have local importance.

An important taxonomic descriptor involves the persistence/deciduous nature of the male bracts, buds, and axis at the terminal end of the raceme (cf. Simmonds 1962; Tezenas du Montcel et al. 1983; IBPGR 1984). This phenotypic change represents a significant development change and has major implications for fruit production (Krikorian et al. 1993).

[1] Department of Biochemistry and Cell Biology, State University of New York at Stony Brook, Stony Brook, New York 11794-5215, USA

Biotechnology in Agriculture and Forestry, Vol. 31
Somatic Embryogenesis and Synthetic Seed II (ed. by Y.P.S. Bajaj)

1.2 Significance of Somatic Embryogenesis and Tissue Culture-Related Strategies

Bananas and plantains are excellent candidates for all tissue culture strategies (Krikorian and Cronauer 1984b; Krikorian 1990). The corms or suckers are large and inconvenient for germplasm movement, frequently harbor pathogens, and are often in short supply – especially for large-scale plantings. Germplasm maintenance in the field is expensive and space-consuming, and on a commercial scale germplasm production is usually relegated to "seed" nurseries. Breeding of the essentially seed-sterile clones of interest to plantation agriculture is fraught with many difficulties (Stover and Buddenhagen 1986; Novak 1992; Rowe and Rosales 1995).

A number of diseases seriously affect banana and plantain plantings throughout the world. The amount of time and money that must be invested in control measures using chemicals are beyond subsistence farmers, usually not within reach of ordinary people, and pose financial problems even for large commercial operations. Breeding strategies to generate tolerance or resistance to important diseases such as *Fusarium* wilt (Panama Disease) caused by *F. oxysporum* f. sp. *cubense* (various races), Sigatoka leaf spot caused by *Mycosphaerella musicola*, Black Sigatoka caused by *Mycosphaerella fijiensis* var. *difformis* and to the burrowing nematode *Radopholus similis* and the corn weevil *Cosmopolites sordidus* have thus far been the main breeding objectives, but these have yet to lead to commercially successful clones. The improvement programs are rendered all the more difficult because the potential to breed plantains and bananas is seriously handicapped by the exceedingly high sterility of the clones of primary interest (cf. Simmonds 1962; Stover and Buddenhagen 1986; Stover and Simmonds 1987; Ploetz 1990; Vuylsteke and Swennen 1992 and refs. therein).

The reliable application of tissue, cell, somatic embryogenesis, and protoplast procedures to plantains and bananas, as in the case of many other plants, is a necessary and key component in any comprehensive biotechnology plan aimed at improvement via mutation breeding and genetic transformation. It is also important for germplasm management strategies such as cryopreservation or storage and artificial seed or propagule delivery systems (Krikorian and Cronauer 1984b; Krikorian 1987; Murfett and Clarke 1987; Novak 1992).

1.3 Shoot Tip Procedures as a Preliminary to Suspension/Embryogenesis Procedures

Figure 1 provides a diagrammatic representation of some of the various tissue culture procedures for *Musa* and their interrelationships: (1) a mature plant with a ratoon or sucker. Excision of the apex from a vegetative corm (2 through 3), or alternatively a stem tip from the male floral bud (2a through 2c), yields a system where leaf-opposed and adventitious buds are generated (4 through 8). Subdivision of these multiple shoots and transfer to cytokinin-containing culture medium (semisolid or shallow liquid) keeps the system open-ended (9). Further treatment of aseptic shoot tips in vitro with auxins can foster the development of

Fig. 1. Overview of aseptic culture procedures that yield multiplying shoot tip cultures (1–9 and 17), and the use of these cultures to generate organogenically (9–15) and embryogenically competent (9–14a, b and 17) suspensions; see text for details

callus, calloid, and embryogenic masses (10–13) (cf. Krikorian 1987; Novak et al. 1989; Krikorian et al. 1990). These variously sized and variously differentiated units may be further broken down into still smaller dimensions. These can be regenerated into somatic embryos (14a) or organogenically responsive masses (15). Each of these in turn can be reared to full-sized propagules that can be established in soil (14b, 16–17). Protoplast work can follow from any of these points, but competent cell suspensions are preferable for a number of reasons (Krikorian et al. 1988; Panis et al. 1993; Megia et al. 1993).

Various publications summarize methods to make primary explants and to initiate multiplying shoot cultures (cf. Krikorian and Cronauer 1984a; Cronauer and Krikorian 1984, 1985a,b; Cronauer-Mitra and Krikorian 1988a,b; Vuylsteke 1989; Vuylsteke et al. 1990; Israeli et al. 1991; Rowe and Rosales 1995). Even so, there is considerable opportunity for changes to be implemented in these procedures, and refinements are an on-going activity (Krikorian et al. 1993). Technical guidelines for safe movement of germplasm have been delineated (cf. Stover 1977) and most are now aware of the problems of assuming that meristems or in vitro cultures are necessarily "clean" (Frison and Putter 1989). But there are many possibilities of introducing virus diseases if adequate precautions are not taken to see that plants are virus-indexed or quarantined.

Tests for cucumber mosaic virus should be done (cf. e.g. Stein et al. 1979) as well as for bacteria and fungi. Other precautions will depend on the location of the operation. For instance, bunchy top virus is not eliminated through meristem culture (cf. Drew et al. 1989).

The means whereby a shoot tip multiplication system is developed and maintained has fundamental implications for the successful establishment of morphogenetically competent suspension cultures. The presence of multiple foci of meristematic growth in the basal areas of the multiplying shoot system is crucial to the relatively easy production of callus. By multiple foci we mean numerous meristematic areas (cf. Fig. 1, No. 10, and Fig. 2A, B). Unless cultures are vigorous, these multiple areas are not generated. Moreover, their formation seems to be clone-related (Krikorian and Scott unpubl.). Another significant point is that *Musa* readily serves as a repository for cryptic microbial contaminants and these can make it virtually impossible to obtain clean callus and suspensions (Krikorian 1989, 1990).

MS liquid medium (Murashige and Skoog 1962) with 5 mg/l BAP, 100 mg/1 inositol, and 4% sucrose provides an adequate starting point for initiating multiple shoot systems from excised stem tip or floral bud explants. Response in liquid is much faster than on semisolid medium. Rotation on a gyratory shaker is helpful but not critical. Pulsing the explant from liquid to semisolid and back to liquid may speed up growth (Cronauer and Krikorian 1984).

Once the system is well established, multiple shoots may be separated at 3-week intervals and placed in fresh cytokinin-containing medium. Many adventitious buds are produced in vitro from cultured and subdivided shoot tips (Banerjee et al. 1986; Cronauer and Krikorian 1988d), and the potential for multiplication is very high (cf. Cronauer and Krikorian 1984; Krikorian 1990).

1.4 Callus Induction

Callus culture and cell suspension cultures from preclimacteric fruit tissue were carried out in the 1950s and 1960s (cf. Mohan Ram and Steward 1964). Cultures were relatively slow-growing and no morphogenesis was encountered. Since then, the procedures for *Musa* have been steadily improving but callus induction from somatic tissue is still not necessarily fast, prolific, or routine from many clones. Clean, in vitro propagated shoots or plants provide a good source of primary explant material for the initiation of callus and suspension cultures.

Auxins like dicamba and picloram can be used, but the more conventional ones like 2,4,5-T and 2,4-D are active as well. Semisolid medium provides a somewhat better substrate in the initial stages of callus induction for research purposes, since one can follow the explant more easily. The report that Gelrite is nominally pivotal to successful callus culture of banana may be explained on the basis of failure to use adequately washed agar (Huang and Chi 1988). Even so, Gelrite has value in *Musa* tissue culture. It is very helpful, for instance, in detecting subtle contaminants. There is much darkening in explants due to polyphenol oxidation (Palmer 1963) and Gelrite can help minimize this. Figure 1, part 10, is the most important part of the procedure of obtaining vigorous callus.

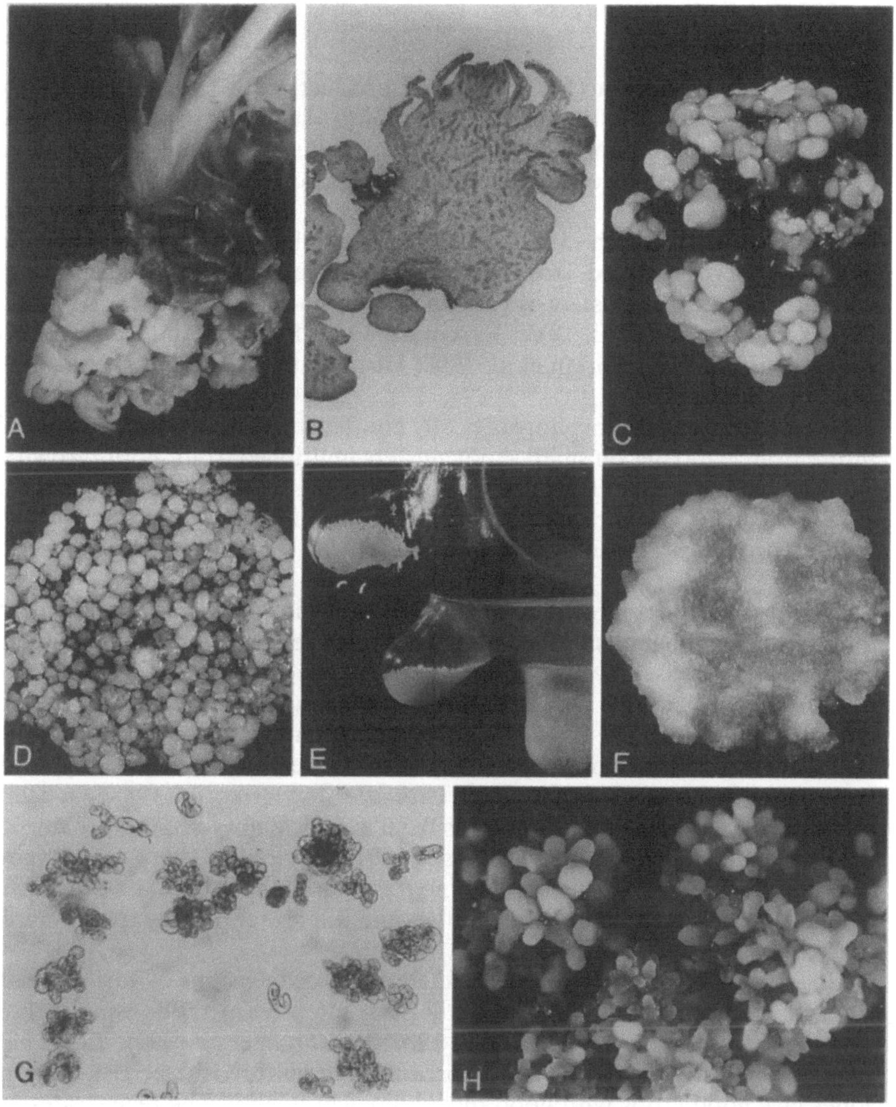

Fig. 2A–H. Photographs of some stages in the process of generating suspensions and embryogenic cultures; see text for details. **A** Nodular growths formed at the base of a micropropagated shoot system of an ABB clone, Cardaba, 0.5x; **B** section of the same, 10x. **C** AAB clone Lady's Finger showing globular masses (lighter color interspersed between tissues which have darkened severely) produced from nodular growths similar to those in **A** and **B**, 2x. **D** View of globular masses separated from units like those shown in **C**, 3x. **E** Partial view of a nipple flask (containing 220 ml of liquid nutrient medium) showing suspended cells, 0. 33x. **F** Low pH-maintained culture of embryogenic masses on a semisolid medium. Note the uniformity and general fineness of the units comprising the culture, 4x; **G** magnification of the embryogenic cells shown in **F** after dispersion in liquid medium, 71x. **H** Somatic embryos generated on semisolid medium, 5x; see text for details

2 Somatic Embryogenesis

Cronauer and Krikorian (1983) reported production of somatic embryos of *Musa* from suspension cultures of triploid clones of cooking bananas. Although their shoot tips were poorly developed, the work showed that very small cell clusters were morphogenetically competent (cf. Cronauer 1986 for histological details). Neomorphs, i.e., structures that were developmentally aberrant, embryonal structures with varying levels of morphological fidelity, and bonafide somatic embryos derived from suspensions have, since then, been readily generated from shoot clusters of various dessert, cooking banana, and plantain clones (see Banerjee et al. 1987; Krikorian 1987; Escalant and Teisson 1988; Novak et al. 1989; Krikorian et al. 1990; Dhed'a et al. 1991; Panis et al. 1992; Escalant et al. 1994).

We have initiated morphogenetically competent globules directly in liquid culture from leaf bases on small shoot clusters multiplied from shoots grown on semisolid medium. 2,4,5-T was used initially in our studies and has remained an auxin that can be used with success. We have also confirmed reports by Novak et al. (1989) that dicamba and picloram work as well (but picloram perhaps a bit better).

The presence of a large number of active cells in the area of the basal and primary thickening meristem makes it an attractive zone from which to secure vigorous cells. Excised zygotic embryos are responsive as well, and yield callus cultures and suspensions which, in turn, yield somatic embryos which develop normally; albeit they are arguably not the choice materials since many of the edible clones of *Musa* do not normally produce seeds and, hence, any model based on zygotic embryos is flawed (cf. Cronauer and Krikorian 1988c for details on the production of somatic embryos from zygotic embryos of an ornamental diploid banana; also Escalant and Teisson 1987, 1988 for work on excised zygotic embryos of a diploid *M. acuminata*). Somatic embryos have been generated from triploid clones using both in vitro plantlets and full-size corms as explant sources (cf. Novak et al. 1989).

Plantlets are routinely regenerated from cell suspensions of various edible *Musa* clones using fairly indirect and still labor-intensive methods using explants from shoot tips since 1985 (Krikorian 1987; Krikorian et al. 1990). Yields have been relatively erratic (cf. Fig. 2 for representative, cultured materials), but some clones give good crops with good morphological fidelity. The regenerates are, however, usually not of the same quality as plantlets obtained from other systems we routinely work on such as daylily, or even equal to the elegant somatic embryos produced from zygotic embryos of diploid *Musa.*

Plantlets have been regenerated by other laboratories as well from nodular, calloid growths occurring at the base of proliferating shoots. These nodular growths are somewhat reminiscent of orchid protocorms and are said to be capable of yielding cultures which can generate somatic embryos (Sannasgala 1989; Dhed'a et al. 1991).

Cronauer and Krikorian (1988b) showed that adventitious growths that grossly resemble somatic embryos can be mistaken for true somatic embryos. It is highly advisable, therefore, to verify the pathway of nominally somatic embryo development rigorously. One problem is that the stem line tissue that yields the cells has been difficult to characterize. The precursors of the somatic embryos obtained by Sannasgala (1989) and Dhed'a et al. (1991) are very much like the globular structures described by Cronauer (1986), but these are perhaps better viewed as aberrant, albeit morphogenetically competent, organogenic globules.

Novak et al. (1989) reported somatic embryogenesis from suspensions generated from callus initiated from the basal areas of shoot tips and leaf bases of bananas and plantains. Critical details as to the origin of the somatic embryos are not provided. Substantial gaps exist in the published papers, particularly in those steps between initiation of a suspension and generation of a plantlet/propagule. For us then, the matter of routine production of somatic embryos in the edible (i.e., nonseeded), triploid clones requires more work and the process is far from resolved for the majority of commercially important clones. There are many discrepancies and shortcomings in the published information and one should not feel comfortable with the presumption that *Musa* somatic embryogenesis has been reduced to a routine or simple process despite suggestions to the contrary (Panis et al. 1991).

2.1 Procedures

Shoot tip cultures exposed in darkness at 28 °C or thereabouts on a shaker rotating at 70 rpm to MS medium with 4% sucrose and 3 mg/l 2,4,5-T and 2.5 mg/l BAP start off morphogenetically competent cell cultures. Such suspensions obtained using 2,4,5-T are, however, better maintained with 2,4-D, and are then followed by NAA to push it to the next level of competence, but growth can be and frequently is, especially in the beginning, slow and considerable selection against organized shoots is necessary at each subculture. The initial growth becomes tighter and more compact and, eventually, compact globular centers of growth predominate. These detach from the parent mass and lead to suspended cell clusters. The cell products of this type of growth can be maintained with 1 mg/l 2,4-D and 1 mg/l BAP using very finely sieved suspensions with less or greater difficulty over the long term, but they are subject to deterioration over consecutive months of subculture. By transferring any of the material from any of the stages just described, and transferring to MS hormone-free medium, plantlets are recovered. The unit sizes required prior to organized growth are frequently not small enough to give a highly discrete response. Slight or substantial variations in the responses are obtainable depending on how precisely cultures are initiated and handled (cf. e.g. Jarret et al. 1985; Banerjee et al. 1987). Auxins such as picloram and dicamba in place of 2,4-D can also be used to initiate and maintain competent cultures (cf. Novak et al. 1989).

We have also used an alternative strategy to achieve a culture which is virtually completely compromised of somatic proembryos. Such cultures retain their ability to proliferate for extended periods of time. This is based on the fact that somatic embryos can bud-off at low pH (pH 4) additional somatic embryos – at the preglobular stage. These multiple embryos, in turn, can release additional embryos. As long as the more recently generated and developed somatic embryos can be separated and subcultured, one can have a sustainable and subculturable culture of somatic embryos. The level or stage of development at which such cultures are maintainable can vary considerably. In some instances, one can have small proembryos or proembryonic globules; in other cases, the somatic embryos can be well advanced and clearly visible to the naked eye and at a stage wherein the next level would be maturation "germination", and further growth or sprouting. And there is, of course, the possibility of achieving and sustaining a culture with all levels of development in between. The ultimate state of a culture condition maintenance strategy or plan for *Musa* depends largely, at this point in the time of our knowledge, on what one can get.

2.2 Strategy to Obtain Somatic Embryos of *Musa*

The strategy we follow involves initiating, maintaining, and manipulating cell suspensions which are competent and capable of undergoing progressive, organized development either by adventitious routes from small callus masses, or via somatic embryogenesis involving pre-proembryonic (i.e., embryogenic cell cluster) and/or proembryonic (i.e., globular stage somatic embryo) suspensions from which continued somatic embryo growth and development can derive.

Figure 2A,B shows the nodular meristematic masses (the source of so-called scalps?) (we prefer not to call them protocorm-like bodies, PLBs, because their morphology is not identical and their presence seems to depend on clone and culture methodology; cf. Matsumoto and Yamaguchi 1989) produced at shoot cluster bases. When subjected to 2,4,5-T in liquid medium (Fig. 2C), these yield compact globular masses. The compact globular masses separate and can grow as units of varying sizes, ranging from discrete units to clusters (Fig. 2D). These units can organize and yield a nodular growth mode from which more units and, hence, plantlets can be proliferated. Competent cells (Fig. 2E) can be derived from the compact globular masses since they release cells more or less readily, and either adventitious or embryogenic pathways may be achieved. In the former case, shoots are generated adventitiously which must be rooted to yield plantlets; in the case of somatic embryogenesis, a process analogous to germination encountered in zygotic embryos occurs (Fig. 3). The means, in varying degrees of details to carry this out, may be found in Krikorian and Cronauer (1984a), Cronauer and Krikorian (1985a,b, 1988d); Cronauer (1986); and Krikorian (1987) and references therein. We emphasize here, however, that it is the strategy that is crucial and there are sufficient differences among clones for us to be confident that the particulars will vary. In some instances in the lierature, it is clear that more than one route of regeneration is being activated or realised – both embryogenic and organogenic.

The work of Bakry and Rossignol (1985) on "callogenesis" and production of "neoformations" (calloids? – i.e., proliferations that "look" callus-like but have an epidermis and hence are not true callus) of various types substantiates the view that many developmental events can go on in a limited space. Adventitious shoot formation from the cultured floral axis of various so-called

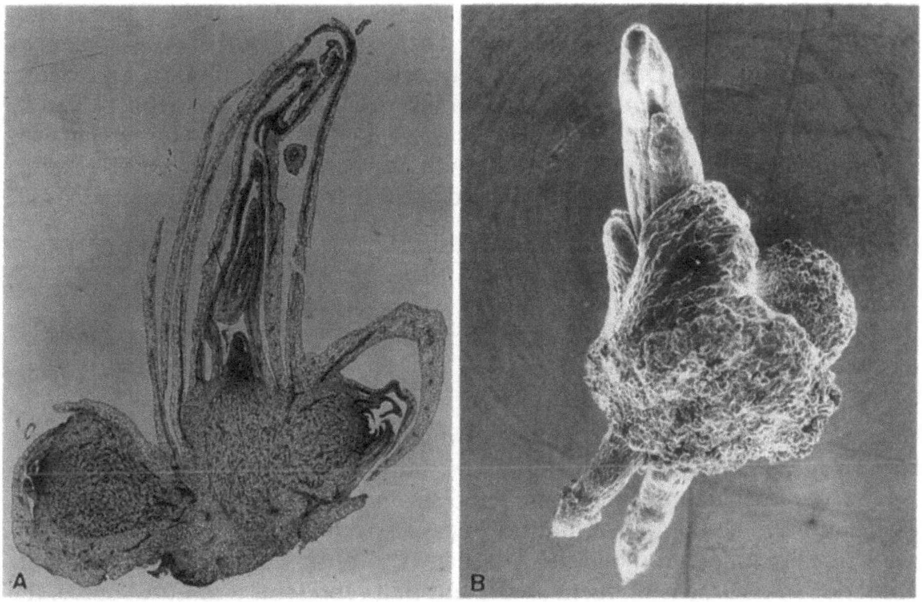

Fig. 3A,B. Photographs taken to emphasize differences in the product of a shoot multiplication system and a somatic embryo-generating one. **A** Unit grown in vitro showing main shoot and some off-shoots of the ABB clone Cardaba in section. There are eight leaves and two newly emergent side shoots visible. The shoots would have to be separated (severed) and rooted in a separate step to obtain a plantlet, ca. 4.7x. **B** A somatic embryo showing well-formed shoot and roots. The discreteness of the structure emphasizes the differences between a somatic embryo with a shoot generated via micropropagation. Somatic embryos with high fidelity offer not only the possibility of being generated in very large numbers, but with sufficient understanding of the mechanisms that control development of somatic embryos along the lines of zygotic embryos, one has hopes for development of management systems involving "artificial seed"; ca. 22x

determinate plantain clones without major callus production has been routinely obtained (cf. Cronauer-Mitra and Krikorian 1988a). This indicates that there is a high potential for differentiation of meristems *de novo* ("meristemoids") from otherwise meristem-free areas.

We have generated competent cell masses of varying sizes from cells routinely sieved through No. 80 sieves (177-μm pore diameter) and even No. 200 sieves (74-μm pore diameter) and their ability to generate plantlets is beyond all doubt (cf. Krikorian 1987; Krikorian et al. 1990). The clones we have worked on most are Cardaba ABB (a cooking banana) and Lady Finger (AAB), a dessert banana. Even so, there are many problems such as understanding why it is that in some cases, small groups of cells seem to need to grow into larger masses before they organize. Is there a requirement for critical mass? The greater the number of cells in a morphogenetically competent mass before full developmental competence is expressed, the greater the chance for introduction of chimeral status to the developing propagule. The implications of this for clonal stability and integrity are obvious (cf. Bergann 1967 and references therein).

In our somatic embryo-producing cultures there seems to be a correlation with the ability to "germinate" and grow further with the level of development of the haustorium (cf. Cronauer and Krikorian 1988c; Krikorian and Scott, unpubl.). Is there a critical nutritional- or information-processing role played by this cotyledon-like structure? Unless one can identify all the requirements for controlled germination and/or retention of dormancy or "quiescence" prior to desired use, the system is only partial and one is thereby limited to restricted and early developmental phenomena.

3 Conclusions

There has been steady and significant progress on the regeneration of bananas and plantains from somatic embryos both on semi-solid medium from embryogenically competent units and from embryogenic or organogenic suspensions in liquid. Still, a number of gaps exist in the papers published so far on *Musa* somatic embryogenesis, particularly in those steps between reliable initiation of a suspension from an explant that represents a mature source, the formation of somatic embryos from progenitor cells, and the generation of a plantlet or propagule. Most work using somatic embryos for field testing has derived so far from shoot tip micropropagation initiated from plantlets which derived from a relatively few somatic embryos. Thus, the phenotypic and genotypic fidelity of somatic embryo-derived progeny from many important clones remains to be adequately evaluated especially from the perspective of yield of marketable fruit. Since many edible *Musa* clones are mixoploid (cf. Krikorian et al. 1993 and references therein), there are sure to be many variants associated with such things as breakdown of chimeral structure (cf. e.g. Bergann 1967).

The matter of routine production of somatic embryos in the full range of edible, triploid, and other clones thus requires more work and is far from satisfactorily resolved. Opportunities to study factors influencing abortion and production of aberrant embryos (cf. Rowe and Richardson 1975; Rowe and Rosales 1995) should yield new insights on developmental controls.

References

Bakry F, Rossignol L (1985) Analyse des capacités de callogenèse et d'organogenèse obtenues à partir de differents tissus de Bananiers (*Musa* sp., Musacées). Fruits (Paris) 40: 697–708

Banerjee N, Vuylsteke D, De Langhe EAL (1986) Meristem tip culture of *Musa*: histomorphological studies of shoot bud proliferation. In: Withers L, Anderson PG (eds) Plant tissue culture and its agricultural applications. Butterworths, London, pp 139–147

Banerjee N, Schoofs J, Hollevoet S, Dumortier F, De Langhe E (1987) Aspects and prospect of somatic embryogenesis in *Musa*, ABB, cv. Bluggoe. Acta Hortic 212: 727–730

Barker WG, Steward FC (1962a) Growth and development of the banana. I. The growing regions of the vegetative shoot. Ann Bot 26: 389–411

Barker WG, Steward FC (1962b) Growth and development of the banana plant. II. Transition from the vegetative to the floral shoot in *Musa acuminata* cv. Gross Michel. Ann Bot 26: 413–423

Bergann F (1967) The relative instability of chimeral clones – the basis for further breeding. Abh Dtsch Akad Wiss Berl Kl Med 2: 287–300

Cronauer SS (1986) In vitro growth responses of *Musa*. PhD Thesis, State University of New York at Stony Brook, Stony Brook, New York, 206 pp. Diss Abstr Int 47: 1816–B. Univ Microfilms Order No DA 8616686

Cronauer SS, Krikorian AD (1983) Somatic embryos from cultured tissues of triploid plantains (*Musa* 'ABB'). Plant Cell Rep 2: 289–291

Cronauer SS, Krikorian AD (1984) Multiplication potential of *Musa* from excised stem tips. Ann Bot 53: 321–328

Cronauer SS, Krikorian AD (1985a) Banana (*Musa* spp.). In: Bajaj YPS (ed) Biotechnology in agriculture and forestry vol 1. Trees I. Springer, Berlin Heidelberg New York, pp 233–252

Cronauer SS, Krikorian AD (1985b) Reinitiation of vegetative growth from aseptically cultured terminal floral apex of banana. Am J Bot 72: 1598–1601

Cronauer-Mitra SS, Krikorian AD (1988a) Determinate floral buds of plantain (*Musa* AAB) as a site of adventitious shoot formation. Ann Bot 61: 507–512

Cronauer SS, Krikorian AD (1988b) Adventitious shoot production from calloid cultures of banana. Plant Cell Rep 6: 443–445

Cronauer SS, Krikorian AD (1988c) Plant regeneration via somatic embryogenesis in the seeded diploid banana *Musa ornata* Roxb. Plant Cell Rep 7: 23–25

Cronauer SS, Krikorian AD (1988d) Temporal, spatial and morphological aspects of multiplication in aseptically cultured *Musa* clones. In: Valentine F (ed) Progress and prospects in crop and forest biotechnology. State University of New York College of Environmental Science and Forestry, Colloquium. April 18–20, 1985. Syracuse. Springer, Berlin Heidelberg New York, pp 45–57

Devos P (1985) The location of lateral buds of banana (*Musa* sp.) clarified by the discovery of two new types of adventitious buds within the ABB plantain group. Fruits (Paris) 40: 471–474

Dhed'a D, Dumortier F, Panis B, Vuylsteke D, De Langhe E (1991) Plant regeneration in cell suspension cultures of the cooking banana cv. 'Bluggoe' (*Musa* spp. ABB group). Fruits (Paris) 46: 125–135

Drew RA, Moisander JA, Smith MK (1989) The transmission of banana bunchy-top virus in micropropagated bananas. Plant Cell Tissue Organ Culture 16: 187–193

Escalant JV, Teisson C (1987) Comportements in vitro de l'embryon isolé du bananier (*Musa* species). Fruits (Paris) 42: 333–342

Escalant JV, Teisson C (1988) Embryogenese somatique chez *Musa* sp. CR Acad Sci Paris Sér III: 306: 277–281

Fisher J (1978) Leaf-opposed buds in *Musa*: their development and comparison with allied monocotyledons. Am J Bot 65: 784–794

Frison EA, Putter CAJ (eds) (1989) FAO/IBPGR technical guidelines for the safe movement of *Musa* germplasm. Food and Agriculture Organisation of the United Nations, Rome/International Board for Plant Genetic Resources, Rome

Huang L-C, Chi D-L (1988) Pivotal roles of picloram and gelrite in banana callus culture. Environ Exp Bot 28: 249–258

IBPGR (1984) International Board for Plant Genetics Resources Revised banana descriptors. AGPG: IBPGR/83/11 April. IBPGR Secretariat, Rome, 31 pp

Israeli Y, Reuveni O, Lahav E (1991) Qualitative aspects of somaclonal variations in banana propagated by in vitro techniques. Sci Hortic 48: 71–88

Jarret RL, Fisher JB, Litz RE (1985) Organ formation in *Musa* tissue cultures. J Plant Physiol 121: 123–130

Krikorian AD (1987) Callus and cell culture, somatic embryogenesis, androgenesis and related techniques for *Musa* improvement. In: Persley GJ, De Langhe EA (eds) Banana and plantain breeding strategies: Pro Int Worksh held at Cairns, Australia, 13–17 Oct, 1986. ACIAR Proc 21, pp 128–135

Krikorian AD (1989) In vitro culture of bananas and plantains: background, update and call for information. Trop Agric (Trinidad) 66: 194–200

Krikorian AD (1990) Baseline tissue and cell culture studies for use in banana improvement schemes. In: Ploetz RC (ed) Fusarium wilt of banana. American Phytopathological Society Press, St. Paul, pp 127–133

Krikorian AD, Cronauer SS (1984a) Banana. In: Sharp WR, Evans DA, Ammirato PV, Yamada Y (eds) Handbook of plant cell culture, vol 2. Macmillan, New York, pp 327–348

Krikorian AD, Cronauer SS (1984b) Aseptic culture techniques for banana and plantain improvement. Econ Bot 38: 322–333

Krikorian AD, Cronaeur-Mitra SS, Fitter Corbin MS (1988) Protoplast culture of perennials. Sci Hortic 37: 277–293

Krikorian AD, Scott ME, Cronauer-Mitra SS, Smith DL (1990) *Musa* callus and cell culture: strategies, achievements and directions. In: In vitro mutation breeding of bananas and plantains I. Report of the First Res Co-ordination Meeting for the FAO/IAEA, 29 May–2 June 1989. Technical Document, IAEA Vienna, 1989. IAEA 312. D2. RC. 411, pp 9–23

Krikorian AD, Irizarry H, Cronauer-Mitra SS, Rivera E (1993) Clonal fidelity and variation in plantain (*Musa* ABB) regenerated from vegetative and floral stem tips in vitro. Ann Bot 71: 519–535

Matsumoto K, Yamaguchi H (1989) Nonwoven materials as a supporting agent for in vitro culture of banana protocorm-like bodies. Tropical Agric (Trinidad) 66: 8–10

Megia R, Haicour R, Tizroutine S, Bui Trang V, Rossignol L, Sihachakr, Schwendiman J (1993) Plant regeneration from cultured protoplasts of the cooking banana cv. Bluggoe *Musa* spp., ABB group). Plant Cell Rep 13: 41–44

Mohan Ram HY, Steward FC (1964) The induction of growth in explanted tissue of the banana fruit. Can J Bot 42: 1559–1579

Mohan Ram HY, Ram M, Steward FC (1962) Growth and development of the banana plant. III. A The origin of the inflorescence and development of the fruit. Ann Bot 26: 657–673

Murashige T, Skoog F (1962) A revised medium for rapid growth and bioassay with tobacco tissue cultures. Physiol Plant 15: 473–497

Murfett J, Clarke A (1987) Producing disease-resistant *Musa* cultivars by genetic engineering. In: Persley GJ, De Langhe EA (eds) Banana ad plantain breeding strategies: Proc Int Worksh held at Cairns, Australia, 13–17 Oct 1986. ACIAR Proc 21, pp 87–94

Novak FJ (1992) *Musa* (bananas and plantains). In: Hammerschlag FA, Litz RE (eds) Biotechnology of perennial fruit crops. CAB International, Wallingford pp 449–488

Novak FJ, Afza R, Van Duren M, Perera-Dallos M, Conger BV, Tang X (1989) Somatic embryogenesis and plant regeneration in suspension cultures of dessert (AA and AAA) and cooking (ABB) bananas (*Musa* spp.). Bio/Technology 7: 154–159

Palmer JK (1963) Banana polyphenoloxidase. Preparation and properties. Plant Physiol 38: 508–513

Panis BJ, Withers LA, De Langhe EAL (1991) Cryopreservation of *Musa* suspension cultures and subsequent regeneration of plants. Cryo Lett 11: 337–350

Panis B, Dhed'a D, Swennen R (1992) Freeze-preservation of embryogenic *Musa* suspension cultures. In: Adams RP, Adams JE (eds) Conservation of plant genes. DNA banking and in vitro technology. Academic Press, San Diego, pp 183–195

Panis B, Van Wauwe A, Swennen R (1993) Plant regeneration through direct somatic embryogenesis from protoplasts of banana (*Musa* spp.). Plant Cell Rep 12: 403–407

Ploetz RC (ed) (1990) Fusarium wilt of banana. APS Press, The American Phytopathological Society, St Paul

Rowe P, Richardson DL (1975) Breeding bananas for disease resistance fruit quality and yield. Tropical Agriculture Research Services (SIATSA) Bull 2. La Lima, Honduras, 44 pp

Rowe P, Rosales F (1995) Breeding bananas and plantains. In: Janick J (ed) Advances in fruit breeding. J. Wiley and Sons, New York

Sannasgala K (1989) In vitro somatic embryogenesis in *Musa*. PhD Thesis, Dissertationes de Agicultura No. 180, Catholic University of Louvain, Belgium, 172 pp.

Simmonds NW (1962) The evolution of bananas. Longmans, London

Simmonds NW, Shepherd K (1955) The taxonomy and origins of cultivated bananas. J Linn Soc Lond Bot 55: 302–3012

Smith DL, Krikorian AD (1990) Somatic embryogenesis of carrot in hormone-free medium: external pH control over morphogenesis. Amer J Bot 77: 1634–1647

Stein A, Loebenstein G, Koenig R (1979) Detection of cucumber mosaic virus and bean yellow mosaic virus in *Gladiolus* by enzyme-linked immunosorbent assay (ELISA). Plant Dis Rep 63: 185–188

Stover RH (1977) Banana (*Musa* spp.). In: Hewitt WR, Chirappa S (eds) Plant health and quarantine in international transfer of genetic resources. CRC Press, Boca Raton pp 71–79

Stover RH, Buddenhagen IW (1986) Banana breeding: polyploidy, disease resistance and productivity. Fruits (Paris) 41: 175–191

Stover RH, Simmonds NW (1987) Bananas, 3rd edn. Longman, Edinburgh; John Wiley, New York

Tezenas du Montcel H, De Langhe E, Swennen R (1983) Essai de classification des bananiers plantain (ABB). Fruits (Paris) 38: 461–474

Vuylsteke DR (1989) Shoot-tip culture for the propagation, conservation and exchange of *Musa* germplasm. International Board of Plant Genetic Resources, Rome

Vuylsteke DR, Swennen R (1992) Biotechnological approaches to plantain and banana improvement at IITA. In: Thottapilly G, Monti L, Mohan Raj DR, Moore AW (eds) Biotechnology; enhancing research on tropical crops in Africa. CTA/IITA co-publication, Ibadan, Nigeria, pp 143–150

Vuylsteke D, Swennen R, De Langhe E (1990) Tissue culture technology for the improvement of African plantains. In: Fullerton RA, Stover RH (eds) Sigatoka leaf spot diseases. Proc Int Worksh San José, Costa Rica, March 28–April 1, 1989. International Network for the Improvement of Banana and Plantain, Montpellier, France, pp 316–337

Section III
Somatic Embryogenesis in Legumes and in Oilseed Crops

III.1 Somatic Embryogenesis in Legumes

W.A. PARROTT[1], R.E. DURHAM[2], and M.A. BAILEY[1]

1 Introduction

Legumes are members of the third largest family (Leguminosae) of flowering plants, and are globally distributed, with as many as 19,000 species. Several species have the ability to undergo a symbiotic association with nitrogen-fixing bacteria of the genera *Rhizobium* or *Bradyrhizobium*. Consequently, legumes tend to have high protein contents, and are an essential source of food, feed, and forage. They are also used as ornamentals, or valued for green manure, timber, gums, and other compounds.

Legumes have traditionally been difficult to regenerate from cell culture (McHughen and Swartz 1984; Huyghe 1990). Regeneration via somatic embryogenesis offers great potential for use in mass propagation or nonconventional genetic improvement programs, including transformation. Unlike shoots derived via organogenesis, somatic embryos consist of complete propagules that do not require a separate rooting regime. As technology develops further, it may become possible to grow large numbers of somatic embryos in bioreactors for use as artificial seed to propagate elite or hybrid genotypes.

Over the past 5 years, great progress has been made in the development of regeneration systems for legumes, and many legumes are now amenable to regeneration via somatic embryogenesis. The purpose of this review is to focus on patterns of regeneration that are becoming evident across species. The study of these patterns and their application could help facilitate both a fundamental understanding of somatic embryogenesis and the development or refinement of regeneration systems for most legumes.

Almost all reports of regeneration in legumes have been confined to members of the Papilionoideae, the subfamily of greatest economic importance. Plants had been regenerated via organogenesis from 54 different legume species, while only 39 species had been regenerated from somatic embryos (Parrott et al. 1992). A listing of species currently regenerated from somatic embryos, may be found in Table 1. In addition, reports of somatic embryo formation without plant recovery are given in Table 2. In some of these instances, plant recovery was never an objective.

[1] Department of Crop and Soil Sciences, The University of Georgia, 3111 Plant Sciences Building, Athens, Georgia 30602-7272, USA
[2] Department of Plant and Soil Sciences, Texas Tech University, Lubbock, Texas, 79409-2122, USA

Biotechnology in Agriculture and Forestry, Vol. 31
Somatic Embryogenesis and Synthetic Seed II (ed. by Y.P.S. Bajaj)

Table 1. Explant source for somatic embryogenesis observed for various species of legumes from which whole plants have been obtained

Species	Explant[a]	Reference
Albizia lebbeck (L.) Benth	H	Gharyal and Maheshwari (1983); Tomar and Gupta (1988)
A. richardiana King	H	Tomar and Gupta (1988)
Arachis hypogaea L.	C, CN, EA, S, ZE	Hazra et al. (1989); Ozias-Akins (1989); Sellars et al. (1990); Baker and Wetzstein (1992); Durham and Parrott (1992); Ozias-Akins et al. (1992)
A. paraguariensis Chod. et. Hassl.	Ca, ZE	Still et al. (1987); Sellars et al. (1990)
Cercis canadensis L.	ZE	Geneve and Kester (1990)
Chamaecytisus purpureus (Scop.) Link	H, L, R	Greinwald and Czygan (1991)
C. austriacus (L.) Link	SC, H, L,R	Greinwald and Czygan (1991)
Cicer arietinum L.	L	Rao and Chopra (1989)
Cladrastis lutea (Michz) K. Koch	C	Weaver and Trigiano (1991)
Clitoria ternatea L.	H,R	Dhanalakshmi and Lakshmanan (1992)
Coronilla Varia L	Ca, L, R	Arcioni and Mariotti (1982); Arcioni et al. (1988); Dušková et al. (1990)
Crotalaria juncea L.	CP	Ramanuja Rao et al. (1982)
Glycine canescens F.S. Herm.	C	Grant (1984)
G. max (L.) Merr.	C, EA	Christianson et al. (1983); Lazzeri et al. (1985, 1987a, b, 1988); Li et al. (1985); Ranch et al. (1985); Barwale et al. (1986); Ghazi et al. (1986); Hammatt and Davey (1987); Tétu et al. (1987); Finer (1988); Finer and Nagasawa (1988); Hartweck et al. (1988); Hepher et al. (1988); Komatsuda and Ohyama (1988); Parrott et al. (1988, 1989); Shoemaker and Hammond (1988); Buchheim et al. (1989); Christou and Yang (1989); Feng et al. (1989); Kien et al. (1989); Komatsuda and Ko (1990); Ferreira et al. (1991); Kiss et al. 1991; Shoemaker et al. (1991)
G. soja Sieb et Zucc.	H	Gamborg et al. (1983)
Lathyrus sativus L.	ST	Gharyal and Maheshwari (1983)
Lens culinaris Medik	EA	Saxena and King (1987)
Lotus corniculatus L.	L	Arcioni and Mariotti (1982); Arcioni et al. (1988)
Lupinus albus L.	C	Nadolska-Orczyk (1992)
L. angustifolius L.	C	Nadolska-Orczyk (1992)
L. mutabilis Sweet	C	Nadolska-Orczyk (1992)
Medicago coerulea Less	LP	Arcioni et al. (1982)

M. difalcata Sinsk. (=*M. sativa* spp. *difalcata*)	LP	Glimour et al. (1987)
M. falcata Arcengeli. (= *M. sativa* spp. falcata)	LP	Glimour et al . (1987); Denchev et al. (1991)
M. glutinosa M.B (= *M. sativa* spp. *glutinosa*)	LP	Arcioni et al. (1982) ; Glimour et al. (1987)
M. hemicycla Grossh. (= *M. sativa* spp. *hemicycla*)	LP	Gilmour et al. (1987)
M. media Pers.	R, H	Nagarajan and Walton (1987)
M. sativa L.	C, H, L, LP, O, P, S, SC, ST	Saunders and Bingham (1972); Dos Santos et al. (1980,1983); Kao and Michayluk (1980, 1981); Reisch and Bingham (1980); Stavarek et al. (1980); Johnson et al. (1981); Walker and Sato (1981); Arcioni et al. 1982; Lu et al. (1982, 1983); Mezentsev et al. (1982); Nóvak and Konecná (1982); Xu et al. (1982); Lupotto (1983, 1986); Hartman et al. (1984); Mitten et al. (1984); Stuart and Strickland (1984a,b); Brown and Atanassov (1985); Skokut et al. (1985); Meijer and Brown (1985, 1987a,b, 1988); Matheson et al. (1990); Chen and Marowitch (1987); Chen et al. (1987); Strickland et al. (1987); Bianchi et al. (1988); Latunde-Dada and Lucas (1988); Meijer and Simmonds (1988); Seitz Kris and Bingham (1988); Walton and Brown (1988); Wan et al. (1988); Feher et al. (1989); Fujii et al. (1989,1990); Hernández-Fernández and Christie (1989); McKersie et al. (1989); Meijer (1989); Ray and Bingham (1989); Nolan et al. (1989), Senaratna et al. (1989, 1990); Slade et al. (1989); Anandarajah and McKersie (1990a,b); Binarová et al. (1990); Song et al. (1990); Yu et al. (1990); Denchev et al. (1991); Frame et al. (1991); Nagy et al. (1991) Nichol et al. (1991)
M. varia Mart. (= *M. sativa* spp. *varia*)	LP	Gilmour et al. (1987)
Phaseolus acutifolius Gray	L	Kumar et al. (1988b)
P. coccineus L.	C, H	Genga and Allavena (1991); Malik and Saxena (1992)
P. wrightii L.	ST	Malik and Saxena (1992)
Pisum sativum L.	LP, ST, ZE	Kysely et al. (1987); Lehminger-Mertens and Jacobsen (1989); Kysely and Jacobsen (1990); Tétu et al. (1990)
Robinia pseudoacacia L.	ZE	Merkle and Wiecko (1989)

(*Contd.*)

Table 1. (*Contd.*)

Species	Explant[a]	Reference
Trifolium arvense L.	H	Bhojwani et al. (1984)
T. alpestre L.	C, H, R	Yamada and Higuchi (1990)
T. amabile Humb., Bonpl. et Kunth	C,H, R	Yamada and Higuchi (1990)
T. apertum Bobrov	C, H, R	Yamada and Higuchi (1990)
T. caucasicum Tausch	C,H, R	Yamada and Higuchi (1990)
T. cherleri L.	C, H,R	Yamada and Higuchi (1990)
T. heldreichanum Haurskn.	C,H, R	Yamada and Higuchi (1990)
T. incarnatum L.	H	Pederson (1986)
T. medium L.	P.	Choo (1988)
T. montanum L.	C,H, R	Yamada and Higuchi (1990)
T. pratense L.	Ca, H P, SC, ZE	Keyes et al. (1980); Phillips and Collins (1980); Collins and Phillips (1982); Bhojwani et al. (1984); Maheswaran and Williams (1986b); Wang and Holl (1988); McGee et al. (1989)
T.repens L.	H, L, ZE	Bhojwani et al. (1984); Maheswaran and Williams (1984, 1986a, 1987); Bond and Webb (1989); Parrott (1991)
T. resupinatum L.	ZE	Maheswaran and Williams (1986b)
T. rubens L.	H	Parrott and Collins (1983); McGee et al. (1989)
T. subterraneum L.	ZE	Maheswaran and Williams (1986b)
T. vesiculosum Savi	H	Pederson (1986)
Trigonella corniculata		Lu et al. (1982)
Vicia faba L.	C	Griga et al. (1987)
V. narbonensis L.	ST	Pickardt et al. (1989)
Vigna acontifolia (Jacq) Meréchal	L, LP	Shekhawat and Galston (1983); Kumar et al. (1988a) ; Eapen and George (1990)
V. glabrescens × *V. radiata* (L.) Wilczek	ZE	Chen et al. (1990)
V. mungo (L.) Hepper	C	Eapen and George (1990)

[a] C= cotyledon; CN = cotyledonary node; CP = cotyledon protoplast; Ca= callus; CaP = callus protoplast; E= epicotyl; EA= embryonic axis; H= hypocotyl; HP= hypocotyl protoplast; I = internode; L= leaf; LP = leaf protoplast; O= ovary; P= petiole; R= root; RA= radicle; S=stem; SC= suspension culture; ST= shoot; tip=stolon; ZE= zygotic embryo.

Table 2. Explant source for somatic embryogenesis observed for various species of legumes from which whole plants were not obtained

Species	Explant[a]	Reference
Acacia koa Gray	H, ST	Skolmen (1986)
Cercis canadensis L.	ZE	Trigiano et al. (1988)
Cyamopsis tetragonoloba L.	SC	McHughen and Swartz (1984)
Glycine gracilis Skvortz	C	Komatsuda (1990); Komatsuda et al. (1991)
G. max (L.) Merr.	C, H, SC	Beversdorf and Bingham (1977); Phillips and Collins (1981); Kerns et al. (1986); Komatsuda (1990); Yang et al. (1991)
G. soja Sieb et Zucc.	H, SC	Beversdorf and Bingham (1977); Phillips and Collins (1981)
G. tabacina	H	Beversdorf and Bingham (1977)
Medicago sativa L.	P, SC, ZE	Walker et al. (1982); Maheswaran and Williams (1984); Stuart et al. (1987, 1988); Binarová and Dolezel (1988)
Phaseolus vulgaris L.	ST	Martins and Sondahl (1984)
Poinciana regia	C	Bajaj (1989)
Trifolium ambiguum Bieb	H	Pederson (1986)
T. arvense L.	H	Bhojwani et al. (1984)
T. pratense L.	ZE	Maheswaran and Williams (1984)
T. repens L	H, ZE	Maheswaran and Williams (1985); Pederson (1986)
T. rubens L.	P	Cui et al. (1988)
Vicia faba L.	C	Griga et al. (1987)
Vigna radiata (L.) Wilczk	C	Eapen and George (1990)

[a] C= cotyledon; H= hypocotyl; P= petiole; SC= suspension culture; ST = shoot tip; ZE = zygotic embryo.

2 Induction of an Embryogenic State

At first glance, somatic embryogenesis in legumes appears to be a collection of diverse mechanisms. Some legumes, such as clovers (*Trifolium* spp.), have been reported to form somatic embryos (Fig. 1A) upon exposure to cytokinins (e.g., Maheswaran and Williams 1984). Others, such as soybean (*Glycine max*), form somatic embryos in response to auxins (Fig. 1B), and regenerate directly from zygotic embryo explant tissues without an intervening callus phase (Lazzeri et al. 1985). Finally, some legumes (Fig. 1C), notably alfalfa (*Medicago sativa*), regenerate well from callus (Bingham et al. 1988). The use of auxins and cytokinins to induce embryogenesis, as well as the presence or absence of a callus phase, can be explained collectively under the concepts put forth by Sharp et al. (1982). These authors used the term PEDC (pre-embryogenic determined cell) to describe cells within very immature zygotic embryos, recognizing that these are expressing an embryogenic developmental program. In such tissues, the presence of a cytokinin to stimulate cell division is all that is required to propagate the embryogenic state and to form new somatic embryos (Fig. 1A). As zygotic embryos mature, the original embryogenic program is replaced by the expression of other developmental programs. In this case, the presence of an auxin is required to return the tissues to their original embryogenic state (Fig. 1C). Once the tissues have been returned to the original embryogenic state, they consist of induced embryogenic cells, or IEDCs (Sharp et al. 1982). An IEDC and a PEDC should be functionally equivalent.

Tissues at different stages of development require different numbers of mitotic divisions in the presence of an auxin before becoming embryogenically determined. These intervening cell divisions are manifested as a callus phase. When few, if any, mitotic divisions are required for embryogenesis, such as that which occurs from zygotic cotyledons, embryogenesis is said to be direct (Fig. 1A,B). When many divisions are required, such as regeneration from leaf-derived callus, somatic embryogenesis is said to be indirect (Fig. 1C). Tissues associated with reproduction, such as ovary or anther tissue, or tissue from young seedlings, may be more easily returned to an embryogenic state than more mature tissue (Williams et al. 1990). Occasionally, some legumes, such as *Trifolium rubens*, can be identified which retain the capacity for direct embryogenesis from mature tissues (Cui et al. 1988). The ability to induce somatic embryogenesis in these plants without an intervening callus phase may make such species ideal for further investigations of the induction process.

The role of auxin for induction of an embryogenic state is not clear. Some evidence from pea (*Pisum sativum*) suggests that soluble auxin-binding proteins play a role in somatic embryogenesis. Picloram (4-amino-3,5,6-trichloropicolinic acid), which does not affect cell wall elongation in pea, but does induce somatic embryogenesis, only binds to cytoplasmic receptors. Furthermore, soluble auxin-binding proteins are found in zygotic embryos only at the developmental stage which will give rise to somatic embryos, and are limited to the axis, the region from which somatic embryos originate (Jacobsen 1991).

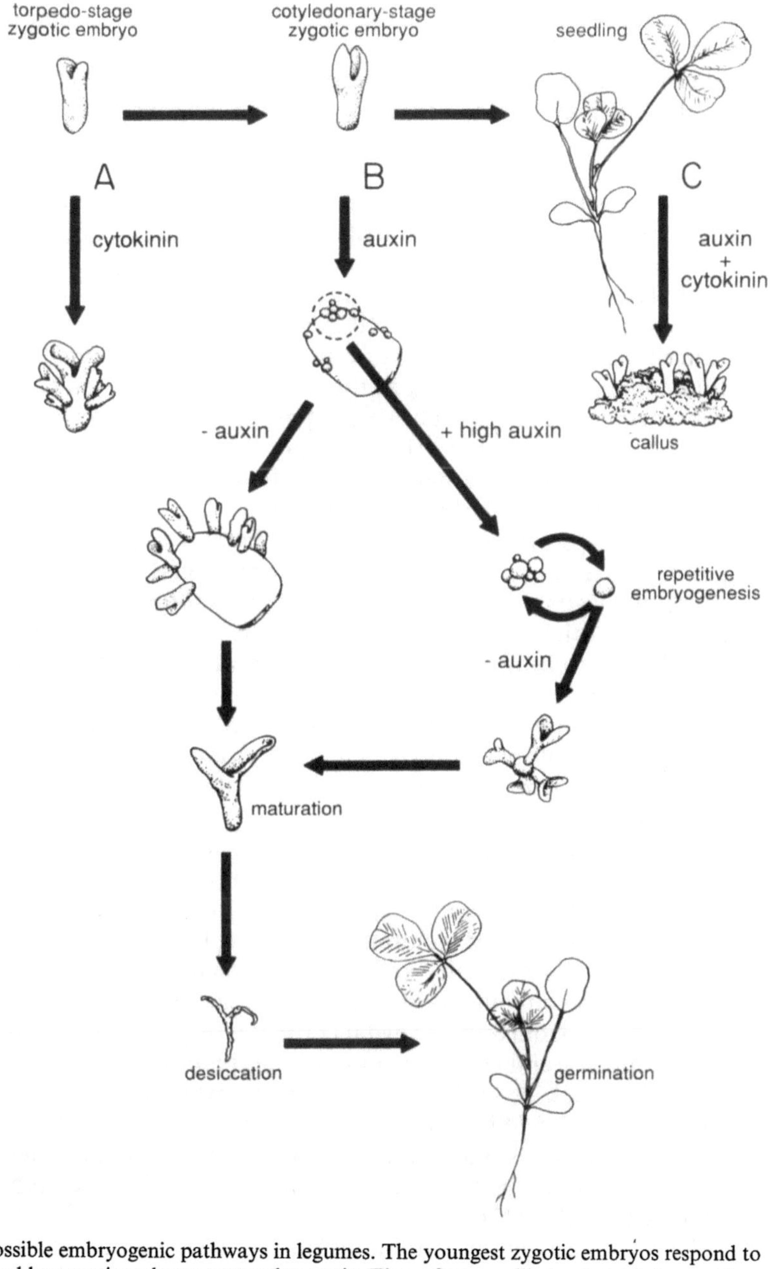

Fig. 1. Some possible embryogenic pathways in legumes. The youngest zygotic embryos respond to cytokinin, while older zygotic embryos respond to auxin. Tissue from seedlings responds to combinations of auxins and cytokinins by forming callus. Somatic embryos may be matured, desiccated, and germinated. Alternatively, enough auxin may be used to induce repetitive embryogenesis, which continues until auxin is withdrawn from the medium, permitting the somatic embryos to resume their development. (Drawing by J. Adang)

Little additional research on the induction of an embryogenic state has been done in legumes. Nevertheless, this process has been reviewed extensively using evidence from families of plants other than legumes (Merkle et al. 1990; Parrott et al. 1991). Briefly, DNA methylation, which may terminate or alter existing gene expression programs, occurs in the presence of auxins. In addition, cells may become isolated from their neighbors through the severance of plasmodesmata or necrosis of surrounding tissues, interrupting cell–cell interactions which, in turn, can reinforce the termination of existing gene expression programs and the reexpression of an embryogenic development program. Embryogenically determined cells of soybean have been found to have ultrastructural features in common with megagametophytes (Bonnelle et al. 1990), highlighting the degree of intracellular changes which may occur during the induction of embryogenesis. Another function of auxins may be the subsequent imposition of polarity within an embryogenic cell or cells. The role of polarity has also been reviewed (Merkle et al. 1990; Parrott et al. 1991), and may help to establish the identity and coordinate the behavior of embryogenic cells separately from that of surrounding tissues (Williams and Maheswaran 1986).

If these steps are correct, then other factors that alter gene expression programs (e.g., stress) or disrupt cell–cell interactions (physical disruption of the tissue) should lead to the formation of somatic embryos, and such examples are available from other plant families. Within the legumes, hypocotyl segments of the East Indies walnut, *Albizia lebbeck* (Gharyal and Maheshwari 1983), and of kardofan, *Clitoria ternatea*, (Dhanalakshmi and Lakshmanan 1992), form somatic embryos on growth regulator-free medium. The application of a low-voltage field to protoplast-derived callus of alfalfa stimulates the formation of somatic embryos, perhaps in response to polarity imposed by the electrical field (Dijak et al. 1986).

The auxin most commonly used to induce embryogenesis is 2,4-dichlorophenoxyacetic acid (2,4-D), but examples may be found where the use of 2,4-D is ineffective (Greinwald and Czygan 1991). Other auxins, such as α-naphthaleneacetic acid (NAA) and picloram are also used. Legumes differ in response to various auxins. For example, soybean responds to both NAA and 2,4-D, with NAA giving fewer but more normal embryos (Lazzeri et al. 1987a). NAA induces somatic embryos from only the distal perimeter of cotyledon explants, while 2,4-D induces somatic embryos from most of the epidermal surface of the cotyledon explant (Hartweck et al. 1988). In contrast, white clover (*Trifolium repens*) responds well to 2,4-D but poorly to NAA (Parrott 1991). The use of 2,4-D in lupins (*Lupinus* spp.) stimulates more embryo formation than the use of NAA, while indole-3-acetic acid (IAA) and indole-3-butyric acid were not effective (Nadolska-Orczyk 1992).

The role of exogenous cytokinins during the induction phase is difficult to assess, as published studies almost never include the appropriate control, namely, an auxin treatment without any cytokinin. The role of a cytokinin may depend on whether somatic embryogenesis is direct or indirect. When somatic embryogenesis is possible from callus, such as in alfalfa (Saunders and Bingham 1975) or in lentil, *Lens culinaris* (Saxena and King 1987), the frequency of somatic embryo formation is apparently enhanced by the presence of cytokinins in the callus induction medium. However, in direct systems, such as in crownvetch, *Coronilla varia,* in which somatic embryos formed directly on roots developed from callus (Arcioni and Mariotti 1982), or in pea (Kysely and Jacobsen 1990) and soybean (Lippmann and Lippmann 1984; Lazzeri et al. 1987a), in which somatic embryos form directly from immature zygotic embryos, addition of cytokinin reduces the frequency of embryo formation. In lupins, addition of a cytokinin does not affect the frequency of somatic embryogenesis if 2,4-D is the inducing auxin, but lowers the frequency if NAA is the inducing auxin (Nadolska-Orczyk 1992).

Some legume species, such as soybean, have not responded to conventional auxin-cytokinin treatments to form somatic embryos from callus tissue. The cocultivation of soybean callus with *Pseudomonas maltophilia* to induce somatic embryo formation (Yang et al. 1991) is evidence that it may eventually be possible to regenerate more species from callus tissue. The role of the *P. maltophilia* in embryo induction is unknown, but raises the intriguing possibility that unidentified compounds exist with the ability to turn on the gene(s) necessary to initiate the embryogenic pathway.

Somatic embryogenesis is influenced by explant source and genotype. The most responsive explant type is species-specific. For those species which form somatic embryos from callus, the choice of explant is less critical, and may consist of tissues as diverse as stolons, petioles, stems, leaves, hypocotyls, and ovaries. Otherwise, immature zygotic embryos are the most responsive explants. Very immature zygotic embryos can form somatic embryos upon exposure to a cytokinin (Fig. 1A). Species tested for cytokinin-induced embryogenesis are limited to alfalfa and a few clover species, all of which have responded (Maheswaran and Williams 1984, 1986b). As the zygotic embryo becomes more mature, the use of auxin becomes necessary to induce somatic embryogenesis (Fig. 1B) and eventually, the embryos of some species mature to a point where somatic embryogenesis can no longer be induced. The entire zygotic embryo may not be embryogenic, such as in pea, in which somatic embryos only form on the axis (Kysely and Jacobsen 1990). In contrast, both axes (Hazra et al. 1989) and cotyledons (Ozias-Akins 1989) of peanut (*Arachis hypogaea*) are embryogenic. The orientation of the cotyledon on the medium can also be significant, In soybean, the greatest number of embryos form when excised cotyldons are placed with the abaxial surface on induction medium (Hartweck et al. 1988).

In some species, embryos may be induced from other tissues, such as shoot apices of pea (Kysely and Jacobsen 1990) or immature leaflets of peanut (Baker and Wetzstein 1992). While pollen-derived embryos have been reported in peanut (Bajaj et al. 1980), there are no verified reports of a haploid plant being obtained from microspore- or pollen-derived embryos of any legume.

Plant genotype can be a critical factor for somatic embryogenesis. For example, soybean genotypes (Komatsuda and Ohyama 1988; Parrott et al. 1989) and alfalfa germplasms and cultivars differ in the frequency of individuals with embryogenic capacity (Mitten et al. 1984; Brown and Atanassov 1985; Meijer and Brown 1985; Chen and Marowitch 1987; Chen et al. 1987; Bianchi et al. 1988). Recurrent selection has been successful for increasing the embryogenic capacity of both tetraploid (Bingham et al. 1975) and diploid (Ray and Bingham 1989) alfalfa. However, different alfalfa germplasms vary in genetic control of regeneration, and consequently respond dissimilarly to different regeneration protocols, manifested as genotype × protocol interactions (Seitz Kris and Bingham 1988). The genetics of regeneration are probably better defined in alfalfa than in any other plant species. Three separate studies have identified two dominant Mendelian genes conditioning embryogenic capacity. These have been called *Rn1* and *Rn2* (Reisch and Bingham 1980), *Rn3* and *Rn4* (Wan et al. 1988), and *Rna* and *Rnb* (Hernández-Fernández and Christie 1989). These gene pairs have not been tested for allelism. In red clover (*Trifolium pratense*), embryogenic capacity has been determined to be under the control of additive genetic factors (Keyes et al. 1980).

3 Histodifferentiation of Somatic Embryos

Zygotic embryo ontogeny has been described as a sequence of three confluent stages: histodifferentiation, maturation, and desiccation (Kermode 1990), and distinct patterns of gene expression are associated with these stages (Hughes and Galau 1989). Histodifferentiation refers to the stage of extensive cell division starting with the zygote and ending with a differentiated, immature, cotyledon-stage embryo. Maturation is marked by cell expansion and accumulation of storage reserves. Development ends when the embryo desiccates and becomes quiescent (Kermode 1990). Here, somatic embryo ontogeny is described as it parallels the stages of zygotic embryogenesis. As the process of somatic embryogenesis is better understood and better regulated, zygotic and somatic embryogenesis will become more similar at the molecular, developmental, and morphological levels.

Histodifferentiation of somatic embryos begins after induction of the embryogenic state. However, the histodifferentiation stage of some species is suppressed until auxins are removed from the medium (Walker et al. 1979) or, at least, until the auxin concentration is lowered to a threshold that will permit somatic embryo development (Fig. 2). In species where embryos are capable of developing in the presence of an auxin the continued presence of an exogenous auxin is detrimental to normal development (Fig. 2). Auxins suppress the development of the apical meristem, probably by the same mechanism involved in the establishment of apical dominance. While the original observation of this phenomenon was made in carrot somatic embryos, the same observation has been made for those of soybean (Parrott et al. 1988) and crownvetch (Dušková et al. 1990).

4 Maturation of Somatic Embryos

Embryo maturation begins after histodifferentiation is complete. In legumes, growth by mitosis stops, and growth through cell expansion begins as cells accumulate storage reserves (Raghavan 1986). There is a traditional tendency in

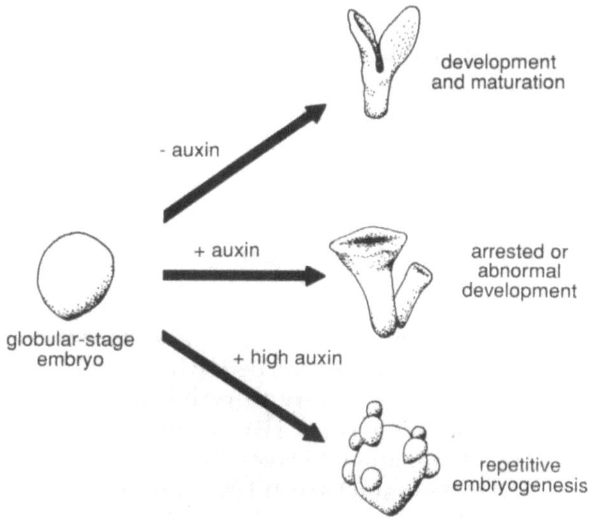

Fig. 2. Effect of exogenous auxins on somatic embryo development. (Drawing by J. Adang)

published reports to provide various growth regulators in the medium during this stage. Yet, there is enough information suggesting that exogenous auxin or cytokinin is not necessary for normal embryo maturation, as evidenced by the normal development of legume zygotic (Stafford and Davies 1979; Hsu and Obendorf 1982) or somatic embryos (e.g., Parrott et al. 1988) on medium devoid of all growth regulators. Poorly developed meristem or swollen hypocotyls may result from the application of exogenous auxins or cytokinins, respectively. Consequently, a treatment that binds and removes auxins, such as the addition of activated charcoal to the maturation medium (Ebert and Taylor 1990), can improve embryo normalcy and enhance germination (Buchheim et al. 1989). However, application of activated charcoal at the wrong stage, such as during the induction phase (instead of the development/maturation phase) suppresses embryogenesis (Genga and Allavena 1991).

In contrast to auxins and cytokinins, abscisic acid (ABA) may be necessary during embryogenesis to initiate the synthesis of storage proteins and proteins involved in dessication tolerance (Sussex and Dale 1979; Barratt 1986; Galau et al. 1990). Alfalfa is notable in that exogenous ABA promotes maturation Denchev et al. 1991), desiccation tolerance (Senaratna et al. 1989), and subsequent conversion (Fujii et al. 1989) of somatic embryos, permitting plant recovery from as many as 80% of somatic embryos. Optimal ABA applications range from 5 μM (Gujii et al. 1990) to 10 μM (Senaratna et al. 1989). Application of ABA has also aided the recovery of plants from somatic embryos of various lupin species (Nadolska-Orczyk 1992).

The time necessary for somatic embryos to achieve physiological maturity is species-specific, mirroring the maturation period of zygotic embryos *in planta.* Studies of zygotic soybean and bean (*Phaseolus vulgaris*) embryos cultured in vitro suggest that seed growth rate and maturation are controlled by the embryo

(TeKrony et al. 1979; Adams and Rinne 1981; Egli et al. 1981; Dyer et al. 1987), and parallel that of embryos *in planta* (Long et al. 1981). Accordingly, current protocols for soybean (Buchheim et al. 1989) or peanut (Durham and Parrott 1992) include a 2–3 month maturation period (Buchheim et al. 1989), while white clover only requires 2–3 weeks (Parrott 1991).

5 Desiccation, Germination, Conversion, and Artificial Seed

Physiological maturity of legume zygotic embryos is followed by a period of desiccation. This desiccation period has been linked with the synthesis of proteins associated with the ability to germinate (Rosenberg and Rinne 1986, 1988). Partial desiccation (Fig. 1B) enhanced conversion of somatic embryos of soybean (Hammatt and Davey 1987; Parrott et al. 1988; Buchheim et al. 1989), peanut (Durham and Parrott 1992), and alfalfa (McKersie et al. 1989; Senaratna et al. 1989,1990; Anandarajah and McKersie 1990). Nevertheless, a desiccation treatment may not be universally required for all legumes. For example, somatic embryos of birdsfoot trefoil (*Lotus corniculatus*), crown vetch (Arcioni and Mariotti 1982), and white clover (Parrott 1991) germinate readily without desiccation.

Together with desiccation, photoperiod can play an important role in the successful conversion of somatic embryos of day length-sensitive species. Yet, the role of photoperiod in the conversion process has received very little attention. In soybean, the use of a 23 h photoperiod (Parrott et al. 1988) is necessary to prevent the premature induction of flowering. Once floral induction occurs, germinating somatic embryos cease all growth, even if the resulting seedling is only a few centimeters tall.

Although published research and development of artificial seed technology in legumes have been limited to alfalfa, advances in somatic embryo technology, including the ability to desiccate somatic embryos, should eventually facilitate their storage and use as artificial seed. Reviews on artificial seed technology in general (Parrott et al. 1991) and alfalfa in particular (Redenbaugh and Walker 1990) have been published recently.

Prior to the discovery that somatic embryos could withstand desiccation like their zygotic counterparts, efforts on the development of artificial seed centered on the encapsulation of somatic embryos within hydrated calcium alginate gels (Redenbaugh et al. 1986). While such gels need to be covered with an artificial seed coat to prevent drying of the capsule and prevent the capsules from sticking to each other (Redenbaugh et al. 1987a), the gel can serve as a delivery agent for nutrients, pesticides, and/or growth regulators (Redenbaugh and Walker 1990). Alternatively, somatic embryos may be desiccated and encapsulated in a water-soluble resin, although this procedure has not been attempted with leguminous somatic embryos.

Naked, desiccated somatic embryos can serve directly as artificial seeds, resulting in conversion rates of 60% for unselected alfalfa somatic embryos, and 90–100% for somatic embryos preselected for high quality (Senaratna et al.

1990). Work in our laboratory with desiccated soybean somatic embryos has resulted in conversion rates for embryos placed directly into soil that range from 25–50%, depending on the genotype. Partially desiccated soybean somatic embryos have been stored at 4 °C for 3 months, while maintaining some viability and conversion ability.

The greatest limitation to artificial seed technology appears to be the requirement for high quality somatic embryos with high conversion rates. As somatic embryo technology has progressed, so have conversion rates, but one limitation continues to be the lack of accumulation of storage reserves in somatic embryos to the extent seen in zygotic embryos. Increased levels of the 11S storage protein of alfalfa correlate with the quality of somatic embryos, but the highest concentration of 11S protein obtained in somatic embryos was still an order of magnitude less than that of zygotic embryos (Stuart et al. 1988). Other obstacles to overcome include the efficient mass production of synchronized somatic embryos, the elimination of somaclonal variation, and the price of artificial seed. Cost estimates for artificial alfalfa seed of $ 0.07 per seed (Redenbaugh and Walker 1990) do not compare favorably with the cost of $ 0.00066 per true alfalfa seed (Redenbaugh et al. 1987b).

The current status of technology will undoubtedly restrict, at least for the foreseeable future, the use of artificial seed to a few specialty applications. Within the confines of a laboratory, artificial seed may serve as a relatively easy way to preserve valuable cell lines or genotypes. Commercially, artificial seed may serve for the mass propagation of genotypes of ornamental or other value, but which are difficult to propagate by other methods.

6 Role of Other Medium Components

The role of other medium components has not received as much attention as growth regulators. Glucose is superior to sucrose for the induction of somatic embryos of scarlet runner bean, *Phaseolus coccineus* (Genga and Allavena 1991). The use of maltose instead of sucrose has been reported to enhance somatic embryo production and conversion in alfalfa (Strickland et al. 1987), and both maltose and glucose are better than sucrose for repetitive embryogenesis in alfalfa (Parrott and Bailey 1993). The most recent protocols for the maturation of soybean (Finer and McMullen 1991) and alfalfa (Denchov et al. 1991) somatic embryos use maltose instead of sucrose. Conversion of white clover somatic embryos is significantly enhanced when 6% maltose is used instead of the more common 3% sucrose (Weissinger and Parrott 1993). While these results suggest that maltose enhances embryo development and/or conversion, it is necessary to test maltose on a wider range of species before it becomes possible to evaluate its full potential as a medium component.

Amino acids, notably proline and its analogs such as alanine, glutamine, lysine, and serine, have also been reported to increase the frequency of somatic embryogenesis, with proline resulting in a threefold increase in embryo number. The use of glutamine, arginine, or alanine increased the size of embryos and improved conversion into plants (Stuart and Strickland 1984a, b). For all these amino acids, except glutamine, enhancing effects were dependent on the presence of NH_4^+ in the medium.

Sufficient ammonium in the basal salt mixture appears to be essential for somatic embryogenesis (Greinwald and Czygan 1991). While induction may occur with as little as 5 mM (Meijer and Brown 1987b), at least 12.5 mM is required for embryo development (Walker and Sato 1981; Meijer and Brown 1987b). If ammonium levels are too high, they may inhibit embryogenesis (Trigiano et al. 1988).

Any basal salt formulation that contains enough ammonium may probably be used, but in general, there has been a lack of critical evaluation of the various media formulations. MS (Murashige and Skoog 1962), B5 (Gamborg et al. 1968), and SH (Schenk and Hildebrandt 1972) formulations have all been used successfully. B5 has been reported to be superior to MS (Saxena and King 1987; Nadolska-Orczyk 1992) and vice versa (Lehminger-Mertens and Jacobsen 1989). Basal salts formulations optimized for specific legumes have also been developed, such as L-2 for red clover (Collins and Phillips 1982), L-6 for the moth bean, *Vigna acontifolia* (Kumar et al. 1988a), and EC6 for immature clover embryos (Maheswaran and Williams 1984). The L-6 formulation later proved to be the best formulation for the culture of tepary bean *Phaseolus acutifolius* (Kumar et al. 1988b). Finally, the possibility exists that embryogenesis may be best optimized by using different basal salts during different stages of the process. For example, while somatic embryogenesis of white clover is best induced and maintained on EC6 medium, MS salts are better for achieving conversion of embryos into plants (Weissinger and Parrott 1993).

7 Somatic Embryogenesis and Somaclonal Variation

Hanna et al. (1984), observing a lack of somaclonal variation in guinea grass regenerated from somatic embryos, postulated that less somaclonal variation is recovered from embryogenic systems than from organogenic systems, plausibly because any severe perturbations caused by somaclonal variation during ontogeny could prevent the recovery of somatic embryos. The actual amount of somaclonal variation recovered probably depends on the regeneration protocol. Somaclonal variation has been reported when somatic embryos are obtained following a callus phase, such as in red clover (Wang and Holl 1988) and alfalfa (Nagarajan and Walton 1987), and this variation increases with time in culture (Bingham et al. 1988). In contrast, somaclonal variation was not found in direct embryogenesis of white clover (Maheswaran and Williams 1987). In soybean, somaclonal variation decreases with increasing amounts of 2,4-D in the induction medium (Shoemaker et al. 1991).

8 Repetitive Embryogenesis and Mass Propagation

A common characteristic of embryogenic tissue is that it can remain embryogenic indefinitely (Terzi and Lo Schiavo 1990), a phenomenon variously termed secondary, recurrent, or repetitive embryogenesis (Figs. 1B and 2). Maintenance of an embryogenic state may require exogenous auxin (Fig. 3A), or it may occur in the absence of exogenous growth regulators (Fig. 3F). The amount of auxin required may vary with culture protocol. For example, on solid medium, soybean and white clover (Weissinger and Parrott, in prep.) require 20–40 mg/l, 2,4-D (Fig. 3B,C). In liquid medium, soybean (Finer and Nagasawa 1988) and peanut (Durham and Parrott 1992) undergo repetitive embryogenesis at 5 mg/l 2,4-D (Fig. 3D,E). Alfalfa (Fig. 3F) undergoes repetitive embryogenesis on solid medium devoid of all growth regulators (Dos Santos et al. 1983; Lupotto 1983,

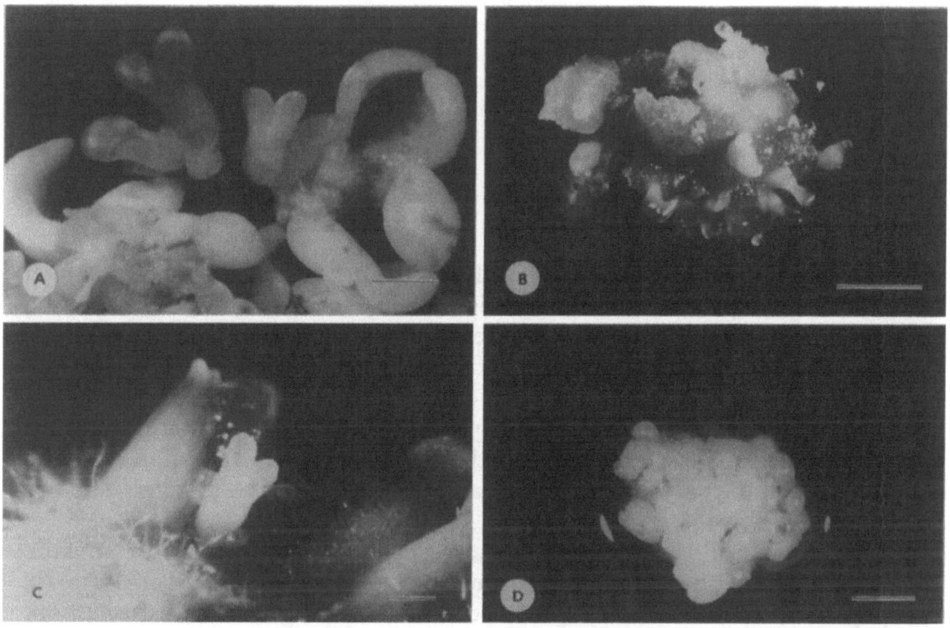

Fig. 3A–D. Somatic embryos of legumes. **A** Embryos arising from cultures of proembryogenic masses of black locust (*Robinia pseudoacacia* L.). The cultures require 3 mg/l 2,4-D to remain embryogenic; *bar* = 2.5 mm (photo courtesy of S.A. Merkle). **B** Primary somatic embryos of white clover following exposure to 40 mg/l 2,4-D; *bar* = 5 mm. **C** Mature secondary somatic embryo of white clover; *bar* = 1 mm **D** Clumps of globular somatic embryos from a suspension culture of soybean. The cultures remain embryogenic when exposed to 5 mg/l 2,4-D; *bar* = 5 mm. **E** Embryogenic clumps from a suspension culture of peanut. The cultures remain embryogenic when exposed to 5 mg/l 2, 4-D; *bar* = 5 mm **F** Recurrent somatic embryogenesis of alfalfa from a late cotyledonary stage embryo plated on growth regulator-free medium; *bar* = 5mm. **G** Development of soybean somatic embryos after removal of auxin and exposure to 6% maltose and 0.5% activated charcoal; *bar* = 5 mm. **H** Maturation of peanut somatic embryos after removal of auxin and 8 weeks of culture on growth regulator-free medium; *bar* = 5 mm

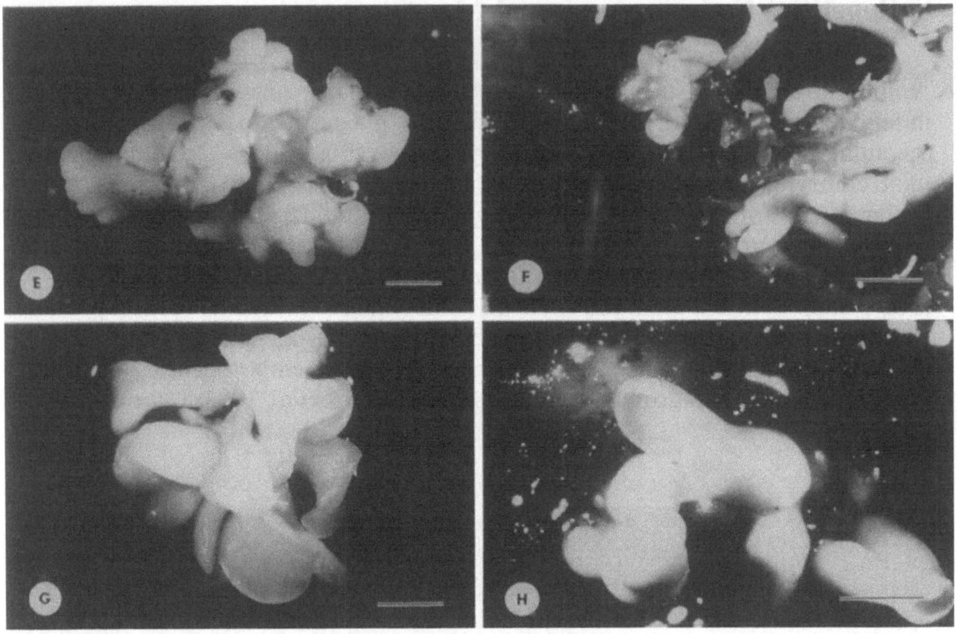

Fig. 3E–H. (*Contd.*)

1986). These particular species also differ in the developmental stage at which they undergo repetitive embryogenesis. White clover and soybean somatic embryos reach a globular stage of development before secondary embryos appear, while peanut somatic embryos reach a heart to torpedo stage before the appearance of new embryos. For white clover, peanut, and soybean, secondary embryogenesis is inhibited and embryo differentiation and maturation are promoted when auxin is removed from the medium (Fig. 3G,H). Alfalfa somatic embryos reach the cotyledonary stage and may even start to germinate before becoming repetitive.

The ability of somatic embryos to become repetitive makes somatic embryogenesis a powerful tool. Aside from its obvious potential for mass propagation, repetitive embryogenesis can be used for genetic transformation, especially in those species that do not regenerate readily from callus (Finer and McMullen 1991). Somatic embryogenic systems may be additionally modified for mass propagation by using micropropagation techniques and exposing somatic embryos to high cytokinin:low auxin ratios to induce multiple shoot formation from the apical meristem, thus increasing the number of plants obtainable from one embryo (Collins and Phillips 1982).

Table 3. Recent references on somatic embryogenesis in legumes

Species	Reference
Arachis hypogaea L.	Gill and Saxena (1992); Saxena et al. (1992)
Cicer arietinum L.	Barna and Wakhlu (1993)
Glycine max (L) Merr.	Dahmer et al. (1992); Komatsuda (1992); Liu et al. (1992); Bailey et al. (1993); Ranch (1993)
Medicago sativa L.	Kepczynski et al. (1992); Kielly and Bowley (1992); Krochko et al. (1992); Lai et al. (1992) ; McElroy and Brown (1992); Pramanik et al. (1992); Varga and Badea (1992); Xu and Bewley (1992); Bowley et al. (1993); McKersie and Bowley (1993)
Phaseolus coccineus L.	Malik and Saxena (1992a)
Pisum sativum L.	Altherr et al. (1993)
Stylosanthes scabra	Dornelas et al. (1992)
Trifolium pratense L .	Quesenberry and Smith (1993)

9 Conclusions

The number of legume species regenerated from somatic embryos is continually increasing, and the phenomenon of somatic embryogenesis is becoming better understood (Table 3). Enough is now known about the process to refine previously developed embryogenic protocols and to define embryogenic protocols for additional species. By manipulating the growth regulators in the medium, it is possible to regulate the ontogeny of somatic embryos and induce repetitive embryogenesis. The latter phenomenon is especially useful for genetic transformation and has permitted the transformation of species which do not regenerate from callus, making them amenable to nonconventional genetic improvement techniques.

10 Protocols for Somatic Embryogenesis

A detailed description of the various protocols for somatic embryogenesis in legumes would be impractical. Rather, some examples of representative regeneration schemes are listed here.

Glycine max (L.) Merr.

Two protocols (A and B) are described below. They are derived from various sources in the literature (Lazzeri et al. 1985; Finer and Nagasawa 1988; Parrott et al. 1988; Finer and McMullen 1991) and our own recent experience. Since the two protocols have much in common they are discussed separately only when they differ. All media are solidified with 0.2% (w/v) Gelrite and adjusted to pH 5.8 prior to autoclaving unless otherwise indicated. Cultures are maintained at 2–27 °C, with a light intensity of 10–75 μ E $m^{-2}s^{-2}$ and a 23-h photoperiod, unless indicated otherwise.

Growth of Source Plants

Any growth conditions which support vigorous plants and flowering are suitable. Photoperiod should be extended initially to allow for substantial vegetative growth prior to the induction of flowering, which will in turn lead to a high yield of immature seeds for explants.

Induction of Embryogenesis

1. Harvest pods when seeds are 3–6 mm long.
2. Surface-sterilize pods with 70% propanol for 30 s, followed by 1% sodium hypochlorite (20% commercial bleach) for 15 min. Rinse three times in sterile water.
3. Excise immature seeds from the pods. Remove the end of the seed containing the embryonic axis. Apply gentle pressure to the narrow end to extrude cotyledons. Select dissected cotyledons 3–5 mm long.

A) 4. Place abaxial surface of each cotyledon on medium containing MS salts, B5 vitamins, 6% sucrose, and 40 mg/l 2,4-D. Adjust medium pH to 7 prior to autoclaving.

B) 4. Same as in A, but replace 2,4-D with 10 mg/l NAA.

Maintenance of Embryogenesis

A) 1. After 4–6 weeks on induction medium transfer clumps of globular-stage embryos to 10A40N medium (Finer and Nagasawa 1988).

A) 2. Subculture monthly. Select rigorously for nodular, compact, green clumps of embryogenic tissue. Alternatively, clumps of globular stage embryos may be subcultured monthly on solid induction medium with the 2,4-D level reduced to 20 mg/l and pH adjusted to 5.8

Development of Somatic Embryos

A) After 4–6 weeks on induction medium, or at any time on maintenance medium, transfer globular-stage somatic embryos to a growth regulator-free medium consisting of MS salts, B5 vitamins, 6% maltose, and 0.5% activated charcoal (MS M6AC).

B) After 10 days on induction medium, transfer whole explants to growth regulator-free medium with MS salts, B5 vitamins, and 3% sucrose (MS0).

Maturation of Somatic Embryos

1. After 4 weeks on development medium (MS0 or MS M6AC), transfer cotyledon stage somatic embryos to growth regulator-free maturation medium with MS salts, B5 vitamins, and 6% maltose.
2. After 4 weeks on maturation medium, transfer mature somatic embryos to a desiccation chamber consisting of a dry Petri plate containing a 1-cm^3 block of maturation medium to maintain high humidity. Seal plate with Nescofilm. Some genotypes require desiccation at 85% RH, which may be obtained using saturated KCl.

Germination and Conversion of Somatic Embryos

1. After 7 days in a desiccation chamber, transfer embryos to MS0.
2. Transfer seedling with roots and primary foliage leaves (usually after 1–3 weeks) to a GA-7 Magenta vessel containing MS0.
3. Transfer healthy plants to a 6.35-cm pot containing a sterilized 1:1 mixture of sand and commercial potting mix. Place potted plant inside sterile GA-7 vessel with GA-7 coupler (Magenta).

4. After 1 week in soil acclimatize plant by progressively loosening the GA-7 coupler.
5. Transfer hardened plants to a greenhouse.

Medicago sativa L.

Here, we report only on repetitive somatic embryogenesis, as modified from the protocol of Lupotto (1986). Other protocols for somatic embryogenesis of alfalfa have been described elsewhere (Saunders and Bingham 1972; Meijer and Brown 1987a; Denchov et al. 1991).

Callus Induction

1. Harvest the top 5 cm of 10-cm plant.
2. Surface-sterilize in 70% 2-propanol for 30s followed by immersion in 1.05% sodium hypochlorite for 12 min, and three rinses in sterile, deionized water.
3. Place petiole and /or leaf sections on B1INK medium, consisting of Blaydes salts and vitamins (Blaydes 1966), 3% sucrose, 3 g/l Gelrite as solidifying agent, pH adjusted to 5.8, and supplemented with 2 mg/l each of IAA, NAA, and kinetin.
4. Subculture monthly.

Initiation of Repetitive Somatic Embryogenesis

1. Transfer cotyledon-stage somatic embryos which develop on the surface of the callus to MS0 medium.
2. Secondary embryos develop from the surface of primary embryos. Excise and subculture new embryos at monthly intervals.

Conversion of Secondary Somatic Embryos

1. Some embryos germinate and develop foliage leaves prior to becoming embryogenic. Transfer these seedlings directly to 6.35-cm pots containing a sterilized mixture of 1 : 1 sand and commercial potting mix.
2. Acclimatize and transfer plants to the greenhouse by the soybean protocol (above).

Trifolium repens L.

A) Modified protocols for the cytokinin-based induction of primary (Maheswaran and Williams 1984) and secondary (Maheswaran and Williams 1986a) embryogenesis are described. Genetic fidelity of plants regenerated by both protocols has been documented (Maheswaran and Williams 1987).
B) Regeneration via auxin-induced embryogenesis (Parrott 1991; Weissinger and Parrott 1993).

Growth of Source Plants

Source plants are cross-pollinated by hand to produce embryos for dissection.

Initiation of Cultures

A) 1. Developing seed pods are harvested 5–7 days after pollination, depending on temperature and growth conditions, to yield torpedo-stage embryos 0.4–0.8 mm in length.
B) 1. Developing seed pods are harvested 7–8 days after pollination, depending on temperature and growth conditions, to yield cotyledon-state somatic embryos. The endosperm has solidified at this point.
2. The pods are surface-sterilized for 30 s in a solution of 70% 2-propanol (v/v), followed by 12 min in a 1% sodium hypochlorite (20% commercial bleach) solution, then washed thoroughly with sterile water and dissected aseptically.
A) 3. Seed coats are removed, and embryos are placed on EC6 medium containing 0.05 mg/l 6-benzylaminopurine (BAP) (Maheswaran and Williams 1984).
B) 3. Seed coats and embryonic axes are removed, and cotyledons placed adaxial side up, on EC6 basal medium (Maheswaran and Williams 1984) supplemented with 40 mg/l 2,4-D for 10 days.

Embryo Development and Conversion

A) 1. Somatic embryos arise from the hypocotyl portion of the zygotic embryo 7–9 days after culture initiation.
B) 1. Transfer cotyledons to MS0 Medium (growth regulator-free MS salts, B5 vitamins and 3% sucrose).
A) 2. Somatic embryos germinate approximately 3 weeks after culture initiation. Transfer germinated embryos to growth regulator-free EC6 medium.
B) 2. Somatic embryos mature after about 3 weeks. Separate from explant, and transfer individual embryos to MS0 till germination occurs.
3. After 1–2 weeks, roots are well developed allowing plantlets to be transferred to soil and hardened off.
Note. Somatic embryos have also been obtained from zygotic embryos *T. pratense, T. resupinatum, T. subterraneum*, and *Medicago sativa* using cytokinin-stimulated embryogenesis.

Initiation of Cultures and Conversion of Secondary Embryos

A) 1. Primary somatic embryos are initiated as described above.
B) 1. Primary somatic embryos are initiated as described above, but the explant cotyledons remain on medium with 40 mg/l 2,4-D for 30 days.
A) 2. Primary embryos at the torpedo to cotyledonary stage are removed from the zygotic embryo and placed on EC6 medium with 2 mg/l BAP.
B) 2. Primary embryos at the globular to heart stage are removed from the zygotic cotyledon and placed on EC6 medium with 20–40 mg/l 2,4-D.
A) 3. After 2 weeks, secondary embryos at various stages of development are apparent on all parts of the primary embryo that directly contact the medium . Transfer new embryos to EC6 medium with BAP to repeat the cycle up to one additional time, or transfer to germination medium (below).
B) 3. New somatic embryos form after 2–4 weeks. Transfer the new embryos to EC6 medium with 2,4-D to repeat the cycle or transfer to development / maturation medium (below).
A) 4. Upon germination, transfer secondary embryos to growth regulator-free EC6 medium for root development, then transfer to soil.
B) 4. Transfer embryos to growth regulator-free medium consisting of MS salts, B5 vitamins, and 6% maltose for germination.

References

Adams CA, Rinne RW (1981) Seed maturation in soybeans (*Glycine max* L. Merr.) is independent of seed mass and of the parent plant, yet is necessary for production of viable seeds. J Exp Bot 32: 615–620

Allen ON, Allen EK (1981) The Leguminosae. A source book of characteristics, uses, and nodulation. The University of Wisconsin Press Madison

Altherr S, Stirn S, Jacobsen H-J (1993) Immunobiochemical analysis of a nuclear protein marker for regeneration potential in higher plants. J Plant Physiol 141: 415–422

Anandarajah K, McKersie BD (1990a) Enhanced vigor of dry somatic embryos of *Medicago sativa* L. with increased sucrose. Plant Sci 71: 261–266

Anandarajah K, McKersie BD (1990b) Manipulating the desiccation tolerance and vigor of dry somatic embryos of *Medicago sativa* L. with sucrose, heat shock and abscisic acid. Plant Cell Rep 9: 451–455

Arcioni S, Mariotti (1982) Tissue culture and plant regeneration in the forage legumes *Onobrychis viciaefolia* Scop., *Coronilla varia* and *Lotus corniculatus* L. In: Fujiwara A (ed) Plant tissue culture 1982. Maruzen, Tokyo, 11: 707–708

Arcioni S, Davey MR, Dos Santos AVP, Cocking EC (1982) Somatic embryogenesis in tissues from mesophyll and cells suspension protoplasts of *Medicago coerula* and *M. glutinosa*. Z Pflanzenphysiol 106: 105–110

Arcioni S, Mariotti D, Damiani F, Pezzotti M (1988) Birdsfoot trefoil (*Lotus corniculatus* L.) crownvetch (*Coronilla varia* L.) and sainfoin (*Onobrychis viciifolia* Scop.). In Bajaj YPS (ed) Biotechnology in agriculture and forestry vol. 6. Crops II. Springer, Berlin Heidelberg New York, pp 548–572

Bailey MA Boerma HR, Parrott WA (1993) Genotype effects on repetitive embryogenesis and plant regeneration of soybean. In Vitro Cell Dev Biol 29P:102–108

Bajaj YPS (1989) Some Indian ornamental trees: *Cassia fistula* Linn., *Poinciana regia* (Boj.) and *Jacaranda acutifolia* auct. In: Bajaj YPS (ed) Biotechnology in agriculture and forestry, vol 5. Trees II. Springer, Berlin Heidelberg New York, pp 469–476

Baker CM, Wetzstein HY (1992) Somatic embryogenesis and plant regeneration from leaflets of peanut, *Arachis hypogaea*. Plant Cell Rep 11: 71–75

Barna KS, Wakhlu AK (1993) Somatic Embryogenesis and plant regeneration from callus cultures of chickpea (*Cicer arietinum* L.). Plant Cell Rep 12: 521–524

Barratt DHP (1986) Regulation of storage protein accumulation by abscisic acid in *Vicia faba* L. cotyledons cultured in vitro. Ann Bot 57: 245–256

Barwale UB , Kerns HR, Widholm JM (1986) Plant regeneration from callus cultures of several soybean genotypes via embryogenesis and organogenesis. Planta 167: 473–481

Beversdorf WD, Bingham ET (1977) Degrees of differentiation obtained in tissue cultures of *Glycine* species. Crop Sci 17: 307–311

Bhojwani SS, Mullins K, Cohen D (1984) Intra-varietal variation for in vitro plant regeneration in the genus *Trifolium*. Euphytica 33: 915–921

Bianchi S, Flament P, Dattee Y (1988) Somatic embryogenesis and organogenesis in alfalfa; genotypic variation in regeneration ability. Agronomie 8: 121–126

Binarová P. Dolezel J (1988) Alfalfa embryogenic cell suspension culture: growth and ploidy level stability. J Plant Physiol 133: 561–566

Binarová P, Nedelník J, Nedbálková B (1990) Selection for resistance to filtrates of *Fusarium* spp. in embryogenic cell suspension culture of *Medicago sativa* L. Pant Cell Tissue Organ Cult 22: 191–196

Bingham ET, Hurley LV, Kaatz DM, Saunders JW (1975) Breeding alfalfa which regenerate from callus tissue culture. Crop Sci 15: 719–721

Bingham ET, McCoy TJ, Walker KA (1988) Alfalfa tissue culture . In: Hanson AA, Barnes DK, Hill RR, Jr (eds) Alfalfa and alfalfa improvement. ASA-CSSA-SSSA, Madison WI, pp 903–929

Blaydes DF (1966) Interaction of kinetin and various inhibitors in the growth of soybean tissue. Physiol Plant 19: 748–753

Bond JE, Webb KJ (1989) Regeneration and analysis of plants from stolon segments of *Trifolium repens* (white clover). Plant Sci 61: 119–125

Bonnbelle C, Lejeune F, Fournier D, Tourte Y (1990) Remaniements infrastructuraux des cellules cotylédonaires en culture chez deux légumineuses; relations avec l'acquisition de potentialités embryogènes. CR Acad Sci Paris 310: 657–664

Bowely SR, Kielly GA, Anandarajah K, McKersie BD, Senaratna T (1993) Field evaluation following two cycles of backcross transfer of somatic embryogenesis to commercial alfalfa germplasm. Can J Plant Sci 73: 131–137

Brown DCW, Atanassov A (1985) Role of genetic background in somatic embryogenesis in *Medicago*. Plant Cell Tissue Organ Cult 4: 111–122

Buchheim JA, Colburn SM. Ranch JP (1989) Maturation of soybean somatic embryos and the transition to plantlet growth. Plant Physiol 89: 768–775

Chen THH, Marowitch J (1987) Screening of *Medicago falcata* germplasm for in vitro regeneration. J plant Physiol 128: 271–277

Chen THH, Marowitch J, Thompson BG (1987) Genotypic effects on somatic embryogenesis and plant regeneration from callus cultures of alfalfa. Plant Cell Tissue Organ Cult 8: 73–81

Chen HK, Mok MC, Mok DWS (1990) Somatic embryogenesis and shoot organogenesis from interspecific hybrid embryos of *Vigna glabrescens* and *V. radiata*. Plant Cell Rep 9: 77–79

Choo TM (1988) Plant regeneration in zigzag clover (*Trifolium medium* L.). Plant Cell Rep 7: 246–248

Christianson ML, Warnick DA, Carlson PS (1983) A morphogenetically competent soybean suspension culture, Science 222: 632–634

Christou P, Yang N-S (1989) Developmental aspects of soybean (*Glycine max*) somatic embryogenesis. Ann Bot 64: 225–234

Collins GB, Phillips GC (1982) In vitro tissue culture and plant regeneration in *Trifolium pratense* L. In: Earle ED, Demarly Y (eds) Variability in plants regenerated from tissue culture. Praeger, New York, pp 22–34

Cui D, Myers JR,Collins GB, Lazzeri PA (1988) In vitro regeneration in *Trifolium*. 1. Direct somatic embryogenesis in *T. rubens* (L.). Plant Cell Tissue Organ Cult 15 : 33–45

Dahmer ML, Hildebrand DF and Collins GB (1992) Comparative protein accumulation patterns in soybean somatic and zygotic embryos. In Vitro Cell Dev Biol 28P: 106–114

Denchev P, Velcheva M, Atanassov A (1991) A new approach to direct somatic embryogenesis in *Medicago*. Plant Cell Rep 10: 338–341

Dhanalakshmi S, Lakshmanan KK (1992) In vitro somatic embryogenesis and plant regeneration in *Clitoria ternatea*. J Exp Bot 43: 213–219

Dijak M, Smith DL, Wilson TJ, Brown DCW (1986) Stimulation of direct embryogenesis from mesophyll protoplasts of *Medicago sativa* Plant Cell Rep 5: 468–470

Dornelas MC, Vieira MLC, Appezzato-da-Gloria B (1992) Histological analysis of organogenesis and somatic embryogenesis induced in immature tissues of *Stylosanthes scabra* Ann Bot 70: 477–482

Dos Santos AVP, Outka DE, Cocking EC, Davey MR (1980) Organogenesis and somatic embryogenesis in tissues derived from leaf protoplasts and leaf explants of *Medicago sativa*. Z Pflanzenphysiol 99: 261–270

Dos Santos AVP, Cutter EG Davey MR (1983) Origin and development of somatic embryos in *Medicago sativa* (alfalfa). Protoplasma 117: 107–115

Durham RE, Parrott WA (1992) Repetitive somatic embryogenesis in liquid cultures of peanut. Plant Cell Rep 11: 122–125

Dušková J, Opantrný Z, Sovová M, Dušek J (1990) Somatic embryogenesis and plant regeneration in *Coronilla varia* L. (crowvetch) long term tissue cultures Biol Plant 32: 8–13

Dyer DJ, Cotterman CD, Cotterman JC (1987) Comparison of in situ and in vitro regulation of soybean seed growth and development. Plant Physiol 84: 298–303

Eapen S, George L (1990)Ontogeny of somatic embryos of *Vigna aconitifolia*, *Vigna mungo* and *Vigna radiata*. Ann Bot 66: 219–226

Ebert A, Taylor HF (1990) Assessment of the changes of 2,4-dichlorophenoxyacetic acid concentrations in plant tissue culture media in the presence of activated charcoal. Plant Cell Tissue Organ Cult 20: 165–172

Egli DB, Fraser J, Leggett JE, Poneleit CG (1981) Control of seed growth in soya beans [*Glycine max* (L.) Merrill]. Ann Bot 48: 171–178

Feher F, Tarczy MH, Bocsa I, Dudits D (1989) Somaclonal chromosome variation in tetraploid alfalfa. Plant Sci 60: 91–99

Feng X-H, Jiang X-C, Shao Q-Q (1989) Plant regeneration from immature cotyledonous tissue cultures of soybean by somatic embryogenesis . Sci Chi Ser B 32: 58–63
Ferreira AG, Ching Y-H, Romanato Santarém E (1991) Somatic embryogenesis of soybean (*Glycine max* (L.) Merrill), Brazilian cultivars Ivorá and IAS-5 Øyton 51: 139–144
Finer JJ (1988) Apical proliferation of embryogenic tissue of soybean [*Glycine max* (L.) Merrill]. Plant Cell Rep 7: 238–241
Finer JJ , McMullen MD (1991) Transformation of soybean via particle bombardment of embryogenic suspension culture tissue. In Vitro Cell Dev Biol 27P: 175–182
Finer JJ Nagasawa A (1988) Development of an embryogenic suspension culture of soybean (*Glycine max* Merrill.). Plant Cell Tissue Organ Cult 15: 125–136
Frame B, Yu K-Y, Christie BR, Pauls KP (1991) In vitro selection for resistance to verticillium wilt in alfalfa (*Medicago sativa* L.) using a fungal culture filtrate. Physiol Mol Plant Pathol 39: 325–340
Fujii JAA, Slade D, Redenbaugh K (1989) Maturation and greenhouse planting of alfalfa artificial seeds. In Vitro Cell Dev Biol 25: 1179–1182
Fujii JAA, Slade D, Olsen R, Ruzin SE, Redenbaugh K (1990) Alfalfa Somatic embryo maturation and conversion to plants. Plant Sci 72: 93–100
Galau GA, Jakobsen KS, Hughes DW (1990) The controls of later dicot embryogenesis. Physiol Plant 81: 280–288
Gamborg OL, Miller RA, Ojima K (1968) Nutrient requirements of suspension cultures of soybean root cells. Exp Cell Res 50: 151–158
Gamborg OL, Davis BP, Stahlquist RW (1983) Somatic embryogenesis in cell cultures of *Glycine* species. Plant Cell Rep 2: 209–212
Geneve RL, Kester ST (1990) The initiation of somatic embryos and adventitious roots from developing zygotic embryo explants of *Cercis canadensis* L. cultured in vitro . Plant Cell Tissue Organ Cult 22: 71–76
Genga A, Allavena A (1991) Factors affecting morphogenesis from immature cotyledons of *Phaseolus coccineus* L. Plant Cell Tissue Organ Cult 27: 189–196
Gharyal PK, Maheshwari SC (1983) In vitro differentiation of somatic embryoids in a leguminous tree – *Albizzia lebbeck* L. Naturwissenschapen 68: 379–380
Ghazi TD, Cheema HK, Nabors MW (1986) Somatic embryogenesis and plant regeneration from embryogenic callus of soybean, *Glycine max* L. Plant Cell Rep 5: 452–456
Gill R, Saxena PK (1992) Direct somatic embryogenesis and regeneration of plants from seedling explants of peanut (*Arachis hypogaea*): promotive role of thidiazuron. Can J Bot 70: 1186–1192
Gilmour M, Davey MR, Cocking EC (1987) Plant regeneration from cotyledon protoplasts of wild *Medicago* species. Plant Sci 48: 107–112
Grant JE (1984) Regeneration from cotyledonary tissue of *Glycine canescens*, a perennial relative of the soybean. Plant Cell Tissue Organ Cult 3: 169–173
Greinwald R, Czygan F-C (1991) Regeneration of plantlets from callus cultures of *Chamaecytisus purpureus* and *Chamaecytisus austriacus* (Leguminosae). Bot Acta 104: 64–67
Griga M, Kubaláková M, Tejklová E (1987) Somatic embryogenesis in *Vicia faba* L. Plant Cell Tissue Organ Cult 9: 167–171
Hammatt N, Davey MR (1987) Somatic embryogenesis and plant regeneration from cultured zygotic embryos of soybean (*Glycine max* L. Merr). J Plant Physiol 128: 219–226
Hanna WW, Lu C, Vasil IK (1984) Uniformity of plants regenerated from somatic embryos of *Panicum maximum* Jacq. (guinea grass) . Theor Appl Genet 67: 155–159
Hartman CL, McCoy TJ, Knous TR (1984) Selection of alfalfa (*Medicago sativa*) cell lines and regeneration of plants resistant to the toxin(s) produced by *Fusarium oxysporum* f. sp. *medicaginis*. Plant Sci Lett 34: 183–194
Hartweck LM, Lazzeri PA, Cui D, Collins GB Williams EG (1988) Auxin-orientation effects on somatic embryogenesis from immature soybean cotyledons. In Vitro Cell Dev Biol 24: 821–828
Hazra S, Sathaye SS, Mascarenhas AF (1989) Direct somatic embryogenesis in peanut (*Arachis hypogaea*). Bio/Technology 7: 949–951
Hepher A, Boulter ME, Harris N, Nelson RS (1988) Development of a superficial meristem during somatic embryogenesis from immature cotyledons of soybean (*Glycine max* L.). Ann Bot 62: 513–519
Hernández-Fernández MM, Christie BR (1989) Inheritance of somatic embryogenesis in alfalfa (*Medicago sativa* L.). Genome 32: 318–321

Hsu FC, Obendorf RL (1982) Compositional analysis of in vitro matured soybean seeds. Plant Sci Lett 27:129–135

Hughes DW, Galau GA (1989) Temporally modular gene expression during cotyledon development. Genes Dev 3: 358–369

Huyghe C (1990) Les cultures in vitro chez les légumineuses à grosses graines. Agronomie 10: 29–49

Jacobsen H-J (1991) Somatic embryogenesis in seed legumes: The possible role of soluble auxin receptors. Isr. J Bot 40: 139–143

Johnson LB, Stuteville DL, Higgins RK, Skinner DZ (1981) Regeneration of alfalfa plants from protoplasts of selected Regen S clones, Plant Sci Lett 20: 297–304

Kao KN, Michayluk MR (1980) Plant regeneration from mesophyll protoplasts of alfalfa. Z Pflanzenphysiol 96: 135–141

Kao KN, Michayluk MR (1981) Embryoid formation in alfalfa cell suspension cultures from different plants . In Vitro Cell Dev Biol 17: 645–648

Kepczynski J, McKersie BD, Brown DCW (1992) Requirement of ethylene for growth of callus and somatic embryogenesis in *Medicago sativa* L. J Exp Bot 43: 1199–1202

Kerns HR, Barwale UB Meyer MM Jr, Widholm JM (1986) Correlation of cotyledonary node shoot proliferation and somatic embryoid development in suspension cultures of soybean (*Glycine max* (L.) Merr.). Plant Cell Rep 5: 140–143

Keyes GJ, Collins GB, Taylor NL (1980) Genetic variation in tissue-cultures of red clover. Theor Appl Genet 58: 265–271

Kermode AR (1990) Regulatory mechanism involved in the transition from seed development to maturation Crit Rev Plant Sci 9: 155–195

Kielly GA, Bowely SR (1992) Genetic control of somatic embryogenesis in alfalfa. Genome 35: 474–477

Kein TD, Mehandjiev A, Phuc CB (1989) Factors influencing the somatic embryogenesis in soybean (*Glycine max*. L. Merr.). CR Acad Bull Sci 42: 91–93

Kiss E, Heszky E, Gyulai G, Horváth Z, Csillag A (1991) Neomorph and leaf differentiation as alternative morphogenetic pathways in soybean tissue culture. Acta Biol Hung 42: 313–321

Komatsuda T (1990) Ability of soybean (*Glycine max* L. Merr.) to produce somatic embryos on a medium containing a low concentration of sucrose, Jpn J Breed 40: 371–375

Komatsuda T (1992) Research on somatic embryogenesis and plant regeneration. Bull Nastl Inst Agrobiol 7: 1–78

Komatsuda T, Ko S-W (1990) Screening of Soybean [*Glycine max* (L.) Merrill] genotype for somatic embryoproduction from immature embryo, Jpn J Breed 40: 249–251

Komatsuda T, Ohyama K (1988) Genotypes of high competence for somatic embryogenesis and plant regeneration in soybean *Glycine max*. Theor Appl Genet 75: 695–700

Komatsuda T, Kaneko K, Oka S (191) Genotype × sucrose interactions for somatic embryogenesis in soybean . Crop Sci 31: 333–337

Krochko JE, Pramanik SK, Bewley JD (1992) Contrasting storage protein synthesis and messenger RNA accumulation during development of zygotic and somatic embryos of alfalfa (*Medicago sativa* L.) Plant Physiol 99:46–53

Kumar AS, Gamborg OL, Nabors MW (1988a) Plant regeneration from cell suspension cultures of *Vigna aconitifolia*. Plant Cell Rep 7: 138–141

Kumar AS, Gamborg OL, Nabors MW (1988b) Regeneration from long-term cell suspension cultures of tepary bean (*Phaseolous acutifolius*). Plant Cell Rep 7 : 322–325

Kysely W, Jacobsen H-J (1990) Somatic embryogenesis from pea embryos and shoot apices. Plant Cell Tissue Organ Cult 20: 7–14

Kysely W, Myers JR, Lazzeri PA, Collins GB, Jacobsen H-J (1987) Plant regeneration via somatic embryogenesis in pea (*Pisum sativum* L.). Plant Cell Rep 6: 305–308

Lai F, Senaratna T, McKersie BD (1992) Glutamine enhances storage protein synthesis in *Medicago sativa* L. somatic embryos. Plant Sci. 87: 69–77

Latunde-Dada AO, Lucas JA (1988) Somaclonal variation and resistance to *Verticillium* wilt in lucerne, *Medicago sativa* L., plants regenerated from callus. Plant Sci, 58: 111–119

Lazzeri PA, Hilderbrand DF, Collins GB (1985) A procedure for plant regeneration from immature cotyledon tissue of soybean. Plant Mol Biol Rep 3: 160–167

Lazzeri PA. Hildebrand DF, Collins GB (1987a) Soybean somatic embryogenesis: effects of hormones and culture manipulations. Plant Mol Biol Rep 10: 197–208

Lazzeri PA, Hildebrand DF, Collins GB (1987b) Soybean somatic embryogenesis: effects of nutritional, physical and chemical factors. Plant Mol Biol Rep 10: 209–220

Lazzeri PA, Hildebrand DF, Sunega J, Williams EG, Collins GB (1988) Soybean somatic embryogenesis: interactions between sucrose and auxin. Plant Cell Rep 7: 517–520

Lehminger-Mertens R, Jacobsen H-J (1989) Plant regeneration from pea protoplasts via somatic embryogenesis. Plant Cell Rep 8: 379–382

Li BJ, Langridge WHR, Szalay AA (1985) Somatic embryogenesis and plantlet regeneration in the soybean, *Glycine max*. Plant Cell Rep 4: 344–347

Lippmann B, Lippmann G (1984) Induction of somatic embryos in cotyledonary tissue of soybean , *Glycine max* L. Merr. Plant Cell Rep. 3: 215–218

Liu W, Moore PJ, Collins GB (1992) Somatic embryogenesis in soybean via somatic embryo cycling. In Vitro Cell Dev Biol 28P: 153–160

Long SR, Dale RMK, Sussex IM (1981) Maturation and germination of *Phaseolus vulgaris* embryonic axes in culture. Planta 153: 405–415

Lu DY, Pental D, Cocking EC (1982) Plant regeneration from seedling cotyledon protoplasts. Z Pflanzenphysiol 107: 59–63

Lu DY, Davey MR, Cocking EC (1983) A comparison of the cultural behavior of protoplasts from leaves, cotyledons and roots of *Medicago sativa*. Plant Sci Lett 31: 87–99

Lupotto E (1983) Propagation of an embryogenic culture of *Medicago sativa* L. Z Pflanzenphysiol 111: 95–104

Lupotto E (1986) The use of single somatic embryo culture in propagating and regenerating lucerne (*Medicago sativa* L.). Ann Bot 57: 19–24

Maheswaran G, Williams EG (1984) Direct somatic embryoid formation on immature embryos of *Trifolium repens*, *T. pratense* and *Medicago sativa*, and rapid clonal propagation of *T. repens*. Ann Bot 54: 201–211

Maheswaran G, Williams EG (1985) Origin and development of somatic embryoids formed directly on immature embryos of *Trifolium repens* in vitro. Ann Bot 56: 618–630

Maheswaran G, Williams EG (1986a) Direct secondary somatic embryogenesis from immature sexual embryos of *Trifolium repens* cultured in vitro. Ann Bot 57: 109–117

Maheswaran G, Williams EG (1986b) Clonal propagation of *Trifolium pratense*, *Trifolium* resupinatum and *Trifolium subterraneum* by direct somatic embryogenesis on cultured immature embryos. Plant Cell Rep 3: 165–168

Maheswaran G, Williams EG (1987) Uniformity of plants regenerated by direct somatic embryogenesis from zygotic embryos of *Trifolium repens*. Ann Bot 59: 93–97

Malik KA, Saxena PK (1992a) *In vitro* regeneration of plants: a novel approach. Naturwiss enschaften 79: 136–137

Malik KA, Saxena PK (1992b) Somatic embryogenesis and shoot regeneration from intact seedling of *Phaseolus acutifolius A. P. aureus* (L.) *Wilczek, P . coccineus* L., and *P. wrightii* L. Plant Cell Rep 11: 163–168

Martins IS, Sondahl MR (1984) Early stages of somatic embryo differentiation from callus cells of bean (*Phaseolus vulgaris* L.) grown in liquid medium. J Plant Physiol 117: 97–103

Matheson SL, Nowak J, MacLean NL (1990) Selection of regenerative genotypes from highly productive cultivars of alfalfa. Euphytica 45: 105–112

McElroy AR, Brown DCW (1992) A transplant plug technique for production of alfalfa (*Medicago sativa* L.) plants from somatic embryos. Can J Plant Sci 72: 483–485

McGee JD, Williams EG, Collins GB, Hildebrand DF (1989) Somatic embryogenesis in *Trifolium*: protein profiles associated with high- and low-frequency regeneration. J Plant Physiol 135: 306–312

McHughen A, Swartz M (1984) Embryo formation in cell cultures of the legume *Cyamopsis tetragonoloba* (guar). J Plant Physiol 117: 163–167

McKersie BD, Bowley SR (1993) Synthetic seeds of alfalfa. In: Redenbaugh K (ed) Synseeds. Applications of synthetic seeds to crop improvement. CRC Press, Boca Raton, pp 231–253

McKersie BD, Senaratna T, Bowley SR, Brown DCW, Krochko JE, Bewley JD (1989) Application of artificial seed technology in the production of hybrid alfalfa (*Medicago sativa* L.) . In Vitro Cell Dev Biol 25: 1183–1188

Meijer EGM (1989) Developmental aspects of ethylene biosynthesis during somatic embryogenesis in tissue cultures of *Medicago sativa* J Exp Bot 40: 479–484

Meijer EGM, Brown DCW (1985) Screening of diploid *Medicago sativa* germplasm for somatic embryogenesis. Plant Cell Rep 4: 285–288

Meijer EGM, Brown DCW (1987a) A novel system for rapid high frequency somatic embryogenesis in *Medicago sativa*. Physiol Plant 69: 591–596

Meijer EGM, Brown DCW (1987b) Role of exogenous reduced nitrogen and sucrose in rapid high frequency somatic embryogenesis in *Medicago sativa*. Plant Cell Tissue Organ Cult 10: 11–20

Meijer EGM, Brown DCW (1988) Inhibition of somatic embryogenesis in tissue cultures of *Medicago sativa* by aminoethoxyvinylglycine, amino-oxyacetic acid, 2,4-dinitrophenol and salicylic acid at concentrations which do not inhibit ethylene biosynthesis and growth. J Exp Bot 39: 263–270

Meijer EGM, Simmonds J (1988) Polyamine levels in relation to growth and somatic embryogenesis in tissue cultures of *Medicago sativa* L. J Exp Bot 39: 787–794

Merkle AS, Wiecko AT (1989) Regeneration of *Robinia psuedoacacia* via somatic embryogenesis. Can J For Res 19: 285–288

Merkle SA, Parrott WA, Williams EG (1990) Applications of somatic embryogenesis and embryo cloning. In: Bhojwani SS (ed) Plant tissue culture: applications and limitations. Elsevier Amsterdam, pp 67–102

Mezentsev AV, Lyubavina LA, Karelina NA (1982) Cell cultures in the breeding of clover and alfalfa. Sov Agric Sci 7: 7–10

Mitten DH, Sato SJ, Skokut TA (1984) In vitro regenerative potential of alfalfa germplasm sources. Crop Sci 24: 943–945

Murashige T, Skoog F (1962) A revised medium for rapid growth and bioassays with tobacco tissue cultures. Physiol Plant 15: 473–497

Nadolska-Orczyk A (1992) Somatic embryogenesis of agriculturally important lupin species (*Lupinus angustifolius, L. albus, L. mutabilis*). Plant Cell Tissue Organ Cult 28: 19–25

Nagarajan P. Walton PD (1987) A comparision of somatic chromosomal instability in tissue culture regenerants from *Medicago media* Pers. Plant Cell Rep 6: 109–113

Nagy B, Fehér F, Bócsa I, Dudits D (1991) The genetical transmission of embryogenity in lucerne (*Medicago sativa* L.). Növénytermelés 40: 11–20

Nichol JW, Slade D, Viss P, Stuart DA (1991) Effect of organic acid pretreatment on the regeneration and development (conversion) of whole plants from callus cultures of alfalfa *Medicago sativa* L. Plant Sci 79:181–192

Nolan KE, Rose RJ, Gorst JR (1989) Regeneration of *Medicago truncatula* from tissue culture: increased somatic embryogenesis using explants from regenerated plants. Plant Cell Rep 8: 278–281

Novåk FJ, Konecná D (1982) Somatic embryogencsis in callus and cell suspension cultures of alfalfa (*Medicago sativa* L.). Z Pflanzenphysiol 105: 279–284

Ozias-Akins P (1989) Plant regeneration from immature embryos of peanut. Plant Cell Rep 8: 217–218

Ozias-Akins P, Anderson WF, Holbrook CC (1992) Somatic embryogenesis in *Arachis hypogaea* L.: genotype comparison. Plant Sci 83: 103–111

Parrott WA (1991) Auxin-stimulated somatic embryogenesis from immature cotyledons of white clover. Plant Cell Rep 10: 17–21

Parrott WA, Bailey MA (1993) Characterization of recurrent somatic embryogenesis in alfalfa. Plant Cell Tissue Organ Cult 32: 69–76

Parrott WA, Collins GB (1983) Callus and shoot-tip culture of eight *Trifolium* species in vitro with regeneration via somatic embryogenesis of *T. rubens*. Plant Sci Lett 28: 189–184

Parrott WA, Dryden G, Vogt S, Hildebrand DF, Collins GB, Williams EG (1988) Optimization of somatic embryogenesis and embryo germination in soybean. In Vitro Cell Dev Biol 24: 817–820

Parrott WA, William EG, Hildebrand DF, Collins GB (1989) Effect of genotype on somatic embryogenesis from immature cotyledons of soybean. Plant Cell Tissue Organ Cult 16: 15–21

Parrott WA, Merkle SA, Williams EG (1991) Somatic embryogenesis: potential for use in propagation and gene transfer systems. In: Murray DR (ed) Advanced methods in plant breeding and biotechnology. CAB International, Wallingford, pp 158–200

Parrott WA, Bailey MA, Durham RE, Mathews HV (1992) Tissue culture and regeneration of legumes. In: Moss JP, (ed) Proc Int Worksh on Biotechnology and crop improvement in Asia. ICRISAT, Hyderabad, India pp 115–148

Pederson GA (1986) In vitro culture and somatic embryogenesis of four *Trifolium* species . Plant Sci 45: 101–104

Phillips GC, Collins GB (1980) Somatic embryogenesis from cell suspension cultures of red clover. Crop Sci 20: 323–326

Phillips GC, Collins GB (1981) Induction and development of somatic embryos from cell suspension cultures of soybean. Plant Cell Tissue Organ Cult 1: 123–129

Pickardt T, Huancaruna Perales E, Schieder O (1989) Plant regeneration via somatic embyogenesis in *Vicia narbonensis*. Protoplasma 149: 5–10

Pramanik SK, Krochko JE, Bewley JD (1992) Distribution of cytosolic mRNAs between polysomal and ribonucleoprotein complex fractions in alfalfa embryos. Plant Physiol 99: 1590–1596

Quesenberry KH, Smith RR (1993) Recurrent selection for plant regeneration from red clover tissue culture. Crop Sci 33: 585–589

Raghavan V (1986) Embryogenesis in angiosperms. A developmental and experimental study. Cambridge University Press, New York

Ramanuja Rao IV , Mehta U, Mohan Ram HY (1982) Whole plant regeneration from cotyledonary protoplasts of *Crotalaria juncea* In : Fujiwara A (ed) Plant tissue culture 1982. Maruzen, Tokyo, pp 595–596

Ranch JP (1993) The potential for synthetic soybean seed. In: Redenbaugh K (ed) Synseeds. Applications of synthetic seeds to crop improvement. CRC Press, Boca Raton, pp 329–350

Rach JP, Oglesby L, Zielinski AC (1985) Plant regeneration from embryo-derived tissue cultures of soybean by somatic embryogenesis. In Vitro Cell Dev Biol 21: 653–657

Rao BG Chopra VL (1989) Regeneration in chickpea (*Cicer arietinum* L.) through somatic embryogenesis . J Plant physiol 134: 637–638

Ray IM, Bingham ET (1989) Breeding diploid alfalfa for regeneration from tissue culture .Crop Sci 29: 1545–1548

Redenbaugh K, Walker K (1990) Role of artificial seeds in alfalfa breeding. In: Bhojwani SS (ed) Plant tissue culture: applications and limitations. Elsevier, Amsterdam, pp 102–135

Redenbaugh K, Paasch BD, Nichol JW, Kossler ME, Viss PR, Walker K (1986 Somatic seeds encapsulation of asexual plant embryos. Bio/Technology 4: 797–801

Redenbaugh K, Slade D, Viss P, Fujii JA (1987a) Encapsulation of somatic embryos in synthetic seed coats. HortScience 33: 803–809

Redenbaugh K, Viss P, Slade D Fujii JA (1987b) Scale up: Artificial seeds. In: Green CE, Somers DA Hackett WP, Biesboer DD (eds) Plant tissue and cell culture. Alan R Liss, New York, pp, 473–493

Reisch B, Bingham ET (1980) The genetic control of bud formation from callus cultures of diploid alfalfa. Plant Sci Lett 20: 71–77

Rosenberg LA, Rinne RW (1986) Moisture loss as a prerequisite for seedling growth in soybean seeds (*Glycine max* L. Merr.). J Exp Bot 37: 1663–1674

Rosenberg LA, Rinne RW (1988) Protein synthesis during natural and precocious soybean seed (*Glycine max* [L.] Merr.) maturation. Plant Physiol 87: 474–478

Saunders JW, Bingham ET (1972) Production of alfalfa plants from callus culture. Crop Sci 12: 804–808

Saunders JW, Bingham ET (1975) Growth regulator effects on bud initiation in callus cultures of *Medicago sativa*. Am J Bot 62: 850–855

Saxena PK, King J (1987) Morphogenesis in lentil: plant regeneration from callus cultures of *Lens culinaris* Medik. via somatic embryogenesis . Plant Sci 52: 223–227

Saxena PK, Malik KA, Gill R (1992) Induction by thidiazuron of somatic embryogenesis in intact seedlings of peanut. Planta 187: 421–424

Schenk RU, Hildebrandt AC (1972) Medium and techniques for induction and growth of monocotyledonous and dicotyledonous plant cell cultures. Can J Bot 50: 199–204

Seitz Kris MH, Bingham ET (1988) Interactions of highly regenerative genotypes of alfalfa (*Medicago sativa*) and tissue culture protocols. In Vitro Cell Dev Biol 24: 1047–1052

Sellars RM, Southward GM, Phillips GC (1990) Adventitious somatic embryogenesis from cultured immature zygotic embryos of peanut using soybean as a model system. Crop Sci 30: 408–414

Senaratna T, McKersie BD, Bowley SR (1989) Desiccation tolerance of alfalfa (*Medicago sativa* L.) somatic embryos. Influence of abscisic acid, stress pretreatments and drying rates. Plant Sci 65: 253–259

Senaratna T, McKersie BD, Bowley SR (1990) Artificial seeds of alfalfa (*Medicago sativa* L.). Induction of desiccation tolerance in somatic embryos. In Vitro Cell Dev Biol 26: 85–90
Sharp WR, Evans DA, Sondahl MR (1982) Application of somatic embryogenesis to crop improvement. In: Fujiwara A (ed) Plant tissue culture 1982. Maruzen, Tokyo, pp 759–762
Shekhawat NS, Galston AW (1983) Isolation , culture, and regeneration of moth bean *Vigna aconitifolia* leaf protoplasts. Plant Sci Lett 32: 43–51
Shoemaker RC, Hammond EG (1988) Fatty acid composition of soybean (*Glycine max* (L.) Merr.) somatic embryos. In Vitro Cell Dev Biol 24: 829–832
Shoemaker RC, Amberger LA, Palmer RG, Oglesby L, Ranch JP (1991) Effect of 2, 4-dichlorophenoxyacetic acid concentration on somatic embryogenesis and heritable variation in soybean [*Glycine max* (L) Merr.]. In Vitro Cell Dev Biol 27P: 84–88
Skolmen RG (1986) Acacia (*Acacia koa*) In: Bajaj YPS (ed) Biotechnology in agriculture and forestry vol 1. Trees I. Springer Berlin Heidelberg New York pp, 375–384
Skokut TA, Manchester J, Schaeffer J (1985) Regeneration in alfalfa tissue culture. Plant Physiol 79: 579–583
Slade D, Fujii JA, Redenbaugh K (1989) Artificial seeds : a method for the encapsulation of somatic embryos. J Tissue Cult Methods 12: 179–184
Song J, Sorensen EL, Liang GL (1990) Direct embyogenesis from single mesophyll protoplasts in alfalfa (*Medicago sativa* L.). Plant Cell Rep 9: 21–25
Stafford A, Davies DR (1979) The culture of immature pea embryos. Ann Bot 44: 315–321
Stavarek SJ, Croughan TP, Rains DW (1980) Regeneration of plants from long-term cultures of alfalfa cells. Plant Sci Lett 19: 253–261
Still PE, Plata MI, Campbell RJ, Bueno LC, Chichester EA, Niblett CL (1987) Regeneration of fertile *Arachis paraguariensis* plants from callus and suspension cultures. Plant Cell Tissue Organ Cult 9: 37–43
Strickland SG, Nichol JW, McCall CM, Stuart DA (1987) Effect of carbohydrate source on alfalfa somatic embryogenesis. Plant Sci 48: 113–121
Stuart DA, Strickland SG (1984a) Somatic embryogenesis from cell cultures of *Medicago sativa.* 2. The interaction of amino acids with ammonium. Plant Sci Lett 34: 175–182
Stuart DA, Strickland SG (1984b) Somatic embryogenesis from cell cultures of *Medicago sativa.* 1. The role of amino acid additions to the regeneration medium. Plant Sci Lett 34: 165–174
Stuart DA, Strickland SG, Walker KA (1987) Bioreactor production of alfalfa somatic embryos. HortScience 22(5): 800–803
Stuart S, Nelson J, Nichol J (1988) Expression of 7S and 11S alfalfa seed storage proteins in somatic embryos. J Plant Physiol 132: 134–139
Sussex IM Dale RMK (1979) Hormonal control of storage protein synthesis in *Phaseolus vulgaris*. In: Rubenstein I, Phillips RL, Green CE, Gengenbach BG (eds) Plant seed: development, preservation and germination. Academic press, New York, pp 129–141
TeKrony DM, Egli DB, Balles J, Pfeiffer T Fellows RJ (1979) Physiological maturity in soybean. Agron J 71: 771–775
Terzi M, Lo Schiavo F (1990) Somatic embryogenesis. In: Bhojwani SS (ed) Plant tissue culture: applications and limitations. Elsevier, Amsterdam, pp 54–66
Tétu T, Sangwan-Norreel B, Sangwan R (1987) In vitro somatic embryogenesis and plant regeneration in three precocious varieties of soybean. CR Acad Sci Paris 305: 613–617
Tétu T, Sangwan RS, Sangwan-Norreel BS (1990) Direct somatic embryogenesis and organogenesis in cultured immature zygotic embryos of *Pisum sativum* L. J Plant Physiol 137: 102–109
Tomar UK, Gupta SC (1988) Somatic embryogenesis and organogenesis in callus cultures of a tree legume–*Albizia richardiana* King. Plant Cell Rep 7: 70–73
Trigiano RN, Beaty RM, Graham ET (1988) Somatic embryogenesis from immature embryos of redbud (*Cercis canadensis*). Plant Cell Rep 7: 148–150
Varga P, Badea EM (1992) *In vitro* plant regeneration methods in alfalfa breeding. Euphytica 59:119–123
Walker KA, Sato SJ (1981) Morphogenesis in callus tissue of *Medicago sativa*: the role of ammonium ion in somatic embryogenesis. Plant Cell Tissue Organ Cult 1: 109–121
Walker KA, Wendeln ML, Jaworski EG (1979) Organogenesis in callus tissue of *Medicago sativa.* The temporal separation of induction processes from differentiation processes. Plant Sci Lett 16: 23–30

Walker KA, Yu PC, Sato SJ, Jaworski EG (1982) The hormonal control of organ formation in callus of *Medicago sativa* L. cultured in vitro . Am J Bot 65: 654–659

Walton PD, Brown DCW (1988) Screening of *Medicago* wild species for callus formation and the genetics of somatic embryogenesis. J Gen 67: 95–100

Wan Y, Sorensen EL, Liang GH (1988) Genetic control of in vitro regeneration in alfalfa *Medicago sativa* L. Euphytica 39: 3–10

Wang H, Holl FB (1988) In vitro culture and the incidence of somaclonal variation in regenerated plants of *Trifolium pratense* L. Plant Sci 55: 159–167

Weaver LA, Trigiano RN (1991) Regeneration of *Cladrastis lutea* (Fabaceae) via somatic embryogenesis. Plant Cell Rep 10: 183–186

Weissinger AK II, Parrott WA (1993) Repetitive somatic embryogenesis and conversion to plants in white clover. Plant Cell Rep 12: 125–128

Williams EG, Maheswaran G. (1986) Somatic embryogenesis: factors influencing coordinated behaviour of cells as an embryogenic group. Ann Bot 57: 443–462

Williams EG, Collins GB, Myers JR (1990) Clovers (*Trifolium* spp.). In: Bajaj YPS (ed) Biotechnology in agricultural and forestry vol 10. Legumes and oilseed crops. I. Springer Berlin Heidelberg New York, pp. 242–287

Xu N, Bewley JD (1992) Contrasting pattern of somatic and zygotic embryo development in alfalfa (*Medicago sativa* L.) as revealed by scanning electron microscopy. Plant Cell Rep 11: 279–284

Xu Z-H , Davey MR, Cocking EC (1982) Organogenesis from root protoplasts of the forage legumes *Medicago sativa* and *Trigonella foenum-graecum*. Z Pflanzenphysiol 107: 231–235

Yamada T, Higuchi S (1990) In vitro culture of genus *Trifolium* germplasm and plant regeneration . J Jpn Grassl Soc 36: 47–45

Yang Y-S, Wada K, Goto M, Futsushara Y (1991) In vitro formation of nodular calli in soybean (*Glycine max* L.) induced by cocultivated *Pseudomonas maltophilia*. Jpn J Breed 41: 595-604

Yu K, Christie BR, Pauls KP (1990) Effects of *Verticillium albo-atrum* culture filtrate on somatic embryogenesis in alfalfa. Plant Cell Rep 8: 509–511

III.2 Somatic Embryogenesis in Peanut (*Arachis hypogaea* L.)

A.H. McKently[1]

1 Introduction

1.1 Importance and Morphology

Arachis hypogaea L., the cultivated peanut, also called groundnut, is an important source of oil and protein and makes a major contribution to human nutrition. Peanut is native to South America with its primary center of origin believed to occur in southern Bolivia or northern Argentina (Gregory and Gregory 1976). Commercial production is found primarily between 40 °N and 40 °S latitude (Cobley 1985). Approximately 20.1 million ha are devoted to peanut production worldwide, and the major producers are India, China, the United States, and Senegal. Annual world production for the 1991–1992 marketing year is 22.7 million metric tons (USDA Official Data). Most of the world's peanut crop is crushed for the extraction of edible oil and protein meal. It is also an important whole seed food and ingredient of processed foods.

Peanut is an allotetraploid (2n = 40) (Stalker and Dalmacio 1986) annual legume crop, and is quite variable morphologically. There are many distinct cultivars grown commercially, and these are classified into two botanical groups according to their branching pattern (Gregory et al. 1951). This primary division of the species separates the subspecies *hypogaea* (alternately branched) and *fastigiata* (sequentially branched). In ssp. *hypogaea* (Virginia), alternating pairs of vegetative and reproductive branches (inflorescences are borne on the cotyledonary and other lateral branches. The first two branches of each lateral are vegetative and the main axis produces vegetative branches only. In ssp. *fastigiata* (Spanish-Valencia), inflorescences are borne on successive nodes of the cotyledonary and other lateral branches. Inflorescences are also borne directly on the main axis at higher nodes. Cultivar distinction within botanical varieties is based upon growth habit – prostrate, erect, or intermediate between the two – pod characters, seed number, and testa color (Gibbons et al. 1972).

The plant develops a tap root system with abundant branching, and is usually heavily supplied with nitrogen-fixing nodules (Duke 1981). Peanut is principally self-pollinating, and its papilionaceous flowers are sessile and borne in axillary inflorescences. Following fertilization, the ovary elongates by means of an intercalary meristem located just below the ovarian cavity (Smith 1950).

[1] Science and Technology, The Land, EPCOT Center, P.O. Box 10, 000, Lake Buena Vista, Florida 32830, USA

Biotechnology in Agriculture and Forestry, Vol. 31
Somatic Embryogenesis and Synthetic Seed II (ed. by Y.P.S. Bajaj)

Fig. 1. A peanut plant bearing fruits. (McKently et al. 1990)

As the young fruit elongates, it grows geotropically and penetrates the soil, whereupon pod development ensues (Fig. 1). The fruit is a netted, functionally indehiscent legume, containing one to five ovoid seeds.

1.2 Significance of Somatic Embryos

Somatic embryogenesis is emerging as an efficient and essential tool for peanut crop improvement programs. The promise of recombinant DNA technology for cultivar enhancement is being realized as reliable regeneration systems are perfected. Regeneration via somatic embryogenesis increases the probability of achieving transformed plants due to the potentially large numbers of regenerates originating from single or several cells. Chimeric transformants can be cycled in repetitive systems as a means of obtaining wholly transformed plants (Durham and Parrott 1992)

Somatic embryogenic systems would potentially facilitate peanut breeding programs by providing a means for mass producing F_1 hybrids, haploids, and other rare plants carrying traits desirable for improved lines. Somatic embryogenesis also has application in the production of interspecific hybrids. Wild *Arachis* species possess many agronomically desirable characters not found in peanut (Bajaj 1984; Stalker and Moss 1987). Such variability, however, has not been fully exploited in breeding programs due to frequent postfertilization incompatibilities leading to embryo abortion. Such interspecific hybrid embryos can be rescued prior to degeneration and cultured for the induction of multiple somatic embryos and subsequent hybrid plant production (Ozias-Akins 1989; Ozias-Akins et al. 1992b).

1.3 Literature Review of Work

Initial observations of somatic embryogenesis in peanut were made in cultures of leaflet tissue (Pittman et al. 1983) and cotyledonary node tissue (Banerjee et al. 1988). Embryos, ranging from globular to early heart-shaped stages, were reported in both studies, and secondary embryo development was documented in the second.

Somatic embryogenic systems for peanut have been developed using mature zygotic embryo axes (McKently 1991), immature zygotic embryos (Hazra et al. 1989; Ozias-Akins 1989; Sellars et al. 1990), and young leaflets (Baker and Wetzstein 1992). The somatic embryos are auxin-induced, with a wide variety of auxin types and concentrations being utilized. Secondary embryogenesis is often observed, and a long-term repetitive system has been developed using immature zygotic embryo cotyledons as the initial explant material (Durham and Parrott 1992).

2 Somatic Embryogenesis

2.1 Culture of Mature Zygotic Embryo Axes

Plant regeneration via somatic embryogenesis has been obtained in peanut from the axes of mature zygotic embryos (McKently 1991). The area of greatest embryogenic activity was a 2-mm region adjacent to and encircling the epicotyl. Somatic embryo production was achieved on a wide range of auxin types and concentrations, which will be described below. Embryogenic cultures, transferred to media devoid of hormones, were capable of germination and plant development.

Explant Sterilization. Effective disinfection of zygotic embryo explant material can be a concern, as high contamination rates are inherent in explants harvested from the soil. A protocol has been developed which provides 98 to 100% contaminant-free cultures, without damaging explant epicotyls. The legumes are harvested, soaked in water for 2 h, and washed to remove surface debris. Seeds are removed from the fruits, and the embryos separated from the seed coats. The embryos are surface-sterilized two times in 2.6% sodium hypochlorite for 6 min each. Embryo axes are excised and further sterilized in 1.6% sodium hypochlorite for 3 min and 70% ethyl alcohol for 1 min. The explants are rinsed three times with sterile, deionized water.

Methodology. Seed from the peanut cultivar Florigiant were harvested from greenhouse-grown plants at the R8 stage of development (Boote 1982), in which the inner pericarp and testa are colored. Embryo axes 8 to 10 mm long were excised, sterilized, and cultured on Murashige and Skoog (MS) medium (1962). Sixteen auxin treatments [1,3,5, and 10 mg/l naphthaleneacetic acid (NAA), 4-amino-3,5,6-trichloropicolinic acid (PIC), 2,4-dichlorophenoxyacetic acid (2,4-D), and 2,4,5-trichlorophenoxyacetic acid (2,4,5-T)] were evaluated.

Somatic Embryo Formation. Direct somatic embryogenesis was observed within 30 days of culture initiation from a 2-mm band of hypocotyledonary tissue surrounding explant epicotyls (Figs. 2 and 3). Initial investigation found this band of tissue to be highly embryogenic in mature explants and much less so in younger explants. The embryos initially had the appearance of trichomes that were 0.8 to 1.5 mm in length. Approximately 10% of these structures developed into embryos with a suspensor. The others ceased growth and withered. This observation is similar to the findings of Maheswaran and Williams (1985) who described the development of structures with the appearance of epidermal hairs from immature zygotic embryos of *Trifolium repens*, which subsequently developed into somatic embryos.

In this study, the effects of auxin type and concentration were evaluated and found to influence this developmental process. Auxin type was significant in eliciting a response from explant material. Although all four of the tested auxins did induce somatic embryogenesis, explant response rates varied among them, with the 2,4,5-T treated explants displaying the highest response (Table 1). Auxin concentration effects on response rates were insignificant. Auxin concentrations

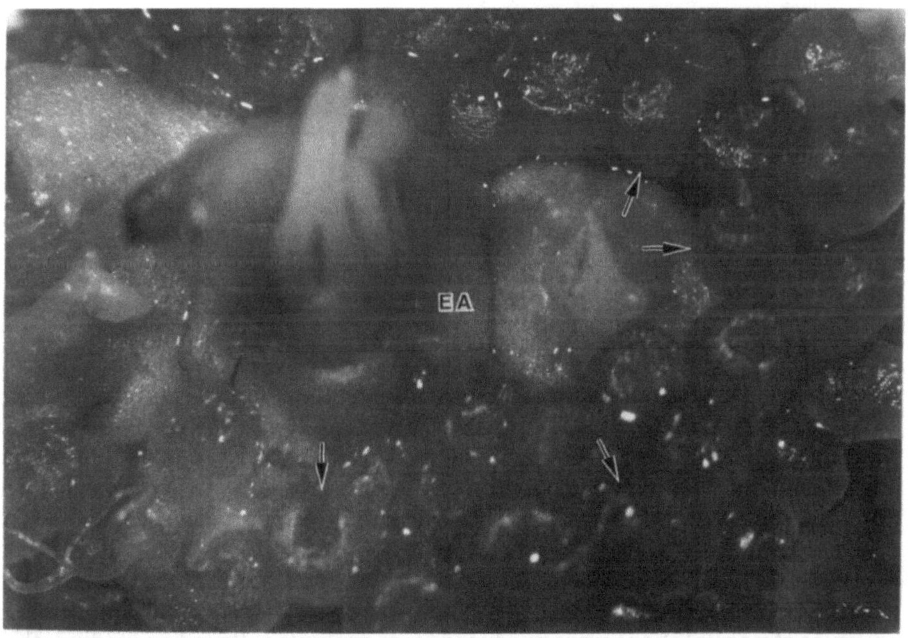

Fig. 2. Somatic embryos (*arrows*) originating from a zygotic embryo axis (*EA*) of Florigiant peanut cultured on 3 mg/l 4-amino-3,5,6-trichloropicolinic acid. (McKently 1991)

media supplemented with 1 to 6 mg/l 2,4-D. Embryo production varied among the tested concentrations, with a maximum of 8 to 15 embryos developing per explant on 3 mg/l 2,4-D. Induction and maturation of the embryos occurred on the same medium, providing a one-step process for plantlet regeneration. Globular-stage embryos developed into fully mature embryos within 20 days, and germinated within 7 days of transfer to half-strength basal medium.

Ozias-Akins (1989) obtained somatic embryos from immature embryo cotyledons 4 to 9 mm in length on B5 medium (Gamborg et al. 1968) plus 0.5 mg/l PIC. Plants developed from the somatic embryos only after extended culture on MS basal medium. Shoots were excised from thick embryonic roots and rerooted on half-strength MS medium.

Using whole embryos 2 to 5 mm in length, Sellars et al. (1990) obtained primary and secondary somatic embryogenesis from three genotypes during a 150-day culture period. Primary embryos arose directly from explant axes and cotyledons on L2 medium (Phillips and Collins 1979) supplemented with 0.02 mg/l PIC, with continuing secondary embryo development on 0.001 mg/l PIC. A subsequent study of seven cultivars (Ozias-Akins et al. 1992a) revealed that most genotypes are competent for somatic embryogenesis, but may vary in their optimal culture requirements. In this study, immature embryos, separated into cotyledon and embryo axis explants, were cultured on MS medium plus 0.5 mg/l PIC. The zygotic embryos were harvested and cultured at various stages of development ranging from early cotyledonary to morphologically mature, corresponding to stages 4 through 7 of Pattee et al. (1974). The percentage of cultures producing somatic embryos and the number of somatic embryos produced per responding explant were higher for all genotypes with embryo axis than with cotyledon explants. Maturity of the embryo explants did not show a significant effect on response, although there was a trend toward reduced somatic embryo formation from older cotyledons.

2.3 Repetitive Somatic Embryogenesis

A regeneration system based on repetitive somatic embryogenesis has been developed for peanut using individual somatic embryos initially induced from immature cotyledons (Durham and Parrott 1992). The cotyledon explants were removed from 10- to 12-mm zygotic embryos and cultured on MS medium containing 40 mg/l 2,4-D for 30 days. Upon transfer to liquid MS medium supplemented with 5 mg/l 2,4-D, the primary somatic embryos (Fig. 4) produced masses of secondary and tertiary embryos (Fig. 5) which continued to proliferate following manual separation and subculture. The cultures exhibited exponential growth for over 1 year without loss of embryogenic potential. Further embryo development, germination, and conversion to plants (Fig. 6) were achieved by placing embryo clumps onto solid basal medium. The inclusion of a 10-day desiccation period after embryo maturation resulted in a fourfold increase (from 3.8 to 15.3%) in the percentage of embryos which converted into plants. The establishment of repetitive embryogenic cultures in liquid medium allows large-scale plant regeneration while avoiding the extensive labor associated with

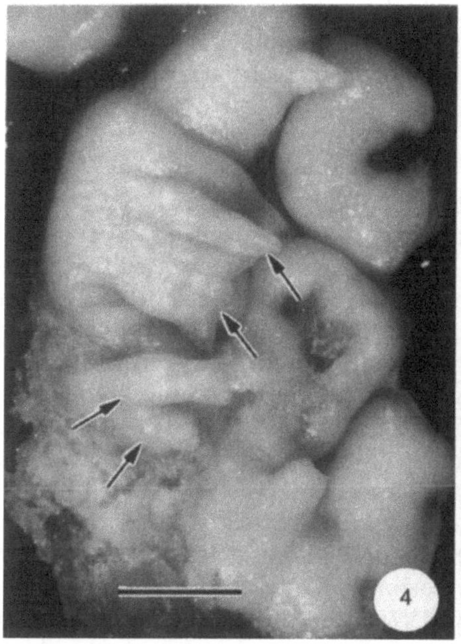

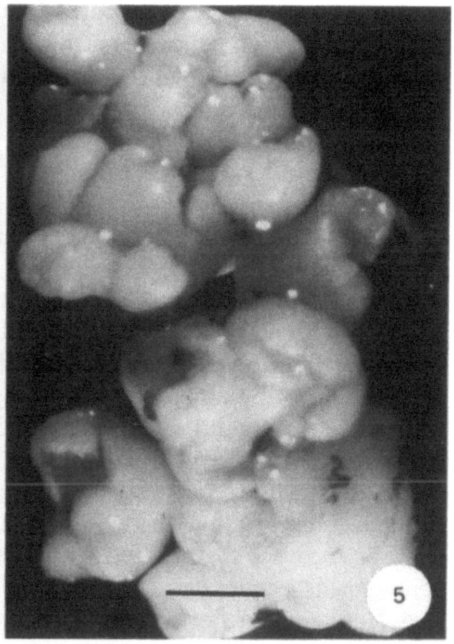

Fig. 4. Primary somatic embryos forming from immature peanut cotyledons on 40 mg/l 2,4-dichlorophenoxyacetic acid. The embryos designated by *arrows* are used to initiate repetitive embryogenic cultures in liquid medium; *bar* = 5 mm. (Durham and Parrott 1992)

Fig. 5. Embryogenic masses harvested from liquid culture supplemented with 5 mg/l 2,4-dichlorophenoxyacetic acid; *bar* = 5 mm. (Durham and Parrott 1992)

Fig. 6. Peanut plant, cv. AT127, regenerated from embryogenic suspension culture and established in soil. (Durham and Parrott 1992)

continuous initiation of cultures from fresh explants. In addition, genetically altered somatic embryos resulting from transformation efforts should be easily produced in quantity via repetitive culture of the primary transformant.

2.4 Culture of Leaflets

A leaf-derived somatic embryogenic system has been developed for peanut (Baker and Wetzstein 1992). Leaflets 5 to 8 mm long were excised from aseptically germinated embryo axes, cut in half perpendicular to the midvein, and cultured with adaxial surfaces in contact with the medium. Maximum embryogenesis of 14.6% was obtained from the leaflets after a 15-day culture period on MS medium supplemented with 40 mg/l 2,4-D plus 0.2 mg/l kinetin followed by transfer to a secondary medium with 5 mg/l 2,4-D plus 0.2 mg/l kinetin. Primary somatic embryos developed and were fused along the axes with no distinct cotyledons, while secondary embryogenesis was observed with the formation of single axis embryos with two cotyledons. Approximately 2% of the secondary somatic embryos germinated and developed into plants. With over 14% embryogenesis, this system demonstrates the feasibility for further development into a high frequency leaflet system (Baker and Wetzstein 1992). Such a system would provide some advantage over the use of zygotic embryo explants. Aseptic cultures with little or no contamination are readily obtained, while bypassing the need to maintain excessive greenhouse plantings for exact staging of embryo explants. A leaf-derived regeneration system also provides an additional avenue for the development of transformation protocols. Leaf tissue is a favored explant type for transformation studies, as medium components for selection readily penetrate tissue sections and effectively suppress growth of nontransformed cells (Fraley et al. 1985).

3 Summary and Conclusions

In the last decade significant advances have been made in developing tissue culture techniques for peanut. Regeneration systems via organogenesis and somatic embryogenesis are now available, and are being perfected for specific crop improvement applications. Explants giving rise to somatic embryos have included mature zygotic embryo axes (McKently 1991), various immature embryo tissues (Hazra et al. 1989; Ozias-Akins 1989; Sellars et al. 1990), and young leaflets (Baker and Wetzstein 1992).

A wide variety of auxin types and concentrations i.e., PIC (0.001–3.0 mg/l), 2,4-D (0.5–40 mg/l), NAA (1.0–10 mg/l), and 2,4,5-T (1–10 mg/l) have been utilized for the induction of somatic embryos. Secondary embryogenesis has been observed in a number of these studies, and a long-term, repetitive system has been developed. Somatic embryogenesis has been successful with 17 genotypes of peanut to date, although explant response rates, embryo production numbers, germination capacity, and conversion into whole plants have varied among them. Investigation of factors, such as medium components, explant selection, and culture conditions that influence embryo initiation through plant development, may be necessary on an individual basis to optimize somatic embryogenesis for specific genotypes. Somatic embryos will facilitate genetic manipulation in peanut.

4 Protocol for the Induction of Somatic Embryos from the Axes of Mature Peanut Embryos

Harvest seed from mature plants at the R8 stage of development according to Boote (1982). Remove embryos from washed fruits and surface sterilize (without testa) in 2.6% sodium hypochlorite for 12 min, changing the solution after 6 min. Excise embryo axes and further sterilize in 1.6% sodium hypocholorite for 3 min and 70% ETOH for 1 min. Rinse the explants 3 times with sterile, deionized water and place in 100 × 15-mm Petri plates containing 25 ml medium. The medium consists of MS salts, 3 mg/l PIC, 25 g/l sucrose, 2 mg/l glycine, 0.5 mg/l nicotinic acid, 0.1 mg/l pyridoxine hydrochloride, and 8 g/l Phytagar (GIBCO Laboratories, Grand Island NY). The culture medium, pH adjusted to 5.8, is autoclaved at 121 °C, 15 psi for 15 min. Maintain cultures at 25 °C with a 16-h photoperiod. Isolate mature somatic embryos from explant material and place on basal medium for germination and subsequent plantlet development. Transfer young plants to a 1 peat: 2 perlite: 1 vermiculite potting medium that has been autoclaved for 30 min at 15 psi and acclimatize with 50% shade and 90% humidity for 30 days.

References

Bajaj YPS (1984) Peanut. In: Ammirato PV, Evans DA, Sharp WR, Yamada Y (eds) Handbook of plant cell culture, Vol. 3. MacMillan, New York, pp 193–225

Baker CM, Wetzstein HY (1992) Somatic embryogenesis and plant regeneration from leaflets of peanut, *Arachis hypogaea*. Plant Cell Rep 11: 71–75

Banerjee S, Bandyopadhyay S, Ghosh PD (1988) Cotyledonary node culture and multiple shoot formation in peanut: evidences for somatic embryogenesis. Curr Sci 57 (5): 252–255

Boote KJ (1982) Growth stages of peanut (*Arachis hypogaea* L.). Peanut Sci 9: 35–40

Cobley LS (1985) The legumes. In: Cobley LS, Steele WM (eds) An introduction to the botany of tropical crops. Longman, London, pp 80–85

Duke JA (1981) *Arachis hypogaea* L. In: Duke JA (ed) Handbook of legumes of world economic importance. Plenum Press, New York, pp 18–22

Durham RE, Parrott WA (1992) Repetitive somatic embryogenesis from peanut cultures in liquid medium. Plant Cell Rep 11: 122–125

Fraley RT, Rogers SG, Horsch RB (1985) Genetic transformation in higher plants. CRC Crit Rev Plant Sci 4: 1–46

Gamborg OL, Miller RA, Ojima K (1968) Nutrient requirements of suspension cultures of soybean root cells. Exp Cell Res 50: 151–158

Gibbons RW, Bunting AH, Smartt J (1972) The classification of varieties of groundnut (*Arachis hypogaea* L). Euphytica 21: 78–85

Gregory WC, Gregory MP (1976) Groundnut, *Arachis hypogaea* (Leguminosae-Papilionacae). In: Simmonds NW (ed) Evolution of crop plants. Longman, London, pp 151–154

Gregory WC, Smith BW, Yarbrough JA (1951) Morphology, genetics and breeding. In: The peanut – the unpredictable legume. Natl Fertile Assoc, Washington, DC

Hazra S, Sathaye SS, Mascarenhas AF (1989) Direct somatic embryogenesis in peanut (*Arachis hypogaea*). Biotechnology 7: 949–951

Maheswaran G, Williams EG (1985) Origin and development of somatic embryos formed directly on immature embryos of *Trifolium repens* in vitro . Ann Bot 56: 619–630

McKently AH (1991) Direct somatic embryogenesis from axes of mature peanut embryos. In Vitro Cell Dev Biol 27P: 197–200

Murashige T, Skoog F (1962) A revised medium for rapid growth and bioassays with tobacco tissue cultures. Physiol Plant 15: 473–497

Ozias-Akins P (1989) Plant regeneration from immature embryos of peanut. Plant Cell Rep. 8: 217–218

Ozias-Akins P, Anderson WF, Holbrook CC (1992a) Somatic embryogenesis in *Arachis hypogaea* L: genotype comparison. Plant Sci 83: 103–111

Ozias-Akins P, Singsit C, Branch WD (1992b) Interspecific hybrid inviability in crosses of *Arachis hypogaea* × A. *stenosperma* can be overcome by in vitro embryo maturation or somatic embryogenesis. J Plant Physiol 140: 207–212

Pattee HE, Johns EB, Singleton JA, Sanders TH (1974) Composition changes of peanut fruit parts during maturation. Peanut Sci 1: 57–61

Phillips GC, Collins GB (1979) In vitro tissue culture of selected legumes and plant regeneration from callus cultures of red clover. Crop Sci 19: 59–64

Pittman RN , Banks DJ, Kirby JS, Mitchell ED, Richardson PE (1983) In vitro culture of immature peanut (*Arachis* spp.) leaves: morphogenesis and plantlet regeneration. Peanut Sci 10: 21–25

Sellars RM, Southward GM, Phillips GC (1990) Adventitious somatic embryogenesis from cultured immature zygotic embryos of peanut and soybean. Crop Sci 30 : 408–414

Smith BW (1950) *Arachis hypogaea*, aerial flower and subterranean fruit. Am J Bot 37: 802–815

Stalker HT, Dalmacio RD (1986) Karyotype analysis and relationship among varieties of *Arachis hypogaea* L. Cytologia 51: 617–629

Stalker HT, Moss JP (1987) Speciation, cytogenetics, and utilization of *Arachis* species. Adv Agron 41: 1–39

III.3 Somatic Embryogenesis in Soybean (*Glycine* Species)

T. KOMATSUDA[1]

1 Introduction

1.1 Importance and Distribution

The genus *Glycine*, to which soybean belongs, has been divided into two subgenera, *Soja* and *Glycine*. The subgenus *Soja* contains *G. max* (L.) Merr., the cultivar, and its wild progenitor *G. soja* Sieb. & Zucc. Both species are annual and carry 2n = 40 chromosomes. A weedy, intermediate semiwild form between *G. max* and *G. soja* with 2n = 40 also occurs in the wild, named *G. gracilis* Skvortz, although this species is sometimes included in *G. max* in view of classical taxonomy (Singh and Hymowitz 1989). The cultivated soybean is responsible for a wide range of useful products and as a source of human nutrition and animal feed. In Japan, soybean is traditionally utilized in the fermented form for the production of miso, soy sauce, natto; and in the nonfermented form for tofu, abura-age, yuba, boiled beans, soy milk, okara, kinako, and edamame. The wild and semiwild soybeans harbor several useful traits such as high protein and high methionine contents that would be worth transferring to the cultivated soybean. The wild and semiwild species thus have been exploited in soybean breeding programs for improving protein quantity and quality.

The subgenus *Glycine* comprises 12 perennial species indigenous to Australia; *G. arenaria* Tind., *G. argyrea* Tind., *G. canescens* F.J. Herm., *G. clandestina* Wendl., *G. curvata* Tind., *G. cyrtoloba* Tind., *G. falcata* Benth., *G. latifolia* (Benth.) Newell & Hymowitz, *G. latrobeana* (Meissn.) Benth., and *G. microphylla* (Benth.) Tind. are diploid (2n = 40), while *G. tabacina* (Labill.) Benth. and *G. tomentella* Hayata are diploid (2n = 40) and tetraploid (2n= 80), and *G. tomentella* aneuploid (2n = 38, 78) (Singh and Hymowitz 1988). According to Marshall and Broué (1981), the perennial *Glycine* species possess many characteristics such as tolerance to drought, heat and cold, apparent day length insensitivity, and disease resistance. Perennial *Glycine* species also possess disease resistance to soybean cyst nematode (Riggs and Hamblen 1962), soybean yellow mosaic virus and soybean rust (Singh et al. 1974; Burdon and Marshall 1981a,b), and powdery mildew (Mignucci and Chamberlain 1978). The perennial *Glycine* species hence represent a potentially valuable source of germplasm for soybean breeding.

[1] Department of Cell Biology, National Institute of Agrobiological Resources, Tsukuba, Ibaraki 305, Japan

Biotechnology in Agriculture and Forestry, Vol. 31
Somatic Embryogenesis and Synthetic Seed II (ed. by Y.P.S. Bajaj)

1.2 Significance of Somatic Embryos

In the past decade, several investigators reported successful intersubgeneric crosses between the soybean and wild perennial *Glycine* species such as *G. tomentella*, *G. canescens*, *G. argyrea*, and *G. clandestina*. The intersubgeneric crosses, however, have a quite low efficiency of hybridization, which resulted in inviable plants, and therefore have never been applied to practical soybean breeding programs. Protoplast fusion and genetic transformation techniques can transfer foreign chromosomes and genes into a desired plant cultivar. These new techniques are valuable tools for plant genetics and breeding. Studies on somatic embryogenesis lead to an understanding of the basis of plant regeneration and provide a practical culture system for protoplast fusion and genetic transformation.

1.3 Review of Work on Somatic Embryogenesis

Somatic embryos have been successfully induced directly from immature embryo tissue (Table 1) and embryogenic suspension cultures initiated from immature embryos (Table 2). In early work on somatic embryogenesis of the soybean and other *Glycine* species, seedling tissues such as hypocotyl and epicotyl sections were used for callus initiation, and from the calli cell suspension cultures were initiated (Beversdorf and Bingham 1977; Phillips and Collins 1981). Soybean suspension cultures induced embryo-like structures, however, these structures differentiated into roots, but none developed into a plantlet. Only one leafy shoot was induced from the somatic embryos of *G. soja* by the same method (Phillips and Collins 1981). Similar embryo-like structures were obtained in hypocotyl-derived cell suspension cultures of *G. soja* and *G. max* (Gamborg et al. 1983), but the structures again did not show continued development.

Investigators have now learned that the tissue of immature embryos is an appropriate material to initiate embryogenic callus and suspension cultures resulting in plant differentiation. Christianson et al. (1983) succeeded in regenerating plants from a morphogenetically competent cell suspension culture derived from the axis tissue of immature embryos, although abnormal, small plants were produced from a single cultivar. Li et al. (1985) reported somatic embryogenesis from single-cell suspension cultures initiated from low temperature-treated immature embryos, but again abnormal, small plants were produced. Finer and Nagasawa (1988) generated a maintainable embryogenic suspension culture. The suspension culture consisted of clumps of proliferating globular embryos. The embryogenic callus was initially induced from immature cotyledons on solid MS medium supplemented with B5 vitamins, 6% sucrose, and a very high concentration of 2,4-D (40 mg/l). The embryogenic callus was placed into suspension culture medium consisting of MS salts, B5 vitamins, 6% sucrose, and 5 mg/l 2,4-D to initiate the embryogenic suspension culture. The embryos matured and germinated on hormone-free solid media, resulting in consistent plant regeneration.

Solidified MS media are more efficient for routine production of functional somatic embryos possessing regeneration competence. Lippmann and Lippmann (1984) reported that direct somatic embryogenesis was induced from immature embryo tissue of *G. max* on a solid culture medium containing a high concentration of auxin such as 2,4-D, MCPA (2-methyl-4-chlorophenoxyacetic acid), NAA, 2,4,5-T, IAA, and IBA. This kind of somatic embryogenesis was reproducible when 2,4-D (Ranch et al. 1985; Lazzeri et al. 1985) and NAA (Lazzeri et al. 1985; Barwale et al. 1986) were used. Lazzeri et al. (1987a) reported that NAA-induced embryos were more closely related to zygotic embryos than 2,4-D-induced embryos. Komatsuda and Ohyama (1988) found a large genotypic difference in the embryogenic response to 2,4-D and NAA, and suggested a significant interaction between auxin type and genotype on embryogenesis.

Effects of sucrose concentration on embryo induction have been well studied. Lazzeri et al. (1987b) showed that the efficiency of embryogenesis increased as sucrose concentration decreased from 120 to 15 g/l. Kageyama et al. (1990) reported that the sucrose concentration affected not only the efficiency of embryogenesis, but also the shape of the embryos. The 5 g/l sucrose medium produced 21 times as many two-cotyledon type embryos as did the 30 g/l sucrose medium. Komatsuda et al. (1991) showed a significant interaction between sucrose concentration and genotype in somatic embryogenesis.

The sucrose concentration also played an important role in the maturation and germination of somatic embryos. According to the maturation study by Buchheim et al. (1989), the somatic embryos which were produced by 19.2 mg/l 2,4-D and 96 g/l sucrose assumed a globular shape, and then developed on a medium containing a high sucrose concentration (100 g/l) to the torpedo stage, resembling zygotic embryos. The somatic embryos then developed into plantlets through a germination-like process on a low sugar medium. The maturation study by Komatsuda et al. (1992) dealt with a different type of somatic embryo which was induced by 10 mg/l NAA and 5 g/l sucrose; the embryos were small (less than 1 mm in length), and had 0.6 mg in fresh weight. However, almost all of the somatic embryos showed a morphologically normal shape, having a well-defined hypocotyl, cotyledon, apical meristem, and radicle (Fig. 1A). The NAA-induced small somatic embryos matured on media containing a sucrose concentration that was gradually increased from 30 to 60 g/l (Fig. 1B), and then germinated upon transfer to a medium containing 10 g/l sucrose (Fig. 1C,D). The processes of induction, maturation and germination of the somatic embryos were thus affected by sucrose concentration; this will be described below in more detail. Plant regeneration from 2,4-D-induced somatic embryos was described in detail for a solid culture medium by Ranch et al. (1986) and Buchheim et al. (1989), as well as for a suspension culture by Finer and Nagasawa (1988).

Table 1. Conditions for somatic embryo induction, maturation, and germination on solid media in soybean

Species	Stage and tissue of immature embryo	Medium for embryogenesis	Maturation	Germination	Reference
G. max	Cotyledons	Blaydes (1966) salts and vitamins, sucrose (3%), 2,4-D (2 mg/l), NAA (2 mg/l), kinetin (2 mg/l), agar (1%) Transfer to Miller (1965) salts and vitamins, sucrose (3%), 2,4-D (0.05 mg/l), kinetin (0.3 mg/l), agar (1%)			Beversdorf and Bingham (1977)
G. max	4-5-mm-long cotyledons	L2 medium (Phillips and Collins 1979), sucrose (1%) 2,4-D (1 mg/l), agar (0.7%)	L2, sucrose (1%), zeatin (0.1mg/l), agar (0.7%)		Lippmann and Lippmann (1984)
G. canescens	Cotyledons of 2–3-week-old embryos	MS major salts, MS × 10 minor salts, B5 vitamins, sucrose (3%), NAA (0.018 mg/l), zeatin (2.2 mg/l), agar (0.8%)	MS major salts, MS × 5 minor salts, B5 vitamins, sucrose (3%), NAA (0.018 mg/l), zeatin (2.2 mg/l), agar (0.8%)		Grant (1984)
G. max, *G. soja*	Heart to early mature stage, 0.5–4-mm-long	MS (Murashige and Skoog 1962) salts and vitamins, sucrose (3%), 2,4-D (5 mg/l), agar (0.8%)	B5 (Gamborg et al. 1968) salts and vitamins, sucrose (2%), IBA (0.12 mg/l), agar (0.8%)	B5, sucrose (2%), GA_3 (0.1 mg/l), agar (0.8%)	Ranch et al. (1985)
G. max, *G. soja*	15–20 days past anthesis, 4 ± 1mm long	MS salts, B5 vitamins, sucrose (3%), 2,4-D (5 mg/l) or NAA (10 mg/l), agar (0.65%)	MS salts, B5 vitamins, sucrose (3%), NAA (0.15 mg/l), BAP, kinetin and zeatin (0.033 mg/l for each), agar(0.65%)		Lazzeri et al. (1985)
G. max	Embryos 4–7 mm long	MS, sucrose (3%), NAA (10 mg/l), thiamine (5 μM), nicotinic acid (0.03 m M), agar (0.6%)	MS sucrose (3%), IBA (0.05 mg/l), BAP (0.4 mg/l), agar (0.6%) or MS, sucrose (3%), IBA (2 mg/l), GA_3 (1 mg/l), agar (0.6%)		Barwale et al. (1986)

G. max	Embryo from 3–5-mm-long-seeds	MS salts, B5 vitamins, sucrose (1.5%), NAA (10 mg/l), agar (0.65%)	MS salts, B5 vitamins, sucrose (3%), BAP, kinetin and zeatin (0.1 mg/l for each), agar (0.65%)		Lazzeri et al. (1987 a,b)
G.max *G. gracilis*	15–20 days post anthesis	MS salts, B5 vitamins, sucrose (3%), NAA (10 mg/l), gelrite (0.18%), or MS salts B5 vitamins, sucrose (3%), 2,4-D (2 mg/l), agar (0.8%)	B5, sucrose (1%), IBA (0.5 mg/l), BA (0.2 mg/l), gelrite (0.4%) Transfer to B5, sucrose (1%), liquid medium Transfer to MS (1/2 major salts), NAA (0.1 mg/l), BA (0.001 mg/l), gelrite (0.18%)		Komatsuda and Ohyama (1988)
G. max	Embryos from 3–5-mm-long seeds	MS salts, B5 vitamins, sucrose (1.5%), NAA (10 mg/l), gelrite (0.2%)	LS salts, B5 vitamins, sucrose (5%), glutamine(6.25 mM), gelrite (0.2%)	MS salts, B5 vitamins, sucrose (1.5%), gelrite (0.2%)	Parrott et al. (1988)
G. max	Cotyledon 4–5-mm long, 15–21 days after flowering	MS, sucrose (3%), 2, 4-D (19mg/l), agar (0.8%)	MS, sucrose (10%), activated chacoal (0.5%), agar (0.8%)	Schenk and Hildebrandt (1970) medium, sucrose (1%), gelrite (0.2%)	Buchheim et al. (1989)
G. max *G. gracillis*	Embryos half the length of Mature dry seeds	MS salts, B5 vitamins, sucrose (0 5%), NAA (10 mg/l), gelrite (0.22%)	MS salts, B5 vitamins, sucrose (3%), NAA (1 mg/l), gelrite (0.22%) Transfer to MS salts, B5 vitamins, sucrose (6%), NAA (1 mg/l), gelrite (0.22%)	MS salts, B5 vitamins, sucrose (1%), GA_3 (0.001 mg/l), gelrite (0.22%)	Komatsuda et al. (1992)

Table 2. Conditions for somatic embryo induction in suspension culture, maturation, and germination in soybean

Explants	Medium for induction of embryogenic callus	Suspension culture	Regeneration	Reference
Hypocotyl sections of *G. max, G. soja*, and *G. tabacina*	B5 medium, sucrose (2%), 2,4-D (2 mg/l), agar (1%)	Miller (1965) medium, sucrose (2%), 2,4-D (0.3–1 mg/l), IAA (0–1mg/l)	Miller medium sucrose (2%), 2, 4-D (0.5 mg/l), kinetin (0.1 or 0.5 mg/l)	Beversdorf and Bingham (1977)
Hypocotyl and epicotyl sections of 7-day-old *G. soja* seedlings	L2 (Phillips and Collins 1979), sucrose (2.5%), picloram (0.06 mg/l), BA (0.1 mg/l), agar (0.8%)	SL2 (Phillips and Collins 1980) basal, sucrose (2.5%), 2,4-D (0.5 mg/l), adenine (15 μM), ABA (2 μg/l), Amo 1618 (1 μM), CH (100 mg/l)	L2, sucrose (2.5%), 2,4-D (0.5 mg/l), adenine (15 μM), ABA (50 μg/l), Amo 1618 (10 nM), glutathione (30 μ M) CH (100 mg/l), agar (0.8%)	Phillips and Collins (1981)
Hypocotyl sections of 8–10-day-old *G. max* and *G. soja*	L2 medium, sucrose (2.5%), picloram (0.06 mg/l), BA (0.1 mg/l) agar (1%)	L2 medium, sucrose (2.5%), picloram (0.06 mg/l), BA (0.1 mg/l) Transfer to L2 major salts, B5 micro-nutrients and vitamins, casamino acid (10 mg/l), adenine sulfate (15 μg/l), picloram (0.006 mg/l), AMO 1618 (0.25 μM)	L2 medium, sucrose (2.5%), picloram (2.4 μg/l), GA (3.5 μg/l), zeatin (0.1 mg/l)	Gamborg et al. (1983)

20-day-old *G. max* embryos	MS, sucrose (2%), 2,4-D (1–2mg/l), gelrite (0.22%)	MS sucrose (2%), 2,4-D (0.5 mg/l), coconut milk (5%)	MS, sucrose (2%), 2,4-D (1mg/l), kinetin (1mg/l), coconut milk (5%), gelrite (0.22%), Transfer to MS, sucrose (2%), BA (0.2mg/l), IAA (0.01 mg/l), gelrite (0.22%) Further transfer to MS, sucrose (2%), 2,4-D (0.5 mg/l), coconut milk (5%)	Li et al. (1985)
Embryo axes from 2.5–3-cm-long *G. max* pods	MS salts and vitamins, sucrose (2%), 2,4-D (5 mg/l), agar (0.9%)	MS salts (nitrogen replaced by 20 mM dibasic ammonium citrate) and vitamins, sucrose (2%), 2,4-D (5 mg/l)	MS salts and vitamins, sucrose (2%),IBA (0.005 mg/l), BA (0.2 mg/l), agar (0.9%)	Christianson et al. (1983)
7–14-day-old *G. max* embryos	MS salts, B5 vitamins, sucrose (6%), 2,4-D (40 mg/l), agar (0.8%)	MS salts (10 m) M NH_4NO_3, 30mM KNO_3), B5 vitamins, sucrose (6%), 2,4-D (5 mg/l) asparagine (5 mM)	MS salts, B5 vitamins, sucrose (3%), agar (0.8%)	Finer and Nagasawa (1988)

Fig. 1A–D. Regeneration of plants via somatic embryogenesis in soybean variety Masshokutou Kou 502. **A** Somatic embryos produced from a cotyledon explant on MS medium containing B5 vitamins. 10 mg/l NAA, and 5 g/l sucrose 5 weeks after culture initiation. **B** Developed somatic embryos 2 weeks after transfer onto MS medium containing B5 vitamins, 1mg/l NAA and 30 g/l sucrose. **C** Elongation of hypocotyls of somatic embryos on 1/2 MS medium containing 1 µg/l GA_3. **D** A regenerated plantlet at the trifoliolate stage. (**A** and **B,** Kageyama et al. 1990; **C** and **D**, Komatsuda et al. 1992)

2 Somatic Embryogenesis

Immature zygotic embryos were selected as a source of explant, with regard to their competence in inducing somatic embryos. The embryogenesis medium contained MS salts, B5 vitamins, various concentrations of sucrose, and 10 mg/l NAA, it was solidified with 2.2 g/l gelrite (modification of Lazzeri et al. 1985). The culture system has been well established and simplified (see protocol, points 1 and 2, Sect. 5).

2.1 Explants

The age of immature embryos clearly influences the efficiency of somatic embryogenesis. Immature embryos at an optimal developmental stage were obtained from developing seed; they were half the length of the dry mature seeds (Fig. 2). The optimal stage was about 18 days after flowering, which is probably the most active period of cell division (Meinke et al. 1981). This is only a few days before the time of transition from a period of cell division to a period of cell enlargement,

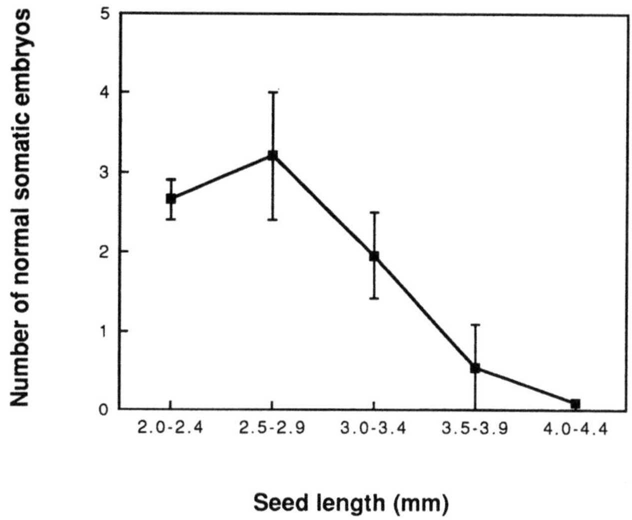

Fig. 2. Effect of developmental stage of explants on the production of normal somatic embryos. Each point is a mean of 20 pairs of cotyledons. Induction medium contained 5 g/l sucrose; cultivar Masshokutou Kou 502

and 7 to 10 days before the time when considerable accumulation of seed storage proteins occurs (Meinke et al. 1981; Komatsuda 1985).

The efficiency of embryogenesis decreased in the following 7–10 days, presumably due to the change in the developmental phase of the immature embryos. The reason for the rapid loss of morphogenetic competence of immature embryos during development is unclear.

Immature soybean embryos obviously possess a relatively narrow range of development stages capable of somatic embryogenesis in comparison with that of alfalfa and carrot. Practical study on tissue culture will require the preferential use of immature embryos at the optimal developmental stage. Although the use of immature embryos is very time-consuming, immature embryos are far more competent than other explants, and their high regeneration ability compensated for the minor inconvenience.

2.2 Sucrose

Lower sucrose concentrations considerably improved the ratio of somatic embryos of two-cotyledon types. Sucrose at 5 g/l produced 21 times more normal-type embryos compared to sucrose at 30 g/l (Fig. 3). A sucrose concentration of 30 g/l possibly promotes the development of somatic embryos just after embryogenesis, and this causes the abnormal shape in almost all of the somatic embryos. It is therefore important to separate the induction and maturation phases of the somatic embryo. The 5–10 g/l sucrose concentration promoted the overall production of somatic embryos (Fig. 4) as well as the morphology. The 5–10 g/l sucrose medium enabled the induction of somatic embryos in certain

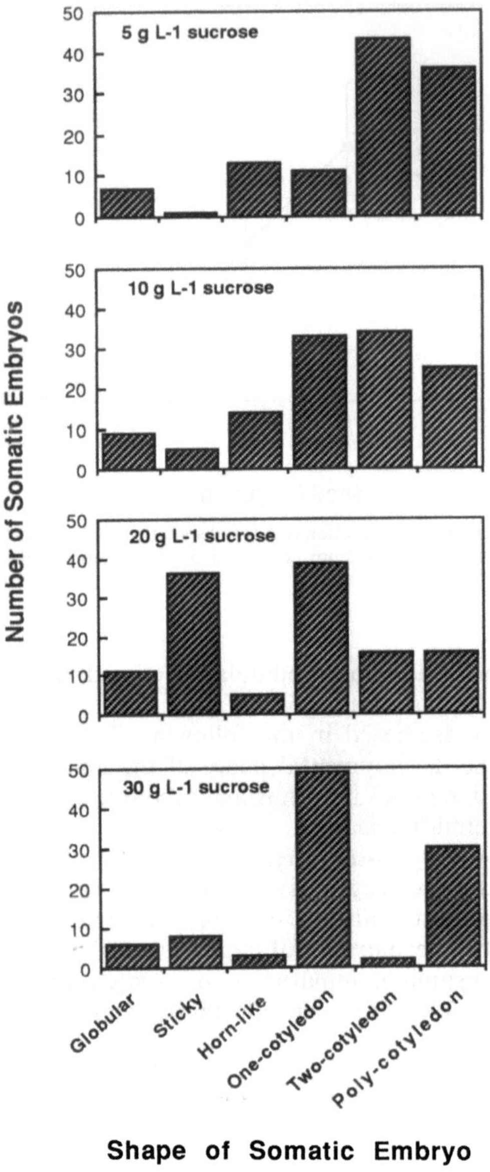

Fig. 3. Shape of somatic embryos affected by sucrose concentration in soybean cultivar Masshokutou Kou 502. Somatic embryos obtained from 40 pairs of 2.5–3.4-mm-long cotyledons were classified into six shapes. (Kageyama et al. 1990)

cultivars such as Tachisuzumari which did not produce somatic embryos with 30 g/l sucrose (Komatsuda et al. 1991). The lower sucrose medium produced an average of about ten normal somatic embryos from a single, isolated pair of cotyledons (Komatsuda 1990). Somatic embryos with normal shapes have a much higher germination frequency. Thus, the minor revision of the sucrose concentration made plant regeneration much more efficient.

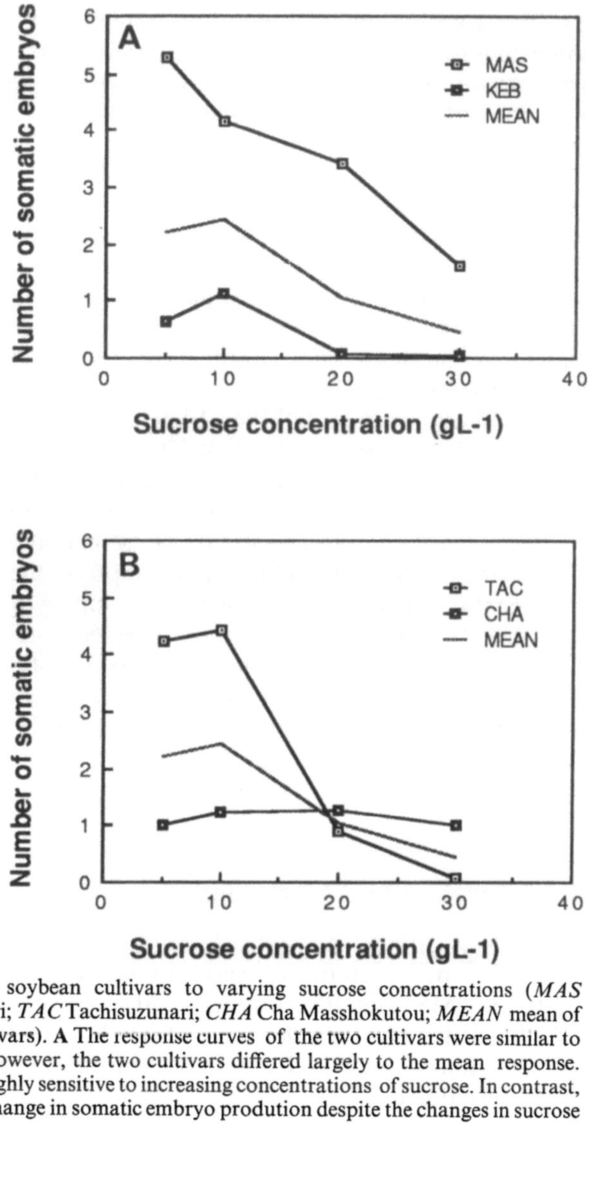

Fig. 4A,B. Various responses of soybean cultivars to varying sucrose concentrations (*MAS* Masshokutou Kou 502; *KEB* Keburi; *TAC* Tachisuzunari; *CHA* Cha Masshokutou; *MEAN* mean of 16 cultivars including the four cultivars). **A** The response curves of the two cultivars were similar to that of the mean of 16 cultivars. However, the two cultivars differed largely to the mean response. **B** The cultivar Tachisuzunari was highly sensitive to increasing concentrations of sucrose. In contrast, Cha Masshokutou exhibited little change in somatic embryo prodution despite the changes in sucrose concentration

2.3 Genotype

Screening studies were made on *G. max*, *G. gracilis* and *G. soja* with regard to somatic embryogenesis from immature cotyledons (Komatsuda and Ko 1990; Komatsuda 1992). Some varieties of *G. gracilis* and *G. soja* were more capable of somatic embryogenesis than those of *G. max*. However, other members of *G. gracilis* showed an intermediate or low capability of somatic embryogenesis. *G. gracilis* may have a wide range of variation in the genetic potential for somatic embryogenesis.

Plants with a high regeneration potential are adapted to a wide range of conditions for tissue culture, whereas recalcitrant plants have a narrow range of appropriate conditions. The ability of 16 soybean genotypes to produce somatic embryos with 5–30 g/l sucrose was tested (Komatsuda et al. 1991). The effect of the genotype × sucrose concentration interaction proved to be highly significant. This interaction was ascribed to the differences between the sensitivities of genotypes to varying sucrose concentrations (Fig. 4B). The embryogenic ability of each cultivar could hence be dependent on two factors: the cultivar mean across all the sucrose concentrations tested and the dose response to sucrose. Masshokutu kou 502, which belongs to *G. gracilis*, produced the highest number of somatic embryos among all the cultivars at each sucrose concentration, thus indicating that this variety was the most embryogenically competent one among the tested cultivars (Fig. 4A).

Masshokutou kou 502 has been used to determine the optimal culture conditions (Kageyama et al. 1990). The data on the somatic embryo production of this genotype showed a near-normal distribution, thus making it possible to compare the effects of several treatments statistically. Masshokutou kou 502 has also been used as a positive control in screening studies (Komatsuda and Ko 1990; Komatsuda 1990). One way to improve the embryogenic response is by transferring the genes responsible for embryogenesis to other genotypes. Masshokutou kou 502 has already been exploited in a program to develop a soybean with a superior regeneration capacity (Bailey et al. 1990). The cultivar also provided enough somatic embryos for other experiments, such as embryo maturation and germination (Komatsuda et al. 1992) and somaclonal variation (Komatsuda 1992).

3 Maturation and Germination

Induced somatic embryos were transferred onto the maturation medium containing MS salts, B5 vitamins, 5–90 g/l sucrose, 1 mg/l NAA, solidified with 2.2 g/l gelrite, and then germinated on the same medium with 10 g/l sucrose (Komatsuda et al. 1992). Refer to Section 5 points (3–6) for precise instructions for tissue culture.

3.1 Sucrose Concentration

The maturation of somatic embryos required gradually increasing sucrose concentrations from 30 to 60 g/l to obtain a gain in fresh weight (Fig. 5). The somatic embryos gradually changed color to yellow, and finally to white, during maturation. The higher sucrose concentration accelerated not only the increase in fresh weight, but also the accumulation of storage proteins. So far, a high sucrose concentration was required when the somatic embryos were in the period of maturation.

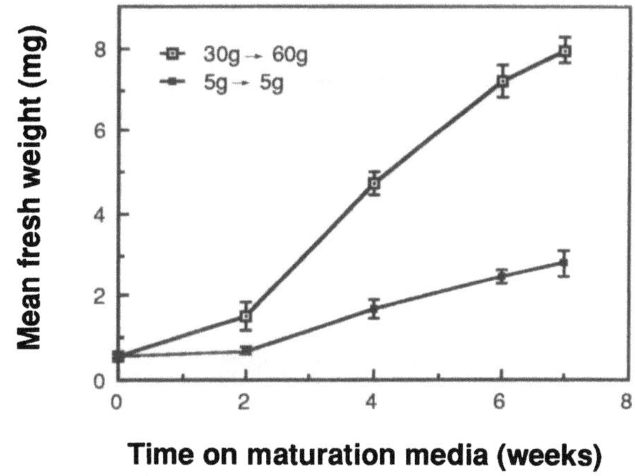

Fig. 5. Increase in fresh weight with maturation of somatic embryos of soybean cultivar Viking (B). The somatic embryos were grown for 2 weeks on the first maturation medium containing 5 or 30 g/l sucrose, and subcultured for 5 weeks on the next maturation medium containing 5, 30, 60, 90 g/l sucrose. The fresh weight of ten somatic embryos was determined, and the most efficient and inefficient curves are shown

These observations are in partial agreement with those of Parrott et al. (1988), who reported a positive effect of desiccation of somatic embryos on their germination. Somatic embryos were either exposed to the high osmotic pressure of the maturation medium (this study) or incubated in empty Petri dishes (Parrott et al. 1988), thereby achieving a degree of desiccation. The developmental process of soybean somatic embryos is partly analogous to that of zygotic embryos, which also needs a period of desiccation. The high sucrose concentration in the maturation medium may cause an elevated osmolarity, which promotes normal maturation (Ammirato 1989) and, at the same time, inhibits precocious germination (Crouch and Sussex 1981).

The germination of somatic embryos required a lower sucrose concentration (10 g/l). The mature somatic embryos turned green on the medium, and the converted embryos germinated to differentiate both roots and shoots. No plant growth regulator was necessary, and hence the sucrose concentration was considered to be the main factor controlling the maturation and germination of soybean somatic embryos. Embryo development without exogenous hormones is more advantageous than hormone-regulated development for several reasons. First, hormone-free medium rarely produces epigenetic malformations, thus promoting normal development of embryos, good regeneration, and rapid acclimatization of vigorous plantlets. Secondly, genetic variation will be minimized in the absence of hormones which are mutagenic. Thirdly, rather simple compositions of culture medium can be used conveniently for cell manipulation.

3.2 Genotype

G. gracilis and *G. soja* showed a higher percentage of germination of somatic embryos (53.0 ± 2.2%) than *G. max* (19.1 ± 1.8%) even on the second-best culture system (Komatsuda and Ohyama 1988). Germination of somatic embryos requires both the organization of shoot meristem at the base of the cotyledons (Lippmann and Lippmann 1984; Barwale and Widholm 1990) and primary leaf outgrowths. The germination process thus includes an organogenic event, and

G. gracilis and *G. soja* were hence considered to possess a higher competence than *G. max* with respect to both embryogenesis and organogenesis. The reason for the genotypic differences is unclear, although it is apparent that organogenesis is partially under genetic control. The genetic difference in germination efficiency, however, could be minimized by optimizing the culture system for maturation and germination. The efficiency of somatic embryo germination was 77% in *G. gracilis*, and 60–64% in *G. max* (Komatsuda et al. 1992).

4 Summary and Conclusion

A system for inducing somatic embryogenesis has been established. Immature embryos, about 18 days after flowering, when cultured on MS-based solid medium containing 10 mg/l NAA produced the most somatic embryos. A sucrose concentration of 5–10 g/l improved the ratio of normal-type somatic embryos compared to the higher concentrations of 20–30 g/l. Some wild and semiwild soybeans gave a better response to somatic embryo production, showing that they are useful positive controls in screening studies and are competent gene stocks for developing a soybean with a superior regeneration capacity. Single, isolated pairs of immature cotyledons could produce, on average, more than ten somatic embryos.

For maturation, somatic embryos were grown for 2 weeks on MS-based medium containing 30 g/l sucrose and then for another 5 weeks on MS-based medium containing 60 g/l sucrose. The maturation of somatic embryos showed some analogy with that of zygotic embryos with respect to the changes in morphology, color, fresh weight, and accumulation of seed storage proteins. For germination, mature somatic embryos were transferred to the sucrose-reduced (10 g/l) MS-based medium. There was 60–77% germination.

In conclusion, a reliable system for regenerating plants has been established in which 600–700 plants can be regenerated from 100 pairs of immature embryos. Soybeans have an outstanding capacity for both oil and protein production, and nitrogen fixation is also very effective. The regeneration system described here is applicable to genetic manipulation in soybean.

5 Protocol

1. Grow soybean plants in the field or greenhouse. After 14–21 days of flowering, harvest immature pods, wash them, and open them by hand to find young seeds. Take the seeds which are half the length of the mature dry seeds. Remove the seed coat and the embryonic axis to isolate a pair of cotyledons.
2. Place the cotyledons with abaxial side down (Hartweck et al. 1988) on solid medium containing MS salts, B5 vitamins, 10 mg/l NAA, 5–10 g/l sucrose, and 2.2 g/l gelrite at pH 7 before autoclaving for 15 min at 120 °C. Culture the cotyledons in 25 × 120 mm test tubes or in 90 × 20 plastic Petri dishes left unsealed unless otherwise noted. Incubate the cultures at 25 °C under continuous dim light

(0.6 µmol $m^{-2}s^{-1}$). Somatic embryos are generally visible 14 days after culture initiation, but 14 more days of culture may provide more embryos.

3. Transfer the somatic embryos connected with the cotyledon explant onto solid maturation medium containing MS salts, B5 vitamins, 1 mg/l NAA (if necessary), 30 g/l sucrose, and 2.2 g/l Gelrite at pH 7 before autoclaving for 15 min at 120 °C. Culture the cotyledons in 90 × 20 plastic Petri dishes left unsealed. Incubate the cultures for 14 days at 25 °C with a 19-h photoperiod under cool white fluorescent lights (34 µmol $m^{-2}s^{-1}$).
4. Transfer the half-mature embryos onto another maturation medium of the same composition supplemented with 60 g/l sucrose, and incubate cultures under the same environmental conditions.
5. After 21–28 days when somatic embryos change color to yellow or white, transfer the embryos onto germination medium containing 1/2 × major and 1 × minor MS salts, B5 vitamins, 1µg/l GA_3 (if necessary), 10 g/l sucrose, and solidified with 2.2 g/l Gelrite. Incubate the cultures in 90 × 20 plastic Petri dishes at 25 °C with a 19-h photoperiod under cool white fluorescent lights (34 µmol $m^{-2}s^{-1}$).
6. When somatic embryos have turned green and germinated to develop the cotyledons, hypocotyl, primary root, and trifoliolate, transfer them to an aqueous solution containing 1 mg/l IBA. Plantlets that have formed new root systems can be transplanted to soil and acclimatized by gradually decreasing the humidity.

Acknowledgements. The author expresses his gratitude to Dr. K. Ohyama for encouragement throughout the investigation, Dr. K. Nakajima and Dr. S. Oka for continuing interest, and Professor K. Hinata (Tohoku University) for stimulating discussions. The research presented here was supported by funds from the Ministry of Agriculture, Forestry and Fisheries (Japan).

References

Ammirato PV (1989) Recent progress in somatic embryogenesis. IAPTC News 57: 2–16

Bailey MA, Parrott WA, Boerma HR (1990) Development of soybean germplasm with superior regeneration capacity and susceptibility to *Agrobacterium*. Proc 3rd Biennial Conf Molecular and cellular biology of the soybean. Iowa State University, Ames, Iowa, p 34

Barwale UB, Widholm JM (1990) Soybean: plant regeneration and somaclonal variation. In: Bajaj YPS (ed) Biotechnology in agriculture and forestry, vol 10. Legumes and oilseed crops I. Springer, Berlin Heidelberg New York, pp 114–133

Barwale UB, Kerns HR, Widholm JM (1986) Plant regeneration from callus cultures of several soybean genotypes via embryogenesis and organogenesis. Planta 167: 473–481

Beversdorf WD, Bingham ET (1977) Degrees of differentiation obtained in tissue cultures of *Glycine* species. Crop Sci 17: 307–311

Blaydes DF (1966) Interaction of kinetin and various inhibitors in the growth of soybean tissue. Physiol Plant 19: 748–753

Buchheim JA, Colburn SM, Ranch JP (1989) Maturation of soybean somatic embryos and the transition to plantlet growth. Plant Physiol 89: 768–775

Burdon JJ, Marshall DR (1981a) Evaluation of Australian native species of *Glycine* for resistance to soybean rust. Plant Dis 65: 44–45

Burdon JJ, Marshall DR (1981b) Inter- and intra-specific diversity in the disease response of *Glycine* species to the leaf-rust fungus *Phakopsora pachyrhizi*. J Ecol 69: 381–390

Christianson ML, Warnick DA, Carlson PS (1983) A morphogenetically competent soybean suspension culture. Science 222: 632–634

Crouch ML, Sussex IM (1981) Development and storage-protein synthesis in *Brassica napus* L. embryos in vivo and in vitro. Planta 153: 64–74

Finer JJ, Nagasawa K (1988) Development of an embryogenic suspension culture of soybean (*Glycine max* Merrill). Plant Cell Tissue Organ Cult 15: 125–136

Gamborg OL, Miller RA, Ojima K (1968) Nutrient requirement of suspension cultures of soybean root cells. Exp Cell Res 50: 151–158

Gamborg OL, Davis BP, Stahlhut RW (1983) Somatic embryogenesis in cell cultures of *Glycine* species. Plant Cell Rep 2: 209–212

Grant JE (1984) Plant regeneration from cotyledonary tissue of *Glycine canescens*, a perennial wild relative of soybean. Plant Cell Tissue Organ Cult 3: 169–173

Hartweck LM, Lazzeri PA, Cui D, Collins GB, Williams EG (1988) Auxin-orientation effects on somatic embryogenesis from immature soybean cotyledons. In Vitro 24: 821–828

Kageyama C, Komatsuda T, Nakajima K (1990) Effects of sucrose concentration on morphology of somatic embryos from immature soybean. Plant Tissue Cult Lett 7: 108–110

Komatsuda T (1985) Genetic mechanism of soybean seed protein biosynthesis. Master Thesis, Iwate University, Faculty of Agriculture, Morioka, Japan

Komatsuda T (1990) Ability of soybean (*Glycine max* L. Merr.) genotypes to produce somatic embryos on a medium containing a low concentration of sucrose. Jpn J Breed 40: 371–375

Komatsuda T (1992) Research on somatic embryogenesis and plant regeneration in soybean. Bull Natl Inst Agrobiol Resour 7 : 1–78

Komatsuda T, Ko S (1990) Screening of soybean [*Glycine max* (L.) Merrill] genotypes for somatic embryo production from immature embryo. Jpn J Breed 40: 249–251

Komatsuda T, Ohyama K (1988) Genotypes of high competence for somatic embryogenesis and plant regeneration of soybean *Glycine max*. Theor Appl Genet 75: 695–700

Komatsuda T, Kaneko K, Oka S (1991) Genotype × sucrose interactions for somatic embryogenesis in soybean. Crop Sci 31: 333–337

Komatsuda T, Lee WB, Oka S (1992) Maturation and germination of somatic embryos as affected by sucrose and plant growth regulators in soybeans *Glycine gracilis* Skvortz and *Glycine max* (L.) Merr. Plant Cell Tissue Organ Cult 11: 3–11

Lazzeri PA, Hildebrand DF, Collins GB (1985) A procedure for plant regeneration from immature cotyledon tissue of soybean. Plant Mol Biol Rep 3: 160–167

Lazzeri PA, Hildebrand DF, Collins GB (1987a) Soybean somatic embryogenesis: effects of hormones and culture manipulations. Plant Cell Tissue Organ Cult 10: 197–208

Lazzeri PA, Hildebrand DF, Collins GB (1987b) Soybean somatic embryogenesis: effects of nutritional physical and chemical factors. Plant Cell Tissue Organ Cult 10: 209–220

Li BJ, Langridge WHR, Szalay AA (1985) Somatic embryogenesis and plant regeneration in the soybean *Glycine max*. Plant Cell Rep 4: 344–347

Lippmann B, Lippmann G (1984) Induction of somatic embryos in cotyledonary tissue of soybean *Glycine max* L. Merr. Plant Cell Rep 3: 215–218

Marshall DR, Broué P (1981) The wild relatives of crop plants indigenous to Australia and their use in plant breeding. J Aust Inst Agric Sci 47: 149–154

Meinke DW, Chen J, Beachy RN (1981) Expression of storage-protein genes during soybean seed development. Planta 153: 130–139

Mignucci JS, Chamberlain DW (1978) Interactions of *Microsphaera diffusa* with soybeans and other legumes. Phytopathology 68: 169–173

Miller CO (1965) Evidence for the natural occurrence of zeatin and derivatives: compounds from maize which promote cell division. Proc Natl Acad Sci USA 54: 1052–1058

Murashige T, Skoog F (1962) A revised medium for rapid growth and bioassay with tobacco tissue cultures. Physiol Plant 15: 473–497

Parrott WA, Dryden G, Vogt S, Hildebrand DF, Collins GB, Williams EG (1988) Optimization of somatic embryogenesis and embryo germination in soybean. In Vitro 24: 817–820

Phillips GC, Collins GB (1979) In vitro tissue culture of selected legumes and plant regeneration from callus cultures of red clover. Crop Sci 19: 59–64

Phillips GC, Collins GB (1980) Somatic embryogenesis from cell suspension cultures of red clover. Crop Sci 20: 323–326

Phillips GC, Collins GB (1981) Induction and development of somatic embryos from cell suspension cultures of soybean. Plant Cell Tissue Organ Cult 1: 123–129

Ranch JP, Oglesby L, Zielinski AC (1985) Plant regeneration from embryo-derived tissue cultures of soybeans. In Vitro Cell Dev Biol 21: 653–658

Ranch JP, Oglesby L, Zielinski AC (1986) Plant regeneration from tissue culture of soybean by somatic embryogenesis. In: Vasil IK (ed) Cell culture and somatic cell genetics of plants, vol 3. Academic Press, Orlando, pp. 97–110

Riggs RD, Hamblen ML (1962) Soybean-cyst nematode host studies in the family Leguminosae. Agric Exp Stn, Division of Agriculture, University of Arkansas, Fayetteville, Rep Ser 110

Schenk RU, Hildebrandt AC (1972) Medium and techniques for induction and growth of monocotyledonous and dicotyledonous plant cell cultures. Can J Bot 50: 199–204

Singh BB, Gupta SC, Singh BD (1974) Sources of field resistance to rust and yellow mosaic diseases of soybean. Ind J Genet Plant breed 34: 400–404

Singh RJ, Hymowitz T (1988) Further data on the gonomic relationships among wild perennial species (2n = 40) of the genus *Glycine* Willd. Genome 30: 166–176

Singh RJ, Hymowitz T (1989) The genomic relationship among *Glycine soja* Sieb. and Zucc., *G. max* (L.) Merr. and '*G. gracilis*' Skvortz. Plant Breed 103 : 171–173

III.4 Somatic Embryogenesis in Cotton (*Gossypium* spp.)

N.J. GAWEL[1] and C.D. ROBACKER[2]

1 Introduction

1.1 Importance and Distribution of Cotton

Whether cultivated or wild, all the cottons of the world belong to the genus *Gossypium*, and four domesticated species of cotton are currently under cultivation. Two are "Old World" species: *Gossypium arboreum* is cultivated in India, and remains a fairly important crop there; *G. herbaceum* is grown only as a local crop in the drier areas of Africa and Asia. Both these species are diploid (2n = 26). The domesticated "New World" cottons, *G. hirsutum* and *G. barbadense*, are allotetraploid, characterized by 26 pairs of chromosomes (2n = 52) (Lee 1984). Cultivation of *G. hirsutum*, often referred to as Upland cotton, makes up about 90% of the current world production of 80 million bales, and about 99% of the US cotton crop. *G. barbadense* (commonly referred to as Egyptian or Pima cotton) accounts for most of the remaining production, and is used in the manufacture of luxury fabrics and sewing thread. The economic impact of commercial cotton production is significant; the value of the US crop is roughly $4 billion (US$) annually (USDA Agricultural Statistics 1991).

1.2 Significance of Somatic Embryos in Cotton

Considerable advances in cotton yield and fiber quality have been realized over the past 50 years. Improved production practices account for part of these advances, but a significant portion has been due to the development of superior cultivars (Niles and Feaster 1984). The application of biotechnological techniques in breeding programs provides alternative avenues for cotton improvement through the direct manipulation of genes, and can produce improvements not possible through classical breeding programs. A basic prerequisite for the utilization of in vitro methods for crop improvement is the development of a reliable method to regenerate whole plants from single, genetically transformed

[1] Cooperative Agricultural Research Program, Tennessee State University, Nashville, Tennessee 37209, USA
[2] Department of Horticulture, University of Georgia, Georgia Station, Griffin, Georgia 30223, USA

Biotechnology in Agriculture and Forestry, Vol. 31
Somatic Embryogenesis and Synthetic Seed II (ed. by Y.P.S. Bajaj)

cells (either through organogenesis or somatic embryogenesis). Somatic embryogenesis has been reported in the wild diploid *G. klotzschianum* (Price and Smith 1979; Finer and Smith 1984), and plants have also been regenerated (Abdukarimov et al. 1987). Genetic transformation of *G. barbadense* has been reported through the direct injection of DNA into immature bolls (Zhou et al. 1983), but this method was not confirmed molecularly. Transformation of *G. hirsutum* has been demonstrated (Firoozabady et al. 1987; Umbeck et al. 1987; Finer and McMullen 1990; Bayley et al. 1992), and field testing of genetically engineered cottons is now underway (Deaton et al. 1991). This chapter will provide an overview of the requirements for somatic embryogenesis and plant regeneration in cultivated *G. hirsutum*.

2 Somatic Embryogenesis

2.1 Choice of Species, Cultivar, and Genotype

An examination of 17 *G. hirsutum* cultivars revealed extreme differences in the ability to induce callus, and somatic embryogenesis was only found to occur in two cultivars (Coker 201 and Coker 315) (Shoemaker et al. 1986). A comparison of the cultivars Acala SJ-5, Coker 312, and Paymaster 303 found Coker 312 to be the only cultivar to exhibit embryogenesis (Gawel et al. 1986). Similarly, Trolinder and Xhixian (1989) tested 38 *Gossypium* species, cultivars, strains, and races and reported a wide range of responses. No embryogenesis was found in *G. barbadense*, *G. arboreum*, or *G. thurberi.* Seven Coker cultivars exhibited the greatest embryogenic response; however, embryogenesis was also noted in cultivars from Acala, Deltapine, Paymaster, and Lankart. No embryogenesis was found in the Stoneville cultivars tested.

Considerable variability for somatic embryogenesis also exists within cultivars and lines. Shoemaker et al. (1986) found that only 73% of the Coker 201 and 64% of the Coker 315 seed they tested produced embryogenic callus . Trolinder and Xhixian (1989) noted that specific plants within the Coker 312 cultivar had a higher embryogenic response than others. An examination of individual plants of Coker 312 and line T25 of the race punctatum showed considerable genotypic specifically for somatic embryogenesis (Gawel and Robacker 1990a). Intracultivar variation for somatic embryogenesis should not be unexpected, as cotton cultivars are somewhat heterozygous and heterogeneous (Poehlman 1979). Furthermore, somatic embryogenesis is not a trait expressed at the whole-plant level, and hence, not selected during cultivar development. The process of self-pollination during development of these cultivars will have pushed the embryogenesis genes towards homozygosity, but not homogeneity. Therefore, for unselected trait such as somatic embryogenesis, each plant is a potentially unique genotype.

2.2 Callus Initiation

2.2.1 Explant Type

A comparison of immature and mature explants found that, in general, immature tissues (embryo axis, 3-day-old cotyledon, 3-day-old hypocotyl) produced higher levels of somatic embryogenesis than did mature tissues (leaf portions, stem segments) (Trolinder and Goodin 1988a). Furthermore, growth regulator type and concentration were more critical to achieving embryogenesis from mature tissues than from immature tissues. The most common immature explants are either cotyledons (Davidonis 1983; Davidonis and Hamilton 1983; Firoozabady et al. 1987; Finer 1988) or hypocotyl segments (Shoemaker et al. 1986; Umbeck et al. 1987; Trolinder and Goodin 1987, 1988b), usually from 3-day-old seedlings. In experiments requiring the use of mature tissue explants, leaf disks (Gawel et al. 1986) and petiole segments (Gawel and Robacker 1990a,b) have each been used in the establishment of embryogenic cultures, however, petiole segments produce a greater percentage of embryogenic cultures. Because immature tissue explants generally give more favorable results, mature tissues explants should only be used when the perpetuation of the explant genotype is necessary, i.e., in genetic studies.

2.2.2 Medium Composition

To initiate embryogenic callus, explants are cultured on basal salt MS medium (Murashige and Skoog 1962). The most common vitamin formulations used are either the B5 formulation (1 g/l nicotinic acid, 1 g/l pyridoxin, 10 g/l thiamine) of Gamborg et al. (1968) or the addition of 0.4 g/l thiamine (Linsmaier and Skoog 1965). Myo-inositol at a concentration of 100 mg/l is included in the callus initiation medium, as is glucose (30 g/l).

Many different growth regulator combinations and concentrations have been used for the induction of embryogenic callus. The auxins used for callus initiation are almost exclusively 2,4-dichlorophenoxyacetic acid (2,4-D) or naphthaleneacetic acid (NAA). These are usually used in combination with a cytokinin: either (2-isopentyl)-adenine (2iP), 6-furfurylaminopurine (kinetin, Kn), or benzyladenine (BA). Optimal concentrations (mg/l) range as follows: NAA, 1–4; 2,4-D, 0.05–1; Kn, 0.1–1; 2iP, 0.1–5; BA, 0.1–1. Embryogenic callus has been induced using 2,4-D alone (Trolinder and Goodin 1987) and 2iP alone (4 mg/l, mature petiole tissues, N. Gawel, unpubl.).

A comparison of NAA (0.5–2 mg/l) and 2,4-D (0.1–0.5 mg/l) showed no significant difference in the ability to induce embryogenic callus from immature tissue explants (Trolinder and Goodin 1988a). Of the cytokinins tested, Kn (0.1–1 mg/l) was most efficient, followed by 2iP (0.1–1 mg/l). BA (0.1–1 mg/l) gave undesirable results, often producing hard green callus.

Culture of mature petiole tissue on various auxin/cytokinin combinations gave results differing from those observed using immature tissue (N. Gawel and C. Robacker, unpubl.). The combination of 2iP (0.1–4 mg/l) and NAA (1–9 mg/l)

Table 1. Percentage of Coker 312 and line T25 petiole callus cultures exhibiting embryogenesis when initiated on media containing various concentrations of auxins and cytokinins (mg/l). Basal medium consisted of MS salts plus (per liter) 0.5 mg pyridoxine, 0.5 mg nicotinic acid, 10 mg thiamine, 30 g glucose, 100 mg myo-inositol, 2 g Gelrite, and 0.75 g $MgCl_2$.(unpubl.)

	T25				Coker 312			
NAA	1	3	6	9	1	3	6	9
2iP								
0.1	85.7	88.9	77.8	87.5	87.5	100.0	55.6	77.7
1.0	100.0	85.7	62.7	55.6	85.7	55.6[a]	14.3[a]	0
2.0	20.0	77.8	66.7	83.3	0	55.6	66.7[a]	0
4.0	[b]	25.8	28.6	44.4	0	11.1[a]	25.0[a]	0
Kn								
0.0	23.0	28.6	7.1	0	7.1[a]	27.3	33.3[a]	20.0
2.0	0	0	0	6.6[a]	6.6[a]	6.25[a]	0	0
4.0	0	0	0	0	0	0	0	0
6.0	0	0	0	0	–	–	0	0
2,4-D	0.0	0.1	0.5	1.0	0.0	0.1	0.5	1.0
2iP								
0.1	0	40.0	0	0	0	0	29.6[a]	0
1.0	0	16.7[a]	0	0	0	0	0	0
2.0	0	0	0	0	0	12.5[a]	0	0
4.0	20.0	0	0	0	66.7	0	0	0

[a] Mean of < 3 embryos/culture.
[b] Treatment contaminated; – no callus produced; n = 10 to 15 petioles per treatment.

was far superior to other auxin/cytokinin combinations (Table 1). Very little embryogenesis occurred on NAA/BA, 2,4-D/Kn, or 2,4-D/BA combinations.

Either agar or Gelrite can be used to gel the media. Most published work has used Gelrite, however, successful protocols have been published which use agar (Shoemaker et al. 1986; Finer 1988). A direct comparison of agar vs. Gelrite showed that Gelrite produced superior rates of callus initiation, though these calli were not evaluated for embryogenic potential (Zimmerman et al. 1985).

2.2.3 *Characteristics of Embryogenic Callus*

Generally, after an explant is placed in culture it is incubated for 6–16 weeks without subculture. Embryogenic *G. hirsutum* callus is very friable and brown to light gray-green in appearance. Hard, bright green, dark green, or white (snow-like) callus will usually not produce embryos. Organogenic (root or shoot-producing) callus will rarely be profusely embryogenic. Visual inspection of embryogenic callus will sometimes reveal somatic embryos on the surface of the callus in smooth, gray-white globular clusters (Fig. 1). Embryogenic tissues can also be evaluated by dispersing an embryogenic callus in a water-filled Petri dish, and examining the slurry under a dissecting microscope. Embryogenic tissues are rounded, dense, and highly cytoplasmic (Fig. 2). In an advanced culture,

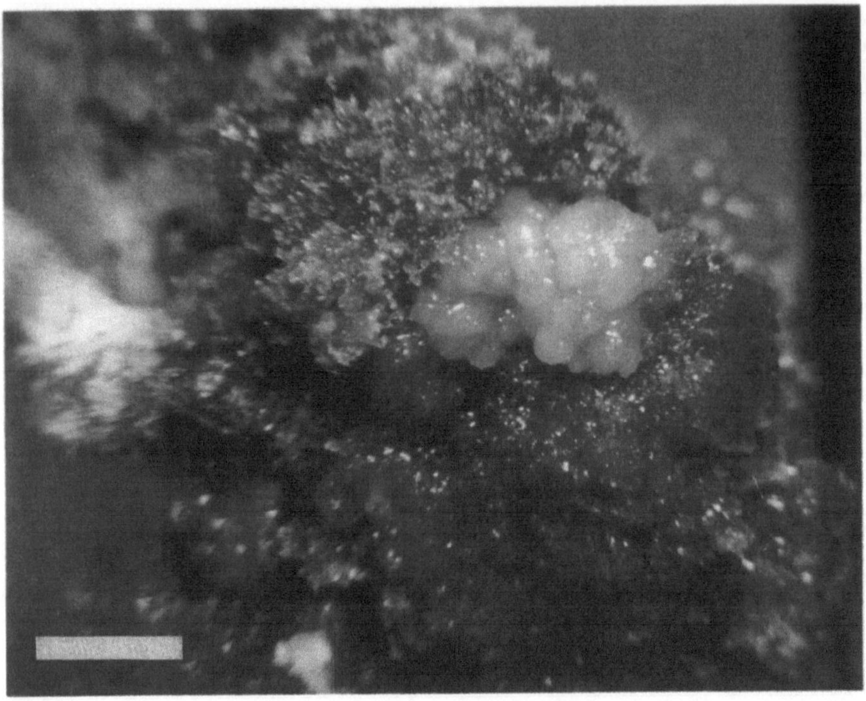

Fig. 1. Globular stage embryos growing on *G. hirsutum* callus: *bar* = 0.8 mm. (Gawel et al. 1986)

many stages of embryonic development (globular through bipolar) can be discerned (Fig. 3).

2.3 Proliferation and Maturation of Embryogenic Callus

Various procedures for the multiplication and maturation of *G. hirsutum* somatic embryos have been published. Although these protocols share similarities, a single, standardized protocol has not been defined. Generally, after an embryogenic callus has been identified, embryo proliferation is achieved by transferring the callus to either on MS-based semisolid or liquid medium, sometimes containing low levels of growth regulators. Comparisons of semisolid vs. liquid proliferation media on embryogenic callus derived from petioles indicate that significantly more somatic embryos are produced in liquid culture than on semisolid media (Table 2; Gawel and Robacker 1990b). Finer (1988) also obtained a high rate of embryo proliferation using suspension cultures. Increased proliferation in liquid media may be attributable to a number of factors. The slurry of cells formed in liquid cultures has a higher level of accessibility to nutrients, and is exposed to more gradual pH changes than that cultured on a semisolid medium. Additionally, any toxic metabolites produced by the callus tissues will be diluted by the liquid medium (Stevenson et al. 1982; George and

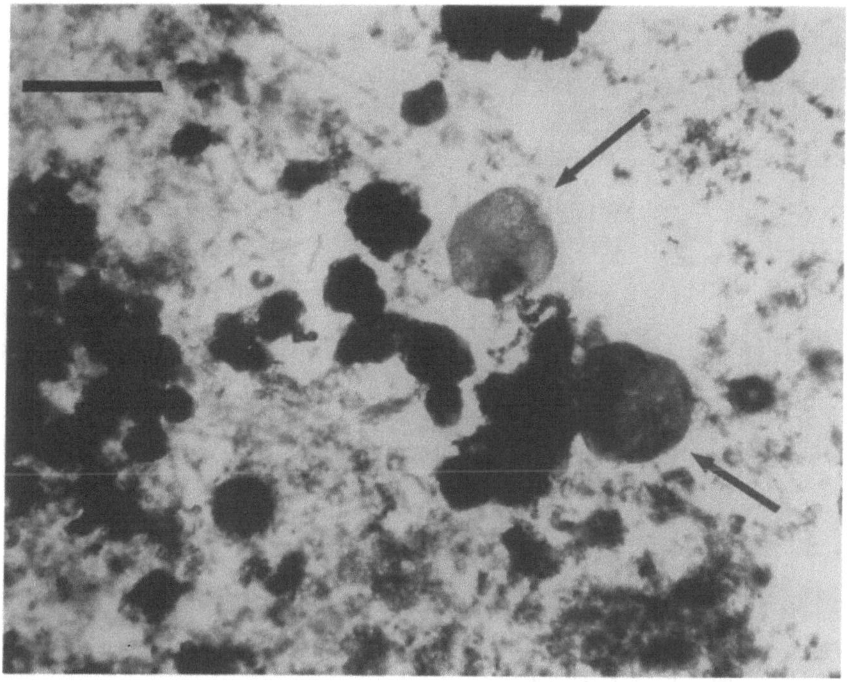

Fig. 2. Early embryonic and globular embryo tissue (*arrows*) in cell suspension culture; *bar* = 0.5 mm. (N. Gawel and C. Robacker, unpubl.)

Sherrington 1984) and endogenous (embryogenesis inhibiting) hormones may be leached more quickly. When propagated on semisolid medium, embryos generally develop in pockets and are not evenly distributed through the callus. This may indicate the presence of nutrient or hormone gradients which are not present in liquid cultures.

The presence of growth regulators in the multiplication medium is not an absolute necessity for embryo proliferation (Davidonis and Hamilton 1983; Umbeck et al. 1987; Trolinder and Goodin 1987); but some researchers report more favorable results when the proliferation medium is supplemented with growth regulators. Finer (1988) reported that the addition of 5 mg/l 2,4-D reduced the size of the embryogenic tissues, which then proliferated as

Table 2. Mean number of Coker 312 and line T25 embryos formed per culture on liquid and semisolid proliferation medium consisting of MS salts supplemented with (per liter) 100 mg myo-inositol, 0.5 mg nicotinic acid, 0.5 mg pyridoxine, and 10 mg thiamine. (Gawel and Robacker 1990b)

	Mean embryos / culture
Liquid proliferation medium	227.5
Semisolid (Gelrite) proliferation medium	134.6

P > T (paired) 0.0002, n = 72 pairs

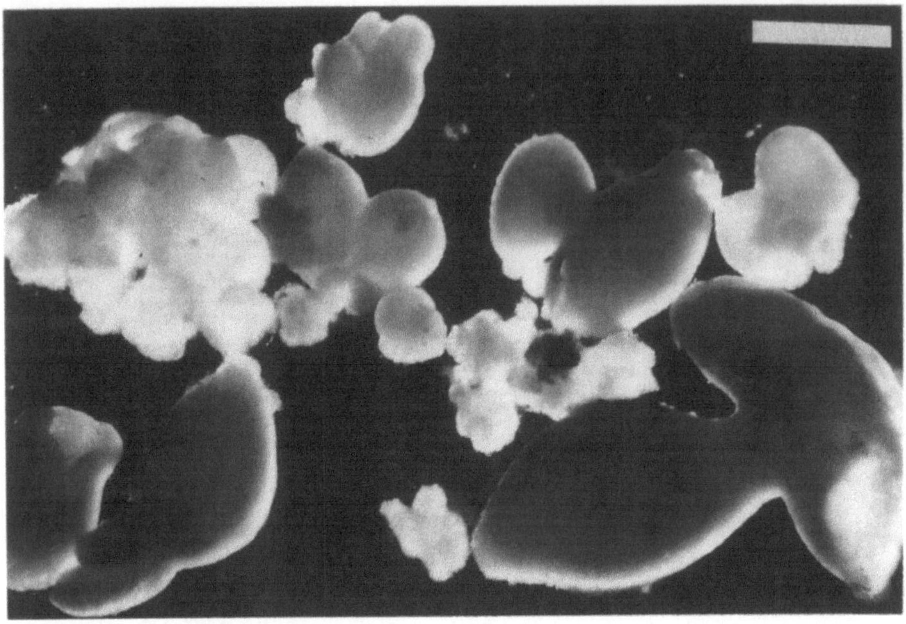

Fig. 3. Stages of somatic embryo development, globular through cotyledonary stage; *bar* = 0.3 mm. (N. Gawel and C. Robacker, unpubl.)

preglobular clumps of tissue. Following proliferation, the embryos were matured in MS medium, B5 vitamins, 15 mM glutamine, and 2% sucrose (this procedure also uses a suspension culture initiation step following the initial callus induction step). Shoemaker et al. (1986) maintained embryogenic callus on MS salts, 2mg/l NAA, 1 mg/l, Kn, and 3% sucrose. These researchers noted the loss of embryogenic potential when callus was subcultured onto glucose-containing media; however, this observation has not been reported by others.

When embryos have differentiated into late heart-shaped to torpedo stages, they are removed from proliferation medium and cultured under conditions to induce germination. Globular or early heart-stage embryos moved to the germination step will rarely develop into plants.

2.4 Embryo Germination

Like the embryo proliferation step, preferred methods of embryo germination also vary. Trolinder and Goodin (1988b) transferred embryos from proliferation medium to semisolid MS-based maturation medium. They found that the addition of 0.1 mg/l GA_3 (gibberellic acid) and/or NAA and/or Kn to the maturation medium generally decreased the number of embryos produced, but increased the number of plants recovered. Also, they determined an additional 1.9 g/l KNO_3 in the maturation medium had a beneficial effect on embryo development. Mature, cotyledonary embryos were removed from the maturation medium and placed on vermiculite moistened with Beaseley and Ting (BT) (1973), Stewart and Hsu (SH) (1977), or MS media. A variety of growth regulator types and concentrations were added to the vermiculite-stage media; no single treatment produced consistently superior results. Umbeck et al. (1987) used vermiculite saturated with SH medium plus GA and Kn (0.1 mg/l each) to induce germination. The addition of growth regulators is not a prerequisite for embryo maturation; mature, bipolar embryos germinated and produced plants when removed from growth regulator-free proliferation medium and placed on vermiculite moistened with growth regulator-free LS medium (Gawel et al. 1986).

Germination has also been demonstrated on semisolid medium (instead of vermiculite). Firoozabady et al. (1987) induced germination from mature embryos by culturing them on Gelrite-solidified SH medium supplemented with 0.1 mg/l GA, 0.01 mg/l NAA, and 0.5% glucose. Mature embryos placed on agar-solidified MS medium plus 1 mg/l Kn and 1.5% sucrose also germinated and produced plants (Shoemaker et al. 1986). Finer (1988) used MS (no NH_4NO_3, 2x KNO_3), B5 vitamins, 1% sucrose, and 0.2% Gelrite for embryo germination. Rinsing the embryos for 48 h in sterile distilled water prior to placing them in the germination treatment has also improved germination rates. It is obvious that much variability exists for maturation and germination techniques. All of the procedures described here have produced plants from somatic embryos. The choice of a particular technique is best determined by the individual researcher. The percentage of plants recovered varies, though it is generally less than 40%. Many forms of abnormal germination can be encountered: radicle development with no apical meristem, apical meristem development with no radicle, callus formation from various parts of the embryo, vitrification, etc. Regenerated plants have exhibited various morphological abnormalities, including prolific branching, weak stems, small leaves. Inability to flower, and floral abnormalities. The degree of fertility is variable, ranging from 15% (Trolinder and Goodin 1987) to 100% (Firoozabady et al. 1987; Finer 1988).

3 Summary and Conclusions

Somatic embryogenesis of cotton has been readily induced from callus initiated on hypocotyls, cotyledons, and petioles. However, regeneration in *G. hirsutum* remains very cultivar/genotype specific. Explants are initially cultured on semi-solid MS salts with B5 vitamins or thiamine, an auxin, cytokinin, glucose, and myo-inositol. Embryo proliferation is achieved by culturing the embryogenic callus on MS-based liquid medium. Embryo germination is the most difficult step in cotton regeneration; the germination percentage is often low and morphological abnormalities and infertility of regenerated plants have frequently been reported. In spite of these difficulties, cotton cells have been successfully transformed and fertile, transformed plants have been regenerated via somatic embryogenesis.

4 Protocol for Somatic Embryogenesis in *G. hirsutum*

As explained in the preceding text, no standard method for regenerating cotton through somatic embryogenesis has been established. The following procedure is a consensus protocol for plant regeneration in *G. hirsutum*. Modifications to this protocol may be required to optimize regeneration in individual circumstances.

1. Surface disinfect explant source. Dip seeds or petioles in 70% ethanol for 10–20 s, then soak in 0.525% sodium hypochlorite (10% commercial bleach) for 20 min. Rinse three times with sterile water. For petiole explants, a 1–2 h rinse in running tap water prior to disinfection is useful to reduce contaminants.
2. Germinate seeds aseptically on moistened paper towel or MS medium with no growth regulators.
3. Excise hypocotyls (or cotyledons) from 3–5-day-old seedlings. Culture on MS basal salts, B5 vitamins, 0.2% Gelrite, pH 5.6, 0.1–0.5 mg/l 2,4-D + 0.1–0.5 mg/l Kn. For petiole explants, split the petiole lengthwise and cut into 1.5–2 cm lengths. Place petiole sections epidermis down on the medium described above, except use 0.1–1 mg/l 2iP + 1 mg/l Kn. Culture conditions should be 28–30 °C, 16-h photoperiod, 60–120 $\mu Em^{-2} s^{-1}$. Grow callus 9–12 weeks without subculturing.
4. Subculture 50–200 mg of embryogenic callus (friable, grey-green to light brown) in 10–20 ml liquid medium (same as callus initiation media, but devoid of growth regulators). The optimal callus/media ratio varies for different cultivars. Maintain cultures at 100–120 rpm for 4 weeks. Embryos in various stages of development should be clearly evident.
5. Filter the cell/embryo suspension to remove large clumps of tissue. One layer of cheesecloth or a similarly sized filtering apparatus can be used for this purpose. Culture the filtrate on semisolid callus initiation medium (devoid of growth regulators) in Petri dishes. After 4–8 weeks mature embryos (developed cotyledons and hypocotyl) should be evident (Fig. 4).

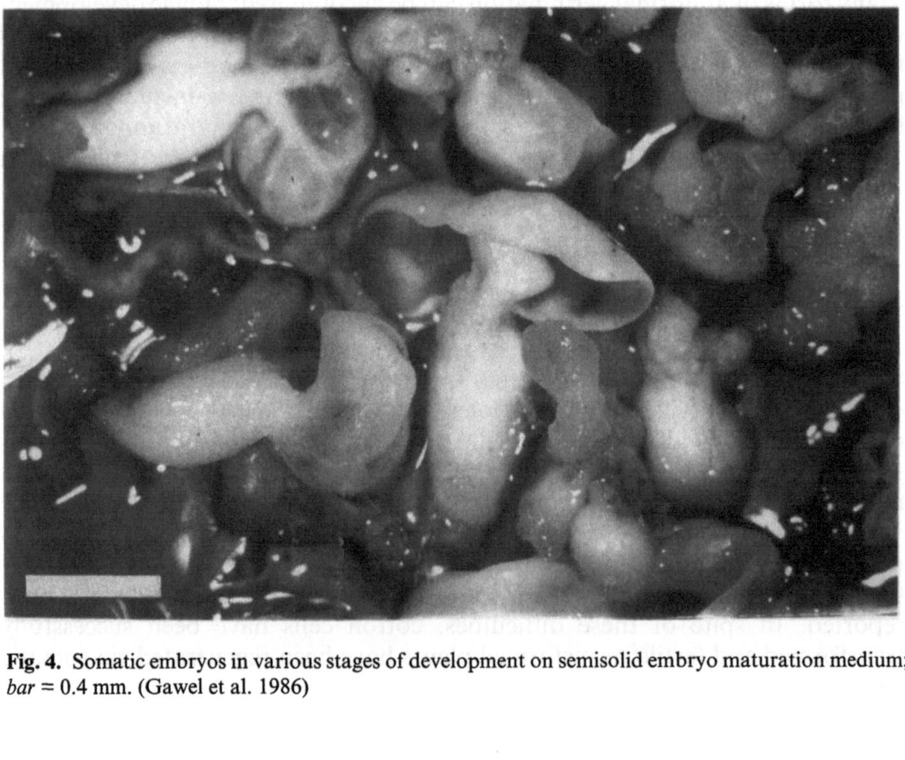

Fig. 4. Somatic embryos in various stages of development on semisolid embryo maturation medium; *bar* = 0.4 mm. (Gawel et al. 1986)

6. Germination can be achieved by culturing mature embryos in test tubes filled with sterile vermiculite saturated with SH media. Alternatively, mature embryos can germinate on semisolid media: MS with no growth regulators and SH with 0.1 mg/l GA_3 + 0.01 mg/l NAA have each been successfully used.
7. When the root and apical shoot are actively growing, transfer the plant to soil and acclimatize through the use of an intermittent mist system or other method of gradually exposing the plants to ambient temperature and humidity conditions.

References

Abdukarimov A, Kuzenetsova NN, Grigina In, Dzhataev SA, Sadykov AS (1987) Regeneration of plants from callus tissue of the cotton *Gossypium klotzschianum*. Dokl Akad Nauk SSSR 292 (2): 502–505

Bayley C, Trolinder N, Ray C, Morgan M, Quisenberry JE, Ow DW (1992) Engineering 2,4-D resistance into cotton. Theor Appl Genet 83: 645–649

Beasley CA, Ting IP (1973) Effects of plant growth substances on in vitro development from fertilized cotton ovules. Am J Bot 60: 130–139

Davidonis G (1983) Embryoid formation and plant regeneration from cotton. Proc Annu Meet Growth Regulator Soc Am, Triangle Park NC, pp 161–164

Davidonis G, Hamilton R (1983) Plant regeneration from callus tissue of *Gossypium hirsutum* L. Plant Sci Lett 32: 89–93

Deaton WR, Jenkins JN, Parrott WL, McCarty JC, Benedict JH, Altman D, Micinski S, Caldwell WD, Gannaway JR, Rummel DR, Wilson FD, Flint HM, Williamson DR (1991) Field performance of cotton genetically modified to express the insecticidal protein from *Bacillus thuringiensis*. In Vitro 27: 89A (Abstr)

Finer J (1988) Plant regeneration from somatic embryogenic suspension cultures of cotton (*Gossypium hirsutum* L.). Plant Cell Rep 7: 399–402

Finer J, McMullen M (1990) Transformation of cotton (*Gossypium hirsutum* L.) via particle bombardment. Plant Cell Rep 8: 586–589

Finer J, Smith R (1984) Initiation of callus and somatic embryos from explants of mature cotton (*Gossypium klotzschianum* Anderss). Plant Cell Rep 3: 41–43

Firoozabady E, DeBoer D, Merlo D, Halk E, Amerson L, Rashka K, Murray E (1987) Transformation of cotton (*Gossypium hirsutum* L.) by *Agrobacterium tumefaciens* and regeneration of transgenic plants. Plant Mol Biol 10: 105–116

Gamborg OL, Miller RA, Ojima K (1968) Nutrient requirements of suspension cultures of soybean root cells. Exp Cell Res 50: 151–158

Gawel N, Rao AP, Robacker C (1986) Somatic embryogenesis from leaf and petiole callus tissues of *Gossypium hirsutum* L. Plant Cell Rep 5: 457–459

Gawel N, Robacker C (1990a) Genetic control of somatic embryogenesis in cotton petiole callus cultures. Euphytica 49: 249–253

Gawel N, Robacker C (1990b) Somatic embryogenesis in two *Gossypium hirsutum* genotypes on semi-solid versus liquid proliferation media. Plant Cell Tissue Organ Cult 23: 201–204

George EF, Sherrington PD (1984) Plant propagation by tissue culture. Exegetics, Eversly, England, pp 487–498

Lee JA (1984) Cotton as a world crop. In: Kohel RJ, Lewis CF (eds) Cotton. American Society of Agronomy, Madison WI pp 1–25

Linsmaier EM, Skoog F (1965) Organic growth factor requirements of tobacco tissue cultures. Physiol Plant 18: 100–127

Murashige T, Skoog F (1962) A revised medium for rapid growth and bioassay with tobacco tissue culture. Physiol Plant 15: 473–497

Niles GA, Feaster CV (1984) Cotton breeding. In: Kohel RJ, Lewis CF (eds) Cotton. American Society of Agronomy, Madison, WI, pp 201–231

Poehlman JM (1979) Breeding Field Crops, 2nd edn. AVI, Westport

Price HJ, Smith R (1979) Somatic embryogenesis in suspension cultures of *Gossypium klotzscianum* Anderss. Planta 145: 305–307

Shoemaker RC, Couche LJ, Galbraith DW (1986) Characterization of somatic embryogenesis and plant regeneration in cotton (*Gossypium hirsutum* L.). Plant Cell Rep 43: 178–181

Stevenson JH, Harris RE, Monette PL (1982) A comparison of liquid and semi-solid culture media for the in vitro proliferation of *Nicotiana tabacum* cv. Xanthi-nc. Plant Propagat 28: 12–14

Stewart J McD, Hsu C (1977) In-ovule embryo culture and seedling development of cotton (*Gossypium hirsutum* L.) Planta 137: 113–117

Trolinder N, Goodin J (1987) Somatic embryogenesis and plant regeneration in cotton (*Gossypium hirsutum* L.) Plant Cell Rep 6 : 231–234

Trolinder N, Goodin J (1988a) Somatic embryogenesis in cotton (*Gossypium*). I. Effects of source of explant and hormone regime. Plant Cell Tissue Organ Cult 12: 31–42

Trolinder N, Goodin J (1988b) Somatic embryogenesis in cotton (*Gossypium*). II. Requirements for embryo development and plant regeneration. Plant Cell Tissues Organ Cult 12: 43–53

Trolinder N, Xhixian C (1989) Genotype specificity of the somatic embryogenesis response in cotton. Plant Cell Rep 8: 133–136

Umbeck P, Johnson G, Barton K, Swain W (1987) Genetically transformed cotton (*Gossypium hirsutum* L.) plants. Bio/ Technology 5: 263–266

USDA (1991) Agricultural statistics. United States Government Printing Office, Washington, DC

Zhou Gy, Weng J, Zeng Y, Huang J, Qian S, Liu G (1983) Introduction of exogenous DNA into cotton embryos. Methods Enzymol 101: 433–481

Zimmerman TW, Robacker CD, Plowman R, Gawel NJ, Rao AP (1985) In vitro callus growth and shoot proliferation of several species on defined media solidified by either agar or gelrite. HortScience 20 (3): 576

III.5 Somatic Embryogenesis and Pattern of Peroxidases in White Mustard (*Sinapis alba* L.syn. *Brassica hirta* Moench.)

Y.P.S. BAJAJ[1]

1 Introduction

Sinapis alba L. (syn. *Brassica hirta* Moench., 2n = 24), commonly known as white mustard or European mustard, is an important oilseed crop, also used as a condiment and high protein meal. The plant shows resistance to leaf blight (*Alternaria brassicae*) and the beet cyst nematode (*Heterodera schachtti*); it is thus a useful source for incorporation of resistance into various oilseed *Brassica* crops.

Since the first publication on the induction of growth and organ formation in *Sinapis alba* by Bajaj and Bopp in 1972, extensive studies have been carried out on various aspects of in vitro culture of this plant. Bajaj et al. (1973) induced somatic embryogenesis, compared the pattern of peroxidases in various tissues, and reported the reversal of 5-fluorodeoxyuridine-caused growth inhibition by thymidine in in vitro grown excised segments of hypocotyls (Bopp et al. 1972). Haploids have also been obtained by culturing anthers (Klimaszewka and Keller 1983), pollen embryogenesis has been induced, and chromosomal variability has been reported in anther-derived callus (Bajaj and Mohapatra 1990). There has also been considerable interest in using this plant in hybridization programs. Mohapatra and Bajaj (1987) successfully obtained interspecific hybrids between *Brassica juncea* × *B. hirta* by culturing excised ovaries and ovules, and later Ripley and Arinson (1990) used embryo rescue to obtain hybrids between *Sinapis alba* × *B. napus*. Recently, protoplast fusion and somatic hybridization studies have been conducted on *Sinapis turgida* × *B. oleracea* (Toriyama et al. 1987) and *Sinapis alba* × *B. napus* (Primard et al. 1988; Lelivelt et al. 1993).

This chapter, primarily based on the author's published work, deals with the establishment of callus cultures, induction of organogenesis and somatic embryogenesis, and the pattern of peroxidases in various organs and differentiating callus cultures.

[1] Former Professor of Tissue Culture, Punjab Agricultural University, Ludhiana (Punjab), India. Present address: A-137 New Friends Colony, New Delhi 110065, India

Biotechnology in Agriculture and Forestry, Vol. 31
Somatic Embryogenesis and Synthetic Seed II (ed. by Y.P.S. Bajaj)

2 Material and Methods

Callus Cultures. The seeds of *Sinapis alba* L. were surface sterilized with 1% solution of Chloramine T for 5 min, and germinated aseptically on modified White's (1943) medium. Cotyledons, hypocotyl, and roots were excised from 1–3 week-old light-grown seedlings, and cultured on Murashige and Skoog's (1962) medium (hereafter referred to as MS medium) supplemented with 1 mg/l 2,4-D + 0.1 mg/l kinetin and 10% coconut water. In other cultures, seeds were germinated directly on this medium. The following supplements were added to MS, either singly or in various combinations: coconut water (CW), dichlorophenoxyacetic acid (2,4-D), indole acetic acid (IAA), yeast extract (YE), and kinetin. The pH of each medium was adjusted to 5.8. The cultures were maintained at 25 ± 2 °C in diffuse daylight. For histological studies, the material was fixed in formalin acetic alcohol. The customary method of dehydration and embedding was followed; 10–12 μ thick sections were cut and stained with hematoxylin. The callus stocks were maintained on MS + 1 mg/l 2,4-D + 0.1 mg/l kinetin + 10% CW.

Electrophoresis. Peroxidase patterns were studied by both starch gel electrophoresis and polyacrylamide disk electrophoresis. Seedlings and callus tissue cultures were weighed, and macerated with equal volumes of 15% sucrose solution in a mortar and pestle in an ice bath. The homogenate was centrifuged at 3000 rpm for 15 min and the supernatant was then used for electrophoresis. For polyacrylamide gel electrophoresis, the procedure used was basically that of Ornstein and Davis (1962) with minor modifications. The gel column was composed of two sections, a 7.5 separation gel and a concentration gel. The electrophoresis tubes were filled 6 cm high with the deaerated separation gel and layered carefully with a drop of water. They were then polymerized for 30 min. When polymerization was complete, the gel formed a sharp line under the water layer. The water was then removed with a syringe and 1 cm of the concentration gel was layered on the separation gel. Then a drop of bromophenol blue was put on top of the concentration gel as a marker, and the gel polymerized using a blue-light lamp. On completion of polymerization the bromophenol blue layer was removed, and 0.1 ml of the enzyme extract put on top of the gel. Electrophoresis was then performed in a Tris-glycine buffer at a current of 4 mA per tube. The completion of electrophoresis could be observed when the bromophenol blue layer had migrated to the bottom of the tubes (about 45 min). The gels were then removed from the tubes and flooded with a mixture of guaiacol and hydrogen peroxide.

Peroxidase Activity. For quantitative measurements the method of Birecka and Galston (1970) was followed. The tissues were ground with 2 vol per unit weight 0.2 M phosphate buffer, pH 5.8, and centrifuged at 3000 rpm for 20 min. A 0.1 ml aliquot of the supernatant was diluted with 5 ml of the phosphate buffer. The peroxidase was measured with 3 ml 15 mM guaiacol-hydrogen peroxide system in phosphate buffer. The optical density readings were taken at 470 mμ and expressed as Δ OD/30s in terms of fresh weight.

3 Results and Discussion

3.1 Callus Differentiation and Somatic Embryogenesis

Explants grown on MS + 1 mg/l 2,4-D + 0.1 mg/l kinetin proliferated within 2 weeks of culturing; the callus was whitish to cream-colored, and mostly hard in texture. This callus when subcultured on MS + 0.2 mg/l 2,4-D + 0.1 mg/l kinetin + 10% CW, showed various morphological patterns: (1) compact, white, round masses of tissue; (2) cabbage-like, folded, leafy patterns; and (3) loose and semifriable callus. Callus consisted mostly of rounded or elongated parenchymatous cells. At places, the callus occasionally showed green or pink pigmentation. The callus tissue, unless subcultured periodically, turned brown and collapsed. A culture obtained from hypocotyl callus grown on medium supplemented with 0.2 mg/l 2,4-D + 2 mg/l kinetin + 15% CW showed differentiation. This culture was passed through six successive subcultures over a period of about 1 year. Twenty to 25 weeks after the first subculture, there appeared, occasionally, on the surface of the slow-growing callus, masses of islets of actively growing tissue. The meristematic zones had small cells with dense protoplasm. These zones differentiated into shoot buds and leafy appendages, irregular rosettes with leafy structures, or abnormal, green, teratoma-like structures (Bajaj et al. 1973).

In some cultures, the surface of the callus tissue was covered with round bodies – comparable to globular embryos. Later, various stages of development of the somatic embryos were observed (Figs. 1 and 2). In many cases, the embryos showed more than two cotyledons. Some showed lobed, leafy, and folded cotyledons. Various early stages of embryo development, studied anatomically, are shown in Fig. 2.

The embryos, if allowed to remain on the same medium, proliferated further to form patches of callus/multiple buds/shoots/accessory embryos. The maturing embryos, when isolated and transferred to a basal medium, started to germinate (Fig. 1D) and grew into plantlets (Bajaj et al. 1973).

3.2 Peroxidases in Relation to Differentiation/Embryogenesis

A differentiating cell undergoes a complex series of alterations involving the accumulation of new enzymes and the loss of old ones. This differential appearance of new patterns can be considered as the elementary process of cell differentiation (Shannon 1968). Morphogenetic roles have also been suggested by the action of peroxidases by producing (Riddle and Mazelis 1964) and by inactivating auxin (Galston and Davies 1969). These enzymes have been shown to alter both quantitatively and qualitatively during growth and development, their patterns can be altered by growth regulators, and they are involved in many of the biochemical processes (see Bopp and Bajaj 1972; Scandalios and Sorenson 1977; Penel et al. 1993).

Peroxidases are known to be organ-specific, and similar observations have been made in different tissues, organs, and seedlings of *Sinapis alba* (Bajaj et al. 1973). The polyacrylamide zymograms of the leaves showed two or three bands,

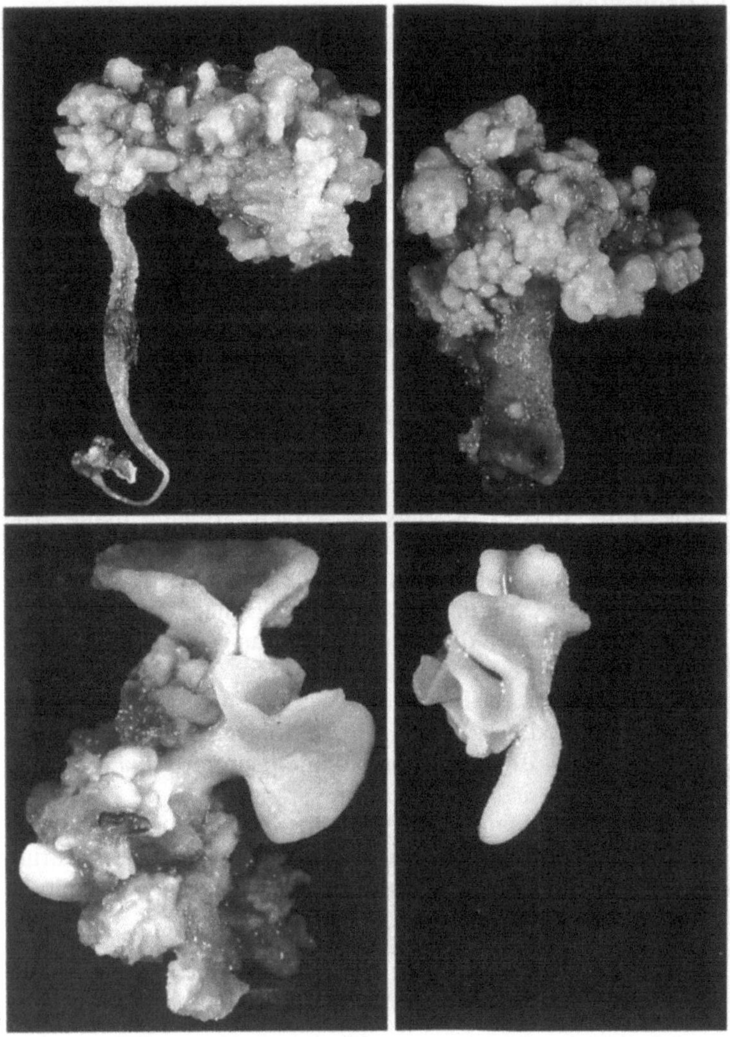

Fig. 1A–D. In vitro induction of somatic embryogenesis in *Sinapis alba*. **A**. 6-week-old culture of a seedling and **B** excised cotyledon grown on MS + 1 mg/l 2,4-D + 0.1 mg/l kinetin. Note the all-round callusing and the differentiation of embryoids. **C** 25-week-old callus tissue originally established on MS +1 mg/l 2,4-D + 0.1 mg/l kinetin, and then transferred to MS + 0.2 mg/l 2,4-D + 2 mg/l kinetin + 10% coconut water. Note the formation of leafy cotyledons of a maturing somatic embryo. **D** A somatic embryo with well-developed cotyledons about to germinate. (Bajaj and Bopp 1972)

one comparatively slow moving and the other faster. However, in the stem, the zymogram pattern was significantly different, and it contained only one band (Fig. 3A,B). This band was faint or absent altogether in the leaves, but peroxidases were present in small quantities in the callus tissue cultures. McCown et al. (1970) observed changes in the stem and callus tissue cultures of *Dianthus*, and attributed this, in part, to repression and induction by the substrate.

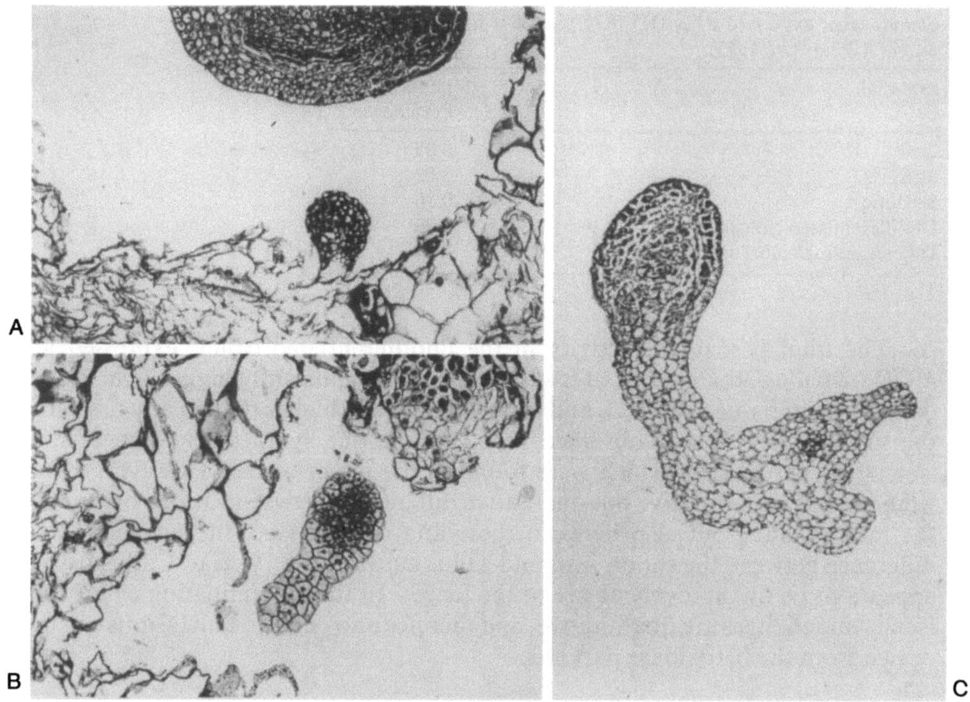

Fig. 2A–C. Anatomy of somatic embryos, from 25-week-old differentiating callus, at various stages of development. (Bajaj et al. 1973)

The undifferentiated and differentiating/embryogenic callus tissue exhibited practically identical isoenzyme patterns consisting of three bands (Fig. 3B). In the differentiated callus the slow-moving band persisted longer on the gel than in the undifferentiated callus tissue where it was faint, and later disappeared altogether (Fig. 3A).

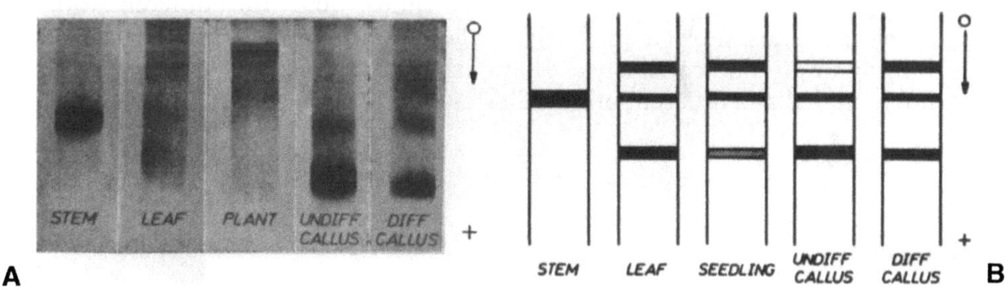

Fig. 3A,B. Polyacrylamide gel electrophoretic patterns of peroxidases. **A** Peroxidases in stem, leaf, 4-week-old plantlet, and undifferentiated and differentiating/embryogenic callus cultures. Electrophoresis was performed in Tris-glycine buffer at a current of 4 m per tube for 45 min; the gels were flooded with guaiacol and hydrogen peroxide. **B** Diagrammatic representation of the zymogram from **A**. *Dark bar* show high peroxidase activity, hatched bar represents comparatively faint bands, and open bar indicates the disappearance of a band. (Bajaj et al. 1973)

Table 1. Peroxidase activity in various organs and callus cultures of *Sinapis alba*, expressed as Δ OD (470 mμ)/30 s in terms of fresh weight. (Bajaj et al. 1973)

Material	Units Peroxidase
Leaf	0.05
Stem	0.02
Seedling	0.10
Undifferentiated callus	0.36
Differentiating/embryogenic callus	0.24

The total peroxidase activity in the undifferentiated callus, expressed as Δ OD (470 mμ)/30 s in terms of fresh weight, was significantly higher than in the differentiated tissue (Table 1), and in both it is much higher than in any tissue of the intact plant. It was observed that callus tissue synthesizes some of the isoenzyme peroxidases which were found in the intact tissues. However, there were not only quantitative but qualitative differences between various organs of the intact plants, but also between these and callus tissue. In particular, the difference between the shoot, leaf, and callus cultures is very clear. Callus tissue appears to be qualitatively closer to the leaves. In the differentiating callus the isoenzyme changes are quantitative, and therefore no definite conclusions can be drawn from the peroxidase patterns.

4 Summary

Callus cultures raised from the segments of cotyledons, hypocotyl, and roots of *Sinapis alba* on MS medium supplemented with 1 mg/l 2,4-D and 0.1 mg/l kinetin, when transferred to MS + 0.2 mg/l 2,4-D + 2 mg/l kinetin and 10% coconut water, underwent differentiation to form shoots, buds/roots, and somatic embryos. The somatic embryos showed various stages of development and, on transfer to a basal medium, germinated to form plants. Although the peroxidases, compared in leaf, stem, seedling, undifferentiated callus, and the differentiating/embryogenic callus, showed qualitative and quantitative changes in the patterns of zymograms, no conclusive relationship was observed between the undifferentiated and differentiating callus cultures.

References

Bajaj YPS, Bopp M (1972) Growth and organ formation in *Sinapis alba* tissue cultures. Z Pflanzenphysiol 66: 378–381

Bajaj YPS, Mohapatra D (1990) Pollen-embryogenesis and chromosomal variability in anther culture of *Brassica hirta* Moench (*Sinapis alba* L.). Cell Res (China) 2: 191–196

Bajaj YPS, Bajaj S, Bopp M (1972) Peroxidases in relation to differentiation in plant tissue cultures. Am J Bot 59: 668

Bajaj YPS, Bopp M, Bajaj S (1973) Patterns of peroxidases and differentiation in *Sinapis alba* L. Phytomorphology 23: 43–52

Birecka H, Galston AW (1970) Peroxidase ontogeny in a dwarf pea stem as affected by gibberellin and decapitation. J Exp Bot 21: 735–745

Bopp M, Bajaj YPS (1972) Enzymic patterns in plant tissue culture. Biochem J 128: 32–33

Bopp M, Bajaj YPS, Ackermann CE, Gapp S (1972) Reversal of 5-fluorodeoxyurodine-caused growth inhibition in intact and excised etiolated hypocotyls of *Sinapis alba* L. by thymidine. Planta 107: 247–256

Galston AW, Davies PJ (1969) Hormonal regulation in higher plants. Science 163: 1288–1297

Klimaszewska K, Keller WA (1983) The production of haploids from *Brassica hirta* Moench (*Sinapis alba* L.) anther culture. Z Pflanzenphysiol 109: 235–241

Lelivelt CLC, Leunissen EHM, Frederiks HJ, Helsper JPFG, Krens FA (1993) Transfer of resistance to the beet-cyst nematode (*Heterodera schachtti* Schm.) from *Sinapis alba* L. (white mustard) to the *Brassica napus* L. gene pool by means of sexual and somatic hybridization. Theor Appl Genet 85: 688–696

McCown BH, McCown DD, Beck GE, Hall TC (1970) Isoenzyme complements of *Dianthus* callus cultures: influence of light and temperature. Am J Bot 57: 148–152

Mohapatra D, Bajaj YPS (1987) Interspecific hybridization in *Brassica juncea* × *Brassica hirta* using embryo rescue. Euphytica 36: 321–326

Murashige T, Skoog F (1962) A revised medium for rapid growth and bioassays with tobacco tissue cultures. Physiol Plant 15: 473–497

Ornstein L, Davis BJ (1962) Disc electrophoresis. Distillation Products Industries, Eastman Kodak, New York

Penel C, Gaspar Th, Greppin H (eds) (1993) Plant peroxidases 1980–1990. Topics and detailed literature on molecular, biochemical, and physiological aspects. University of Geneva.

Primard C, Vedel F, Mathieu C, Pelletier G, Chevre AM (1988) Interspecific somatic hybridization between *Brassica napus* and *B. hirta* (*Sinapis alba* L.). Theor Appl Genet 75: 546–552

Riddle VM, Mazelis M (1964) A role for peroxidase in the biosynthesis of auxin. Nature 202: 391–392

Ripley VL, Arinson PG (1990) Hybridization of *Sinapis alba* and *Brassica napus* L. via embryo rescue. Plant Breed 104: 26–33

Scandalios JG, Sorenson JC (1977) Isozymes in plant tissue culture. In: Reinert J, Bajaj YPS (eds) Applied and fundamental aspects of plant cell, tissue, and organ culture. Springer, Berlin Heidelberg New York, pp 719–730

Shannon LM (1968) Plant isoenzymes. Annu Rev Plant Physiol 19: 187–210

Toriyama K, Kameya T, Hinata K (1987) Selection of a universal hybridizer in *Sinapis turgida* Del. and regeneration of plantlets from somatic hybrids with *Brassica* species. Planta 170 : 308–313

White PR (1943) A handbook of plant tissue culture. Jaques Cattell, Tempe, AZ

III.6 Somatic Embryogenesis and Clonal Propagation in Meadowfoam (*Limnanthes* Species)

D. SOUTHWORTH and S. KWIATKOWSKI[1]

1 Introduction

1.1 Morphology, Importance, and Distribution

Meadowfoam (*Limnanthes* spp.) is a spring-flowering annual, endemic to moist habitats in the Pacific Northwest from California to British Columbia. The Limnanthaceae includes two genera, *Limnanthes* with nine species, and *Floerkea* with one species (Mason 1952; Ornduff and Crovello 1968). The plants grow as low-flowering rosettes with pinnate and bipinnatifid leaves. Flowers are bisexual with 3–6 petals, 6–10 stamens, and 3–5 carpels with a fused style and separate stigma lobes.

The pollen grains of the family are unique (Buchner et al. 1990). An apertural band encircles the grain below the equator. On drying, the smaller portion collapses into the larger, forming a C-shaped pollen grain, Further, the sculpturing pattern on the exine is dimorphic with striations in the zones bordering the apertural band and spinules in the polar regions.

Limnanthes is an oilseed plant with potential for development as a new crop (Jolliff 1981; Jolliff and Snapp 1988). Seeds contain oils with an unusually high proportion of long-chain fatty acids C-20 to C-22 (Princen and Rothfus 1984). Because the wax esters from these oils are stable at high temperatures (160 °C), they are considered to have industrial potential in production of polymers, lubricants, coatings, and adhesives. *Limnanthes* oils also have uses in cosmetics, and may substitute for sperm whale oil.

Agronomic studies have investigated growth parameters such as germination conditions, cultivation, fertilizers, and harvesting methods to optimize seed and oil yields (Jain 1979; Franz and Jolliff 1989; Jahns and Jolliff 1990; Fiez et al. 1991).

Meadowfoam seeds contain glucolimnanthin (Rodman 1991a,b), a plant glucosinolate found to benefit animals by increasing metabolic inactivation of the carcinogen benzo[*a*]pyrene (Baird et al. 1988).

[1] Department of Biology, Southern Oregon State College, Ashland, Oregon 97520-5071, USA

Biotechnology in Agriculture and Forestry, Vol. 31
Somatic Embryogenesis and Synthetic Seed II (ed. by Y.P.S. Bajaj)

1.2 Significance of Somatic Embryos in *Limnanthes*

Somatic embryos in *Limnanthes* serve two purposes: clonal propagation and secondary embryo production. Clonal multiplication of a single genotype allows study of phenotypic variation, especially in oil production. It creates a population of genetically identical organisms on which to test environmental effects. In addition to experimental uses of clones, secondary embryos, produced from zygotic embryos, offer the possibility of perpetuating the differentiated state. Because seeds, cotyledons in particular, produce important oils, somatic embryos offer the possibility of industrial-scale harvesting of oil-rich plant parts from a genetically uniform source.

1.3 Review of Work Done by Others

Rosette plants of *L. floccosa*, *L. alba*, and their hybrid were brought into culture for micropropagation by Reed (1991). Nodal stem explants multiplied on semisolid MS medium with 0.5 mg/l 6-benzyladenine (BA) and 3% sucrose. IAA did not promote multiplication. Plants rooted after transfer to media without BA and were successfully transferred to potting soil.

2 Somatic Embryogenesis

2.1 Secondary Embryos from Immature Embryos

Immature embryos gave diverse growth responses depending on size of the embryo and on growth conditions (Southworth and Kwiatkowski 1991). Globular and heart-shaped embryos < 0.1 mm long (Fig. 1) did not survive well on MS media supplemented with coconut milk, but with no other exogenous phytohormones. Addition of 0.5 μM 2,4-dichlorophenoxyacetic acid (2,4-D) improved survival and growth, but not to the response level of larger (0.1–0.9 mm) embryos. Embryos of late heart-shaped or early cotyledons stage, 0.1–0.9 mm long (Fig. 2), survived and grew. Cotyledonary stage embryos (1–3 mm) often were unresponsive with growth occurring as swelling of cotyledons or occasionally as callus or embryogenic masses arising from the shoot meristem or hypocotyl.

Older embryos, while less sensitive to handling, were more committed to differentiated pathways, and may have been in early stages of dormancy. The middle stages were sufficiently determined to avoid callus formation, and retained sufficient developmental flexibility to respond to altered environmental stimuli.

Sucrose concentration was an important factor in survival. After 2 days on media containing 0–80 mM sucrose, embryos turned white, while embryos on media with 120–290 mM sucrose remained green. On 170–290 mM sucrose, embryos remained green and increased in tissue diameter 10- to 20-fold in 26 days.

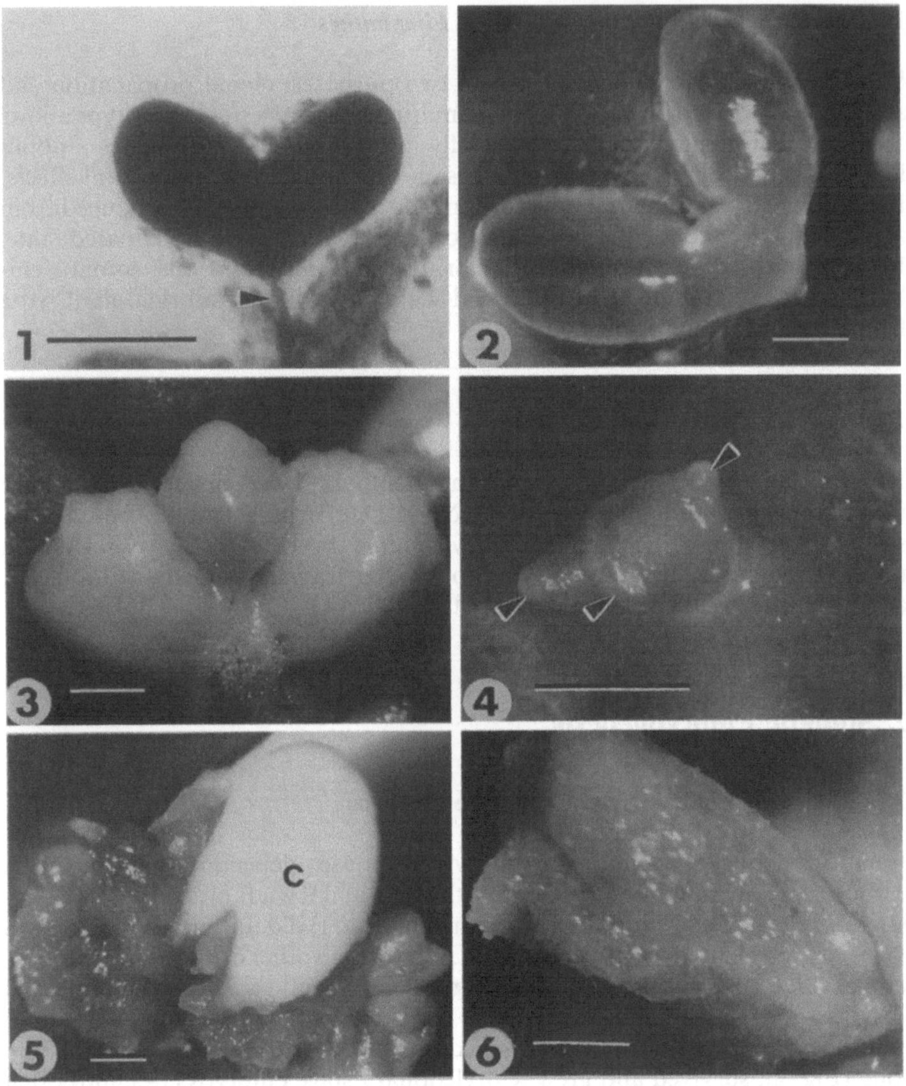

Figs. 1–6. Morphogenesis in cultures of immature zygotic embryos of meadowfoam. **Fig. 1.** Early heart-shaped embryo with suspensor (*arrow head*). **Fig. 2.** Early cotyledonary-stage embryo with shoot and root apex. **Fig. 3.** Growth of cotyledons and shoot apex on 0 2,4-D from a 2-mm embryo. **Fig. 4.** Adventitious shoot formation; apex with leaf primordia (*arrow heads*) grown without 2,4-D, from a 0.6-mm embryo at 22 days. **Fig. 5.** White cotyledon (*C*) with somatic embryos growing from cotyledonary node and hypocotyl region on 0.5 μM 2,4-D. **Fig. 6.** Bipolar embryo with callus-type surface layer on 0.1 μM 2,4-D, from a 0.4-mm embryo at 22 days. Scale *bars* (Figs. 1, 2, 4) = 0.1 mm; (Figs. 3, 5, 6) = 0.5 mm. (Southworth and Kwiatkowski 1991)

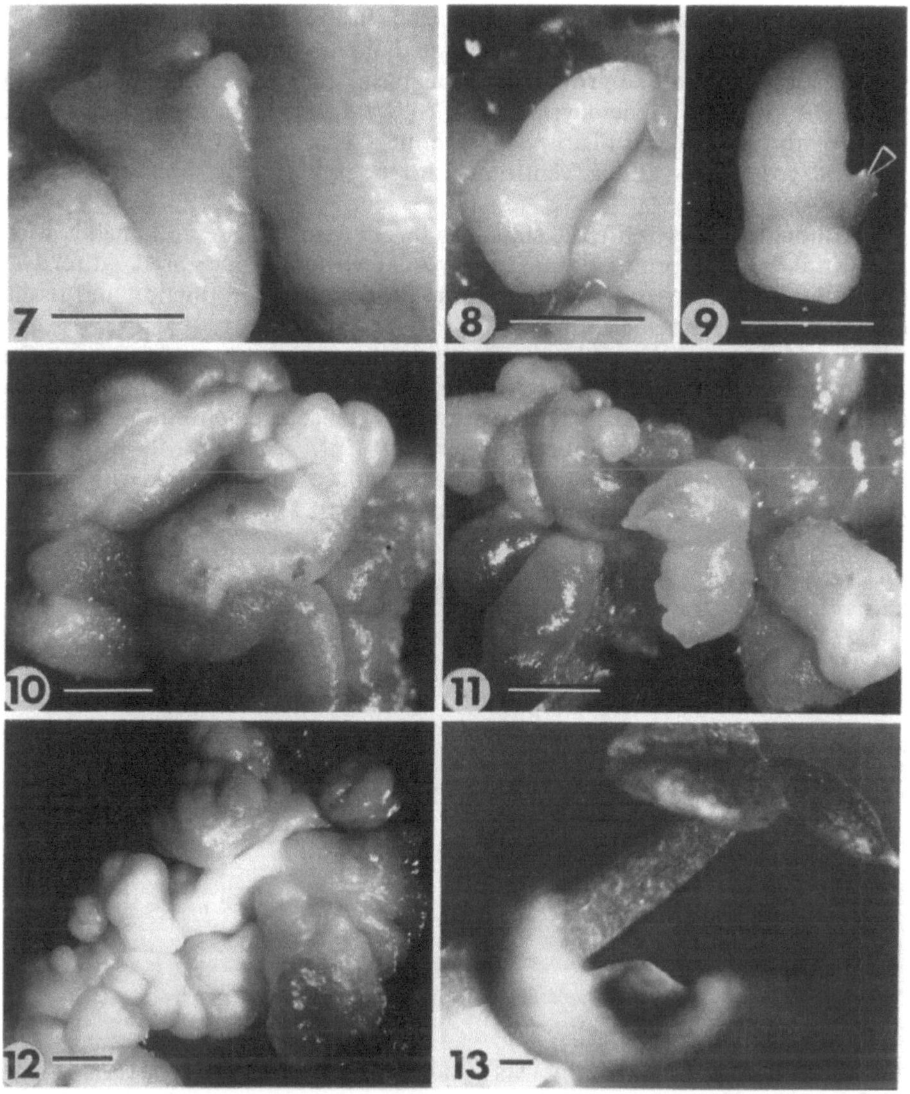

Figs. 7–13. Somatic embryogenesis in cultures of immature zygotic embryos of meadowfoam. **Fig. 7.** Bipolar embryo on 0.1 μM 2,4-D from a 0.5-mm embryo at 22 days. **Fig. 8.** Cotyledonary-stage somatic embryo with unequal cotyledons, on 0.1 μM 2,4-D, from a 0.6-mm embryo. **Fig. 9.** Cotyledonary-stage embryo (Fig. 8.) dissected to show suspensor-like attachment (*arrow head*). **Fig. 10.** Embryogenic culture with organized masses, 0.1 μM zeatin at 20 days. **Fig. 11.** Lobed masses on embryogenic culture, 100 μM zeatin at 20 days. **Fig. 12.** Mass of somatic embryos on 1 μM zeatin at 20 days. **Fig. 13.** Germinating somatic embryo. Scale *bars* = 0.5 mm. (Southworth and Kwiatkowski 1991)

Of the auxins tested, only 2,4-D affected the amount and form of growth. Naphthaleneacetic acid and indoleacetic acid did not induce embryos. Embryogenesis and a high rate of growth of tissue mass occurred on 0.1–4.5 μM 2,4-D (Figs. 5–9; Table 1). Higher levels of 2,4-D (45 μM) also induced embryogenesis, but growth was reduced (Table 1). Without 2,4-D, growth took the form of enlarged cotyledons and adventitious shoots rather than somatic embryos (Figs. 3,4).

Although embryos of all sizes produced tissue masses on media with and without 2,4-D, production of embryo-like structures increased with 2,4-D (Table 1). For embryos 0.5–0.9 mm long, both embryo production and tissue growth were greatest on 0.1 μM 2,4-D.

Cytokinins stimulated tissue growth and somatic embryo formation. On zeatin, embryos produced larger tissue masses with more somatic embryos than on BA (Figs. 10,11; Table 2). The greatest growth occurred on 1 μM zeatin, and the greatest frequency of embryogenesis on 0.1 μM zeatin in MS media containing 0.5 μM 2,4-D. Production of somatic embryos on MS media with 9.1 μM zeatin and low 2,4-D levels was also good (Tables 1, 2).

Table 1. The effect of 2,4-D in MS media containing 9.1 μM zeatin on growth of cultured embryos

Embryo size (mm)	2,4-D (μM) 0	0.1	0.5	4.5	45.0
	Size of cultured tissue (mm)[a]				
0.05–0.4	2.1 ± 3.4	0.3 ± 0.2	3.7 ± 3.1	1.5 ± 1.8	1.8 ± 2.0
0.5–0.9	9.5 ± 3.7	9.6 ± 3.2	8.1 ± 3.9	7.4 ± 3.1	4.8 ± 2.0
1.0–3.0	9.0 ± 1.1	9.5 ± 3.0	6.8 ± 1.9	10.0 ± 3.4	8.7 ± 1.7
	Cultures producing somatic embryos				
0.05–0.4	0/5	0/5	2/5	1/5	0/5
0.5–0.9	0/8	7/8	5/8	6/8	5/8
1.0–3.0	0/4	0/4	2/4	2/4	2/4

[a] Average maximum dimension (mean ± SD) of tissue after 23 days.

Table 2. The effects of zeatin and BAP in MS media containing 0.5 μM 2,4-D on growth of embryos 0.1–0.6 mm at excision

Cytokinin (μM)	0.1	1	10	100
	Size of cultured tissue (mm)[a]			
Zeatin	5.8 ± 4.4	7.4 ± 4.8	4.0 ± 3.9	3.8 ± 3.6
BAP	5.8 ± 5.0	2.8 ± 1.6	4.2 ± 2.4	1.8 ± 0.8
	Cultures producing somatic embryos			
Zeatin	4/5	3/5	2/5	2/5
BAP	2/5	2/5	2/5	0/5

[a] Average maximum dimension (mean ± SD) of tissue after 25 days.

The growth mass derived from immature zygotic embryos consisted of green to yellow, hard, nonfriable tissue with organized protrusions (Figs. 5–12). Somatic embryos arose as fused masses that later separated by differential growth into distinct structures attached to the callus at one side or end by a thin, suspensor-like structure (Figs. 1, 9). Somatic embryos exhibited root-shoot polarity and lateral asymmetry (Figs. 6–12). Normal heart-shaped and cotyledonary-stage embryos were not formed. Some embryos were slightly heart-shaped, but larger than zygotic heart-shaped embryos.

In excised cotyledonary-stage embryos, cotyledons turned from green to yellow or white under all treatments; occasionally, somatic embryos formed in the hypocotyl region (Fig. 5).

Somatic embryos lacked the precise form of zygotic embryos. They were often bipolar with cotyledons of unequal size, with multiple cotyledons or with fasciated structures that may have been abnormal cotyledons. They had a distinct epidermal surface with closely appressed cells that formed a smooth, flat surface. Cotyledonary-stage embryos proliferated somatic embryos from the hypocotyl region of the zygotic embryo. Smaller embryos became globular before proliferation occurred.

2.2 Subculture and Germination of Secondary Embryos

Subcultures of embryogenic masses continued to produce additional embryos that were in turn subcultured and produced further somatic embryos. Some somatic embryos germinated in the absence of exogenous 2,4-D on media containing 88 mM sucrose (Fig. 13). Zeatin was necessary for survival of somatic embryos under these conditions.

2.3 Meristem Culture

Apical meristems from rosette plants and lateral meristems from developing floral stalks could be cultured. Meristems were aseptically excised and cultured on a liquid medium consisting of MS salts, Nitsch and Nitsch vitamins, indole-acetic acid (0.6 μM), gibberellic acid (0.6 μM), kinetin (0.2 μM) and sucrose (88 mM) at pH 5.7 (Southworth and Kwiatkowski 1991).

The meristem itself appeared free of microorganisms. Outer leaf petiole bases were contaminated with soil particles, but the innermost six to eight leaf bases surrounding the meristem were free of soil particles. Surface contamination deep within the rosette indicated difficulty in surface sterilization for shoot culture, as noted by Reed (1991).

After 10 days on liquid medium, meristems were subcultured to semisolid medium containing 250 mg/l carbenicillin for control of bacteria. Antibiotics were unnecessary when lateral meristems from floral stalks were cultured. These lateral meristems were easily excised and cultured, and rapidly produced aseptic plants.

Meristems grew rapidly from an excised length of 0.3–1.0 mm to lengths of 3–15 mm in 9 days. Subculture to semisolid medium from liquid medium was necessary after 10 days. After an additional 55 days on medium with 4.6 μm zeatin and 250 mg/l carbenicillin, seven out of ten shoots survived and produced an average of seven shoots per clump.

2.4 Shoot Multiplication

Cytokinins were tested to determine which best promoted shoot multiplication. Clumps of plants from meristem culture were dissected to individual shoots and transferred to media containing BA, zeatin, kinetin or N^6-(Δ^2-isopentenyl)-adenine (2iP). Tissues were grown under an 8-h photoperiod (19 μE $m^{-2}s^{-1}$) and an 18/12 °C day/night temperature regime.

Shoot multiplication of meadowfoam also occurred on media containing zeatin, kinetin, or 2iP, but not on media containing BA. Multiplication was best on zeatin (Table 3), but kinetin and 2iP gave satisfactory results.

For shoot multiplication and plantlet production, shoots derived from meristem culture were placed on semisolid MS medium containing 4.6 μM zeatin and 250 mg/l carbenicillin in flasks and placed in the growth chamber as above. Shoots that were transferred to zeatin-free medium grew roots, and when transplanted to soil, flowered and produced seed in the greenhouse.

2.5 Clonal Propagation from Bisected Seeds

Seeds of *Limnanthes* can be bisected so that a portion of the seed can be analyzed for oil content and the remainder of the seed germinated (Southworth and Kwiatkowski 1992). High germination rates (97%) and uniform plants were obtained from seeds cut transversely. Cut half-seeds were stored frozen for 25 days without a decrease in viability.

Seeds were sliced transversely across the center, halfway between the attachment scar and seed tip. This cut off the distal portion of both cotyledons, but did not cut the embryonic axis. A sterile technique was not needed. Following frozen storage, seeds were placed on moist filter paper in sealed Petri dishes at 10 °C in darkness. After 9 days at 10 °C, bisected seeds or whole seeds were planted in flats of soil.

Table 3. Shoot multiplication in response to cytokinins

Cytokinin	Shoots per clump[a]
Zeatin (4.7 μM)	9.8 ± 2.8
Kinetin (4.6 μM)	6.7 ± 1.1
N^6-(Δ^2-isopentenyl)-adenine (3 μM)	6.7 ± 2.7
Benzyladenine (4.4 μM)	1.0 ± 0

[a] After 43 days in culture (mean ± SD).

The number of plants established in the greenhouse after 8 days of incubation at 10 °C and 12 days in the greenhouse was greatest for transversely bisected dry seed, of which 100% gave rise to plants. Of bisected frozen seed, 97% formed plants. In dry seeds, cotyledons shattered or cracked when bisected; however, this did not interfere with plant establishment. Seedlings from dry, transversely bisected seed, whether germinated immediately or stored, grew uniformly and flowered.

The high percentage of plants established following frozen storage implies that the oil content of individual seeds can be potentially correlated with the oil content of the progeny of those seeds. The ability to store bisected seeds and later to regenerate and clonally increase plants from those half-seeds should enhance the effort to breed a superior oil-producing meadowfoam.

3 Summary and Conclusions

Somatic embryos of meadowfoam, from immature zygotic embryos on semisolid MS media containing sucrose, 2,4-D, and zeatin, show root-shoot polarity and can germinate or proliferate new somatic embryos. Meadowfoam responds favorably to clonal increase techniques of meristem culture, shoot multiplication, and half-seed propagation.

4 Protocol for Induction of Somatic Embryos

One week after anthesis, when ovaries measure 1–4 mm rinse whole flowers in 0.5% sodium hypochlorite on a magnetic stirrer for 5 min and place uncovered on sterile filter paper. Excise heart-shaped embryos (0.5–0.9 mm long) aseptically using a No.11 scalpel blade and a 26-gauge hypodermic needle.

Culture on media containing MS salts, myo-inositol (0.6 mM), Nitsch and Nitsch vitamins, Noble agar (0.7%), 200 mM sucrose, 0.1 μM 2,4-D, and 0.1–1 μM transzeatin (Southworth and Kwiatkowski 1991). After addition of components, adjust medium to pH 5.7 and autoclave at 110 °C for 15 min. Add sterile coconut water (100 ml/l) to media after autoclaving. Dispense media (25 ml/dish) into 15 × 100 mm polystyrene Petri dishes. Seal dishes with Parafilm and puncture to provide air to the cultures.

Place cultures in a growth chamber at 26 °C (day) and 20 °C (night) with a 16-h photoperiod from cool-white fluorescent lights (19 $\mu E\ m^{-2}\ s^{-1}$).

After 30 days, divide cultures with organized growths and somatic embryos into four or five segments and subculture onto media containing 0.1, 1, or 5 μm 2,4-D and 2 μM zeatin.

For maturation and germination, eliminate 2,4-D and reduce sucrose to 88 mM in medium with zeatin at 0.1 μm.

Acknowledgments. This research was supported by NSF grant DCB-8801902 and Supplement DCB-9144446 through Research in Undergraduate Institutions.

References

Baird WM, Zennie TM, Ferin M, Chae YH, Hatchell J, Cassady JM (1988) Glucolimnanthin, a plant glucosinolate, increases the metabolism and DNA binding of benzo[*a*]pyrene in hamster embryo cell cultures. Carcinogenesis 9: 657–660

Buchner R, Halbritter H, Pfunder D, Hesse M (1990) Pollen of *Limnanthes douglasii*: a reinvestigation. Grana 29: 207–211

Fiez TE, Norberg OS, Jolliff GD (1991) Yield components in three meadowfoam lines. Agron J 83: 598–602

Franz RE, Jolliff GD (1989) Temperature effects of megagametophyte development in meadowfoam. Crop Sci 29: 133–141

Jahns TR, Jolliff GD (1990) Survival rate and reproductive success of *Osmia lignaria propinqua* Cresson (Hymenoptera:Megachilidae) in caged meadowfoam, *Limnanthes alba* Benth. (Limnanthaceae). J Kans Entomol Soc 64: 95–106

Jain SK (1979) Response to mass selection for flowering time in meadowfoam. Crop Sci 19: 337–339

Jolliff GD (1981) Development and production of meadowfoam (*Limnanthes alba*). In: Pryde EH, Princen LH, Mukherjee KD (eds) New sources of fats and oils. Am Oil Chem Soc Monogr 9, Champaign, IL, pp 269–285

Jolliff GD, Snapp SS (1988) New crop development: opportunity and challenges. J Prod Agric l: 83–89

Mason CT (1952) A systematic study of the genus *Limnanthes* R. Br. Univ Calif Publ Bot 25: 455–512

Ornduff R, Crovello TJ (1968) Numerical taxonomy of the Limnanthaceae. Am J Bot 55: 173–182

Princen LH, Rothfus JA (1984) Development of new crops for industrial raw materials. J Am Oil Chem Soc 61: 281–289

Reed BM (1991) Micropropagation of meadowfoam (*Limnanthes* spp.). Plant Cell Rep 10: 94–96

Rodman JE (1991a) A taxonomic analysis of glucosinolate-producing plants. Part 1 Phenetics. Syst Bot 16: 598–618

Rodman JE (1991b) A taxonomic analysis of glucosinolate-producing plants. Part 2 Cladistics. Syst Bot 16: 619–629

Southworth D, Kwiatkowski S (1991) Somatic embryogenesis from immature embryos in meadowfoam (*Limnanthes alba*). Plant Cell Tissue Organ Cult 24: 193–198

Southworth D, Kwiatkowski S (1992) Clonal propagation of meadowfoam (*Limnanthes* sp.) from bisected seeds. Northwest Sci 66: 40–43

Section IV
Ornamental, Medicinal, and Miscellaneous Plants

IV.1 Somatic Embryogenesis in Daylily (*Hemerocallis*)

A.D. Krikorian[1], R.P. Kann[1], and D.L. Smith[1,2]

1 Introduction

1.1 Morphology and Importance

Daylilies (*Hemerocallis* species and cultivars) are herbaceous perennials much sought after for their showy albeit very short-lived blooms. The mature plant is comprised of fans (ramets) consisting of an underground stem (rhizome), an expansive fibrous or fibrous-tuberous root system, and strap-shaped leaves. Axillary buds can be forced to yield vegetative separations. The deciduous or evergreen foliage consists of sheathed-at-the-base, heavily ribbed leaves which when mature (Fitter and Krikorian 1985; Smith et al. 1989) are long and two-ranked. Flowers are borne on a scape, often branched and largely leafless but with bracts, and shoot propagules ("proliferations"). Flower buds per scape may vary from around 5 to 20 (Wilkins 1985).

Flowers are perfect and fairly large, some as large as ca. 7 in., but last only a day (Bielski and Reid 1992). Fertilization (24–48 h after pollination) yields an embryo which when fully mature is bullet-shaped (cf. Arisumi 1973). Some 42 days or so after pollination the fruits and seeds are ripe. Some 80 to 90% germination occurs in the diploids; roughly 60 to 70% in the tetraploids. The plants are genetically heterozygous and evaluation takes about 3–4 years for varieties that are "fast" and longer for those that are "slow" (Voth et al. 1968; Stout 1986; Munson 1989).

Many clones show reduced fertility (Stout and Chandler 1933) or, even once selected, are quite slow to respond to vegetative multiplication techniques (Darrow and Meyer 1968), but there is a vigorous nursery trade based on traditional clones and a steady release of clonally multiplied diploid or tetraploid varieties.

[1] Dep ... y Brook, Stony
Broc
[2] Pre: ... 5W. University
Park

Biotechnology in Agriculture and Forestry, Vol. 31
Somatic Embryogenesis and Synthetic Seed II (ed. by Y.P.S. Bajaj)

Biotechnology in Agriculture and Forestry, Vol. 31
Somatic Embryogenesis and Synthetic Seed II (ed. by Y.P.S. Bajaj)
©Springer-Verlag Berlin Heidelberg 1995

1.2 Significance of Somatic Embryogenesis in Daylilies

Because of their tendency towards slow multiplication and their frequently reduced fertility, daylilies are ideal candidates for the application of any in vitro cloning techniques. We have adopted the view, however, that embryogenic cells which are grown and managed at least partly in liquid hold the greatest promise for clonal multiplication on a massive scale (Krikorian and Kann 1979b; Krikorian et al. 1986) and would also facilitate mutation breeding and genetic engineering (Krikorian et al. 1988b, 1990).

For example, one commonly used technique to produce tetraploid (4n = 44) daylily plants from diploids (2n = 22, cf. Zadoo et al. 1976) is through use of colchicine on whole-plant growing points (Arisumi 1964). In daylily, as in other plants, this is fraught with problems, such as chimera production (Arisumi 1972). Callus is perhaps more readily rendered polyploid (Chen and Goeden-Kallemeyn 1979) and cells in suspension are still more so. Conventional hybridization may also be performed between tetraploids but due to the few seeds produced, the process is often inefficient (Arisumi 1964, 1987). Colchicine treatment of developing microspores in diploid floral buds via syringe injection can induce pollen with doubled chromosome numbers (i.e., $2n$ sperm with 22 chromosomes instead of an n of 11), and this has its value for subsequent pollination of diploid egg cells in the embryo sac of tetraploids (Wallace 1984). With the potential for production of tetraploid somatic embryos from single cells, potential for chimerism is essentially eliminated (cf. also Griesbach 1963). Protoplast fusion (controlled or spontaneous) using embryogenic cells of diploids is yet another way to generate tetraploids (Fitter and Krikorian 1988).

1.3 Review of Published Work

Daylilies can be multiplied from morphogenetically competent callus masses (Chen and Holden 1972; Apps and Heuser 1975; Heuser and Apps 1976; Heuser and Harker 1976; Meyer 1976; Krikorian and Kann 1979a, 1980; Griesbach 1988), cell suspensions (Krikorian and Kann 1981), and cell suspension-derived protoplasts (Fitter and Krikorian 1981, 1988; Krikorian et al. 1988a), but there are clonal differences. The possibility of achieving satisfactory clonal fidelity has been demonstrated (Meyer 1976; Chen and Goeden-Kallemeyn 1979; Krikorian and Kann 1980; Krikorian et al. 1981, 1986; Griesbach 1989), provided certain precautions are followed. However, these precautions, applicable to all plants under in vitro conditions, are based on concepts that derive from rationalized supposition rather than from extended experimentation (cf. e.g. Karp 1989 and refs. cited therein). In the case of daylilies, most of the in vitro procedures involves generation, selection, proliferation, and multiplication of "callus" or callus-derived cells. This means there is potential for generation of mutant off-types which may be similar to or different from those known to exist in the genus (cf. Stout 1956; Krikorian et al. 1982, 1988b; Karp 1991).

2 Somatic Embryogenesis

Our work on somatic embryogenesis has concentrated on the use of the diploid clone Autumn Blaze as a model. The principles to be enunciated apply to other clones, but experience has shown a wide diversity of responsiveness. In all cases, the most important part of the process is the initiation of embryogenic "callus" cultures.

2.1 Initiation of "Callus" Cultures

Bud explants which include the proximal half of the ovary with 1–2 mm of pedicel or the distal half of the ovary with ca. 2 mm of style provide starting material. Alternatively, excised vegetative shoot tips may be used. Although explants from leaves, leaf bases, scapes, and even petals can respond, those from ovaries from flower buds about 1.5–4 cm long and shoot apices have, in our hands, yielded the most consistent results. The basal MS medium (Murashige and Skoog 1962) supplemented with BAP and 2,4-D, each varied logarithmically from 0.001–10 mg/l has proved to be a satisfactory way to design an effective medium for a given clone. Relatively high levels of 2,4-D (e.g., 10 mg/l) in combination with low levels of BAP (e.g., 0.001–0.1 mg/l) generally foster proliferation within 1 month. Explants may then require transfer once or twice prior to separation of the callus from the primary explant as a preliminary to subculture. Lower levels of 2,4-D (e.g., 0.001–0.01 mg/l) with higher levels of BAP (e.g., 0.1–1 mg/l) also yield callus. Equal ratios of auxin to cytokinin (1:1) are usually less satisfactory in inducing callus. Material initiated this way is readily subcultured every 3 to 4 weeks, provided relatively large (up to 1 cm diameter) inocula are used. After an additional 4 weeks, portions of the callus may be transferred to media with lowered levels of BAP and 2,4-D (ratio on the order of 10:1). Shoots and roots form adventitiously within a few weeks and these may be separated if large enough and placed on semisolid medium in vessels containing half-strength MS basal medium with 1.5% sucrose. After an additional 60 days or so they are normally ready for transfer to soil.

Like all tissue culture protocols, these procedures are flexible, however, and no claim is made here that there has been significant progress towards development of a consensus medium that is applicable to a great many clones. For example, Meyer (1976) reported the use of inflorescence slices on a semisolid MS medium supplemented with 10 mg/l NAA and 0.1 mg/l kinetin in darkness for callus initiation, and subsequent subculture and transfer to light on that same medium albeit with NAA lowered to 0.5 mg/l light yielded plantlets. To obtain callus, Heuser and Harker (1976) used small floral bud explants (more specifically, the petals and sepals) and a semisolid MS medium supplemented with 1000 mg/l casein hydrolysate, 2,4-D and kinetin each at 1 mg/l. (If NAA is used instead of 2,4-D to minimize, as these authors claim, the potential for aberrant or off-type plants, it is used at 0.5 mg/l). Eliminating the 2,4-D and lowering the kinetin to 0.1 mg/l, while increasing the sucrose to 4.5%, yielded shoots which in turn could be multiplied on the same medium, and root initiation from such shoots using IAA (anywhere from 1 to 10 mg/l) yielded a good multiplication system.

2.2 Use of Liquid Medium

To increase further the yield and to reduce the dimensions of the embryogenic units required to yield organized growth, a step utilizing liquid was introduced some time ago (Krikorian and Kann 1981). Callus initiated on semisolid medium from primary explants generally needs to be transferred to liquid at the higher auxin and cytokinin levels a few times before a sustainable subculture comprised of fragments of about 1–2 mm can be achieved. Such cultures, which we term "nubbin" cultures (Krikorian et al. 1986,1988b), are maintainable readily through repeated subculture and yield morphogenetically competent units. These units, which may be still further reducible in size down to the level of cell suspensions, retain their embryogenicity and can, when exposed to the correct medium sequences, ultimately give rise to plantlets. In the case of the diploid clone Autumn Blaze, maintenance medium is MS with coconut water (CW) 10% v/v and 2 mg/l 2,4-D (1 mg/l kinetin serves as a satisfactory substitute for CW). Here, the first step in the progression of organization in a suspension is to remove the 2,4-D. After a 3-week subculture in 2,4-D-free medium, the material is cultured on CW-free medium; then the material is transferred to Schenk and Hildebrandt (1972) medium (albeit with Murashige and Skoog iron, cf. Singh and Krikorian 1980) without growth regulators. After about 3 weeks on basal Schenk and Hildebrandt medium, early shoot development is apparent. Continued development is assured on a 50% dilution of the same, generally in the light and in an Erlenmeyer flask. Further growth into plantlets occurs when units are selected and transferred to semisolid medium with 1% sucrose and exposure to light (cf. Krikorian and Kann 1986). Autumn Blaze. Suspensions may also be generated directly in liquid without being started on semisolid medium, which speeds up the process leading to embryogenic cell cultures.

Griesbach (1988) has reported the use of MS basal medium with vitamins and 3% sucrose, 2000 mg/l casein amino acids, 2 mg/l 2,4-D, and 0.25 mg/l kinetin to obtain callus from excised shoots on a semisolid medium, followed by transfer to liquid medium to get suspensions. When the suspensions were plated onto a semisolid regeneration medium, consisting of the same salts and vitamins and casein amino acids but with lowered sucrose (2%), and replacement of the 2,4-D with 10 mg/l NAA and kinetin with 1 mg/l BAP, plantlets were regenerated.

2.3 Generation of Somatic Embryos

2.3.1 Embryogenesis from Cultures Grown on Medium Lacking Hormones

Study of the relationship between habituation or loss of the requirement for external hormones and somatic embryogenesis of carrot (cf. Smith and

Table 1. DS-5a "salts" or base formulation minus nitrogen. Note: Nitrogen source[a] and buffering to be carried out as needed (cf. also Krikorian and Smith 1992)

	(mH)	(mg/l)
$Ca(H_2PO_4).H_2O$	1.0	252.10
$MgSO_4 \cdot 7\,H_2O$	1.0	246.49
NaCl	0.5	29.23
H_3BO_3	0.1	6.18
$MnSO_4 \cdot H_2O$	0.1	16.9
$ZnSO_4 \cdot H_2O$	0.01	2.88
KI	0.005	0.83
$Na_2MoO_4 \cdot 2H_2O$	0.0001	0.024
$CuSO_4 \cdot 5H_2O$	0.0001	0.025
$CoCl_2 \cdot 6H_2O$	0.0001	0.024
$Na_2FeEDTA \cdot 2H_2O$ } $FeSO_4 \cdot 7H_2O$	0.1	b
Thiamine HCl	0.0037	1.0
Sucrose	58.4	20 000.0

[a] Ammonium chloride is a good source of nitrogen.
[b] 27.8 mg/l $FeSO_4.7H_2O$; 37.23 mg/l Na_2 EDTA.$2H_2O$; or 33.6 mg/l Na_2 EDTA. See Singh and Krikorian (1980) for details of chelated iron solutions.

Krikorian 1991) has led to the disclosure of some interesting facts about somatic embryogenesis in the daylily Autumn Blaze. Embryogenic cells in suspensions ordinarily initiated and maintained with growth regulators such as 2,4-D and CW or kinetin (see above for details) may be multiplied in the absence of hormone on a semisolid medium buffered with (10–25 mM MES 2(N-morpholino) ethanesulfonic acid) between pH 4 and 4.5, provided they have been in culture for an appropriate period of time. Hormone-free medium containing ammonium as the nitrogen source (cf. Table 1 for DS 5a medium formula) will sustain multiplication of the cells in the preglobular embryogenic mode, i.e., without progression to later stages of embryogenesis, more or less indefinitely (Smith and Krikorian 1991), whereas increase in pH to 5.8 permits the continued development of the earliest stage embryos all the way to small plantlets that can ultimately be established in soil.

The production of daylily somatic embryos requires the use of reduced nitrogen (cf. Halperin and Wetherell 1965). By using NH_4^+Cl (in the order of 2.5 or 5 mM) morphogenetically competent suspensions embark upon a course of embryogenic development that shows (in its later stages in particular) a high level of similarity to zygotic embryos (Smith and Krikorian 1991).

Cell suspension fractions express embryogenic potential to various degrees. The quality of the resultant somatic embryos, i.e., their degree of morphological fidelity and tendency to the single rather than multiple, varies with the dimension of the units from which they derive. Larger sized fractions generally yield more multiple somatic embryos than finer ones, and not unexpectedly, the larger fractions respond more quickly than the smaller ones. Therefore, depending on the objective, specific fractions should be used. Methods for sieving of cell suspensions through screens of known pore size have been reported in detail (cf. e.g. Giuliano et al.1983; Krikorian and Smith 1992).

2.3.2 *Use of Activated Charcoal*

As in the case with carrot (Smith and Krikorian 1990), daylily somatic embryo development is facilitated by using activated charcoal. Analytical paper containing activated charcoal (Schleicher & Schuell, Keene, New Hampshire #508, 9-cm size) has been useful. These are placed on a semisolid surface of a Petri dish. This avoids the need to prepare charcoal-containing media and provides the added advantage of enabling the convenient handling of tissues, since tissues resting on the papers can be transferred and manipulated with relative ease. However, these papers have to be washed and autoclaved twice to assure sterility (Smith and Krikorian 1990).

2.3.3 *Embryogenesis from Cultures Initiated and Maintained with Hormones*

The procedure of advancing somatic embryos from suspensions that have been initiated with hormones and maintained through serial subculture on media with hormones is essentially the same as that using cells that have been initiated with hormone but maintained without. Cells are washed thoroughly to free them of hormone and inoculated onto DS 5a medium (see Table 1).

Since our work on Autumn Blaze somatic embryogenesis is being done from the viewpoint of attempting to understand the mechanisms of somatic embryogenesis, efforts have not concentrated on massive multiplication. Nevertheless, because we frequently grow suspensions in 220 ml volume, there is potential for operations on a reasonable scale. Figure 1A–C shows a maintenance medium suspension at different levels of magnification. A gross view of the suspension in a so-called nipple flask with 220 ml medium can be seen in Fig. 1A; while Fig. 1B shows a view of a rigorously sieved suspension the unit size of which ranges between 140 and 74 μm and Fig. 1C shows a view under an inverted microscope. Once an appropriate cell fraction has been obtained, appropriate numbers of cells in suspension are applied to charcoal paper on an appropriate medium that allows embryogenesis to progress. Amounts used have generally been on the order of around 0.05 to 0.1 ml settled cell volume per milliliter inoculum (see Fig. 1D for a view of part of a freshly inoculated Petri dish). Most embryos have developed to at least the globular stage by the end of the first 2-week culture period. Figure 1E,F presents views of somatic embryos at 20 and 31 days, respectively. Somatic embryos are moved to fresh medium of identical composition at least once for an additional 2 weeks (total of 4 weeks). These will grow out when placed on a medium containing ammonium and nitrate under continuous low light (about 450 ft-c) (cf. Fig. 1G). Not all the embryos will grow out properly under these conditions, however. After 4 weeks in the light, the vast majority have done so but some appear dormant, others make callus, and still others produce aberrant forms that we call "neomorphs" (cf. Smith and Krikorian 1991).

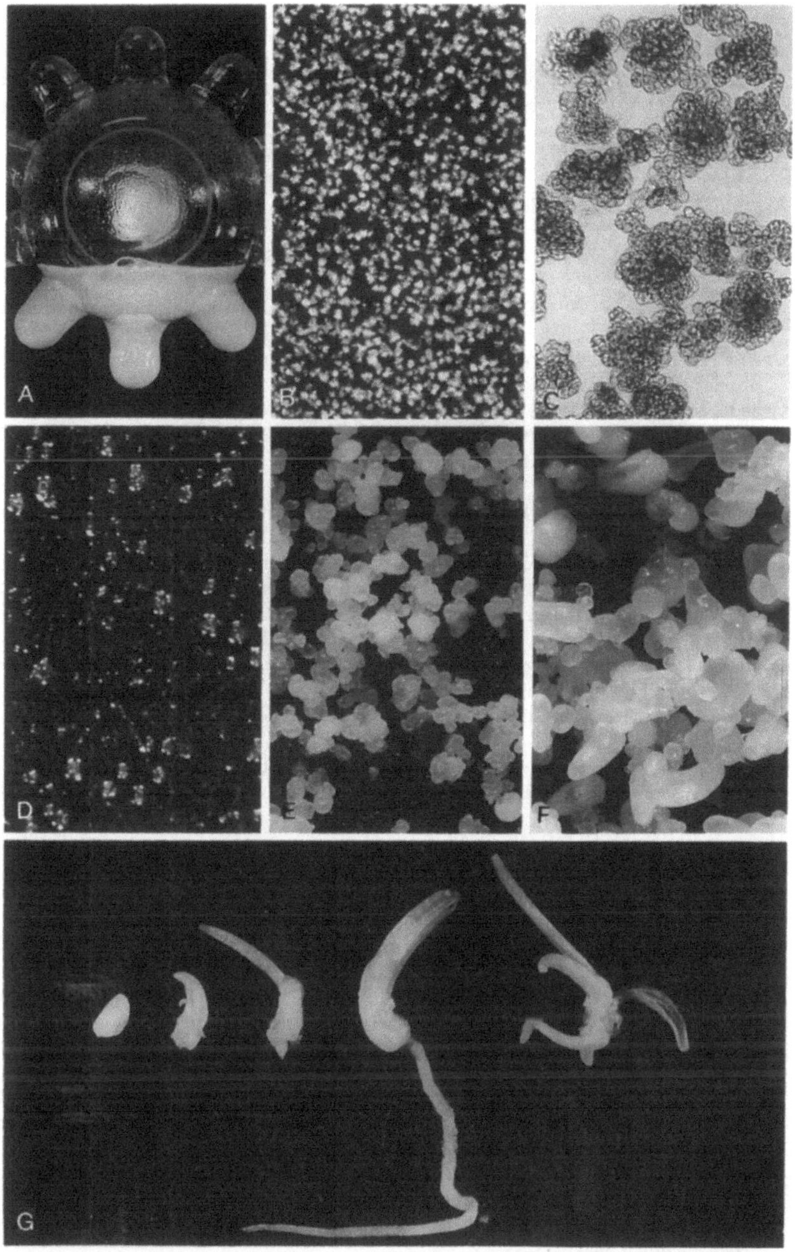

Fig. 1A–G. Generation of daylily somatic embryos using cells grown in suspension. **A** Nipple flask with cells in liquid medium 23 days after subculture. Note the density and fineness of the culture; 0.2x. **B** 100- to 200-mesh fraction of embryogenic cells obtained by sieving a suspension similar to that in **A** through stainless steel screens; ca 9.5x. **C** Microscopic view of cells like those in **B**; 69x. **D** suspension plated from units like those in **C** onto activated charcoal-impregnated filter paper, which in turn was placed on a hormone-free DS 5a culture medium with 1 mM ammonium chloride as a nitrogen source; 13.9x. **E** Readily recognizable somatic embryos 20 days after inoculation; 4.8x. **F** Similar embryos 31 days after inoculation. **G** "Germination" of somatic embryos showing a range of development in the root and shoot portions; 2.8x

References

Apps DA, Heuser CW (1975) Vegetative propagation of *Hemerocallis* – including tissue culture. Int Plant Prop Soc Comb Proc 25: 326–367

Arisumi T (1964) Colchicine-induced tetraploid and cytochimeral daylilies. J Hered 55: 255–261

Arisumi T (1972) Stabilities of colchicine-induced tetraploid and cytochimeral daylilies. J Hered 63: 15–18

Arisumi T (1973) Embryo development and seed set in crosses for triploid daylilies. Bot Gaz 134: 135–139

Arisumi T (1987) Microsporogenesis in diploid and tetraploid 'Sherwood' daylily. J Am Soc Hortic Sci 112: 722–723

Bielski RL, Reid MS (1992) Physiological changes accompanying senescence in the ephemeral daylily flower. Plant Physiol 98: 1042–1049

Chen CH, Goeden-Kallemeyn YC (1979) In vitro production of tetraploid plants from colchicine-treated diploid daylily callus. Euphytica 28: 705–709

Chen CH, Holden DJ (1972) Organogenesis in daylily. Proc SD Acad Sci 51: 146–149

Darrow GM, Meyer FG (eds) (1968) Daylily handbook. Am Hortic Mag 47(2): 41–272

Fitter MS, Krikorian AD (1981) Recovery of totipotent cells and plantlet production from daylily protoplasts. Ann Bot 48: 591–597

Fitter MS, Krikorian AD (1985) Mature phenotype in *Hemerocallis* plantlets fortuitously generated in vitro. J Plant Physiol 121: 97–101

Fitter MS, Krikorian AD (1988) Daylily protoplasts: isolation, culture and organized development into plants. In: Valentine F (ed) Progress and prospects in crop and forest biotechnology. State University of New York College of Environmental Science and Forestry Colloquium, April 18–20, 1985, Syracuse. Springer, Berlin Heidelberg New York, pp 242–256

Giuliano G, Rosellini D, Terzi M (1983) A new method for the purification of different stages of carrot embryoids. Plant Cell Rep 2: 216–218

Griesbach RJ (1963) Induction of polyploidy in newly-germinated *Hemerocallis* seedlings. *Hemerocallis* J 17: 70–75

Griesbach RJ (1988) *Hemerocallis* improvement through tissue culture. HortScience 23: 804

Griesbach RJ (1989) Selection of a dwarf *Hemerocallis* through tissue culture. HortScience 24: 1027–1028

Halperin W, Wetherell DF (1965) Ammonium requirement for embryogenesis in vitro. Nature 205: 519–520

Heuser CW, Apps DA (1976) In vitro plantlet formation from flower petal explants of *Hemerocallis* cv. 'Chipper Cherry'. Can J Bot 54: 616–618

Heuser CW, Harker J (1976) Tissue culture propagation of daylilies. Int Plant Prop Soc Comb Proc 26: 269–272

Karp A (1989) Can genetic instability be controlled in plant tissue cultures? Int Assoc Plant Tissue Cult Newsl 58: 2–11

Karp A (1991) On the current understanding of somaclonal variation. Oxford Surveys Plant Mol Cell Biol 7: 1–58

Krikorian AD, Kann RP (1979a) Clonal micropropagation of daylilies. In: Sharp WR, Larsen PO, Paddock EF, Raghavan V (eds) Plant cell and tissue culture: principles and application. Ohio State University Press, Columbus, pp 835–836

Krikorian AD, Kann RP (1979b) Micropropagation of daylilies through aseptic culture techniques: its basis, status, problems and prospects. *Hemerocallis* J 33(1): 44–61

Krikorian AD, Kann RP (1980) Mass blooming of a daylily clone reared from cultured tissues. *Hemerocallis* J 34(1): 35–38

Krikorian AD, Kann RP (1981) Plantlet production from morphogenetically competent cell suspensions of daylily. Ann Bot 47: 679–686

Krikorian AD, Kann RP (1986) Regeneration in lilies, irises and tulips. In: Vasil IK (ed) Cell culture and somatic cell genetics of plants, vol 3. Plant regeneration and genetic variability. Academic Press, New York, pp 187–205

Krikorian AD, Smith DL (1992) Somatic embryogenesis in carrot (*Daucus carota*). In: Lindsey K (ed) Plant tissue culture manual: fundamentals and application. Kluwer Dordrecht, pp PTCM-A9 1–32

Krikorian AD, Staicu S, Kann RP (1981) Karyotype analysis of a daylily clone reared from aseptically cultured tissues. Ann Bot 47: 121–131

Krikorian AD, O'Connor SA, Kann RP, Fitter MS (1982) Development of, and karyotype stability in, *Hemerocallis* plants reared via tissue, suspension and protoplast culture. In: Fujiwara A (ed) Plant tissue culture 1982. Maruzen Tokyo, pp 429–430

Krikorian AD, Kann RP, O'Connor SA, Fitter MS (1986) Totipotent suspensions and means of multiplication. In: Zimmerman RH, Griesbach RJ, Hammerschlag FA, Lawson RH (eds) Tissue culture as a plant production system for horticultural crops. US Dept of Agriculture Symp, Oct 20–23, 1985, Beltsville, Maryland. Martinus Nijhoff/Dr W Junk, The Hague, pp 61–72

Krikorian AD, Cronauer-Mitra SS, Fitter Corbin MS (1988a) Protoplast culture of perennials. Sci Hortic 37: 277–293

Krikorian AD, Kann RP, O'Connor SA, Fitter MS, Cronauer SS, Smith DL (1988b) The range of morphogenetic responsiveness in aseptically cultured daylily tissues and cells: significance for multiplication and improvement. In: Valentine F (ed) Progress and prospects in crop and forest biotechnology. State University of New York College of Environmental Science and Forestry, Colloquium, April 18–20, 1985, Syracuse. Springer, Berlin Heidelgerg New York, pp 82–98

Krikorian AD, Kann RP, Fitter MS (1990) Daylilies. In: PV Ammirato, Evans DA, Sharp WR, YPS Bajaj (eds) Handbook of plant cell culture, vol 5. Ornamental. McGraw-Hill, New York, pp 375–412

Meyer MM (1976) Propagation of daylilies by tissue culture. HortScience 11: 485–487

Munson RW Jr (1989) *Hemerocallis*, the daylily. Timber Press, Portland

Murashige T, Skoog F (1962) A revised medium for rapid growth and bioassays with tobacco tissue cultures. Physiol Plant 15: 473–497

Schenk R, Hildebrandt AC (1972) Medium and techniques for induction and growth of monocotyledonous and dicotyledonous cell cultures. Can J Bot 51: 199–204

Singh M, Krikorian AD (1980) Chelated iron in culture media. Ann Bot 46: 807–809

Smith DL, Krikorian AD (1990) Somatic embryogenesis of carrot in horomone-free medium: external pH control over morphogenesis. Amer J Bot 72: 1634–1647

Smith DL, Krikorian AD (1991) Growth and maintenance of an embryogenic cell culture of daylily (*Hemerocallis*) on hormone-free medium. Ann Bot 67: 443–449

Smith DL, Kelly K, Krikorian AD (1989) Ethylene-associated phase change from juvenile to mature phenotype of daylily (*Hemerocallis* in vitro. Physiol Plant 76: 466–473

Stout AB (1956) Variegation and somatic mutations in daylilies. *Hemerocallis* J 10(2): 15–21

Stout AB (1986) Daylilies. The wild species and garden clones, both old and new, of the genus *Hemerocallis*. Introduction and updating by Darrel Apps. Sagapress, Millwood, NY

Stout AB, Chandler C (1933) Pollen-tube behavior in *Hemerocallis* with special reference to incompatibilities. Torrey Bot Club Bull 60: 408–416

Voth PD, Griesbach RA, Yeager JR (1968) Developmental anatomy and physiology in daylily. Am Hort Mag 47: 121–151

Wallace M (1984) Scape conversion. Daylily J 38(1): 112–115

Wilkins HF (1985) *Hemerocallis fulva*. In: Halevy AH (ed) CRC Handbook of flowering vol 3. CRC Press, Boca Raton, pp 130–132

Zadoo SN, Roy RP, Khoshoo TN (1976) Variation in karyotype in *Hemerocallis*. La Cellule 71: 253–271

IV.2 Somatic Embryogenesis in *Freesia refracta*

L. WANG and B. HUANG[1]

1 Introduction

Freesia is a small genus in Iridaceae, comprising 11 species. The common freesia (*Freesia refracta* Klatt.) was originally found in southern Africa (Goldblatt, 1982), but is now distributed in many parts of the world. The plant, normally 25–45 cm in height, has branched stems, slender leaves, and bulb-like tunicated corms. The flower possesses three stamens and an inferior ovary with three carpels. A normal plant has two types of roots: the thick contractile roots function mainly to support the plant, whereas the numerous thin roots transport materials from the soil to the plant (Ruzin, 1979). *Freesia refracta* can be propagated either sexually through seeds or vegetatively through corms. The seeds of freesia are difficult to obtain, and it usually takes years for seeds to develop into mature plants. Therefore, corms are practically used to propagate plants.

Freesia refracta is an ornamental flowering plant. During the flowering season, the plant produces long spike inflorescences with flowers orderly arrayed along one side of the inflorescence. The flowers start to bloom in succession from the base to top of the inflorescence, and produce a very pleasant fragrance. Large-scale cultivation of freesia plants for commercial purposes has been carried out in many countries. In addition, the flowers contain aromatic oils which have a certain commercial value (Wang 1981).

The early work on tissue culture and plant regeneration via organogenesis in *F. refracta* was carried out by Davies and Nichol (1971). Bajaj and Pierik (1974) described the plant recovery from corms, stems, leaves, flower buds, and anthers under in vitro culture conditions in one freesia cultivar. These authors also evaluated the effects of the interaction between auxins and cytokinins on morphogenesis in callus cultures. Using buds and roots as explants, Petru et al. (1976) obtained regenerated plants in several cultivars. Pierik and Steegmans (1977) investigated the exogenous hormonal regulation in plant regeneration from bud explants. Other work dealing with freesia tissue culture can be found in the literature (e.g., Hussey 1975; Mori et al. 1975; Kataeva 1981; Bach 1984; Chen and Lu 1985; Sun et al. 1986; Bajaj 1989). The results of these studies showed that *F. refracta* is very amenable to in vitro culture, and the organs and tissues of various parts of the plant have been demonstrated to be capable of

[1]Institute of Genetics and Cytology, Northeast Normal University, Changchun 130024, People's Republic of China

Biotechnology in Agriculture and Forestry, Vol. 31
Somatic Embryogenesis and Synthetic Seed II (ed. by Y.P.S. Bajaj)

regeneration and propagation. However, plant regeneration via somatic embryogenesis has not been reported in *F. refracta* prior to our work (Wang et al. 1990). Recently, however, embryogenesis has been induced in *F. hybrida* by Bach (1992).

Our study demonstrated that somatic embryos can be induced either directly from the surface of explants such as the young inflorescences, young leaves, and immature and mature embryos, or indirectly from calluses initiated from these explants. Moreover, it was possible to control or regulate the two different pathways of somatic embryogenesis by varying the combinations and levels of exogenous hormones in culture media (Wang et al. 1990). The ease with which the somatic embryos can be induced and handled in a controllable manner makes this plant useful experimental material for studying various processes of somatic embryogenesis, such as the differentiation and redifferentiation of cultured cells, the induction of embryogenic potential, as well as their hormonal regulations. These are the fundamental problems in understanding the mechanisms of somatic embryogenesis, which for many years has been a focus of research attention.

In addition to its use in basic studies, somatic embryogenesis in *Freesia* also has some practical value. Plant regeneration via somatic embryogenesis provides a means of rapid clonal propagation of large quantities of virus-free plants. This is important in commercial production of flowers since, firstly, the vegetative propagation through bulb corms takes a relatively long period of time and, secondly, certain virus infections, e.g., the bean yellow mosaic virus (BYMV), have been a serious problem in freesia cultivation (Nagel et al. 1983).

In this chapter, we describe the procedures for the induction of both direct and indirect somatic embryogenesis and the subsequent plant regeneration from tissue cultures of *F. refracta*, as well as the exogenous hormonal control of these embryogenic processes. In addition, the results obtained with histological and scanning electron microscope studies of the segmentations of embryogenic cells and the development of somatic embryos are discussed in association with the ontogeny and origin of embryogenesis in this plant.

2 Somatic Embryogenesis

2.1 Induction of Somatic Embryogenesis and Its Hormonal Control

The young inflorescences, young leaves and zygotic embryos of *F. refracta*, are all suitable explant sources for the induction of somatic embryos and, as a rule, the juvenile tissues and organs are more responsive to the inducing stimuli than the more mature tissues when used as explants.

Under the culture conditions employed in our experiments, satisfactory induction efficiencies were obtained with the explants of young inflorescences segments. The young inflorescences of freesia can be separated into two parts, the flower stalks and the flower buds. These two components were used separately as explants for tissue culture. Although the flower stalks contained no mitotically active meristems, some of the cells in the epidermis were able to induce direct embryogenesis without callus formation when placed under appropriate culture conditions. The dissected flower buds continued to develop into flowers on culture medium and callus was initiated from the basal part of the flower. The young leaves were less regenerative than young inflorescences for embryo induction. Another ideal explant source was the zygotic embryo which exhibited a high potency of somatic embryo formation. However, it is difficult to obtain sufficient embryo material due to the low seed setting rate of freesia.

Explants of *F. refracta* cultured in vitro can be induced to undergo somatic embryogenesis via two different pathways, i.e., embryos can be formed either directly from the epidermal cells of the explants, or indirectly through an intervening callus stage depending on the exogenous hormones included in the media.

After 3 days in culture on MN_6 medium (modified N_6, Chu 1978) supplemented with 2 mg/l IAA (indole-3-acetic acid) and 3 mg/l BAP (6-benzylaminopurine), the margins of the explanted segments of young inflorescence began to swell slightly. Subsequent growth and differentiation of the swellings resulted in the formation of globular somatic embryos which became visible within 7–14 days in culture. These embryos arose from the competent cells in the epidermis of the explants via a direct somatic embryogenesis pathway, as will be evidenced in the following section. The most apparent characteristic of this embryogenic pathway was that the divisions of the pre-existing embryogenic cells in explants gave rise to the embryos without callus formation.

If, alternatively, the segments of young infloresence were inoculated on MS medium (Murashige and Skoog 1962) containing 2 mg/l IAA, 0.5 mg/l BAP, and 0.5 mg/l NAA (naphthalenacetic acid), a translucent, pale-yellow nodular callus formed on explants after 1 week of culture. When this callus was transferred onto MN_6 medium with 2 mg/l IAA and 3 mg/l BAP, some globular proembryos formed (Fig. 1A). A soft, watery, non-embryogenic callus was induced if the same explant was placed on MS medium with 2 mg/l IAA, 3 mg/l BAP, and 0.5 mg/l, 2, 4-D (2,4-dichlorophenoxyacetic acid). This watery callus did not produce embryos and tended to turn brown after prolonged culture on the same medium

Fig. 1A. Pale-yellow nodular callus initiated from a segment of young inflorescence of *F. refracta* cultured on MN_6 + 2 mg/l IAA + 3 mg/l BAP. Note the globular embryos on the callus surface (*arrow*). **B** Soft watery callus grown on MS + 2 mg/l IAA + 3 mg/l BAP + 0.5 mg/l 2, 4-D. **C** Embryo formation on soft watery callus after transfer to MN_6 + 2 mg/l IAA + 3 mg/l BAP

Table 1. Patterns of somatic embryogenesis in explants of young inflorescences and leaves of *F. refracta* using different media and hormone conditions

Medium mg/l	Experiment[a]	Pattern of Induction[a]	No. of explants	No. of explants forming embryoids	Induction rate (%)
MN_6 + IAA 2	YI	D	50	15	30
+ BAP 3	YL	D	60	6	10
MS + IAA 2 + BAP 0.5 + NAA 0.5 ↓ MN_6 + IAA 2 + BAP 3	YI	I	50	13	26
	YL	I	70	15	21.4

[a] YI, Young inflorescences; YL, young leaf; D, direct embryogenesis; I, indirect embryogenesis

(Fig. 1B). However, when this callus was transferred to MN_6 medium containing 2 mg/l IAA and 3 mg/l BAP, nodules appeared, and some proembryos developed on the surface (Fig. 1C).

Table 1 summarizes the pathways of somatic embryogenesis in tissue cultures and their dependence on the hormonal conditions. It is clear from Table 1 that the same explants were able to undergo different patterns of embryogenesis under different culture conditions. Therefore, it was possible to manipulate the embryogenic pathways by modifying the medium composition and other conditions. Among the various factors influencing in vitro morphogenesis, the effect of exogenously supplied hormones was most decisive and crucial. Table 2 shows the influence of the levels and combinations of various exogenous hormones on the induction patterns of somatic embryogenesis in inflorescence segments cultured on the same basal medium (MN_6). The results presented in Table 2 indicate that (1) use of the combination of IAA and BAP (medium 1) had a positive effect on the induction of somatic embryos; (2) the IAA alone (medium 2), or substitution

Table 2. Influence of exogenous hormones on somatic embryogenesis in young inflorescences explants of *F. refracta* cultured on MN_6 basal medium

	Hormone and conc. (mg/l)				Morphogenetic responses
	IAA	BAP	NAA	2,4-D	
Medium 1	2	3	–	–	Massive embryoid formation directly from epidermis
Medium 2	2	–	–	–	Formation of nodular callus
Medium 3	–	3	–	2	Formation of friable, watery callus, partial browning
Medium 4	2	3	0.5	–	Development of embryoids, faster growth at radical end than at plumuler end
Medium 5	2	5–10	–	–	Maintenance of callus state

of IAA with 2,4-D (medium 3), was not sufficient for the induction; (3) NAA had a positive effect on root development (medium 4); and (4) high levels of BAP (medium 5) prevented the differentiation of embryogenic callus. Moreover, the embryogenic potential of this callus has been maintained for over 1 year.

The somatic embryos induced via both the direct embryogenic pathway from explants and the indirect embryogenic pathway from callus cultures were structurally normal, possessing a plumule apex at one end and a radical apex at the other.

The development of globular proembryos was marked first the appearance of a coleoptile, and then simultaneous elongation of the first leaf out of the coleoptile and growth of the root out of the coleorhiza (Fig. 2A). Regeneration of plants from somatic embryos can be accomplished without difficulty. For the embryos formed via the direct pathway, the germination and development of plants can be completed without changing the medium composition (Fig. 2B). For the embryos derived from callus cultures via the indirect pathway, root development was usually delayed when kept in the same culture. Therefore, embryos were subcultured to MN_6 medium devoid of added exogenous hormones to promote rooting of the regenerated plants (Fig. 2C).

Fig. 2A–D. Plant regeneration through somatic embryogenesis in *F. refracta*. **A** Germinating somatic embryos with roots and leaves. **B** Plantlets developed from embryos induced via direct embryogenesis on a leaf segment. **C** Rooting of the plantlet on hormone-free medium. **D** A somatic embryo-derived plant at the flowering stage

The established plants were successfully transplanted to pots and were acclimatized in a growth chamber for several days before subsequent transfer to the greenhouse (Fig. 2D). Generally, 15–20 plants can be obtained directly from a single inflorescence segment within a 7-week period of culture, and 10–15 plants from leaf tissue within an 8-week period of culture. More than 20 regenerated plants can be obtained through indirect embryogenesis from an explant segment, but the formation of an entire plant from callus cultures took about 12–15 weeks.

2.2 Histology of Embryogenesis

To study the ontogeny of somatic embryogenesis and the development of embryos in both direct and indirect embryogenesis explants and calli undergoing somatic embryogenesis were processed for paraffin embedding and sectioning, and the histological sections were examined under the light microscope.

Direct somatic embryogenesis occurred at the surface of the explanted tissues. The embryos seemed to originate from the discrete, embryogenically competent cells in the epidermal tissue (Fig. 3A). Figure 3B shows a section of a globular proembryo arising from the epidermis of a young leaf segment in culture. At the radical end of this proembryo, where it was still in contact with the epidermal tissue, a suspensor-like structure comprising a string of a few cells can be seen (Fig. 3B, arrow). At a later stage, the globular embryo developed further to form a plumule apex growing upwards and a radical apex growing in the opposite direction (Fig. 3C). Direct somatic embryogenesis originating from the pre-embryogenically determined cells (PEDCs, Sharp et al. 1980) has been commonly observed in many plant culture systems. Normally, PEDCs can be found in epidermal tissues of explants (e.g., Konar et al. 1972), but direct embryo formation from mesophyll cells (Conger et al. 1983) and various parts of cultured leaves (Trigiano et al. 1989) have also been reported. These studies indicate that potentially embryogenic cells may exist in tissues that have undergone a very high

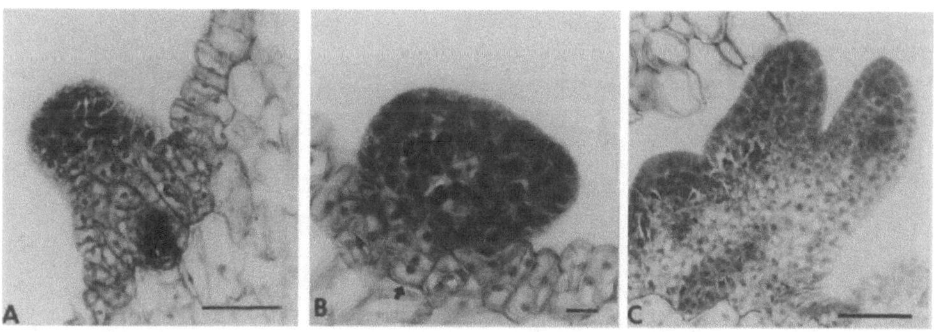

Fig. 3A–C. Sections of somatic embryos induced from leaf segments of *F. refracta* via a direct embryogenesis. **A** A globular proembryo formed from the epidermal cell of the explant, *bar* = 50 μm. **B** A developing globular embryo. Note the suspensor-like structure at the basal part of the embryo (*arrow*), *bar* = 10 μm, **C** A somatic embryo at a later stage. Note the plumule and radical apex at opposite poles of this embryo; *bar* = 50 μm degree of specialization, such as mesophyll tissue. Under conducive culture conditions, these cells may be induced to restore their mitotic activity and may directly undergo embryogenesis.

In direct somatic embryogenesis, the explanted tissues first proliferate to give rise to callus before any morphogenetic processes can be detected. A specific exogenous hormonal regime is required for induction of embryogenically determined cells (IEDCs, Sharp et al. 1980) within the callus. The IEDLs usually appear as clusters of small cytoplasm–rich cells, and were referred to as the "superficial meristems" by Hussey (1978). In *F. refracta*, clusters of embryogenic cells were frequently seen in the peripheral regions of the callus. Indirect embryogenesis commenced with the periclinal division of a single embryogenic cell in callus (Fig. 4A). In another section of similar embryogenic callus, a linear four-celled proembryo can be seen at the periphery of the callus, apparently resulting from two successive periclinal divisions (Fig. 4B). It appears that the periclinal cell divisions at the surface layers of callus may mark the beginning of the somatic embryogenic process. This is consistent with the observation by Maheswaran and Williams (1985) that the first sign of direct somatic embryogenesis was a shift from regular anticlinal divisions to irregular periclinal or oblique divisions of the epidermal cells. Figure 4C shows a longitudinal section of a globular proembryo arising from the meristem-like tissue at the periphery of the callus induced from a young inflorescence. The somatic embryo formed via the indirect pathway lacked a structure equivalent to the suspensor, and it had a relatively broad contact with the maternal tissue (Fig. 4C).

The ontogeny and origin of somatic embryos have been frequently discussed. Generally, there has been ample evidence for the single-cell origin of embryo formation in direct somatic embryogenesis (e.g., Konar et al. 1972; Conger et al. 1983; Trigiano et al. 1989). However, in indirect embryogenesis, the cellular origin of embryos still remains a disputed question. Though the unicellular origin of somatic embryos was reported by some authors (e.g., Haccius 1978), results from many experiments suggested that embryos may develop from groups of embryogenic cells formed in callus or in cell cultures (Halperin and Jensen 1967; Raghavan, 1976; Tisserat et al. 1979). Backs-Husemann and Reinert (1970)

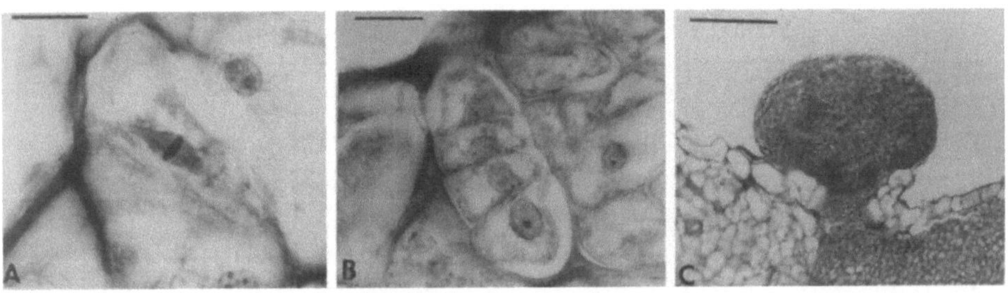

Fig. 4A–C. Sections showing somatic embryo formation from callus of *F. refracta* via indirect embryogenesis. **A** A dividing embryogenic cell at the surface layer of the callus. Note the first division of the embryogenic cells is a periclinal division; *bar* = 10 µm. **B** A linear 4-celled proembryo resulting from two successive cells; *bar* = 10 µm **C** A globular embryo forming at the surface of the callus; *bar* = 100 µm described the formation of embryos from embryogenic cell groups which had a single-cell origin in *Daucus carota* cell cultures. In our sections of embryogenic callus derived from inflorescence explants, the first cell division marking the initiation of somatic embryogenesis and the four-celled proembryo resulting from two divisions of a single embryogenic cell can be clearly seen (Fig. 4A, B). These observations provide evidence for the single-cell origin of indirect somatic embryogenesis in freesia.

2.3 Scanning Electron Microscope (SEM) Study

Under SEM, compact nodular embryogenic callus showed a regular surface morphology, and the superficial cells were arranged in an orderly and uniform manner (Fig. 5A). In contrast, the surface of the soft, friable non-embryogenic callus looked irregular (Fig. 5B). Similar differences in surface structure between embryogenic and non-embryogenic calli were reported in a number of plants (Vasil and Vasil 1982; Wang and Kong 1984; Zhang and Zheng 1988). From these data, it can be assumed that a correlation between the surface structure and the embryogenic potential of the callus may exist. Use of SEM provides an effective means to investigate this correlation.

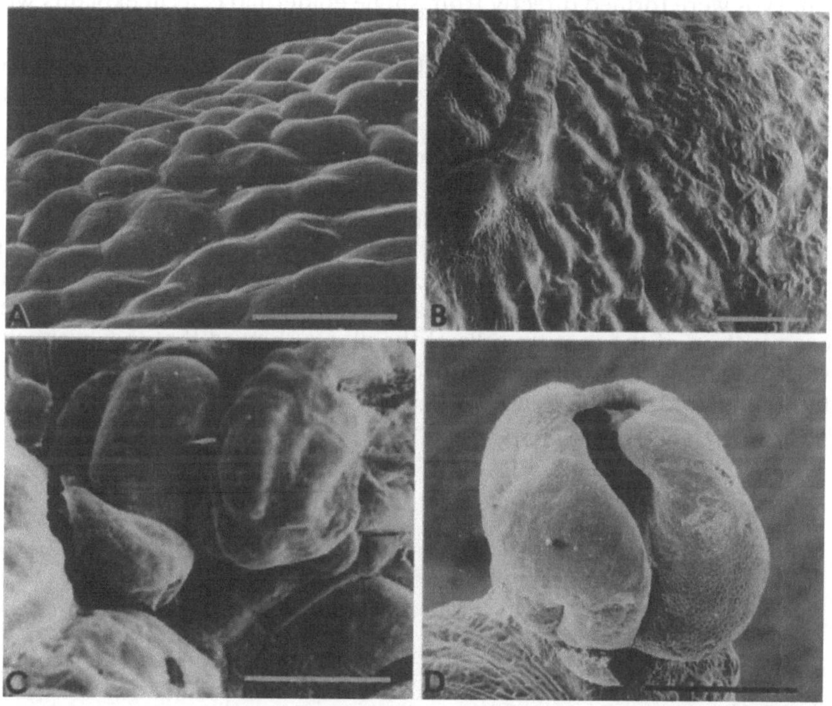

Fig. 5A–D. Scanning electron micrographs of somatic embryogenesis in *F. refracta.* **A** Surface structure of embryogenic callus; *bar* = 49 μm. **B** Surface structure of nonembryogenic callus; *bar* = 100 μm. **C** Globular embryos of different stages; *bar* = 100 μm. **D** A developed somatic embryo with coleoptile; *bar* = 1 μm

Figure 5C shows scanning electron micrograph of an embryogenic callus undergoing somatic embryogenesis. On the previously regular and even surface of the callus numerous nodular protuberances of variable sizes can be seen. At a later stage, highly organized somatic embryos developed on the callus, Fig. 5D shows one of these embryos. As the embryo formation was asynchronous, the embryos on the same callus may be at different developmental stages. The mature somatic embryos were germinated in the same culture and the regenerated plantlets were morphologically similar to the seedlings from the zygotic embryos, except that the regenerated plantlets usually had a conspicuous cotyledon still attached, while the cotyledon was not detected in seedlings (Wang et al. 1991).

2.4 Chromosome Numbers in Regenerated Plants

The diploid chromosome number of the somatic cells of *F. refracta* is 2n = 22, including six pairs of submetacentric chromosomes, four pairs of metacentric chromosomes, and one pair of telocentric chromosomes (Wang and Pu 1988).

Results from cytological studies demonstrated that the freesia plants regenerated from explants through direct embryogenesis had a normal diploid chromosome number of 22 (Fig. 6A). This could be expected, since here, somatic embryos were formed directly from single epidermal cells in explants without a callus culture stage.

The majority of plants obtained through indirect embryogenesis also showed a normal karyotype, provided they were regenerated from callus that had been newly initiated from explants.

Nevertheless, since in this indirect pathway, some of the plants arose from callus that had been kept in culture for up to 1 year, chromosome variations in these plants were most likely to exist.

On the other hand, plants recovered from callus cultures via organogenesis showed a considerable variation in chromosome numbers. In one experiment, a somaclone of dwarf plants from callus via organogenesis was obtained. The examination of the root tip cells of the plants showed that they were chimeric, with about 60% of the somatic cells being diploid, 10.5% haploid, 8.5% tetraploid, and 21% aneuploid (Fig. 6B–D).

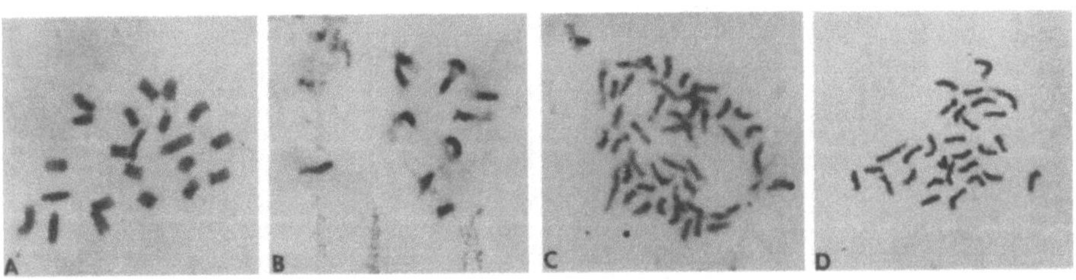

Fig. 6A–D. Root tip cell chromosomes of regenerated *F. refracta* plants. The plant recovered through direct embryogenesis possess a normal diploid chromosome number of 2n = 22 (**A**). Plants regenerated from callus cultures via organogensis are chimerics. In addition to diploid cells, some of the somatic cells possess haploid (**B**), tetraploid (**C**), or aneuploid (**D**) karyotypes

3 Summary

Somatic embryogenesis can be induced in tissue cultures of *F. refracta* either directly from the explants, or indirectly via the intervening callus. These two pathways of embryogenesis can be controlled and regulated by varying the combinations and levels of exogenous hormones in the culture medium. In direct embryogenesis, embryos developed from the pre-embryogenic cells in the epidermis of explants cultured on MN_6 medium with 2 mg/l IAA and 3 mg/l BAP. If the same explants were placed on MS medium containing 2 mg/l IAA, 0.5 mg/l BAP, and 0.5 mg/l NAA, embryogenic callus could be induced. Divisions of the induced embryogenic cells at the periphery of the callus led to the formation of somatic embros on callus. Plants regenerated through direct embryogenesis possessed a normal diploid karyotype, while those recovered from callus cultures through organogenesis showed variable chromosome numbers. The ontogeny and development of somatic embryos of both pathways were studied using histological and scanning electron microscope techniques.

4 Protocol

Young inflorescences of freesia (about 3 cm in length) were an ideal and convenient explant source for both establishment of callus cultures and induction of direct somatic embryogenesis. The MN_6 basal medium (modified N_6) was satisfactory for tissue cultures and somatic embryogenesis.

For the induction of direct embryogenesis, explant segments were placed on MN_6 medium supplemented with 2 mg/l IAA and 3 mg/l BAP, and kept in culture under a 12-h photoperiod of 1800 lx. The cultures were subcultured every 4 weeks, and plants were regenerated from the explants. Transfer of cultures onto a medium either free of added hormones, or with 0.5 mg/l NAA promoted rooting of the regenerated plants.

The procedure for induction of indirect embryogenesis was as follows. Calli were induced from explants cultured on MN_6 (or MS) medium containing 2 mg/l IAA, 0.5 mg/l BAP, and 0.5 mg/l NAA. Nodular embryogenic callus was selected and maintained by subculturing on MN_6 medium with 0.5 mg/l 2,4-D and 5 mg/l BAP at 3-week intervals. To initiate embryo development, nodular callus was transferred to MN_6 medium supplemented with 2 mg/l IAA and 3 mg/l BAP. When the somatic embryos began to germinate, they were transferred to MN_6 medium without hormones or supplemented with 0.5 mg/l NAA for root development.

The regenerated plants were transplanted into soil, and over 90% of the plants survived and grew normally to the flowering stage in the greenhouse.

References

Bach A (1984) The healthiness of *Freesia* × *hybrida* propagated in vitro. In: Novak FJ, Havel L, Dolezel J (eds) Plant tissue and cell culture–application to crop improvement. Czechoslovak Acad Sci Prague, pp 551–552

Bach A (1992) Induction of somatic embryogenesis and regeneration of plants in *Freesia hybrida* cultures. Folia Hortic Ann IV/1 11–21

Backs-Husemann D, Reinert J (1970) Embryobildung durch isolierte Einzelzellen aus Gewebkulturen von *Daucus carota*. Protoplasma 70: 49–60

Bajaj YPS (1989) Freesia. In: Ammirato PV, Evans DA, Sharp WR, Bajaj YPS (eds) Handbook of plant cell culture, vol 5. Ornamental species. McGraw-Hill, New York, pp 413–428

Bajaj YPS, Pierik RLM (1974) Vegetative propagation of *Freesia* through callus cultures. Neth J Agric Sci 22 (3): 153–159

Chen YC, Lu XH (1985) Tissue culture of *Freesia refracta*. Plant Physiol Commun 2: 43–44 (in Chinese).

Chu CC (1978) The N_6 medium and its applications to anther culture of cereal crops. In: Proc Symp Plant tissue culture. Peking, Science Press, Peking, pp 43–50

Conger BV, Hanning GE, Gray DJ, McDaniel JK (1983) Direct embryogenesis from mesophyll cells of orchardgrass. Science 221: 850–851

Davies DR, Nichol MA (1971) In vitro propagation of *Freesia*. Annu Rep John Innes Inst 62:45

Goldblatt P (1982) Systematics of *Freesia* (Iridaceae). J S Afr Bot 48: 39–92

Haccius B (1978) Question of unicellular origin of non-zygotic embryos in callus cultures. Phytomorphology 28: 74–81

Halperin W, Wetherell DF (1965) Ontogeny of adventive embryos in wild carrot. Science 147: 756–758

Halperin W, Jensen WA (1967) Ultrastructural changes during growth and embryogenesis in carrot cell cultures. Jour Unltrastructure Research 18: 428–443

Hussey G (1975) Totipotency in tissue explants and callus of some members of the Liliaceae, Iridaceae and Amaryllidaceae. J Exp Bot 26: 253–262

Hussey G (1978) The application of tissue culture to the vegetative propagation of plants. Sci Prog Oxf 65: 185–208

Kataeva NV (1981) Tissue and organ culture of *Freesia*. Fiziol Rast (Mosc) 28: 1062–1064

Konar RN, Thomas E, Street HE (1972) Origin and structure of embryoids arising from epidermal cells of the stem of *Ranunculus sceleratus* L. J Cell Sci 11: 77–93

Maheswaran G, Williams EG (1985) Origin and development of somatic embryos formed directly on immature embryos of *Trifolium repens* in vitro. Ann Bot 56: 619–630

Mori Y, Hasegawa A, Kano K (1975) Studies on the clonal propagation by meristem culture in *Freesia*. J Jpn Soc Hortic Sci 44: 294–302

Murashige T, Skoog F (1962) A revised medium for rapid growth and bioassays with tobacco tissue cultures. Physiol Plant 15: 473–497

Nagel J, Zettler FW, Hiebert E (1983) Strains of bean yellow mosaic virus compared to clover yellow vein virus in relation to gladiolus (*Gladiolus hortulanus*) production in Florida. Phytopathology 73: 449–454

Petru E, Jirsakova E, Landa Z (1976) Clonal propagation of some *Freesia* cultivars through tissue culture. Biol Plant (Prague) 18: 304–306

Pierik RLM, Steegmans HHM (1977) Vegetative propagation of *Freesia* through the isolation of shoots in vitro. Neth J Agric Sci 24: 274–277

Raghavan V (1976) Adventive embryogenesis: induction of dipoid embryoids. In: Experimental embryogenesis in vascular plants. Academic Press, London, pp 349–381

Ruzin SE (1979) Root contraction in *Freesia* (Iridaceae). Am J Bot 66: 522–531

Sharp WR, Sondahl MR, Caldas LS, Maraffa SB (1980) The physiology of in vitro asexual embryogenesis. Hortic Rev 2: 268–310

Sun LH, Lu XF, Tang BG (1986) Tissue culture of *Freesia refracta* Plant Physiol Commun 4: 55–56 (in Chinese).

Tisserat B, Esan EE, Murashige T (1979) Somatic embryogenesis in angiosperms. Hortic Rev 1: 1–78

Trigiano RN, Gray DJ, Conger BV, McDaniel JK (1989) Origin of direct somatic embryos from cultured leaf segments of *Dactylis glomerata*. Bot Gaz 150: 72–77

Vasil V, Vasil IK (1982) The ontogeny of somatic embryos of *Pennisetum americanum* in cultured immature embryos. Bot Gaz 143: 454–465

Wang DY, Kong Y (1984) Somatic embryogenesis and plant regeneration from inflorescence of *Echinochloa crusgalli* L. Acta Biol Exp Sin 17: 271–279 (in Chinese with English Abstr)

Wang L, Pu XL (1988) Karyotypic study of *Freesia refracta*. J Northeast Normal Univ 3: 93–95 (in Chinese with English Abstr)

Wang L, Huang B, He MY, Hao S (1990) Somatic embryogenesis and its hormonal regulation in tissue cultures of *Freesia refracta*. Ann Bot 65: 271–276

Wang L, Hu AL, Zheng YZ (1991) Scanning electron microscopic observation on somatic embryogenesis in the tissue culture of *Freesia refracta*. J Chin Elect Microsc Soc 10: 16–19 (in Chinese with English Abstr)
Wang YY (1981) *Freesia refracta*. In: Wang YY, Qin GS, Zhong SQ, Huang JL (eds) Ornamental flowering plants. Shanxi Science and Technology Press, Xi, An, pp 103–104 (in Chinese)
Zhang SL, Zheng GC (1988) Scanning electron microscopic observations on somatic embryogenesis in the tissue culture of *Panicum miliaceum* L. J. Chin Elect Microsc Soc 1: 1–4

IV.3 Somatic Embryogenesis and Synthetic Seed in *Apium graveolens* (Celery)

B.L. NADEL[1], A. ALTMAN[2], and M. ZIV[3]

1 Introduction

1.1 Botany and Distribution of Celery

Celery (*Apium graveolens*) is normally a biennial plant, but can complete its life cycle in a year if subjected to low temperatures. In the vegetative phase (the first year) the plant is composed mostly of petioles and leaves (60 cm in height), with a condensed shoot that does not elongate. The stem elongates during the second year, after vernalization, and the plant grows to 1–2 m in height. The stem terminates in a compound umbellate flower. Two types of celery are grown for commercial use: the green and the golden, self-balancing varieties.

Celery is a luxury crop. Its seeds are used in a limited amount as a spice, and it is mainly grown for the stalks. The nutritional value of the stalks in minimal: eaten raw or cooked, the stalks provide only 7–8 cal/100 g. Its only nutritional value is as a source of natural fiber and flavor. According to USDA sources, the yearly acreage of celery has remained consistant from 1976 until the present, between 33000 to 38000 acres (USDA 1991). *Apium graveolens* is grown only in areas with a monthly mean temperature of 16–21 °C. Growth is optimal at 20–22 °C, and it is extremely susceptible to temperature extremes. Temperatures below 14 °C cause stem elongation and early flowering, and those above 25 °C result in stress symptoms, such as leaf burn and pithiness. Consequently, celery production is restricted to mild temperate regions during the summer, and coastal subtropical areas in the winter.

1.2 Previous Studies on Somatic Embryogenesis in Celery (Table 1)

The first successful induction of somatic embryos in celery cultures was reported by Reinert et al. (1966), but these did not develop further. Chen (1976) and Williams and Collin (1976a,b) developed procedures for induction of somatic

[1] AGECOTEC, Moshav Kfar Pines, Post Karkur, 37920 Israel
[2] Dept. of Horticulture, The Otto Warburg Center for Biotechnology in Agriculture, The Hebrew University of Jerusalem, P.O.B. 12, Rehovot 76–100, Israel
[3] Dept. of Agricultural Botany, The Otto working center for Biotechnology in Agriculture, The Hebrew University of Jerusalem, P.O. Box 12, Rehovot 76–100, Israel

Biotechnology in Agriculture and Forestry, Vol. 31
Somatic Embryogenesis and Synthetic Seed II (ed. by Y.P.S. Bajaj)

embryos and plant regeneration. Their procedure involved the following general stages: (1) callus induction from petiole explants, using 2,4-D; (2) maintenance of cultures either as callus or as cell suspensions with 2,4-D; (3) removal of 2,4-D and regeneration of somatic embryos; (4) maturation of globular and heart-shaped embryos into plantlets; (5) establishment of plantlets in vitro and hardening. Thus, somatic embryogenesis is celery similar in principle to that reported previously for carrot and other species (Halperin and Wetherell 1964).

Histological studies to establish the origin of celery somatic embryos were initiated by Al-Abta and Collin (1979) and by Zee and Wu (1979, 1980), They

Table 1. Summary of the work done on somatic embryogenesis in celery

Reference	Explants/culture	Results/remarks
1. Reinert et al. (1966)	Callus	Embryos, but no plant regeneration
2. Williams and Collin (1976a)	Callus	Embryos and regenerated plants
3. Williams and Collin (1976b)	Callus	Plantlet growth and physiology
4. Chen (1976)	Callus	Embryos and regenerated plants
5. Metcalf and Collin (1978)	Suspension	Effects of simazine on growth and respiration
6. Al-Abta and Collin (1979)	Callus	Embryos and changes in cytokinins
7. Zee and Wu (1979)	Petiole/callus	Embryogenesis
8. Zee and Wu (1980)	Leaf/callus	Proembryos, embryogenesis
9. Browers and Orton (1982)	Callus	Chromosomal variability
10. Merrick and Orton (1983)	Callus	Resistance to asulam (herbicide)
11. Murata and Orton (1983)	Cells	Chromosomal analysis (structural changes)
12. Orotn (1983)	Callus	Chromosomal variability
13. Ireland and Lacy (1987)	Callus	Somaclonal variant resistant to *Fusarium*
14. Heath-Pagliuso et al. (1988)	Suspension	Somaclonal variants resistant to *Fusarium*
15. Heath-Pagliuso and Rappaport (1990)	Suspension	Genetic analysis of somaclonal variants
16. Wright and Lacy (1988)	Suspension	Somaclonal variants, disease resistance
17. Altman et al. (1989)	Suspension	Polyamines and somatic embryogenesis
18. Nadel et al. (1988)	Suspension	Embryogenesis and plantlets: 2,4-D, BA, and mannitol
19. Kim and Janick (1989)	Suspension	Desiccated somatic embryos
20. Kim and Janick (1991)	Suspension	Improvement of desiccation tolerance by ABA and proline
21 Nadel et al. (1990a)	Suspension	Embryogenesis and synchronized cultures
22. Nadel et al. (1990b)	Suspension	Cold storage and conversion of embryos
23. Nadel et al. (1990c)	Suspension	Embryogenesis in a bioreactor
24. Altman et al. (1990)	Suspension	Interrelationship of embryogenesis and polyamine biosynthesis
25. Saranga et al. (1992a)	Suspension	Synthetic seeds, desiccation tolerance, and amino acids
26. Saranga et al. (1992b)	Suspension	Synthetic seeds, reduced osmotic potential
27. Van Staden et al. (1992)	Suspension	Embryogenesis and cytokinin metabolism
28. Danin et al. (1993)	Suspension	Embryogenesis and changes in polyamines and cytokinins
29. Jin and Kuo (1988)	Suspension	Embryoids and artificial seeds

traced the origin of celery somatic embryos to globular structures embedded in the callus. Kim and Janick (1989), using similar histological methods, found that somatic embryos of celery formed from single cells on the surface of callus clumps. Secondary somatic embryos occurred on cotyledons and hypocotyls of primary embryos.

Celery somatic embryogenesis has been used as a model system for practical micropropagation, as well as for encapsulation and production of "synthetic seeds" (Kitto and Janick 1985a, Kim and Janick 1989; Saranga et al. 1992a, b). In addition, selection of somaclonal variants of somatic embryos resulted in celery lines which were resistant to herbicides (Metcalf and Collin 1978; Merrick and Collin 1982), or tolerant to the fungal disease *Fusarium oxysporum* F. sp. apii (Health-Pagliuso et al. 1988, 1989).

Summary of the work done on somatic embryogenesis in celery is presented in Table 1.

2 Pathways of Somatic Embryogenesis

Somatic embryogenesis in celery involves several distinct stages and pathways, i.e., establishment of proembryogenic clumps in liquid suspension cultures, formation of clumps of embryos as well as of single, free-floating embryos, culture synchronization, and plant establishment (Nadal et al. 1989, 1990a,b,c), as described briefly in the following.

Initially, induction of somatic embryogenesis was mainly via proembryogenic clumps, composed of both callus and embryo-like structures. To determine their composition, clumps were grown individually or in small groups, on a regeneration medium. Plantlets regenerated abundantly all over the clumps, indicating that the clumps were indeed masses of predetermined cells, organized into proembryos. From these, plantlets were regenerated and maintained in vitro.

Secondary cultures, produced from subcultured in vitro grown plantlets, expressed their embryogenic potential much more rapidly than the original primary cultures. They also produced more somatic embryos, and maintained their embryogenic potential for longer periods. These cultures also maintained, in addition to clumps of embryos, a high rate of single, free-floating somatic embryos. Continuous subculture of these cell lines in a modified regeneration medium (see below) resulted in suspension cultures composed of single free-floating somatic embryos only. The secondary cultures became so regenerative that the concentration of 2,4-D in the medium had to be doubled to prevent regeneration and to maintain the embryogenic suspension cultures.

Five separate pathways of somatic embryogenesis in celery were identified and maintained as indicated in Table 2. Pathway No. 5 seems to be the preferred one for commercial, large-scale production of somatic embryos.

The various pathways of somatic embryogenesis in celery, as affected by the composition of the culture media and sub-culture regimes, are schematically illustrated in Fig. 1.

Table 2. Pathways of somatic embryogenesis in celery

1. Maintenance medium + low 2,4-D[a]	→	**proembryogenic clumps and plantlets**
2. Maintenance medium + high 2,4-D[b]	→	**proembryogenic clumps**
3. Maintenance medium + high 2,4-D+ mannitol	→	**proembryogenic clumps, clumps of embryos, and singular embryos**
4. Maintenance medium + high 2,4-D	→	regeneration medium → **clumps of embryos**
5. Maintenance medium + high 2,4-D+ mannitol	→	regeneration medium + mannitol → **single embryos**

[a] Low 2,4-D: 2.3 μM.
[b] High 2,4-D: 4.6 μM; maintenance and regeneration media: see Protocol (Sect. 8).

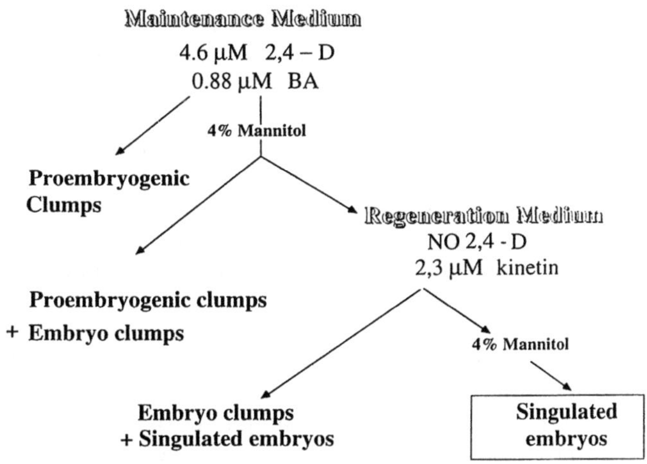

Fig. 1. Pathways leading to different types of somatic embryogenesis in celery

3 Factors Affecting Embryogenesis

Details of the basal nutrient medium are presented in the protocol (Sect. 8). In addition, the following factors were studied.

3.1 Mannitol

During standard subculturing, it was noticed that regeneration occurred mainly from clumps which were composed of both callus and embryo-like structures. These cultures became turbid during standard subculturing, and microscope examination revealed that the turbidity was caused by cell lysis. To alleviate this problem, mannitol (3%, w/v) was added to the medium. It became soon evident that many more singular, free-floating somatic embryos were produced in the presence of mannitol, and turbidity was reduced considerably. The effect of elevated osmolarity, produced by sucrose or hexitols, on somatic embryo development was previously observed in carrot and water parsnip (Ammirato 1983).

Investigations of the effects of mannitol on celery embryogenesis included several parameters, i.e., optimal concentrations, timing and duration of exposure to mannitol, as well as the question of mannitol specificity (Nadel et al. 1989, 1990a,b,c). Different concentrations of mannitol were added to a 250 mesh screen filtrate which contained undifferentiated single cells and small cell clusters, but no preembryonic structures or somatic embryos. Addition of 3% mannitol resulted in the largest number of singular somatic embryos. With the addition of mannitol, embryos were regenerated even in the presence of 2,4-D in contrast to previous studies of somatic embryogenesis in celery, where only in the absence of 2,4-D did embryogenesis occur (Orton 1984). The combination of 3% mannitol and "regeneration medium" resulted in the greatest number of single free-floating somatic embryos, and accelerated normal embryo development, i.e. the greatest number of torpedo embryos. Subsequently, 4% mannitol was found to be the most effective concentration for producing the best somatic embryo cultures. To elucidate the effective duration of exposure to mannitol, cultures were exposed to mannitol for 24 or 72 h, or continuously for 30 days. The results show a fairly rapid response: induction by mannitol occurred between 24 h and 72 h.

Iso-osmotic solutions, either with mannitol alone or in combination with sucrose, were used to determine whether the effects of mannitol were specific, or a result of a general osmotic effect. The data indicated that both growth (cell number) and embryogenesis were accelerated by the combination of mannitol and sucrose. Mannitol or sucrose alone, at concentrations which result in a similar decrease in the osmotic potentials of the medium, was unable to induce somatic embryogenesis. The results show the importance of both sucrose (to provide the necessary energy) and mannitol (contributing to a lowered osmotic potential) for improving somatic embryogenesis in celery.

The effect of mannitol was further investigated with respect to plantlet regeneration. Proembryonic clumps were planted on agar-solidified regeneration medium, with and without 3% mannitol. After 6 days, clumps plated on a mannitol-containing medium produced 12 times more shoots and shoot initials than those without mannitol.

3.2 Auxins

Somatic embryogenesis is controlled by a delicate balance of growth regulators, Auxins, especially 2,4-D, have been shown to affect the early events of induction of somatic embryos in a variety of tissues and plant species (Halperin and Wetherell 1964; Reinert et al. 1966; Ammirato and Steward 1971; Vasil and Vasil 1982; Ammirato 1983; Nomura and Komamine 1985). Following the induction stage, somatic embryogenesis occurs readily only after auxin removal from the medium, whereas the addition of IAA and 2,4-D can inhibit somatic embryogenesis (Halperin and Wetherell 1964; Fujimura and Komamine 1975, 1980).

In celery, somatic embryogenesis was also found to be induced by 2,4-D (Nadel et al. 1989), however, by adjusting the osmotic potential of the suspension cultures with 4% mannitol, it was possible to induce somatic embryogenesis even in the presence of 2,4-D (at the same concentration that was used for induction). In fact, as mentioned above, the level of 2,4-D had to be doubled in order to prevent formation of globular embryos and further regeneration.

3.3 Cytokinins

Unlike auxins, little is known about the specific role of cytokinins during somatic embryogenesis. Fujimura and Komamine (1975) reported that kinetin and BA inhibited carrot somatic embryogenesis, while zeatin promoted it. It has been suggested that cytokinins are required for the initial cell division phase of somatic embryogenesis, but not for the later stages of embryo development and maturation, as found in carrot (Fujimura and Komamine 1980), anise (Ernst et al. 1984), orchardgrass (Wenck et al. 1988), and *Pennisetum* (Rajasekaran et al. 1987), In several of these cases, higher levels of cytokinins were found in non-embryogenic tissues as compared with highly embryogenic cultures.

In a detailed study of cytokinin content and metabolism in celery cultures, it was found that cytokinin levels decreased with increased embryo development (Van Staden et al. 1992; Danin et al. 1993). Most of the detected cytokinin-like activity cochromatographed with benzyladenine and its metabolites, especially ribosylbenzyladenine (BAR), indicating rapid metabolism and/or other ways of removal. Analyses of embryogenic cell cultures, which were fed ^{14}C-benzyladenine, yielded several peaks of radioactive metabolites: BAR was the major metabolite, suggesting that it may play a pivotal role in somatic embryogenesis in celery, but other metabolites were also detected. Thus, more attention should be given to the function of cytokinin metabolites in regulating somatic embryogenesis.

3.4 Abscisic Acid (ABA)

Synchronization of cell and embryo cultures is important for increasing the uniformity of the cultures and, therefore, for improving its efficiency. ABA was found to play a specific role in somatic embryo maturation and germination in caraway, i.e., increasing embryo uniformity (Ammirato 1983), in addition to other effects (Ammirato 1989).

Similar effects were established in celery (Nadel et al. 1990a,c). ABA clearly increased culture synchronization and the uniformity of embryo growth when applied for 1 week. Upon subculture to a regeneration medium without ABA this effect became transient. Additional effects of ABA (0.5 mM) included: (a) inhibition of root elongation, similar to observations in carrot (Kamada and Harada 1981); (2) development of secondary embryos; and (3) ABA was not required for maturation of celery somatic embryos and plantlet formation.

3.5 Polyamines

Several recent studies indicated that putrescine biosynthesis and polyamines play a role in cellular differentiation during somatic embryogenesis in carrot (Montague et al. 1978; Feirer et al. 1984; Altman 1989). The direct involvement of polyamines in somatic embryogenesis in celery was indicated by the fact that the addition of spermidine to the growth medium greatly promoted and enhanced plantlet formation (Altman et al. 1989). Analysis of free polyamine content in globular embryos and in plantlets revealed significant qualitative and quantitative differences in polyamine levels (Altman et al. 1989; Danin et al. 1993). Putrescine content in globular embryos was six fold that in plantlets, whereas the changes in spermidine content were less dramatic, resulting in a high putrescine/spermidine ratio in globular embryos.

The effect and involvement of polyamines were further confirmed by data of polyamine biosynthesis (incorporation of ^{14}C-arginine, ornithine, and putrescine) and by measuring the activity of polyamine biosynthetic enzymes. Several inhibitors of polyamine biosynthesis inhibited cell division and embryo organization. Methylglyoxal-bis-guanylhydrazone (MGBG), an inhibitor of Sadenosyl methionine decarboxylase (and thus of spermindine and spermine synthesis) inhibited growth of cell clumps and globular embryos of celery suspension cultures; addition of spermidine did not relieve this inhibition. MGBG inhibited both growth and normal differentiation, i.e., plantlet formation, in a solidified medium. Root formation, however, was not inhibited, suggesting that shoot formation may be more sensitive to inhibition of SAM-decarboxylase activity (Altman et al. 1989).

3.6 Protein Changes During Embryogenesis

Preliminary studies of ^{14}C-leucine incorporation and the effect of cycloheximide were carried out in globular embryos and plantlets of celery (Altman et al. 1989). Data indicate that total protein synthesis in plantlets was considerably higher than in globular embryos when expressed on a fresh weight basis, and even more so when expressed as specific activity. Cyclohexomide inhibited leucine incorporation by more than 90%. Possible changes in the expression of early and late embryonic genes were studied by comparing SDS-PAGE profiles of proteins extracted from globular embryos and from plantlets. The gels revealed two highly abundant polypeptides in globular embryos (ca. 55 and 14 kDa), which were hardly detected in mature plantlets.

4 Large-Scale Production

Economical large-scale production requires that the following levels of research and development be established: basic biological information on somatic embryo induction and maturation on a laboratory scale (level 1), an intermediary stage where the major biological problems involved in scaling up must be investigated and solved (level II), and technical strategies for scaling up laboratory system to industrial mass production of embryos or plantlets (level III). Many species have been manipulated in vitro to produce somatic embryos, providing the basic biological information on somatic embryogenesis in the laboratory (level I). Strategies on how to scale up the laboratory systems to industrial mass production of embryos or plantlets were usually extrapolated directly from this information. However, level II research, i.e., biological problems involved in scaling up, is most crucial. This intermediary stage involves several distinct problems and is directed at improving the efficiency and quality throughout the entire process, as outlined in the following for celery (Nadal et al. 1989, 1990a,b,c).

4.1 Early Identification of Embryogenic Potential

Early identification of those cell lines which possess good embryogenic potential is most critical. The ability to manipulate and select early expression of embryogenic potential is important in order to include embryogenic cell lines of additional genotypes, and to increase the efficiency of existing cell lines.

All secondary cultures of celery arising from in vitro cultured plantlets proved to be highly embryogenic, and produced more single embryos than primary embryogenic cultures. The embryogenic potential of different cultures changed with time and with the composition of the nutrient medium. A study of the effects of subculture regimes established:

1. Frequent subculturing results in delayed expression of embryogenesis; thus, for early identification of embryogenic potential it is advantageous to reduce the frequency of subculturing of initial explants.
2. Great care must be given to cultures that are maintained for extended periods of time, in order to avoid random selection against regeneration (due to culturing calli with low embryogenic potential).
3. With time in culture, the inhibitory effect of 2,4-D on the development of somatic embryos was greatly reduced, allowing cultures which were grown even at high 2,4-D concentrations to reach maturation. Thus, the response to a growth regulator regime has to be examined regularly during culture.

4.2 Establishment of Single, Free-Floating and Coated Embryos

Production of single embryos, floating freely in the liquid medium, is important for production in bioreactors and for automated dispension. From the several existing pathways leading to celery somatic embryogenesis, the one which yields a large proportion of single embryos (i.e. with mannitol) should be selected.

Automated dispension of embryos from the bioreactor may require a procedure whereby single embryos remain separated and, to a certain degree, more resistant to mechanical handling during the entire production cycle (Levin et al. 1988). One means of achieving this goal is embryo coating. Preliminary studies (Nadel and Altman, unpubl.) indicated that a proper control of the flow rate, of both the alginate and the globular embryos, led to the formation of alginate beads, each containing one to two globular embryos. Further embryo development to the torpedo stage, within the bead, was accomplished by growing the beads in suspension cultures with regeneration medium. Small plantlets developed readily from the alginate beads and were easily separated from the liquid medium by sieving. After 1 month, well-developed plantlets were transferred to a mist chamber, and over 90% of the plantlets developed into healthy, field-grown plants.

4.3 Culture Synchronization and Uniformity

The ability to synchronize and control all stages of somatic embryo development, and to manipulate culture duration, is of great importance for large-scale production. The kinetics of embryo induction and development in celery suspension cultures was determined, and it was established that heart-shaped embryos began to appear within 3 days, and torpedo-shaped embryos within 6 days of subculture. Both embryo induction and conversion from the globular stage to later stages of embryogenesis followed a classic logarithmic curve over an 18-day period. The specific kinetic parameters of embryo development are important for determining the proper strategy, as synchronization should preferably take place at the globular stage after induction (maintaining heart-shaped cultures would be too difficult due to their transient nature).

The ability to mechanically separate the required embryonic stage is another important factor for large-scale automated production. Celery cell cultures were synchronized by sieving through metal mesh screens of different sizes. Embryo free suspension cultures could be produced by collecting the cell fraction that passed through a 250 μ screen. The effect of initial embryo size on further embryo uniformity was examined by comparing the initial size of the sieved embryos to their final size after 8 and 15 days of culture. Embryos of 1400 μ resulted in the best growth in terms of final embryo size, but they were less uniform. The most synchronous growth was found with embryos of 800 and 500 μ, but they evinced a poor growth rate. A synchronized culture with uniformly sized clumps of somatic embryos is presented in Fig. 2 (Nadel and Altman, unpubl.).

4.4 Cold Storage

The ability to store somatic embryos allows for production of a large biomass of embryos at one time, followed by plantlet maturation when desired. Normally, once embryos are induced they have to be converted into plantlets immediately, otherwise they are lost, Cold-storage of celery somatic embryos, followed by plantlet maturation, was found to be most effective (Nadal et al. 1990b). Celery cultures (embryogenic callus, proembryogenic clumps, single embryos) were stored at 4–6 °C for extended periods (up to 24 weeks) with no detrimental reduction of their embryogenic potential (Table 3). The longevity of cultures was

Fig. 2. A synchronized suspension culture of uniform celery embryogenic clumps

Table 3. Cold storage of celery embryogenic cultures. Proembryogenic clumps and single embryos were stored at 4–6 °C as wet cultures (i.e., slightly moistened with regeneration medium), or in liquid (i.e., immersed in liquid regeneration medium), or surface-dried. Callus was stored in Petri dishes with semisolid medium ("agar"). Plantlet growth (G) and regeneration (R) were determined at the end of the indicated storage periods

Storage Period (weeks)	Proembryogenic clumps				Single embryos						Callus	
	Liquid		Wet		Dry		Liquid		Wet		Agar	
	G	R	G	R	G	R	G	R	G	R	G	R
8	+	+	++	++	++	++	+	+	++	++	++	++
10	Dead		++	++	+	+	–	–	++	++	++	++
12			++	++	–	–	–	–	++	++	++	++
14			++	++	Dead		Dead		++	++	++	++
16			++	++					++	++	++	++
18			++	++					++	++	+	+
20			++	++					+	+	–	–
24			++	++					Dead		Dead	

++. + and – indicate good, weak, and no growth, respectively

dependent on the moisture level during cold storage. Proembryogenic clumps and single embryos survived best if stored slightly moistened (i.e., "wet") with regeneration medium (as compared with "liquid", i.e., within the medium, or with surface-dried embryos). Extended cold storage of embryogenic callus, subcultured in semisolid medium, was also investigated. Plantlet regeneration from cold-stored embryogenic cultures was similar to that from cultures, that were not stored and no phenotypic changes were observed.

4.5 Bioreactors

One of the advantages of somatic embryogenesis in adapted liquid culture media is their compliance with mass production in bioreactors. One prevalent problems with bioreactors is the risk of contamination, and the use of antibiotics is one possible way of controlling bacterial contamination. Studies with ampicillin and erythromycin showed that both antibiotics can be used for celery embryogenic cultures, in 1-l bubble bioreactors, at a concentration of up to 100 mg/l each, without causing any damage or interfering with embryo formation and maturation (Nadel and Altman, unpubl.).

Celery culture development and conversion of proembryogenic clumps to plantlets were investigated in 1-l airlift bubble bioreactors (Fig. 3; Nadel and Altman, unpubl.). The results show that high yields of celery plantlets/medium volume i.e., ca. 80 000 minute plantlets per 800 ml, can be obtained in a 14–16 day cycle (Nadal et al. 1990c).

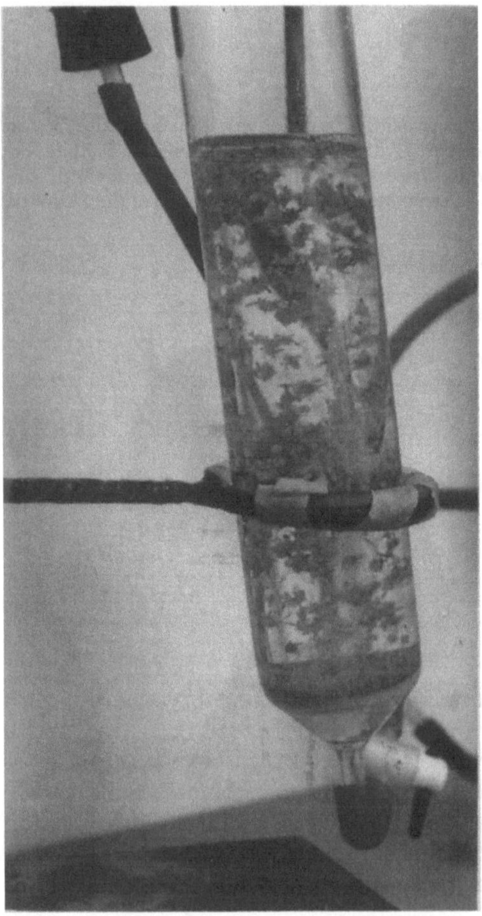

Fig. 3. An air lift bubble bioreactor with celery somatic embryos

4.6 Plant Establishment

The final crucial stage of any somatic embryogenesis system is to determine the success of converting embryos into plantlets and plantlets into hardened, field-growing plants. A very high rate of torpedo-stage embryo conversion to small plantlets in vitro (close to 100%) was achieved by growing them on paper bridges (Nadel et al. 1990a,b). These plantlets were divided according to size, and transferred to a mist chamber for acclimatization. Over 82% of the smaller plantlets (2–4 cm in length) survived, compared with 97% of the larger plantlets (5–7 cm). This represents a high rate of embryo conversion to plants. As an alternative to paper bridges, which are less favorable for large-scale production, other semi-solid supports were sought, e.g., mixture of agar and perlite, perlite and liquid medium (Nadel et al. 1990b). The agar-perlite mix allowed for better aeration and penetration of roots. Field tests established that celery plants regenerated from tissue culture were uniform and horticulturally true to type.

5 Synthetic Seeds and Coated Embryos

Economical application of large-scale micropropagation using somatic embryos is dependent on an efficient conversion of somatic embryos to "artificial seeds". These can then be used for direct planting under field or greenhouse conditions. Three methods have been suggested for conversion of somatic embryos to artificial seeds: (1) encapsulation (coating) in an alginate gel; (2) partial desiccation; (3) partial desiccation plus encapsulation with a water-soluble resin, as applied to celery somatic embryos. Desiccated somatic embryos of carrot and celery have been proposed as a potential analog of true seeds, thus the term "synthetic seeds" (Kitto and Janick 1985a,b; Kim and Janick 1989, 1991; Saranga et al. 1992a,b). Synthetic seed coats, which were first developed for carrot somatic embryos, were applied by mixing equal volumes of embryo suspension and a solution of polyethylene oxide (Polyox) (Kitto and Janick 1985a,b). The resulting drops were dried to form detachable wafers of embryos embedded in Polyox. The ability of somatic embryos to survive this desiccation, yielding normal seedlings, is a key factor in the practical utilization of synthetic seeds. Thus, a high desiccation rate of coated somatic embryos of celery resulted in low survival, but slow desiccation protocols, high humidity, and pretreatment with ABA or with proline or with GABA plus proline greatly increased survival (Kim and Janick 1989, 1991; Saranga et al. 1992a,b). Further studies on the possible mechanisms of survival and desiccation tolerance of celery somatic embryos indicated that ABA increased the accumulation and altered the distribution of fatty acids in the embryos, as well as of proline and GABA (Kim and Janick 1991; Saranga et al. 1992a,b). Treatments having a high osmotic concentration applied during the last stage of embryo production cycle increased embryo survival and conversion after partial desiccation, and markedly increased embryo proline content (Saranga et al. 1992b). Celery somatic embryos were entrapped also in alginate beads, as discussed earlier (see Sect. 4.2 Nadel and Altman unpubl.). To the best of our knowledge, field tests of synthetic seeds have failed so far.

6 Genetic Variability of Somatic Embryos

6.1 Levels of Variation

Genetic variability of regenerated celery plants from somatic embryos was first studied by Williams and Collin (1976b). Over 25% of the regenerated plants were nondiploid (2n = 22), and the aneuploids were indistinguishable horticulturally from the normal diploids. Browers and Orton (1982) conducted a factorial study of chromosomal variability in relation to explant type and culture phase, and found that callus cultures exhibited the largest number of genetic variants. This variability was also observed between cultures arising from different plants of the same variety (Browers and Orton 1986). Using a detailed karyological analysis of a 10-month-old celery suspension culture, Murata and Orton (1983) found that only 33% of the cells was diploid. In another study (Orton), using isozymes as genetic markers, it was found that 85% of the embryogenic celery cultures was karyologically indistinguishable from normal diploids after 6 months in culture. In our studies, all greenhouse and field-planted celery plants were phenotypically identical.

6.2 Use of Variation in Breeding for Resistance

Fungal and Bacterial Diseases. Somaclones from single cells of celery cell cultures were regenerated via somatic embryogenesis, and the resulting plants were evaluated in greenhouses for resistance to *Fusarium oxysporum* f. sp. *apii* race 2 (Ireland and Lacy 1987; Ireland et al. 1987). Of the eight clones selected and selfed, three progenies had significantly lower mean disease ratings, as compared to the parental material. Somaclones regenerated from celery embryoids were screened in greenhouses for their tolerance to three fungal and one bacterial pathogen (*Septoria apiicola*, *Cercospora apii*, *Fusarium oxysporum* f.sp. *apii* race 2 (FOA 2) *Pseudomonas cichorii*). This study showed (Wright and Lacy 1988) that regenerated plants varied greatly in their response to the four pathogens, plants were rarely found with a high resistance to two or more pathogens. Similar selection studies were carried out by another group (Rappaport and Heath-Pagliuso 1987; Health-Pagliuso et al. 1988, 1989; Heath-Pagliuso and Rappaport 1990). S1 progeny of celery somaclones were consistently more resistant to FOA2 than the parent line, in numerous field tests in California. The enhanced resistance persisted in the S2 generation, and when progeny of the S2 were backcrossed to the parent line, the progeny were intermediate in phenotype.

Herbicides. Embryogenic cell cultures of celery have been employed also to select for resistance to herbicides, as discussed by Browers and Orton (1986). Simazine-tolerant cell lines of celery selected, but none of the regenerated plants showed resistance to the herbicide (Metcalf and Collins 1978). Research on selecting for resistance to asulam proved more promising: a resistant cell line was selected and plants were regenerated (Merrick and Collin 1992). The regenerated plants were resistant to the herbicide, but they proved to be sterile, and progeny tests were not carried out. The effect of EMS mutagenization on selection of celery clones resistant to glyphosate (Roundup) was studied, and some resistant field-grown plant were detected (Orton 1983).

7 Summary

Somatic embryogenesis of celery has proven to be an excellent model system for developing a potentially large-scale system for micropropagation of a vegetable crop. Our studies show that a laboratory procedure for tissue culture propagation through somatic embryogenesis can be refined, enhanced, scaledup, and made more efficient for large-scale industrial production. This was achieved by proper and orderly manipulation and adaptation of the induction stage of somatic embryogenesis, plantlet regeneration and acclimatization, culture synchronization and uniformity of single embryos, extended storage of the somatic embryos, use of liquid media, and scaling up in 1-l bubble bioreactors.

8 Protocol

The following protocol is based on our previous studies (Nadel et al., 1989, 1990a,c). A detailed description of the various parameters involved can be found in the preceding sections of this review.

8.1 Parental Plants

Celery seedlings, from commercial nurseries, are maintained in pots, with a peat-moss: perlite: vermiculite (1:1:1) potting mixture, and with drip irrigation (carefully avoiding leaves from getting wet). Pots are maintained in an insect-proof chamber at 22 °C and a 16-h photoperiod (1700 lx). Petioles large enough for explant preparation are removed 4–6 weeks, after planting (new petioles can be harvested every 4–6 weeks, for over 12 months).

8.2 Explant Preparation and Culture Conditions

Petioles are removed from the parental plants, washed thoroughly in a detergent and water for 5–8 min, and petiole and leaf explants, 1–2 × 1–2 cm, are prepared. Surface-sterilization of explants includes: 70% ethanol (90 s) followed by 1% (v/v) sodium hypochlorite with 0.02% Tween 20 (10 min), and 4 successive rinses with sterile water.

The initial percentage of aseptic cultures may be quite low, because of endophytic bacteria (*Erwinia* sp. and *Pseudomonas* sp. were identified). Therefore, for best results, explants are first screened on nutrient agar, and the uncontaminated cultures are plated on growth medium for callus induction.

All cultures are maintained with the appropriate culture medium (see below), at 24 ± 1 °C, and a 16-h photoperiod of 1600 lx.

8.3 Callus Induction and Cell Cultures

Initial callus appears within 3 weeks of culturing on Murashige and Skoog medium (MS) supplemented with 2.3 μM each of 2,4-D and benzyladenine (BA), 3% sucrose, and 1% Bacto agar. After 2 weeks of culture, callused explants are transferred to 250-ml flasks with 90–100 ml liquid MS medium supplemented with 2.3 μM 2,4-D, 0.88 μM BA, 3% sucrose, 500 mg/l casein hydrolysate, and 4% mannitol (i.e., "maintenance medium"). The flasks are maintained on a gyratory shaker at 100 rpm. These initial cultures are referred to as "primary" cultures. The resulting suspension cultures are subcultured by sieving through a 1500 μ metal screen, and 10–50 ml of the sieved suspension is added to 80 ml fresh maintenance medium. Suspension cultures are thus subcultured every 14–18 days, and embryos usually start to appear after 3 subcultures.

8.4 Regeneration of Somatic Embryos

Somatic embryos and plantlet regeneration is accomplished on a "regeneration medium" consisting of MS medium supplemented with 3% sucrose, 2.3 μM kinetin, 500 mg/l casein hydrolysate, and 4% mannitol. The size of inoculum is regulated by sieving suspension cultures (cultured in maintenance medium) through a series of metal screens with varying pore sizes from 2400 to 200 μ. Microscopic examination showed that cultures that had passed a 250 μ screen were embryo-free.

In addition, explants can be removed from aseptic plants which regenerated from primary cultures, and plated on callus induction media. The resulting callus is transferred to liquid maintenance and regeneration media as described above. These cultures, which are designated as "secondary" callus cultures, can become so regenerative that 2,4-D levels higher than 2.3 μM have to be used in the maintenance medium to prevent regeneration.

8.5 Hardening

Torpedo-stage embryos larger than 1 cm are transferred to culture vials with 0.5 strength (of the mineral salts) liquid "regeneration medium" with paper bridges. After 40–45 days, fully developed plantlets with roots are acclimatized in a mist chamber in a peat-moss: perlite: vermiculite (4:4:2) medium, and maintained in a greenhouse, under natural daylight conditions, for a period of 2 weeks.

References

Al-Abta S, Collin HA (1979) Cytokinin changes during embryoid development in celery tissue cultures. New Phytol 82: 29–35

Altman A (1989) Polyamines and plant hormones. In: Bachrach U, Heimer YM (eds) The physiology of polyamines, vol II. CRC Press, Boca Raton, pp 121–145

Altman A, Levin N, Cohen P, Schneider M, Nadel BL (1989) Polyamines in growth and differentiation of plant cell cultures: the effect of nitrogen nutrition, salt stress and embryogenic media. In: Zappia V, Pegg AE (eds) Progress in polyamine research. Plenum Press, New York, pp 559–572

Altman A, Nadel BL, Falash Z, Levin N (1990) Somatic embryogenesis in celery: induction, control and changes in polyamines and proteins. In: Nijkamp HJJ, Van Der Plas LHW, Van Aartrijk J (eds) Progress in plant cellular and molecular biology. Kluwer Academic, Dordecht, pp 454–459

Ammirato PV (1983) Embryogeneis. In: Evans DA, Sharp WR, Yamada Y (eds) Handbook of plant cell culture, vol I. Macmillian New York, pp 82–123

Ammirato PV (1989) Recent progress in somatic embryogenesis. IAPTC Newsletter, pp 2-16

Ammirato PV, Steward FC (1971) Some effects of the environment on the development of embryos from cultured free cells. Bot Gaz 132: 149–158

Ammirato PV, Styer DJ (1985). Strategies for large-scale manipulation of somatic embryos in suspension culture. In: Zaitlin M, Day P, Hollaender A (eds) Biotechnology in plant science: relevance to agriculture in the eighties. Academic Press, New York pp 161–178

Browers MA, Orton TJ (1982) A factorial study of chromosomal variability in callus cultures of celery (*Apium graveolens*). Plant Sci Lett 26: 65–73

Browers MA, Orton TJ (1986) Celery (*Apium graveolens* L). In: Bajaj YPS (ed) Biotechnology in agriculture and forestry 2. Crop I. Springer-Berlin Heidelberg New York, pp 485–519

Chen CH (1976) Vegetative propagation of the celery plant by tissue culture. Proc S Dakota Acad Sci 55: 44–48

Danin M, Upfold SJ, Levin N, Nadel BL, Altman A, Van Staden J (1993) Polyamines and cytokinins in celery embryogenesis cell cultures. Plant Growth Regul 12: 245–254

Ernst D, Oestrehelt D, Schafer W (1984) Endogenous cytokinins during embryogenesis in an anise cell culture (*Pimpinella anisum* L.). Planta 161: 240–245

Feirer RP, Mignon G, Litvay JD (1984) Arginine decarboxylase and polyamines required for embryogenesis in the wild carrot. Science 223: 1433–1435

Fujimura T, Komamine A (1975) Effects of various growth regulators on the embryogenesis in a carrot cell suspension culture. Plant Sci Lett 5: 359–364

Fujimura T, Komamine A. (1980) Mode of action of 2,4-D and zeatin on somatic embryogenesis in a carrot cell suspension culture. Z Pflanzenphysiol 99:1–8

Halperin W, Wethrell DF (1964) Adventive embryony in tissue cultures of the wild carrot, *Daucus carota*. Am J Bot 51: 274–283

Heath -Pagliuso S, Rappaport L (1990) Somaclonal variant UC-T3: the expression of Fusarium wilt resistance in progeny arrays of celery, *Apium graveolens* L. TAG 80: 390–394

Heath-Pagliuso S, Pullman J, Rappaport L (1988) Somaclonal variation in celery: screening for resistance for *Fusarium oxysporum* f. sp. apii. TAG 75: 446–445

Heath-Pagliuso S, Pullman J, Rappaport L (1989) UC-T3: somaclone: celery germplasm resistant to *Fusarium oxysporum* f. sp. *apii*, race 2. HortScience 24: 711–712

Ireland KF, Lacy ML (1987) Greenhouse screening of celery somaclone progeny for resistance to *Fusarium oxysporum* f. sp. *apii* race 2. Phytopathology 77:176

Ireland KF, Elmer WH, Lacy ML (1987) Field evaluation of celery germplasm for resistance to *Fusarium oxysporum* f. sp. *apii* race 2. Phytopathology. 77: 1712

Jin Ji-yi, Kuo Chung-chen (1988) Studies on artificial seeds of celery. In: Recent advances on studies of applied and fundamental aspects of plant cell engineering. Science Periodical Press, Beijing pp 84–89

Kamada H, Harada H (1981) Changes in the endogenous levels and effects of abscisic acid during somatic embryogenesis of *Daucus carota* L. Plant Cell Physiol 22: 1423–1429

Kim YH, Janick J (1989) ABA and polyox-encapsulation or high humidity increase survival of desiccated somatic embryos of celery. Hort Science 24: 674–675

Kim YH, Janick J (1991) Abscisic acid and proline improve desiccation tolerance and increases fatty acid content of celery somatic embryos. Plant Cell Tissue Orgn Cult 24: 83–89

Kitto SL, Janick J (1985a) Production of synthetic seeds by encapsulating asexual embryos of carrot. J Am Soc Hortic Sci 110: 277–282

Kitto SL, Janick J (1985b) Hardening treatment increased survival of synthetically coated asexual embryos of carrot. J Am Soc Hortic Sci 110: 283–286

Levin R, Gaba V, Tal B, Hirsch S, Nola D, Vasil IK (1988) Automated plant tissue culture for mass propagation. Bio/Technology 6: 1035–1040

Merrick MMA, Collin HA (1982) Selection for asulam resistance in tissue cultures of celery, *Apium graveolens* L. var. dulce cv. New Dwarf White. Plant Sci Lett 20: 291–296

Metcalf EC, Collin HE (1978) The effect of simazine on the growth and respiration of a cell suspension culture of celery. New Phytol 81:243–248

Montague MJ, Koppenbrink JW, Jaworski EG (1978) Polyamine metbolism in embryogenic cells of *Daucus carota*. I. Changes in intracellular content and rates of synthesis. Plant Physiol 62: 430–433

Murashige T, Skoog F (1962) A revised medium for rapid growth and bioassays with tobacco tissue cultures. Physiol Plant 15: 473–497

Murata M, Orton TJ (1983) Chromosome structural changes in cultured celery cells. In Vitro 19: 83–89
Nadel BL, Altman A, Ziv M (1989) Regulation of somatic embryogenesis in celery cell suspensions. I. Establishment of embryogenic cultures and the beneficial effect of mannitol. Plant Cell Tissue Organ Cult 18: 181–189
Nadel BL, Altman A, Ziv M (1990a) Regulation of somatic embryogenesis in celery cell suspensions. II. Early detection of embryogenic potential and the induction of synchronous cell cultures. Plant Cell Tissue Organ Cult 20: 119–124
Nadel BL, Altman A, Ziv M (1990b) Cold storage and efficient conversion of somatic celery embryo into transplantable plants. Sci Hortic 44: 9–16
Nadel BL, Altman A, Ziv M (1990c) Regulation of large scale embryogenesis in celery. Acta Hortic 280: 75–82
Nomura K, Komamine A (1985) Identification and isolation of single cells that produce somatic embryos at a high frequency in a carrot suspension culture. Plant Physiol 79: 988–991
Orton TJ (1983) Spontaneous electrophoretic and chromosomal variability in callus cultures and regenerated plants of celery. Theor Appl. Genet 67: 17–24
Orton TJ (1984) Celery. In Sharp WR, Evans DA, Ammirato PV, Yamada Y (eds) Handbook of plant cell culture, vol 2. Macmillan, New York, pp 243-247
Rajasekaran K, Hein MG, Davies GC, Carnes MG, Vasil IK (1987) Endogenous growth regulators in leaves and tissue cultures of *Pennisetum purpeurem* Schum. J Plant Physiol 130: 13–25
Rappaport L, Heath-Pagliuso S (1987) Cloning celery for plant propagation. Annu Rep California Celery Growers Assoc, Binuba, CA, pp 19–22
Reinert J, Backs D, Krosing M (1966) Faktoren der Embryogenese in Gewebekulturen aus Kutuformern von Umbelliferen. Planta 68: 375–378
Rowe WJ (1986) New technologies in plant tissue culture. In Zimmerman RH, Hughes KW, Constantin MJ, Hollaender A (eds) Tissue culture as a plant production system for horticultural crops. Martinus Nijhoff, Dordrecht
Saranga Y, Rhodes D, Janick J (1992a) Changes in amino acid composition associated with desiccation tolerance of celery somatic embryos J Am Soc Hortic Sci 117: 337–341
Saranga Y, Kim YH, Janick J (1992b) Changes in desiccation tolerance and metabolities of celery somatic embryos induced by reduced osmotic potential . J Am. Soc Hortic Sci 117: 342–345
USDA (1991) USDS, Agricultural statistics 1991, p 151
Van Staden J, Upfold SJ, Altman A, Nadel BL (1992) Metabolism of benzyladenine in embryogenic cell cultures of celery . J Plant Physiol 140: 446–469
Vasil V, Vasil IK (1982) Characterization of an embryogenic cells suspension culture derived from cultured inflorescence of *Pennisetum americanum* (pearl millet, Gramineae). Am J Bot 69: 1441–1449
Wenck AR, Conger BV, Trigiano RN, Samms CE (1988) Inhibition of somatic embryogenesis in orchardgrass by endogenous cytokinins. Plant Physiol 88: 990–992
Williams L, Collin HA (1976a) Embryogenesis and plantlet formation in tissue cultures of celery. Ann Bot 40: 325–332
Williams L, Collin HA, (1976b) Growth and cytology of celery plants derived from tissue culture. Ann Bot 40: 333–338
Wright JC, Lacy ML (1988) Increase of disease resistance in celery cultivars by regeneration of whole plants from cell suspension cultures. Plant Dis 72: 256–259
Zee SY, Wu SC (1979) Embryogenesis in the petiole explant of Chinese celery Z Pfanzenphysiol 93: 325–436
Zee SY, Wu SC (1980) Somatic embryogenesis in leaf explants in Chinese celery. Aust J Bot 28: 429–436

IV.4 Somatic Embryogenesis and Artificial Seeds in *Coptis chinensis* F.

S. KE[1], Y. GUI[2], and R.M. SKIRVIN[3]

1 Introduction

Coptis chinensis F. (family Ranunculaceae) is a perennial herbaceous plant, 10 to 13 cm in height. Its stem is rhizomatic and often branched. It is an endemic species distributed in Anhui, Guizhou, Jiangsu, Jiangxi, Hubei, Hunan, Shaanxi, Sichuan, Zhejiang Provinces of China. It is a very important medicinal plant in China. The rhizomes of the plant contain berberine, coptine, and coptisine. These compounds are reputed to result in cooling of the blood, reducing stress and fatigue, removing of toxic substances, and purging sthenic fever and emotions as well as promoting salivation (Wang 1964). Considerable work has been done on the micropropagation and production of medicinal compounds in *Coptis* species (Ikuta et al. 1975; Yamamoto et al. 1981; Ikuta and Itokawa 1989).

Due to its medicinal importance and its indiscriminate collection from the wild, *C. chinensis* is now a rare and endangered plant in China (Din 1984). The crop is seldom cultivated due its small seeds, extended seed dormancy period (1 to 2 years), and low germination rates. Although the economic and pharmaceutical value of *C. chinensis* has been studied, plant propagation and conservation methods for this plant have been largely ignored. The problem of propagation could be overcome by making artificial seeds from somatic embryos.

Many species of *Coptis* are of medicinal importance. Attempts have been made to micropropagate some of these, e.g. *C. japonica* (Syono and Furuya 1972; Staba 1977) and *C. teeta* (Tandon and Rathore 1992). Nakagawa et al. (1982) reported clonal propagation of *C. chinensis* through the induction of somatic embryogenesis. Din (1984) made a similar study and regenerated somatic embryos from seeds. Ke et al. (1990) obtained somatic embryos from leaf cultures of this species. In subsequent studies (Ke et al. 1992), methods to control synchronization of embryogenesis and artificial seed production were investigated. This chapter is based on these studies as well as studies of artificial seed conversion in *C. chinensis*.

[1] Wuhan Institute of Botany, Academia Sinica, Wuhan 430074, People's Republic of China
[2] Institute of Botany, Academia Sinica, Bejing 100044, People's Republic of China
[3] University of Illinois, Plant and Animal Biotechnology Laboratory, 1201 Gregory Drive, Urbana, Illinois 61801, USA

Biotechnology in Agriculture and Forestry, Vol. 31
Somatic Embryogenesis and Synthetic Seed II (ed. by Y.P.S. Bajaj)

2 Somatic Embryogenesis

2.1 Induction and Histology of Somatic Embryos

Callus was observed on leaves cultured on two MS (Murashige and Skoog 1962) media in the dark for 2 weeks at 22 ± 2 °C. The first medium contained MS salts+ 2,4-dichlorophenoxyacetic acid (2,4-D, 2 mg/l); the other medium contained 6-benzylaminopurine (BA, 4 mg/l) and naphthaleneacetic acid (NAA, 0.3 mg/l). The calli induced with BA were milky yellow; those on 2,4-D were light yellow. Calli from both sources were friable and soft.

About 2 weeks after the friable callus had been transferred to MS medium with either reduced levels of 2,4-D and BA (0.5 and 1 mg/l, respectively) or reduced NAA (0.2 mg/l), embryogenic calli were induced (Fig. 1). Compact structures developed later. After 10 weeks on the induction medium, somatic embryos began to form (Figs. 9 and 10). They developed from light-colored calli on 2,4-D medium and from milky colored calli on BA medium. These structures were subcultured for 3 to 4 months in the dark. Somatic embryos developed from these calli on both the 2,4-D and BA media (Figs. 9 and 10). The rate of somatic embryo formation was slightly higher with 2,4-D (87.5%) than with BA (74.2%).

When the cultures were transferred to somatic embryogenesis medium, some of the cells on the callus surfaces differentiated to form meristematic cells (Fig. 2) or cell masses (Fig. 3). These calli gradually became embryogenic cell masses that could be easily separated from each other (Fig. 6). Somatic embryos were of epidermal origin (Figs. 2 and 3). Rapid cell division occurred at the callus surfaces. Some cells developed globular structures after only 2 weeks in culture (Figs. 3 and 4). Some of these structures developed into somatic embryos (Figs. 5–7); others formed more globular structures. Secondary somatic embryos could form on the primary somatic embryos. Suspensors were not obvious at the earliest stages of embryo development (globular and pre heart stages), but the bases of the somatic embryos were embedded in the parental calli tissue (Figs. 6 and 7). There was no marginal cell layer between somatic embryos and calli; so the somatic embryos were in contact with callus (Fig. 7). At later developmental stages, such as the late heart-shaped stage, the suspensors became less prominent and it was easy to isolate heart-shaped embryos from the callus by shaking (Fig. 8).

2.2 Development of Somatic Embryos

2.2.1 Effect of 2,4-D or BA in combination with NAA

Somatic embryos could be induced either by 2,4-D or by BA, but the BA-induced embryos were larger than those from 2,4-D (Table 1). In the 3 to 5-mm size range with BA an average of 48% belonged to this category; with 2,4-D, the value was much less (4.5%). In the 2-mm somatic embryo group, BA produced fewer (27%) than 2,4-D (68%). The size of the somatic embryos was not uniform; embryos transferred from 2,4-D medium to BA medium increased in size.

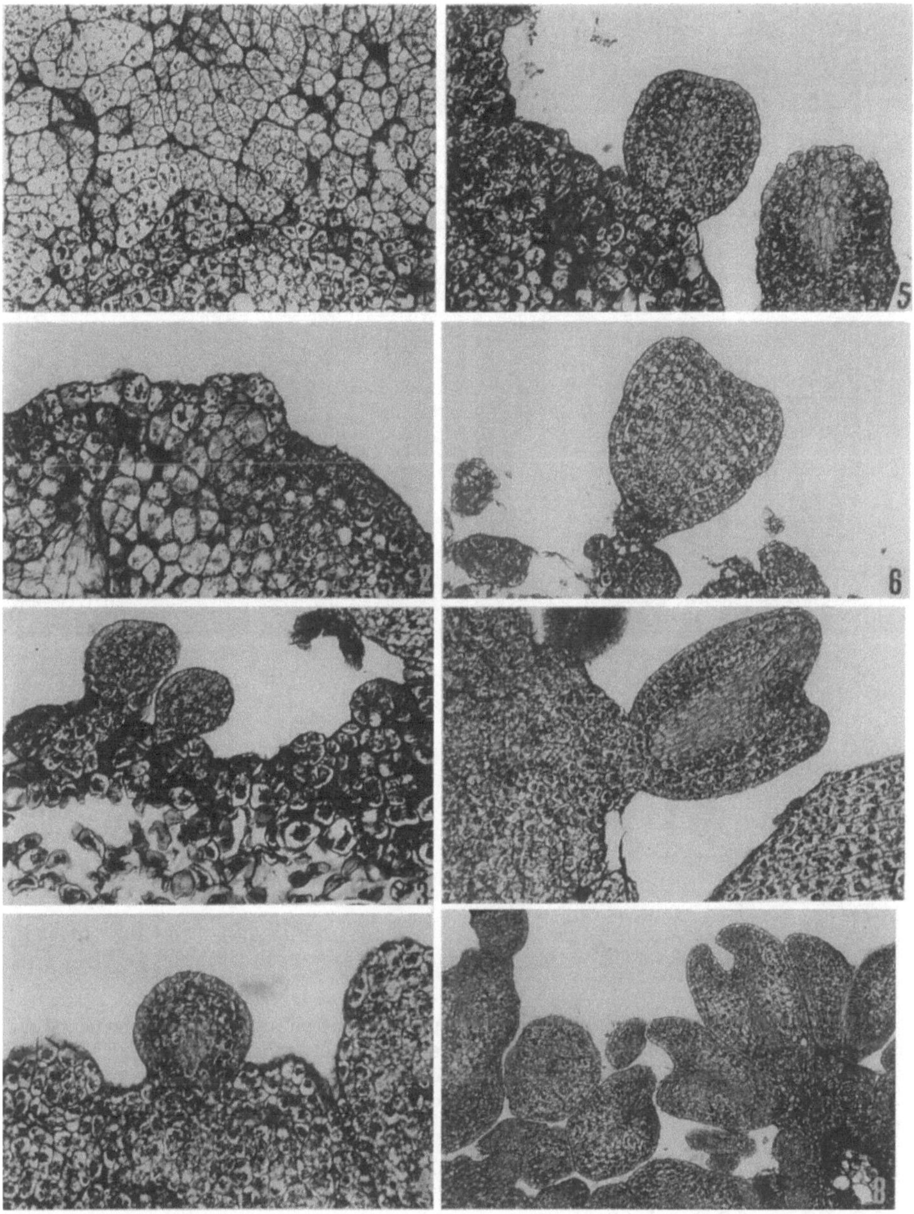

Figs. 1–8. Somatic embryogenesis in *Coptis chinensis*. **1.** Numerous scattered embryogenic cell masses on callus induced from leaves. **2, 3.** Somatic embryos developing from epidermal cells in vitro. **4.** A preglobular somatic embryo. **5.** A late globular somatic embryo. **6.** A heart shaped somatic embryo with its suspensor attached to callus. **7.** A pre cotyledonary stage somatic embryo. **8.** Various stages of somatic embryos

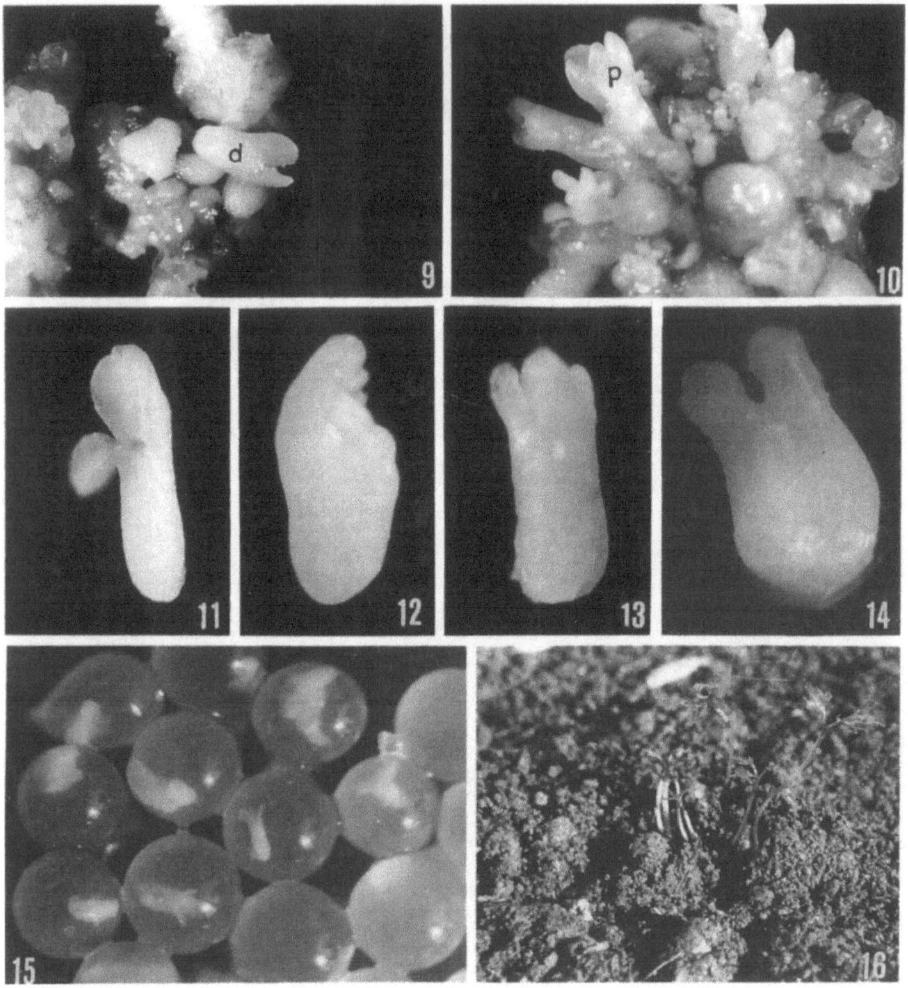

Figs. 9–16. Somatic embryos. *d* Dicotyledonary somatic embryo; *p* polycotyledonary somatic embryo. **10.** also shows secondary embryo formation. **11.** An aberrant dicotyledonary somatic embryo. **12.** An aberrant somatic embryo. **13.** Close-up of a polycotyledonary somatic embryo. **14.** a normal dicotyledonary somatic embryo. **15.** artificial seeds of *C. chinensis*. **16.** Conversion of artificial seeds in soil

2.2.2 Effect of BA and ABA

Somatic embryos were transferred to media with abscisic acid (ABA) in combination with BA and /or NAA. Twenty somatic embryos were incubated for each treatment and the experiments were repeated three times. Somatic embryos were cultured under light (cool white fluorescent light, 20 μmol m^{-2} s^{-1}) or diffuse light conditions (3.5 μmol $m^{-2}s^{-1}$) for a 12/12 h light dark cycle. More dicotyledonous somatic embryos (84.1%; Fig. 14) were observed on basal

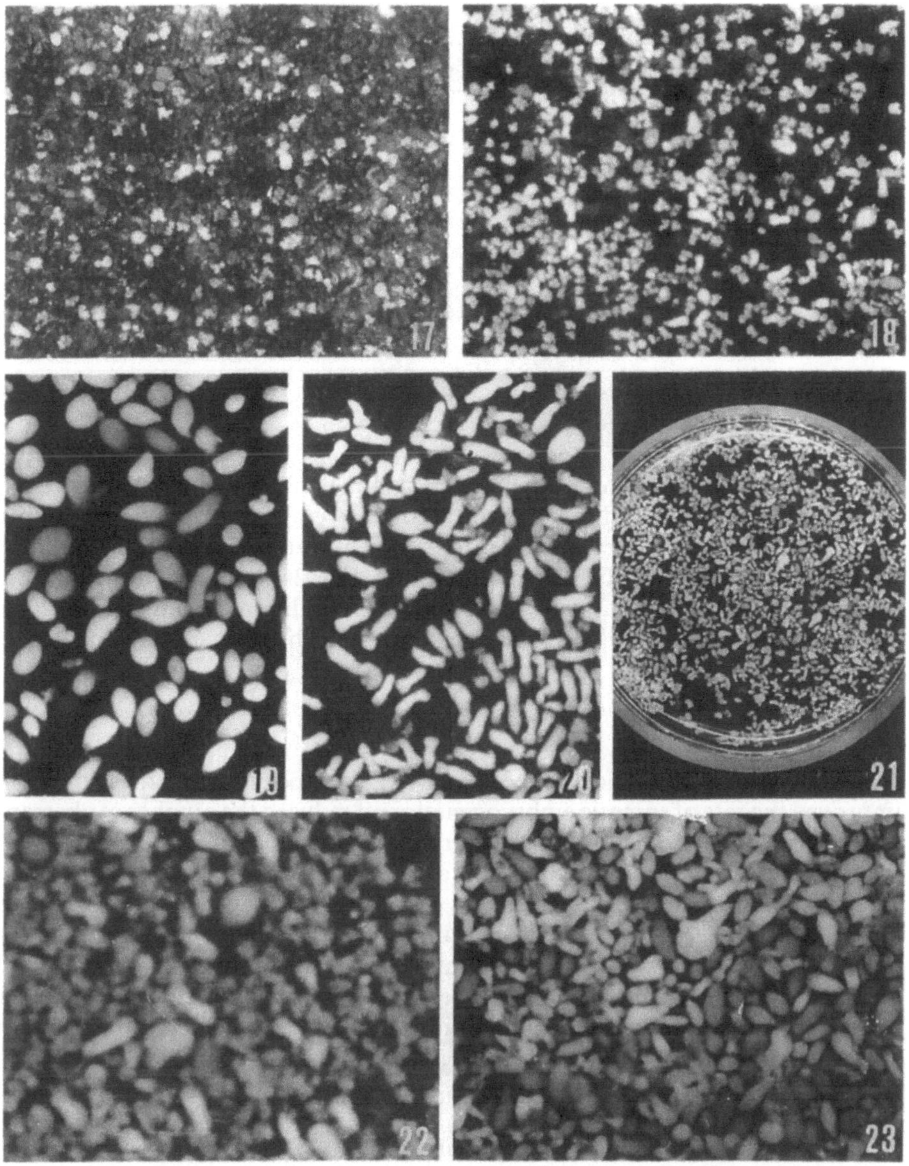

Figs. 17–23. Identically sized embryogenic cell aggregates harvested from 3-week-old suspension cultures using a sieve mesh 0.17 mm diam) **18.** Variously sized embryogenic callus after 2 weeks. **19.** Some globular somatic embryos produced after sieving through a 0.95-mm mesh'. **20.** Uniform torpedo-shaped somatic embryos obtained by sieving callus with a 1.36-mm diam. sieve mesh. **21.** Numerous somatic embryos of nearly uniform size formed 4 weeks after sieving (0.17 mm) in vitro. **22.** Control cells which have developed without being sieved. **23.** Variously sized and staged somatic embryos obtained 6 weeks after sieving and subculturing

Table 1. Effect of 2,4-D and BA on the development of *C. chinensis* somatic embryos (SE) in vitro on MS medium with growth regulators

Growth regulators[a]	Conc. (mg/l)	No. of SEs	SE length (mm)			Cotyledon morphology	
			2	2-3	3–5	Dicot	Aberrant
2,4-D	0.5	179	152 (85)[b]	25 (14)	2 (1)	78 (43)	101 (56)
	1.0	208	103 (50)	96 (46)	9 (4)	142 (68)	66 (32)
BA	1.0	241	51 (21)	54 (22)	136 (56)	178 (74)	63 (26)
	1.5	204	67 (33)	55 (27)	82 (40)	98 (48)	106 (52)

[a] Each medium also contained NAA (0.2 mg/l).
[b] (x)=% of SEs within the specified category.

medium with ABA (0.5 mg/l) than on the control (73.4%) with BA only. Each somatic embryo formed three to seven secondary embryos (Figs. 8 and 10) per subculture. Overall, there was little difference in number of somatic embryos formed under light or diffuse light conditions (Fig. 24). Under diffuse light the somatic embryos maintained their dipolar condition. Even under light conditions, however, cotyledons greened only after the radicle had elongated.

2.2.3 *Effect of Amino acids*

The addition of glycine and alanine (2 and 4 mg/l, respectively) to 2,4-D medium improved the rate of somatic embryogenesis and reduced the number of aberrant somatic embryos (Table 2). These aberrant somatic embryos could represent a type of somaclonal variation (Skirvin et al. 1993). Without growth regulators the rate of somatic embryo production was greatly reduced, even in the presence of amino acids. The addition of glycine improved the rate of somatic embryogenesis, but yielded fewer large (3 to 5 mm) somatic embryos than in media with proline, alanine, and glycine. The number of aberrant somatic embryos (Figs. 11 to 13) was reduced by certain combinations of alanine and proline (Table 2). In general, alanine improved the rate of somatic embryogenesis.

2.3 Synchronization of Somatic Embryogenesis

Populations of somatic embryos typically show a wide range in size and stages of development (Figs. 22 and 23). This is due to the fact that proembryonic cell clusters do not develop simultaneously from all regeneration sites and embryos arise from both cell clusters and maturing embryos. The ability to obtain identical size and developmental stage somatic embryos would be valuable for both theoretical and biochemical analyses of cell differentiation as well as artificial seed production. In this study, growth regulator studies were combined with a graded series of stainless steel meshes to attain uniformly sized somatic

Table 2. Effect of various amino acids on the production and quality of *Coptis chinensis* somatic embryos (SE) after 6 weeks on MS + 0.5 mg/l 2,4-D + 0.2 mg/l NAA

Amino acid	Conc. (mg/l)	No. of SEs		No. of 3–5 mm SEs	No. of cotyledons/SE			
		Cultured	Produced		0	1	2	3
Glycine	2	0	190	125	6	8	150	26
				(66)[b]	(3)	(4)	(79)	(14)
	4	20	179	109	6	12	140	42
				(55)	(2)	(6)	(71)	(21)
Alanine	2	20	204	167	0	15	153	36
				(82)	(0)	(7)	(75)	(18)
	4	20	165	143	8	9	131	17
				(87)	(5)	(5)	(79)	(10)
Proline	2	20	167	154	0	6	143	18
				(92)	(0)	(4)	(86)	(11)
	4	20	183	149	5	4	158	16
				(81)	(3)	(2)	(86)	(9)
Alanine proline	2/4	20	208	153	7	14	164	23
				(74)	(3)	(7)	(79)	(11)
	4/2	20	216	172	6	12	189	9
				(80)	(3)	(6)	(88)	(4)
Control[a] Glycine	2	20	67	26	(Not calculated)			
				(39)				
	4	20	75	24	(Not calculated)			
				(32)				

[a] (x) = % of sample that falls within the specified category.
[b] Control = growth regulator-free medium.

embryos. The embryos used were harvested from suspension cultures grown on basal MS medium with 2,4-D (0.5 mg/l) and NAA (0.2 mg/l). They were transferred to MS medium with ABA (0.5 mg/l) and NAA (0.2 mg/l) and subcultured for 2 weeks. The cultures were screened through mesh (Figs. 17 to 21); 60 to 85% somatic embryos in a similar (synochronous) stage were obtained.

3 Artificial Seeds

Somatic embryos were encapsulated in 2% sodium alginate (Shanghai Chemical Co.) using the methods of Redenbaugh et al. (1987). Each capsule contained a single embryo which was embedded in a bead of sodium alginate dipped in $Ca(NO_3)_2$ (100 mM) for 20 to 30 min followed by a brief rinse in water.

Artificial seeds (Fig. 15) were plated on either MS medium supplemented with gibberellic acid (GA_3, 0.5 mg/l) + indolebutyric acid (IBA, 0.5 mg/l) + BA (1 mg/l) + NAA (0.2 mg/l) or autoclaved soil (peat moss and vermiculite; 1:2 v/v)) in a greenhouse. Germination rates were directly related to the

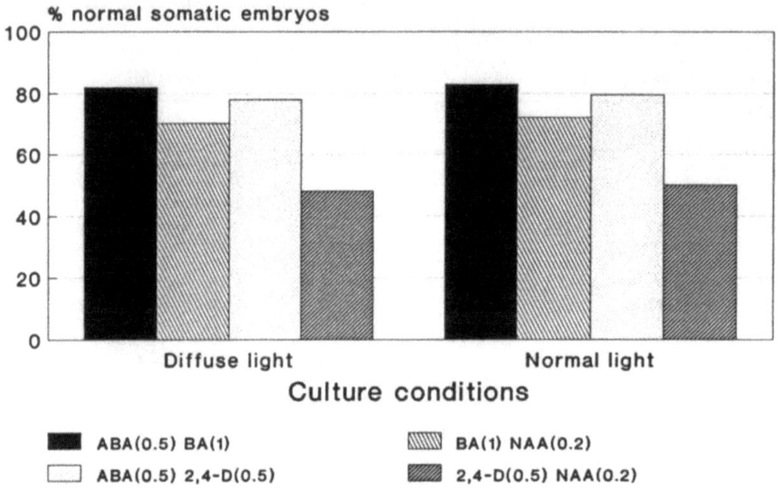

Fig. 24. The effect of BA, 2,4-D, and ABA on somatic embryo development in *Coptis chinensis* under lighted and diffuse light conditions in vitro

developmental stage of the somatic embryo. Among the various developmental stages, the best germination frequency was observed at the cotyledonary stage (Figs. 8 and 14). In vitro germination rates also varied with developmental stage: Cotyledonary (85.7%), torpedo-shaped (41%); heart-shaped (3.2%), and globular (0.5%). only mature embyros (cotyledonary stage) germinated and grew to become whole plants in soil (12.5%; Fig. 16).

The size of the somatic embryo affects its development. In general, the larger the somatic embryo, the higher the conversion frequency observed. The

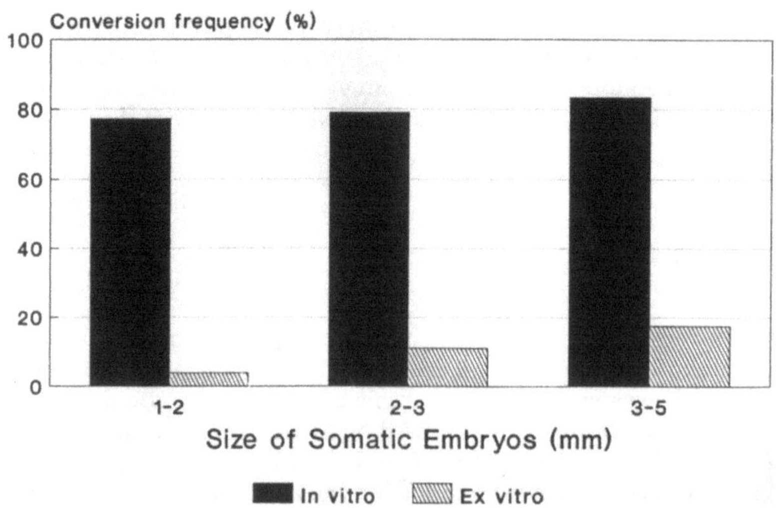

Fig. 25. Conversion rates of variously sized *C. chinensis* somatic embryos to whole plants in vitro and ex vitro

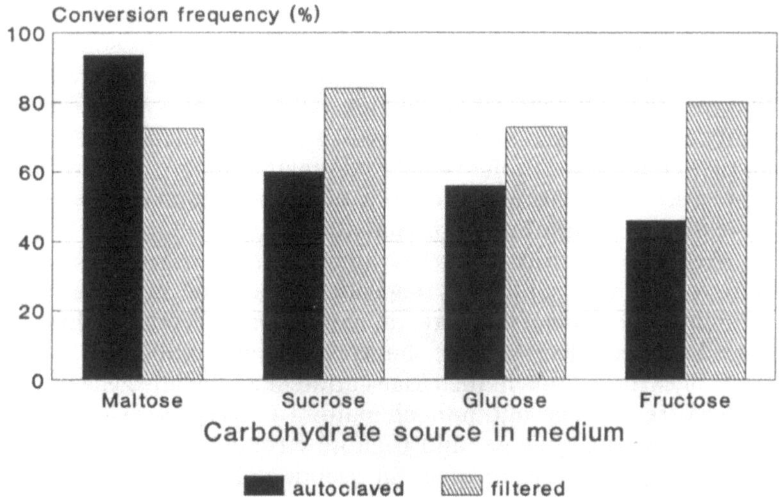

Fig. 26. The effect of various carbohydrates on the conversion frequency of artificial seeds of *Coptis chinensis* in vitro

differences were most pronounced in soil (Fig. 25); the differences were less important under aseptic conditions (Fig. 25). Under in vitro conditions, 77.2 and 83.3% of the "1 to 2" and the "3 to 5" mm somatic embryos, respectively, rooted and grew to whole plants. Somatic embryos of the same size planted in soil (ex vitro) were less likely to survive (3.8 and 17.5%, respectively).

Carbohydrates can be used to regulate osmotic pressure and energy availability in vitro; they also affect conversion rates of somatic embryos (Fig. 26). We found that the addition of either fructose (1.5% w/v) or glucose (1.5% w/v) to autoclaved or filter-sterilized medium (0.5 mg/l MS + IBA) + 0.5 mg/l GA_3 and 1 mg/l BA could serve as artificial endosperm and result in up to 72% and 81.5% conversion rates respectively. The addition of sucrose (1.5%w/v) gave slightly better results; the best results (93.5%) were obtained on medium with maltose (1.5% w/v). Our results are similar to those of Redenbaugh et al. (1987) who studied the conversion of alfalfa somatic embryos with maltose. These results suggest that maltose can serve as a maturation factor to accelerate the development of those embryos that lag behind the more mature embryos. Furthermore, maltose assists embryos that would not convert on the other carbohydrate media. The effect of autoclaving versus filter sterilization of media varied among treatments (Fig. 26).

4 Summary and Conclusions

Embryogenic callus was induced from leaves of *Coptis chinensis* F. on basal Murashige and Skoog medium supplemented with either 1 to 2 mg/l 2,4-D + 0.3 mg/l NAA or 2 to 4 mg/l BA + 0.2 mg/l NAA. These structures were induced to form putative somatic embryos on MS medium with either 0.5 mg/l 2,4-D + 0.2 mg/l NAA or 1 mg/l BA + 0.2 mg/l NAA. The best quality somatic embryos were obtained when 2 to 4 mg/l glycine, alanine, or proline, respectively, were added to the somatic embryo induction medium. All three compounds increased the size of dicotyledonary stage somatic embryos; glycine and alanine increased the number of somatic embryos. Proline improved somatic embryos quality. ABA selectively suppressed the abnormal processes of somatic embryogenesis and permitted more normal maturation of the somatic embryos.

Artificial seed conversion was highly correlated with the size of the somatic embryo and was completely dependent upon the nutrients provided in the medium. The conversion rate of cotyledonary stage embryos was 85.7%, torpedo-shaped 40.6%, heart-shaped 8.2%, and globular 0.5%. The different kinds of carbohydrats in the tissue culture medium affected conversion rates of artificial seeds. The addition of maltose (1.5% w/v) was the most effective (93.5%), sucrose, glucose, and fructose were less effective. Although the conversion rate of artificial seeds in aseptic soil was too low for commercial application (17.5%), the results nevertheless suggest that the application could be useful in the future.

5 Proctocol

Young leaves of *Coptis chinensis* F. were cultured, and embryogenic calli were induced on basal Murashige and Skoog medium supplemented with 2 mg/l 2,4-D + 0.3 mg/l NAA in the dark at 22 ± 2 °C. These cultures were induced to form putative somatic embryos on MS medium with 0.5 mg/l 2,4-D + 0.2 mg/l NAA. The somatic embryos matured on medium with 0.5 mg/l ABA with + 2 mg/l proline. High frequency conversion (85.7%) to artificial seeds was obtained on artificial endosperm medium containing MS salts, 0.5 mg/l IBA, 0.5 mg/l GA_3 0.1 mg/l NAA, 1 mg/l BA and maltose (1.5% w/v). The conversion ratc in soil was lower (17.5%).

References

Din P (1984) Studies on tissue culture of *Coptis chinensis*. Chin Trad Herb Drugs 15: 28–29

Ikuta A, Itokawa H (1989) *Coptis*: in vitro regeneration of plants and the production of berberine. In: Bajaj YPS (ed) Biotechnology in agriculture and forestry, vol. 7. Medicinal and aromatic plants II. Springer. Berlin Heidberg New York, pp 110–128

Ikuta A, Syono K, Furuya T (1975) Alkaloids in plants regenerated from *Coptis* callus cultures. Phytochemistry 14: 1209–1210

Ke S, He Z, Xu L, Hou S, Gui Y, Guo Z (1990) Studies on artificial seeds of *Coptis chinensis*. In: Guo Z, Gui Y (eds) Plant somatic embryogenesis and artificial seeds. Science Press, Beijing, pp 35–42

Ke S, Wu L, Gui Y, Guo Z, He Z, Xu L (1992) Synchronization of somatic embryogenesis in *Coptis chinensis*. J Wuhan Bot Res 10: 1–6

Murashige T, Skoog F (1962) A revised medium for rapid growth and bioassays with tobacco tissue cultures. Physiol Plant 15: 473–497

Nakagawa K, Miura Y, Fukin H, Tabata M (1982) Clonal propagation of medicinal plants through the induction of somatic embryogenesis from the cultured plant cell. In: Fujiwara A (ed) Plant tissue cultures 1982. Maruzen, Tokyo, pp 701–702

Redenbaugh, K, Viss P, Slade D, Fujii TA (1987) Scale-up: artificial seeds. In: Green CE, Somers DA, Hackett WP, Bresboer DD (eds) Plant tissue and cell culture. Alan R Liss, New York, pp 473–493.
Skirvin, RM, Norton M, McPheeters KD (1993) Somaclonal variation: has it proved useful for plant improvement? Acta Hortic 336: 333–340
Staba EJ (1977) Tissue culture and pharmacy. In: Reinert J, Bajaj YPS (eds) Applied and fundamental aspects of plant cell, tissue, and organ culture. Springer, Berlin Heidelberg New York, pp 699–715
Syono K, Furuya T (1972) The differentiation of *Coptis* plants in vitro from callus cultures. Experientia 28: 286
Tandon P, Rathore TS (1992) Regeneration of plantlets from hypocotyl-derived callus of *Coptis teeta*. Plant Cell Tissue organ Cult 28: 115–117
Wang G (1964) Studies on alkaloids of stem and root in *Coptis chinensis*. Acta Pharm Sin 11: 389–392
Yamamoto H, Ishida M, Tomimori T (1981) Studies of the fundamental conditions on growth and alkaloid formation of the callus cultures of *Coptis japonica* var. *Japonica* (III). Effects of cloning and mutagens. Shoyakugaku Zassi 35: 15–21

IV.5 Somatic Embryogenesis and Artificial Seed in Coriander (*Coriandrum sativum* L.)

R.R. CHEN, J.T. ZHANG, B.-PING LI, S.S. GUO, J.P. HAO, and X.M. ZHOU[1]

1 Introduction

Coriander (*Coriandrum sativum* L.), a member of the family Umbellifereae is an annual herb. It is of great economic importance and used as flavoring in Chinese food. It is normally propagated through seed, but micropropagation through somatic embryogenesis would be of great significance. Somatic embryogenesis in this plant was first reported by Steward et al. in 1970, and later studied by Zee (1981) and Schäfer et al. (1986). In this chapter our work on the induction of somatic embryos and their conversion into artificial seed is summarized.

2 Somatic Embryogenesis

2.1 Materials and Media

Stems of coriander seedlings planted in the greenhouse were used as the experimental material. After rinsing with tap water several times, the seedlings were sterilized in 75% (v/v) ethanol for 30 s, then in 10% (w/v) sodium hypochlorite for 15 min, and finally rinsed three times with sterile distilled water.

Callus was induced on basal MS medium of Murashige and Skoog (1962) supplemented with; (1) 1 mg/l 2,4-D; (2) 1 mg/l NAA; (3) 1 mg/l 2,4-D + 0.5 mg/l kinetin (Kn); (4) 2 mg/l NAA + 0.5 mg/l Kn.

2.2 Induction of Callus

Calli were induced with all of four media (Table 1), but induction frequency and structure of calli were quite different in various phytohormones. The efficiency of 2,4-D was inferior to NAA. However, the induction efficiency with the combination of auxins and cytokinins was better than that with auxins only. White, dense, and globular structures occurred on the media containing only 2,4-D. Calli induced on media containing NAA were pale-yellow and loose, and developed globular structures. The calli developed into plantlets through embryoids when transferred to differentiation media. However, the pale-yellow,

[1] Department of Biology, Shanxi University, Taiyuan, 030006 China

Biotechnology in Agriculture and Forestry, Vol. 31
Somatic Embryogenesis and Synthetic Seed II (ed. by Y.P.S. Bajaj)

Table 1. Effect of phytohormones on the induction of coriander callus

Hormone components (mg/l)	No. of explants	No. of calli	Induction rates (%)
2,4-D,1	47	18	38.3
NAA,1	45	20	44.4
2,4-D,1+Kn,0.5	48	23	47.9
NAA, 2 +Kn,0.5	45	24	53.3

Numbers in the table are the average values of three replicates

loose calli did not differentiate, but developed embryoids on the media with higher concentrations of NAA and Kn. Calli formed on 2,4-D and Kn media were hard, and developed very few embryoids on differentiation media.

2.3 Effect of Various Additives on the Differentiation of Somatic Embryos

Globular embryogenic calli developed after subculturing two to three times on induction medium. They were transferred to differentiation medium to induce embryoids. The differentiation medium consisted of basal MS medium supplemented with 1 mg/l NAA or 0.5 mg/l IAA + 3% sucrose + 2000 mg/l CH (casein hydrolysate). Liquid and solid media were used. The former was placed on a gyratory shaker at 70 rpm under light (2000 Ix) for 12 h/day at 25 ± 2 °C; and the latter was cultured under the same conditions, except there was no shaking.

If the embryogenic calli were transferred to differentiation medium or MS solid medium without 2,4-D, they developed successively into globular (Fig. 1), heart-shaped (Fig. 2), and torpedo-shaped embryos (Fig. 3). A large number of mature embryoids were formed after about 20 days (Figs. 4 and 5).

It was necessary to remove 2,4-D in the development process of somatic embryos. When the concentration of 2,4-D was 1 mg/l, the formation of mature somatic embryoids was completely inhibited (Table 2). The optimal concentration of NAA (1–2 mg/l) and IAA (0.5 mg/l) had a positive effect on the development of somatic embryos (Tables 3 and 4). BA had an inhibitory effect on the formation of mature somatic embryos (Table 5).

Table 2. Effect of 2,4-D on the formation of mature somatic embryos

2,4-D (mg/l)	No. of mature somatic embryos	Relative value
0	280	100
0.1	160	57.1
0.5	40	14.3
1.0	0	0

The basic medium is MS + 3% sucrose + 0.8% agar. The number of mature somatic embryos in callus per gram is the average of three replicates (the same in Tables 3,4,5,6, and 7)

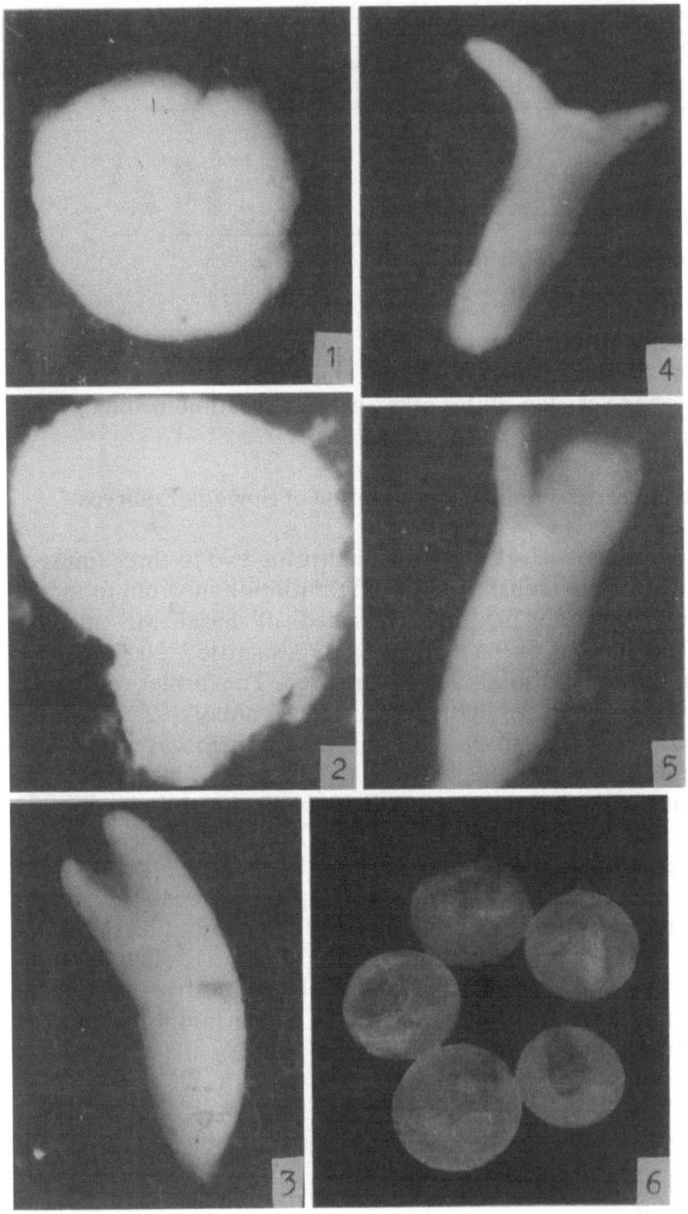

Figs. 1–6. After 3–4 weeks when the embryogenic calli were transferred to MS + 1 mg/l NAA +3% sucrose + 2000 mg/l CH liquid or solid medium. **1** Globular embryoid. **2** Heart-shaped embryoid. **3** Torpedo-shaped embryoid. **4,5** Normal embryoids with cotyledon, used for artificial seeds. **6** Artificial seeds of coriander sown on 1/2 MS + 3% sucrose + 0.8 agar

Table 3. Effect of NAA on the formation of mature somatic embryos

NAA (mg/l)	No. of mature somatic embryos	Relative value (%)
0	35	100
0.1	75	214.3
0.5	91	260.0
1.0	140	400.0
2.0	155	442.9
3.0	138	377.1
5.0	127	362.9

The basic medium is MS + 3% sucrose + 0.8% agar

Table 4. Effect of IAA on the formation of mature somatic embryos

IAA (mg/l)	No. of mature somatic embryos	Relative value (%)
0	35	100
0.1	120	342.9
0.5	128	365.7
1.0	83	273.1
2.0	34	97.1
3.0	33	94.3
5.0	12	34.3

The basic medium is MS + 3% sucrose + 0.8% agar

Table 5. Effect of 6-BA on the formation of mature somatic embryos

6-BA (mg/l)	No. of mature somatic embryos	Relative value (%)
0	140	100
0.1	64	45.17
0.5	45	32.1
1.0	30	21.4
2.0	15	10.7

The basic medium is MS + 1mg/l NAA + 3% sucrose + 0.8% agar

An optimal concentration of CH (1500–2000 mg/l) had a remarkably positive effect on the development of somatic embryos (Table 6). Various sugars also affected the formation and development of somatic embryos (Table 7), 3% sucrose was optimal. If the concentration of sucrose was higher or less than 3%, some inhibition occurred. The higher the concentration of sucrose, the more obvious was the inhibitory effect.

Table 6. Effect of casein hydrolysate on the development of mature somatic embryos

Casein hydrolysate (mg/l)	No. of mature somatic embryos	Relative value (%)
0	80	100
500	180	225.0
1000	262	327.5
1500	386	482.5
2000	432	540.0
3000	240	300.0
5000	156	195.0

The basic medium is MS + 3% sucrose + 1mg/l NAA + 0.8% agar.

Table 7. Effect of sucrose on the formation of mature somatic embryos

Sucrose (mg/l)	No. of mature somatic embryos	Relative value (%)
1.0	280	100.0
3.0	320	177.8
5.0	126	45.0
7.0	36	12.9
9.0	26	9.3
10.0	20	7.1

The basic medium is MS + 1mg/l NAA + 0.8% agar

3 Production of Artificial Seeds

3.1 Encapsulation of Embryoids

The encapsulating agent for artificial seeds of coriander was 3% sodium alginate with 2% $CaCl_2 \cdot 2H_2O$, prepared separately with 1/2 MS.

Normal mature embryoids induced on solid differentiation media could be selected and directly employed for preparing artificial seeds. However, in order to separate mature embryoids from suspension culture, they were filtered through a stainless steel mesh (0.95–1 mm) to remove the larger calli, then mature embryoids (about 2 mm length) were encapsulated. The mature embryoids were mixed with a coating compound (alginate-Na). Then the alginate suspension of embryoids was dropped into $CaCl_2$ solution with a dropper. Encapsulated embryoids were maintained in hardening solution (2% $CaCl_2 \cdot 2H_2O$) for 20–30 min. After the beads were shaped and hardened, they were rinsed twice with sterile distilled water, and artificial seeds were obtained (Fig. 6).

About 400–500 mature embryoids were formed from 1 g of callus on solid media. Most of them developed normally and were used as encapsulating material. In suspension culture, about 1500 mature somatic embryos were formed in each flask containing 50 ml suspension culture, but many embryos, which had the appearance of a glass-like substance, could not regularly develop into seedlings after encapsulation.

3.2 Germination Matrices and Sowing of Artificial Seeds

The germination matrices of artificial seeds were 1/2 MS media containing 3% sucrose, moist filter paper, or vermiculite. Artificial seeds sown on solid 1/2 MS (Fig. 7) showed the best results of the three basic matrices. The germination ratio on solid 1/2 MS, moist filter paper, and moist vermiculite was 86, 62, and 2% respectively. There are two possible reasons: (1) the artificial seeds lost their germination ability due to microbial contamination; (2) the respiration of artificial seeds was inhibited due to poor ventilation. In addition, the germination rate of artificial seeds was also influenced by the storage time. After storage in a refrigerator at 4 °C for a period of time, many of the embryoids in artificial seeds became brown and lost their germination ability (Table 8).

We added sucrose as a nutrient to the encapsulating agents. The results showed that the sucrose could not increase the viability of somatic embryos. On the contrary, there was some inhibition of the germination of artificial seeds (Table 9).

3.3 Transplantation of Seedlings

After the germination of artificial seeds (Fig. 7), most of them developed into normal seedlings. When the seedlings were transplanted into pots in a greenhouse with enough humidity, they could grow to about 5 cm, and their survival rate reached about 83%. Now a number of plants have flowered and set fruit (Fig. 8).

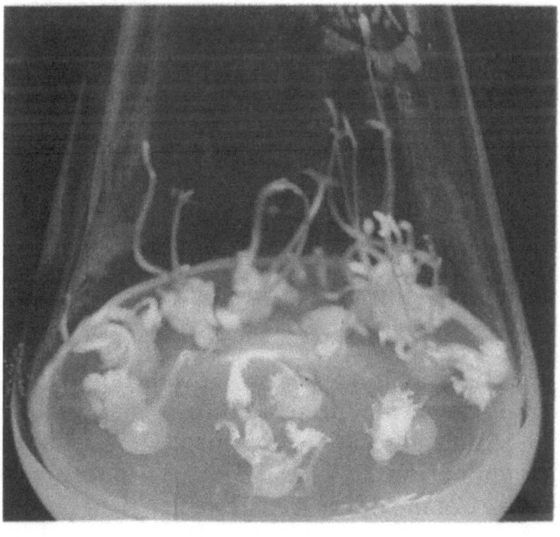

Fig. 7. After 7 days plantlet emerged from artificial seed on agar medium

Table 8. Effect of storage time (4 °C) on germination of artificial seeds

Days Storage	No. of sowings		No. germinated		Germination rate (%)	
	On 1/2 MS medium	On filter paper	On 1/2 MS medium	On filter paper	On 1/2 MS medium	On filter paper
1	200	69	164	41	80.2	59.4
20	220	83	156	38	70.9	45.8
20	60	57	39	23	65.0	40.0

Table 9. Effect of sugar in the encapsulating agent on the germination of artificial seeds of coriander

Sugar and its Concentration	No. of sowings	No. germinated	Germination rate (%)
0	74	64	86.5
3% Sucrose	94	64	48.1
3% Glucose	73	39	53.4
3% Sucrose + 3% glucose	74	41	55.4
3% Sucrose + 6% glucose	74	51	68.9

Numbers in the table are the average values of three replicates

Fig. 8. Plants grown from artificial seeds of coriander after transplantation to a pot in the greenhouse

4 Discussion and Conclusion

By manipulating various concentrations of the growth regulators, somatic embryos were induced in calli originating from seedling stem pieces of coriander, and artificial seeds were formed. The seedling rate of artificial seeds under sterile conditions was 82%, and the survival rate of seedlings of artificial seeds reached 83% after they were transferred to pots.

Auxins were necessary for the induction of embryogenic cells. However, the development of globular embryos to mature embryos was inhibited at a higher concentration of 2,4-D. The calli also developed successively into globular, heart-shaped, and torpedo-shaped embryoids only on NAA media.

Kinetin seems unnecessary for the formation of somatic embryoids. The number of mature somatic embryos formed on media containing Kn was far less than that on medium without Kn. This result was contrary to most plants.

Solid and liquid cultures have both advantages and disadvantages with regard to the production of normal mature embryos. A large quantity of embryoids could be produced from liquid suspension culture each time, but many of them had clonal variations and abnormal embryoids. Gray (1987) and Jin and Kuo. (1988) reported that the viability of somatic embryos in storage increased and the ability to regenerate into plants was partially raised with dry somatic embryos. The embryos induced on solid media were mostly healthy and normal. Redenbaugh et al. (1987, 1991) directly selected healthy embryoids by manual manipulation to prepare artificial seeds, but the number of embryoids obtained by this method was too small to produce a large quantity of artificial seeds.

The research pertaining to artificial seeds is concerned with obtaining a large quantity of normal plants. Therefore, a prerequisite for the preparation of artificial seeds is the culturing of a large number of healthy and normal embryoids.

Artificial seeds encapsulated with alginate could only germinate under aseptic conditions, but they dehydrated easily and died. They could germinate and develop into seedlings only on agar media. Because the poor ventilation of sodium alginate inhibited the normal respiration of embryoids, artificial seeds could not be stored for a long time. We tried using other materials (agarose, gelrite, ethylaminoformate, etc.) as coating compounds, but the results were unsatisfactory. In addition, it was necessary to add other substances to the encapsulating agents. Jin and Kuo (1988) reported that carbohydrates might be necessary for the germination of encapsulated embryoids. However, in our experiments, the coating compound added to sucrose had no stimulatory, but rather an inhibitory effect on the germination of coriander embryoids. Thus, a high sucrose concentration in the sowing matrix (1/2 MS + 3% sucrose) and coating compound may have produced a high osmotic pressure to make the water-absorption ability of embryoids difficult.

5 Protocol for the Induction of Somatic Embryos

1. The stems of coriander seedlings were sterilized in 75% (v/v) ethanol for 30 s, then in 10% (w/v) sodium hypochlorite solution for 15 min, washing three times in sterile distilled water.
2. Cut 1-cm-long stem segments and culture them individually on a solid medium (MS + 1 mg/l NAA + 3% sucrose + 0.8% agar). Incubate the cultures in the dark at 25 ± 2 °C.
3. After 3–4 weeks, transfer the dense, globularly structured calli to fresh medium 2–3 times, A number of globular embryogenic calli occur after 3–6 weeks.
4. To induce somatic embryogenesis, transfer callus to liquid and solid media. The former was a basic MS medium supplemented with 1 mg/l NAA, 3% sucrose, and 2000 mg/l casein hydrolysate place

on a gyratory shaker at 70 rpm under 12-hr light (2000 lx) at 25 ± 2 °C. The solid medium was of the same composition but with 0.8% agar, the conditions were the same as the liquid culture, except shaking.

5. After 3–4 weeks the cultures contain numerous embryos in different stages of development.

References

Gray DJ (1987) Quiescence in monocotyledonous and dicotyledonous somatic embryos induced by dehydration. HortScience 22 (5): 810–814

Jin Ji-yi, Kuo Chung-chen (1988) Studies on artificial seeds of celery. In: Inst Botany, Academia Sinica and Department of Biology Lanzhou University (eds) Recent advances on studies of applied and fundamental aspects of plant cell engineering. Science Periodical Press, Beijing, pp 84–89

Murashige T, Skoog F (1962) A revised medium for rapid growth and bioassays with tobacco cultures. Physiol, Plant 15: 473–497

Redenbaugh K, Viss P, Slade D, Fujii J (1987) Encapsulation of somatic seed coats. HortScience 22(5): 803–809

Redenbaugh K, Fujii J, Slade D, Viss P, Kossler M (1991) Artificial seeds encapsulated somatic embryos. In: Bajaj YPS (ed) Biotechnology in agriculture and forestry, vol 17. High-tech and micropropagation 1. Springer, Berlin Heidelberg New York, pp 395–416

Schäfer F, Li T, Neumann KH (1986) Petiole explants and intact seedlings as sources of somatic embryos of some Apiaceae (*Daucus carota, Anethum graveolens, Pimpinella anisum, Coriandrum sativum, Petroselinum crispum, Carum carvi*). VIth Int Congr Plant tissue and cell culture, Univ Minnesota, Minneapolis pp 346

Steward FC, Ammirato PV, Mapes MO (1970) Growth and development of totipotent cells, Ann Bot 34: 761–787

Zee S-y (1981) Studies on adventive embryo formation in the petiole explants of coriander (*Coriandrum sativum*). Protoplasma 107: 21–26

IV.6 Somatic Embryogenesis in Ginseng (*Panax* Species)

Y. Shoyama[1], H. Matsushita[2], X.X. Zhu[1], and H. Kishira[2]

1 Introduction

Almost all *Panax* spp. (family Araliaceae) have been used in folk medicine. The most famous variety is *P. ginseng*, which was recorded in Chinese Materia Medica 2000 years ago. Ginseng is one of the most important Chinese medicines used in tonics to combat stress and cancer, disturbances of the central nervous system, and hypothermia, and for radio-protection. It contains many dammarane and oleanane saponins of which the biological activity has been studied widely. *P. pseudoginseng* var. *notoginseng* is cultivated in Kumming province in China. It is also an important Chinese crude drug used as an astringent and tonic. *P. japonicus* grows mainly in Japan and China, and is used as a stomachic and a hair-growing tonic in Japan.

A variety of *Panax* species grow widely in the Northern Hemisphere, from the Himalayas through China and Japan to North America. *Panax* spp. consist of *P. ginseng* (Korea, China, Russia), *P. japonicus* (Japan, China), *P. quinquefolium* (North America), *P. pseudoginseng* var. *notoginseng*, *P. pseudoginseng* var. *elegantior*, *P. pseudoginseng* var. *wangiatior*, *P. pseudoginseng* var. *bipinnatifidus*, *P. stipleanathus* (China), *P. pseudoginseng* var. *himalaicus* (Himalaya), and *P. vietnamensis* (Vietnam). The first four species are currently on the market and *P. ginseng* has the biggest market in the world.

P. ginseng C.A. Meyer is a perennial herb and native of Korea, China, and Manchuria. It is erect and grows to a height of about 60 cm. It has many white, five-petaled florets in spherical umbels in early summer. The berry is a flat sphere, which turns red in summer. When 4–6-year-old roots are dried, or steamed and dried, they are used as a crude drug, ginseng. The aerial part of all other *Panax* spp. resembles that of *P. ginseng*, but the morphology of the roots or rhizomes is quite different.

P. japonicus C.A. Meyer is indigenous mainly in Japan and China. Detailed studies on the saponins contained in the rhizome and in the aerial part have been carried out (Tanaka 1989). From these results, it became clear that qualitative and quantitative variations in the saponins of the crude drug are caused by differing growing conditions or strains (Tanaka 1989). Since this plant is not a

[1] Faculty of Pharmaceutical Sciences, Kyushu University, Higashiku, Fukuoka 812, Japan
[2] Nippon Mining Co., Ltd., Bioscience Research Laboratories, 3–17–35, Niizo-Minami, Toda, Saitama 335, Japan

Biotechnology in Agriculture and Forestry, Vol. 31
Somatic Embryogenesis and Synthetic Seed II (ed. by Y.P.S. Bajaj)

cultivated plant, it is difficult to define such factors, hence, it is necessary to breed a homogeneous strain and to establish a propagation system. On the other hand, the embryo-forming rate from callus and the regeneration ratio of normal plants from the embryos gradually decreased and finally became nonexistent after 2 years of subculture (Shoyama et al. 1987). Therefore, there is a need for maintenance of embryogenesis using in vitro culture.

P. ginseng is cultivated in several areas for selecting clones that are suitable for local conditions. However, since it is difficult to propagate this plant by conventional methods of clonal propagation, seedlings are usually used for this purpose. Thus, it is difficult to control the quality of the crude drug and, furthermore, the seeds need pre-treatment for gernination.

Extensive studies have been conducted on various aspects of the in vitro culture of *Panax*.

The callus culture method for *P. ginseng*, and studies on its saponin production have been reported (Furuya et al. 1970,1973,1983a,b; Kim et al. 1980; Yoshikawa and Furuya 1987). Investigations on regeneration via embryogenesis have also been carried out by many workers (see Choi 1988, Asaki et al. 1993). As previous reports indicate, embryogenesis of *P. ginseng* has usually been induced from root tissue and required a long period of time. Our previous works indicate that the flower bud culture of *P. japonicus* is the most effective method for embryogenesis as compared with other organs (Fujioka et al. 1986; Shoyama et al. 1987). In order to throw light on this in vitro propagation method for *P. ginseng* and *P. japonica*, a detailed procedure for somatic embryogenesis is presented.

2 Somatic Embryogenesis in *Panax japonicus* C.A. Meyer

(Fujioka et al. 1986; Shoyama et al. 1987)

2.1 Material and Methods

Flower buds, stems and leaf and rhizome segments of *P. japonicus* were taken from field-grown seedlings 2 weeks after germination. Individual organs were washed with tap water, surface sterilized with 2% sodium hypochlorite solution containing 0.1% Tween 80 for 10 min, then with 70% alcohol for 30 s, and finally thoroughly washed twice with sterilized water. Organs were cultured individually. The basal medium consisted of MS salts (Murashige and Skoog 1962). The cultures were exposed to 16-h light from cool white fluorescent tubes at a temperature of 25 ± 1 °C.

2.2 Results

Table 1 shows that callus was induced from flower buds, and segments of stems, and leaf and rhizome on the MS medium supplemented with 2,4-D and NAA in the dark over 15 weeks. Flower buds and leaf segments produced complete callus

Table 1. Callus formation and embryogenesis (Fujioka et al. 1986)

Origin	Hormone (mg/l)	Number of explants	Number of callusing explants	Fresh weight (g/culture)	Embryogenesis (%)
Flower bud	2,4-D	15	15	1.87	80
	NAA	5	5	0.63	0
Rhizome	2,4-D	42	23	0.36	0
	NAA	22	14	0.53	0
Leaf	2,4-D	9	9	5.29	0
	NAA	9	9	1.20	0
Stem	2,4-D	27	9	2.07	0
	NAA	13	2	0.78	0

Culture conditions: MS medium, 25 ±1 °C, in the dark, 15 weeks

formation when the medium was supplemented with 1 mg/l 2,4-D or 1 mg/l NAA. The callus formation rate from rhizome segments was 55% with 1 mg/l 2,4-D and 64% with 1 mg/l NAA. Callus formation from stem segments was lower compared to the other organs. After 15 weeks the callus segment induced only from flower buds turned partly brown. Eighty percent of the segments produced a few embryoids, whose globular forms predominated on the surface of the callus. It was clear that in order to increase the number of embryoids and also the later stages, such as heart-shaped or cotyledonary forms, further subculture was necessary.

When a callus cluster containing immature embryos was subcultured on MS medium supplemented with NAA and BAP in the dark for 10 weeks, secondary embryogenesis occurred (Fig. 1). The combination of 2 mg/l NAA and 10 mg/l BAP stimulated embryoid propagation (100%). Towards the end of the culture, many heart-shaped embryoids appeared (Fig. 2). The germination from half of the embryoid segments cultured on medium supplemented with 2 mg/l NAA and 2.5, 5, or 10 mg/l BAP produced, on average, one or two shoots after 10 weeks of culture.

It is well known that the addition of GA and cytokinin to the medium stimulates shoot regeneration from embryoids (Chang and Hsing 1980). Furthermore, since normal strength MS medium frequently produced abnormal shoots in a preliminary experiment, 1/2 and 1/4 MS media were investigated. Table 2 shows the effect of GA and/or BAP in the 1/2 and 1/4 MS media. The best result was obtained with the addition of 1 mg/l GA and 1mg/l BAP to the 1/2 MS medium: in this case, 85% of the embryo formed shoots after 6 weeks of culture. The addition of GA alone (1 mg/l) was also effective, as indicated in Table 2 (80% formed shoots). All shoots regenerated from embryoids were without roots. Therefore, the root formation step is necessary at the next stage.

Table 3 indicates the effect of auxins on root formation on shoots. Addition of IAA to the MS medium was ineffective for root formation. When shoots were cultured on the medium supplemented with IBA, rooting occurred, however, the rate was lower, and some callusing appeared. On the other hand, the addition of NAA, particularly 1 mg/l, produced optimal root formation (6.8 roots per shoot).

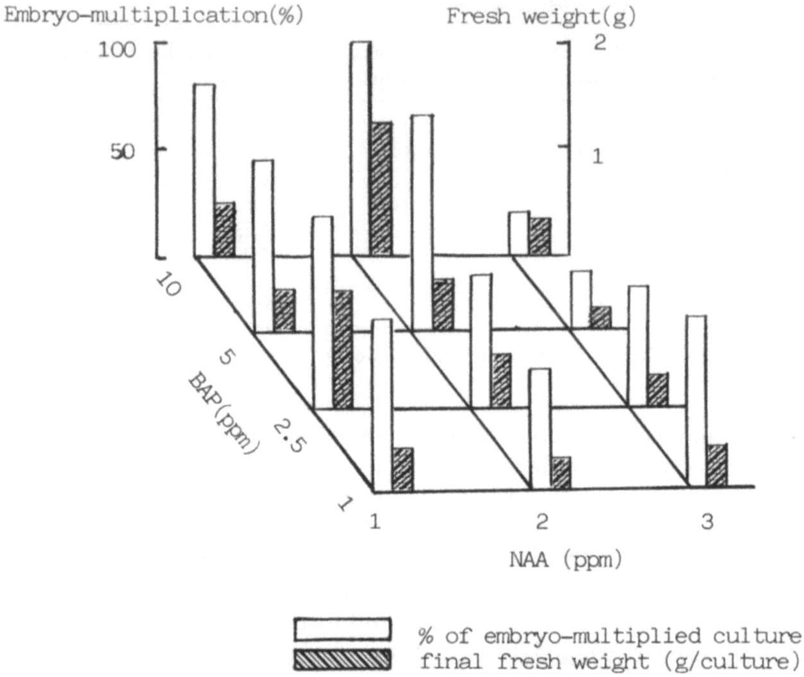

Fig. 1. Embryo multiplication from flower buds. (Fujioka et al. 1986). Culture conditions: MS medium, 24-h light, 25 °C, 10 weeks

Table 2. Shoot formation from embryos. (Fujioka et al. 1986)

Medium	Hormone mg/l		Number of explants	Number of shoots formed
	GA	BAP		
1/2 MS	1	0	60	48
	1	1	40	34
	0	1	35	0
1/4 MS	1	0	33	10
	1	1	20	0
	0	1	20	0

Culture conditions: 25 ± 1 °C, under continuous light, 6 weeks

Since embryo propagation and regeneration from embryos gradually decreased after 2 years of subculture, a multiple shoot formation system using shoots regenerated from embryos has been investigated. Figure 3 shows the effect of BAP and GA, IAA, or NAA in the half-strength MS medium. The most favorable condition for multiple shoot formation was half-strength MS medium supplemented with 1 mg/l GA and 10 mg/l BAP, on which 11 shoots per culture were formed. The 1/2 MS medium containing 5 mg/l GA and 1 or 5 mg/l BAP formed six shoots per culture. Table 4 shows that in the MS medium, almost

Fig. 2A,B. Somatic embryogenesis of *Panax japonicus*. **A** Embryo formation from callus derived from flower bud after 3-month culture on MS medium containing 1 mg/l 2,4-D. **B** Mature embryos formed by subculture of callus supplemented with 1 mg/l 2,4-D for 1 month

A

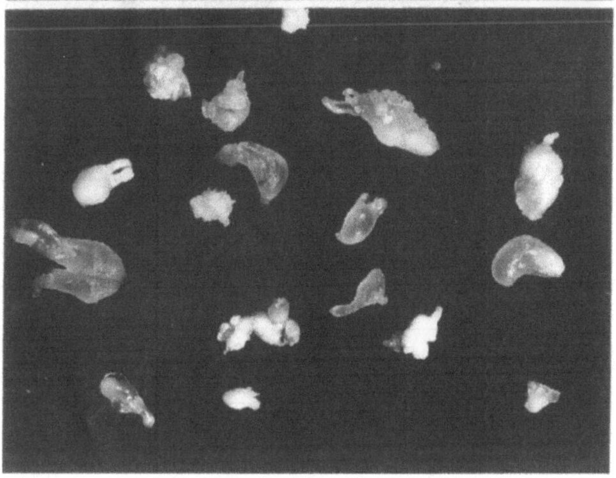

B

Table 3. Root formation from regenerated shoots. (Fujioka et al. 1986)

Hormone	mg/l	Number of explants	Number of rooted cultures	Average number of roots
IAA	1	10	0	–
	2	10	0	–
IBA	1	8	2	6.0
			6[a]	–
	2	8	2	2.0
			6[a]	–
NAA	0.5	10	6	3.7
	1	9	9	6.8

Culture conditions: MS medium, 25 ± 1 °C, under continuous light, 15 weeks
[a] Callusing

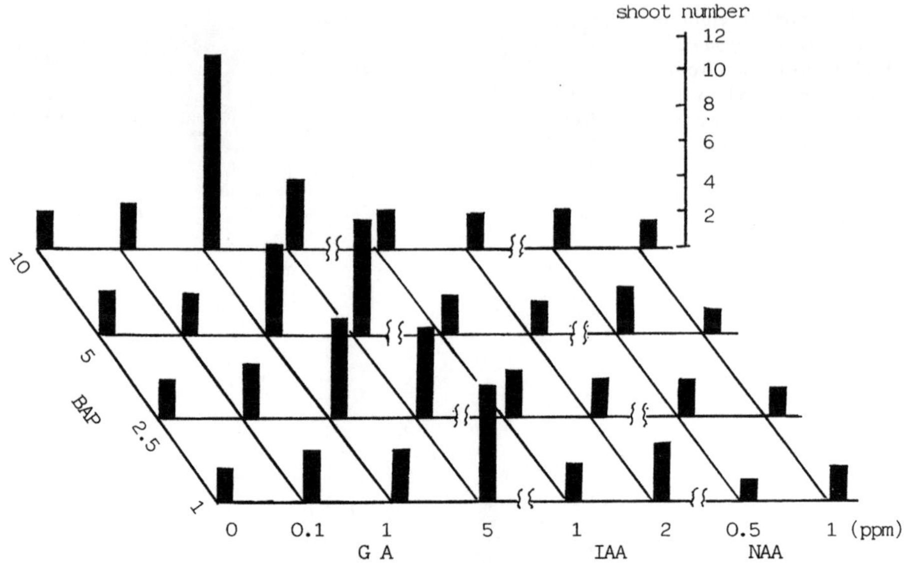

Fig. 3. Multiple shoot formation on 1/2 MS medium supplemented with varying concentrations of GA and BAP. (Shoyama et al. 1987). Culture conditions: 1/2 MS medium, 16-h light, 22 °C, 9 weeks

Table 4. Multiple shoot formation from shoots regenerated via embryogenesis. (Shoyama et al. 1987)

GA mg/l	BAP	No. shoot/culture	Fresh weight (g/culture)
1	1	4.3	1.01
1	2.5	5.4	0.60
1	5	5.4	0.83
1	10	3.6	0.63
5	1	4.5	0.78
5	2.5	5.8	0.98
5	5	5.9	1.03
5	10	4.8	0.75

Culture condition: MS medium, 22 ± 1 °C, 16-h illumination, 9 weeks

any amount of GA-BAP was effective in inducing multiple shoot formation. However, fewer shoots were produced than on the corresponding 1/2 MS media. Therefore, 1/2 MS medium containing 1 mg/l GA and 10 mg/l BAP was routinely used in the multiple shoot formation stage. Occasionally, in vitro flower buds formed in the multiple shoot-forming stage, e.g., *P. ginseng* (Chang and Hsing 1980a).

In the 2-year subculture, embryo propagation ability and the regeneration ratio of normal plantlets gradually decreased. Therefore, embryogenesis from in vitro plant organs was examined. Table 5 shows callus formation and embryo-

Table 5. Callus formation and embryogenesis from in vitro organs. (Shoyama et al. 1987)

Organ	No. of explants	No. callusing explants (embryo-forming explants)
Flower bud	10	10(9)
Leaf	9	9(2)
Stem	9	9(4)
Rhizome-like	15	14(3)
Hairy root	6	3(1)

Culture condition: MS medium, 25 ± 1 °C, in the dark, 4.5 months

genesis from individual in vitro organs on MS medium supplemented with 1 mg/l 2,4-D in the dark for 4.5 months. Flower buds, and leaf and stem segments produced callus. The callus induced from the in vitro flower buds formed embryos after approximately 2 months. Immature and mature embryos were observed on the surface. The best result was obtained by the culture of flower buds; 90% of calli produced embryos.

3 *Panax ginseng* C.A. Meyer

(Shoyama et al. 1988; Kishira et al. 1992)

3.1 Material and Methods

Young flower buds of 5-year-old *P. ginseng* C.A. Meyer were taken from the field-grown collection. They were sterilized as indicated previously. Segments (32–58) were also individually cultured in the dark. In vitro organ segments (6–19) were aseptically separated and were individually cultured in the dark. Embryo-forming cultures were subcultured in MS medium supplemented with 1 mg/l 2,4-D for multiplication and maturation of embryos in the dark at 25±1 °C for 1 month.

Twenty mature embryos were cultured on MS-vermiculite medium supplemented with sucrose, GA, and BAP (0.5 mg/l each) under 16-h irradiation at 16±1 °C for 1 month. The normal plantlets which were longer than 5 mm in length were counted as the germinated plants, and then subcultured on growth regulator-free MS-vermiculite (9:10) medium under 16-h light at 18±1 °C for 3 months.

Twenty mature embryos were cultured in MS liquid medium supplemented with sucrose, and GA (0.5 mg/l) in the dark at 18±1 °C and rotated at 2 rpm for 1 month. The individual lengths of the germinated plantlets were determined, and then they were subcultured on growth regulator-free MS-vermiculite (9:10) medium under 16-h light at 18±1 °C for 3 months.

3.2 Results

When individual tissues of a multiple shoot complex, a flower bud, a leaf segment, and a stem segment, were cultured on MS medium supplemented with 1 mg/l 2,4-D in the dark for 3 months, the embryogenesis induced by the culture of flower buds gave a higher yield (80%) as indicated in Fig 4. The rates of embryogenesis from leaf and stem were 40 and 20%, respectively. It becomes clear that the in vitro flower buds are more capable of embryogenesis than those of the intact 5-year-old plant.

In order to determine the suitable concentration of 2,4-D for callus formation and embryogenesis from the in vitro flower buds, varying concentrations of 2,4-D were tested at different periods of culturing (Fig. 5). From the results, it was evident that the addition of 5 mg/l 2,4-D increased embryogenesis (90%). Therefore, the somatic embryo was routinely induced by culturing in vitro flower buds on MS medium supplemented with 5 mg/l 2,4-D in the dark for 3 months.

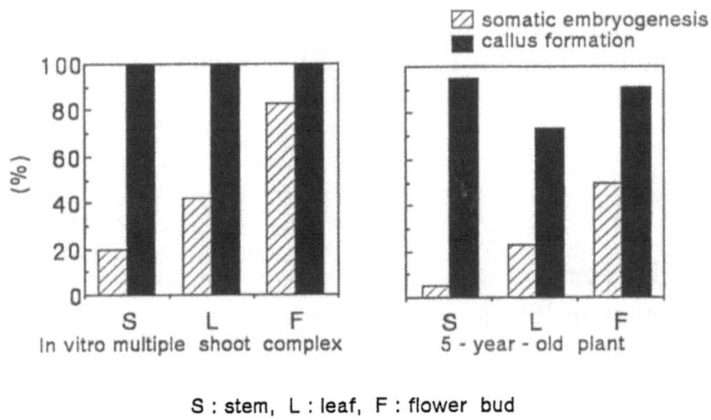

Fig. 4. Callus formation and somatic embryogenesis from various tissues of an in vitro and 5-year-old plant. (Kishira et al. 1992). Culture conditions: S medium, 2,4-D (1 mg/l), sucrose (3%), in the dark, 25 °C, 3 months

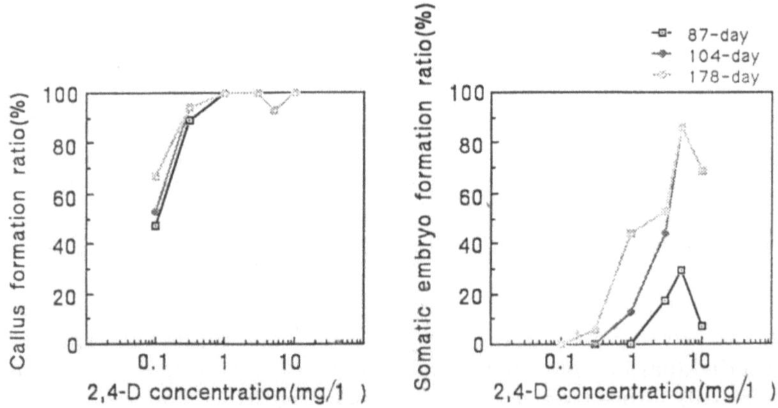

Fig. 5. Effect of 2,4-D concentration on callus formation and somatic embryogenesis from in vitro flower bud. (Kishira et al. 1992). Culture conditions: MS medium, sucrose (3%), in the dark, 25 °C

Although shoot formation occurred in somatic embryos of *P. ginseng*, no root formation was obtained in general (Butenko et al. 1968; Chang and Hsing 1980a, b; Choi et al. 1982; Furuya et al. 1986; Choi 1988). Moreover, in the preliminary investigation, the mature embryos were coated with 3% alginate containing 1/2 MS medium, 0.5 mg/l GA-BAP and 1.5% sucrose, similar to a previously reported procedure (Kitto and Janick 1985), and cultured on MS medium under 16-h light at 25 °C for 1 month, resulting in shoot formation without rooting (Shoyama and Kishira, unpubl.). Therefore, in order to obtain plantlets directly, the mature embryos were incubated in MS-vermiculite medium containing sucrose and BAP-GA under 16-h light for 2 months. Previously, Chang and Hsing (1980) reported that the germination of embryos appeared to be optimal in 1/2 MS medium containing GA and BAP (1 mg/l each). However, since abnormal shoot formation occurred occasionally during culture under these conditions, the authors improved the germination medium to 1/2 MS medium supplemented with 1.5% sucrose and GA-BAP (0.5 mg/l each) (Shoyama et al. 1988). Thus, further aspects regarding the concentrations of MS medium and sucrose were investigated. Figure 6 shows the effect of sucrose concentration on the germination of mature somatic embryos. From these results, it was evident that 3% sucrose was best for germination from mature somatic embryos and growth of plantlets. Figure 7 shows that normal MS medium containing 3% sucrose and GA-BAP (0.5 mg/l each) largely stimulated germination of embryos and the growth of plantlets. Following this study, suitable mixing ratios of MS medium and vermiculite for germination of mature somatic embryos and growth of plantlets were investigated (Fig 8). A combination of MS medium and vermiculite (9:10) was found to be most satisfactory. In this case, germination of embryos was optimal and the average length of whole plantlets was approximately 8 mm (Fig 9 A). This medium was routinely used for the direct production of plantlets from mature embryos. Figure 9B shows the germinated plantlets cultured on the growth regulator-free MS-vermiculite medium after more than 2 months.

In order to obtain rapid elongation of the main root, the effects of liquid medium were tested. When mature embryos were incubated in liquid medium supplemented with sucrose and GA, plantlet formation occurred. The addition

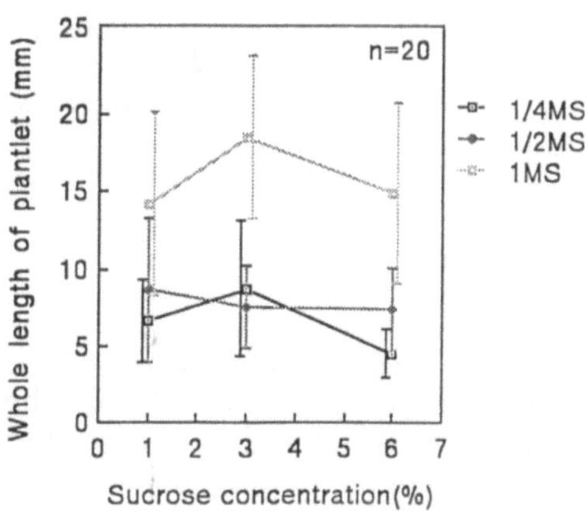

Fig. 6. Effect of sucrose concentration on germination of mature somatic embryos. (Kishira et al. 1992). Culture conditions: GA and BAP (0.5 mg/l each), 16-h light, 16 °C, 2 months

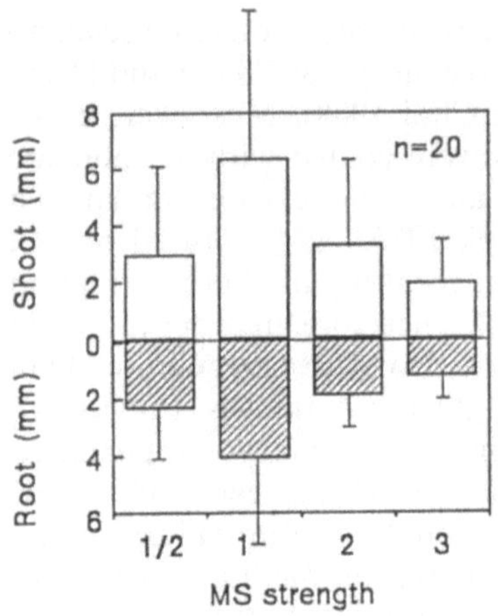

Fig. 7. Effects of MS Strength on germination of embryos and growth of plantlets. (Kishira et al. 1992). Culture conditions: MS medium, GA and BAP (0.5 mg/l each), sucrose (3%), 16-h light, 16 °C, 1 month

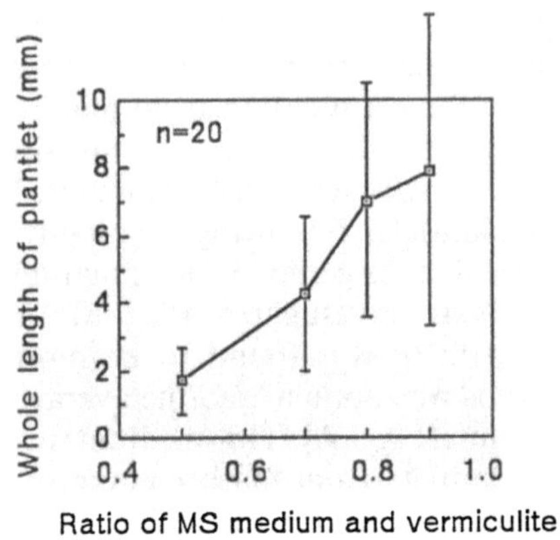

Fig. 8. Effects of mixing ratio of MS medium and vermiculite on germination of somatic embryos and growth of plantlets. (Kishira et al. 1992). Culture conditions: MS medium, GA and BAP (0.5 mg/l each), sucrose (3%), 16-h light, 16 °C, 1 month

of 3% sucrose was effective for germination of somatic embryos. (Fig. 10). Moreover, the most favorable medium was half-strength MS medium, as indicated in Fig. 11. Roots grew more rapidly compared to those grown on MS-vermiculite medium. On the other hand, the growth of shoots was inhibited, resulting in the formation of cotyledon-like plantlets.

Chromosome numbers in the root tip of plantlet regenerated from embryos were determined as 2n=48, which was the same as that of mother plants (Fangyuan et al. 1985).

4 Conclusion

The improved procedure described here can be used as a quick and simple method for germination of somatic embryos, thereby helping to produce a rapid, homogeneous strain of *Panax* spp. Furthermore, after 2 years of subculture, embryo propagation ability and the regeneration ratio of normal plantlets

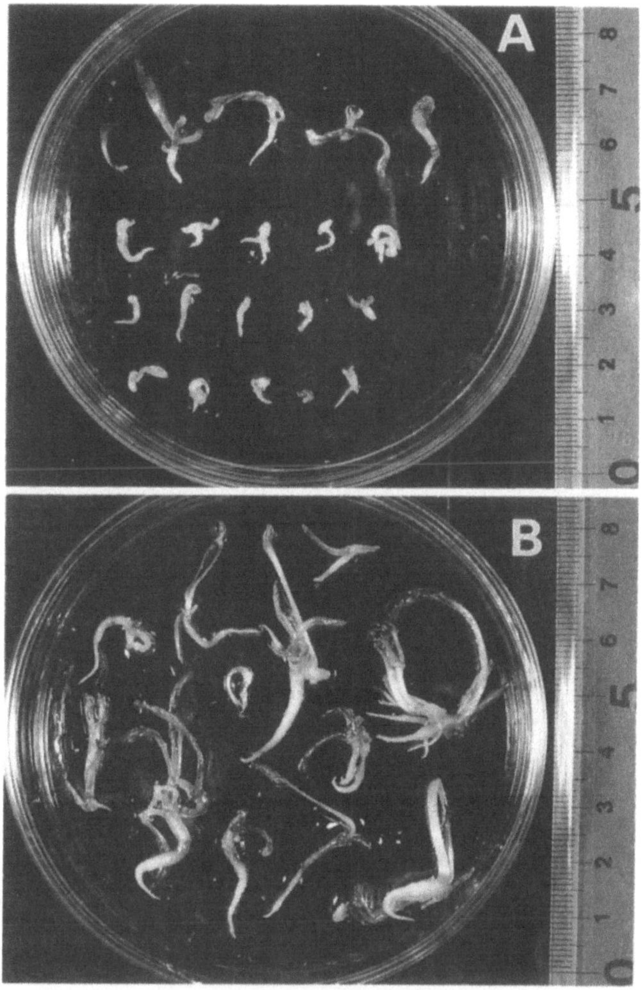

Fig. 9A,B. Germination of mature somatic embryos on half-strength MS-vermiculite medium containing 3% sucrose and GA-BAP (0.5 mg/l each). (Kishira et al. 1992) **A** After 1 month of culture. **B** After 2 month of culture

gradually decreased. The comparison of embryogenic ability of the in vitro plantlet organs and the intact plant organs showed that embryogenesis from the in vitro plant organs is faster than that of intact plants. Especially the callus induced from in vitro flower buds formed embryos at a high frequency after approximately 2 months.

Chromosome numbers of *P. ginseng* in the root tip of plantlets regenerated from embryos were determined to be the same as that of mother plants (Fangyuan et al. 1985). Thus, it is confirmed that propagation via embryogenesis has a place in the clonal propagation system.

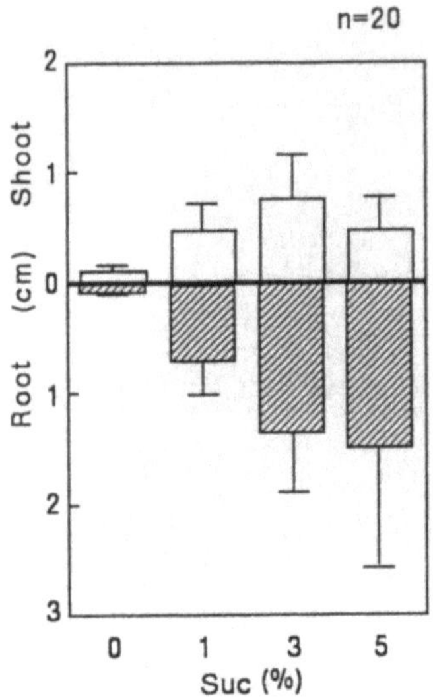

Fig. 10. Effect of sucrose concentration in liquid medium on germination of somatic embryos and growth of plantlets. (Kishira et al. 1992). Culture conditions: 1/2 MS medium, GA (0.5 mg/l), in the dark, 18 °C, 2 rpm, 1 month

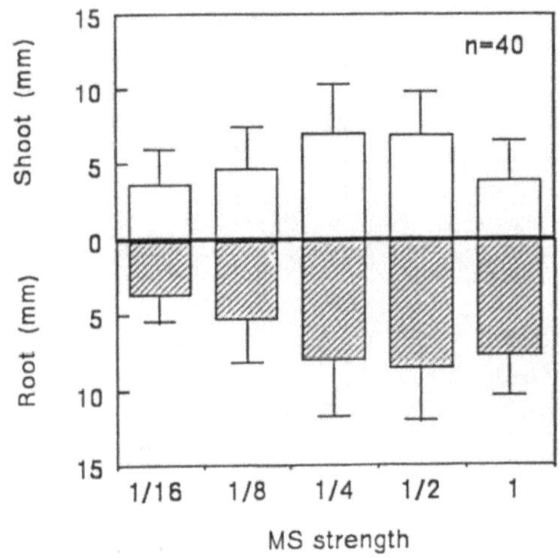

Fig. 11. Effect of MS strength in liquid medium on germination of somatic embryos and growth of plantlets. (Kishira et al. 1992). Culture conditions: GA (0.5 mg/l), sucrose (3%), in the dark, 18 °C, 2 rpm, 1 month

5 Protocol

1. Embryogenesis and regeneration of *Panax japonicus* from young flower buds.
   a) Culture young flower buds on MS medium supplemented with 1 mg/l 2,4-D in the dark at 25 °C for 3 months.
   b) Transfer embryos to 1/2 MS medium supplemented with GA and BAP (0.5 mg/l each) under 16-h light conditions at 25 °C for 1 month to generate shoots.
   c) Culture regenerated shoots on MS medium supplemented with 0.5 mg/l IBA at 25 °C for 1 month to produce plantlets.
2. Embryogenesis and direct plantlet formation of *Panax ginseng*.
   a) Culture flower buds on MS medium containing 1 mg/l 2,4-D in the dark at 25 °C for 3 months to produce embryos.

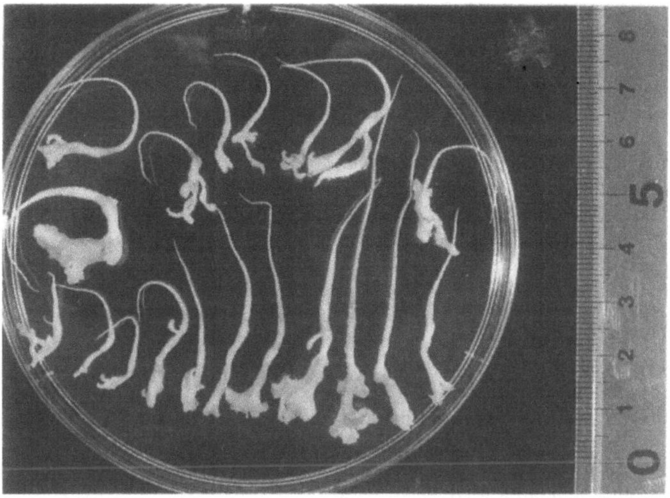

Fig. 12. Germination and growth of somatic embryos on 1/2 MS liquid medium containing 3% sucrose and 0.5 mg/l GA after 2-month culture. (Kishira et al. 1992)

b) Incubate mature embryos in MS liquid medium supplemented with 3% sucrose and 0.5 mg/l GA in the dark at 18 °C rotated at 2 rpm for 1 month to produce plantlets directly.
c) Cultivate regenerated plantlets on MS-vermiculite medium containing 3% sucrose under 16-h light at 18 °C for 3 months to produce plants having main roots.

References

Asaki I, Li I, Yoshikawa T, Hirotani M, Furuya T (1993) Embryoid formation by high temperature treatment for multiple shoots of *Panax ginseng*. Planta Med 59(4): 345–346

Butenko RG, Brushwitzky IV, Stepyan LI (1968) Organogenesis and somatic embryogenesis in the tissue culture of *Panax ginseng* C. A. Meyer. Bot Zh 7: 906–913

Chang WC, Hsing YI (1980a) In vitro flowering of embryoids derived from mature root callus of ginseng (*Panax ginseng*). Nature (Lond) 284: 341–342

Chang WC, Hsing YI (1980b) Plant regeneration through somatic embryogenesis in root-derived callus of ginseng (*Panax ginseng* C. A. Meyer). Theor Appl Genet 57: 133–135

Choi KT (1988) *Panax ginseng* C. A. Meyer: micropropagation and the in vitro production of saponins. In: Bajaj YPS (ed) Biotechnology in agriculture and forestry, vol 4. Medicinal and aromatic plants I. Springer, Berlin Heidelberg New York, pp 484–500

Choi KT, Kim MW, Shin HS (1982) Root and shoot formation from callus and leaflet cultures of ginseng (*Panax ginseng* C. A. Meyer). In: Fujiwara A (ed) Plant tissue culture 1982. Maruzen, Tokyo, pp 171–172

Fangyuan LI, Xian S, Xichen G (1985) The analysis of the chromosomal morphology and giemsa C-banding pattern in ginseng. Sci Agric Sin 5: 31–35

Fujioka N, Kurisu Y, Miyagawa H, Kohda H, Yamasaki K, Shoyama Y, Nishioka I (1986) Studies on the tissue culture of *Panax japonicus* (1). Multiplication by somatic embryogenesis of flower bud and rhizome. Shoyakugaku Zasshi 40: 152–158

Furuya T, Kojima H, Syono K, Ishii T (1970) Isolation of panaxatriol from *Panax ginseng* callus. Chem Pharm Bull 18: 2371–2372

Furuya T, Kojima H, Syono K, Ishii T, Uotani K, Nishio M (1973) Isolation of saponins and sapogenin from callus tissues of *Panax ginseng*. Chem Pharm Bull 21: 98–101

Furuya T, Yoshikawa T, Ishii T, Kajii K (1983a) Effects of auxins on growth and saponin production in callus cultures of *Panax ginseng*. Planta Med 47: 183–187

Furuya T, Yoshikawa T, Ishii T, Kajii K (1983b) Regulation of saponin production in callus cultures of *Panax ginseng*. Planta Med 47: 200–204

Furuya T, Yoshikawa T, Ushiyama K, Oda H (1986) Formation of plantlets from callus cultures of ginseng (*Panax ginseng*). Experientia 42: 193–194

Kim MW, Choi KT, Bae HW, Kang YH (1980) Effects of 2,4-D and kinetin on the production of saponin in ginseng tissue culture. Korean J Bot 23 (3,4): 91–98

Kishira H, Takada M, Shoyama Y (1992) Micropropagation of *Panax ginseng* C. A. Meyer by somatic embryos. Acta Hortic 319: 197–202

Kitto S, Janick J (1985) Production of synthetic seeds by encapsulating asexual embryos of carrot. J Am Soc Hortic Sci 119: 277–282

Murashige T, Skoog F (1962) A revised medium for rapid growth and bioassays with tobacco tissue cultures. Physiol Plant 15: 473–497

Shoyama Y, Nishioka I, Fujioka N, Kohda H, Yamasaki K (1987) Clonal multiplication of *Panax japonicus* by tissue culture. Shoyakugaku Zasshi 41: 333–337

Shoyama Y, Kamura K, Nishioka I (1988) Somatic embryogenesis and clonal multiplication of *Panax ginseng*. Planta Med 54: 155–156

Tanaka O (1989) Saponin composition of *Panax* species. In: Shibata S, Ohtsuka Y, Saito H (ed) Recent advances in ginseng studies. Hirokawa, Tokyo, pp 43–47

Yoshikawa T, Furuya T (1987) Saponin production by cultures of *Panax ginseng* transformed with *Agrobacterium rhizogenes*. Plant Cell Rep 6: 449–453

IV.7 Somatic Embryogenesis from Leaf Protoplasts of *Rauwolfia vomitoria* Afz

J. Trémouillaux-Guiller and J.C. Chénieux

1 General Account

The genus *Rauwolfia* belongs to the family Apocynaceae and includes a very large number of tree and shrub species. Due to the presence of numerous alkaloids, *Rauwolfia* is used extensively for therapeutic purposes, some species are also used as ornamental plants in tropical and subtropical gardens or as forest species (Rao 1956).

1.1 Distribution and Botanical Traits

Rauwolfias are pantropical woody plants from Asia, Central America, tropical Africa and Oceania, found between 30° N and S of the equator. The species grow in moist, warm regions up to an altitude of 1200 m.

Various species are trees or shrubs of varying sizes; the smallest (*R. nana* E.A. Bruce) measures only 15 cm high, whereas other species may reach 30 m (*R. praecox* K. Schum and *R. macrophylla* Stapf.) All species accumulate abundant quantities of latex, especially in the aerial parts (bark, branches, and leaves). The twigs have white lenticels, and the roots are occasionally somewhat tuber-like in appearance. The leaves are ovate, lanceolate, or acuminate at the tip and narrower at the base, and found in whorls of four (or three) or in opposed pairs. The leaf blade contains numerous lateral veins. Leaves of some species are petiolate (*R. vomitoria* Afz. and *R. heterophylla* Willd.). The flowers are small, grouped together in cymes, white, perfumed; the stamen-bearing corolla is initially tubular, then widens. Generally, the indehiscent fruit is a drupe, and contains one or two ovoid, albumen-containing seeds within a fleshy pericarp which becomes red on maturity (Servier 1957). The embryo is erect or curved, and typically dicotyledonous (Woodson et al. 1957).

The most frequently used species are *R. serpentina* and *R. vomitoria. Rauwolfia serpentina* Benth., ex Kurz is an evergreen undershrub 0.5 to 1 m high occurring naturally in the moist, warm regions of India, western Pakistan, Burma, Thailand, Malaysia, Java, and Indonesia. It grows in very shady places, such as deciduous forests, in association with bamboo. The large (50 cm) tap root

Plant Biotechnology Laboratory, Faculty of Pharmacy, University of Tours, 31 Avenue Monge, 37200 Tours Cédex, France

Biotechnology in Agriculture and Forestry, Vol. 31
Somatic Embryogenesis and Synthetic Seed II (ed. by Y.P.S. Bajaj)

grows in a zigzag pattern like a snake, hence the specific name serpentina, from the Latin "serpens", snake; in Sanskrit, the plant's name is sarpagandha: snake root. The slender, 30- to 90-cm stem bears leaves in whorls of three. The flowers are grouped into terminal or axillary cymes, and the fruit is a drupe (Bruneton 1987).

R. vomitoria Afzelius (=*Rauwolfia congolona* Wild), on the other hand, is an African species and has a variety of vernacular names: in the Téké tongue, it is *ompepe*; in N garé, *endouele*; in Bongili, *n douli*; in Koyo, *ondolé*; and in Yoombe, *n doudoudi* (Adjanohoun 1986). The species was first described in 1818 by the Swedish botanist Adam Afzelius (Takenaka 1970). It is a 3- to 6-m-high, tropical African tree with well-developed tap roots (Poisson 1959). The leaves are whorled in fours (or fives), and largely lanceolate or acuminate. The flowers are small, white, perfumed, and grouped together into dense cymes. The fruits are spherical, made up of one or two free carpels, and the berries are orangish or red on maturity (Adjanohoun 1986).

1.2 Medicinal Value and Alkaloid Content

Rauwolfia serpentina, or the Indian sarpagandha, has been used for 3000 years as an antidysenteric, febrifuge, sedative, or snake-bite antidote. The species has been listed in the Indian Pharmacopoeia since 1946 (Paris and Moyse 1971). Chemical isolation of reserpine from roots was performed by Müller et al. (1952). Derived from yohimbane, this alkaloid is antihypertensive and neuroleptic. Between 1931 and 1954, some 17 alkaloids were isolated from *R. serpentina* (Poisson 1959). At this time *R. serpentina* was adopted by Western therapeutic medicine; the WHO classifies reserpine as one of the ten substances of plant origin most commonly used in medicine.

The use of *Rauwolfia vomitoria* in indigenous African medicine has been reported by a number of authors (Takenaka 1970). It is used for mental disorders and infantile convulsions, at higher doses, for dysentery, jaundice, and leprosy (Poisson 1959). Root extracts are used against venereal diseases, and those of the whole plant are considered powerful diuretics and purgatives (Iwu 1980). During the 1970s, chemical research into *Rauwolfia*, especially the African species, resulted in a number of alkaloids being extracted and characterized: from the root of *R. caffra* (Habib and Court 1973); from the root and stem of *R. suavolens* (Majumdar et al. 1973); from the root of *R. macrophylla* (Timmins and Court 1974a); from the root of *R. obscura* (Timmins and Court 1974b, 1976); from the root (Iwu and Court 1978a), stem (Iwu and Court 1978b), and leaves (Iwu and Court 1978c) of *R. cumminsii*; from the root (Iwu and Court 1977), stem (Sabri and Court 1978), and fruit (Iwu 1980) of *R. vomitoria*; and from the stems of *R. monbasiana* (Iwu and Court 1979). Court (1983) has also studied the distribution of indolic alkaloids in the leaves, stems, and roots of ten African *Rauwolfia* species (Fig. 1).

In tropical America, decoctions of *R. tetraphylla* and *R. nitida* Jacquin leaves and bark are used in indigenous medicine against intestinal disorders and syphilis (Rao 1956).

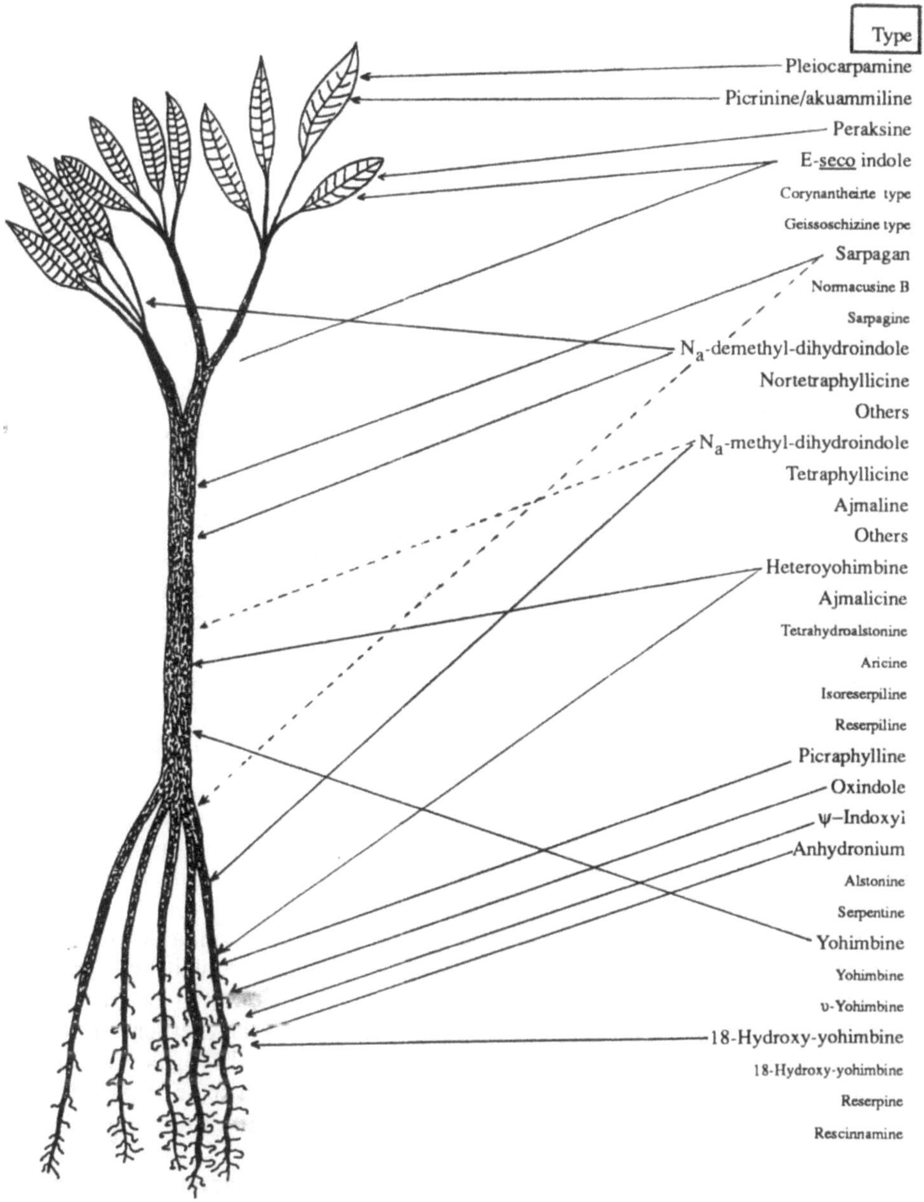

Fig. 1. The distribution of indole alkaloids in plants of the genus *Rauwolfia* in Court (1983)

1.3 Brief Review of Work of Other Authors

The in vitro study of alkaloid production in cell cultures of *R. serpentina* began back in 1964 to find ways of optimizing it. The establishment of such cultures has been reported by numerous workers (Babcock and Carew 1962; Mitra and Kaul

1964; Stöckigt et al. 1981; Kunakh and Alkhimova 1989). Ohta and Yatazawa (1979), for example, started cultures of *R. serpentina* from segments of stem and root and, in the subsequent lines, observed variations in the ajmaline content according to the concentration of the plant growth factors, 2,4-D (2,4-dichlorophenoxyacetic acid) and K (kinetin). Shoot cultures were also developed from *R. serpentina* axillary meristem, and these produced higher levels of major indolic alkaloids than those found in the stem in vivo (Heble et al. 1983). The alkaloid profile of *R. serpentina* shoot cultures demonstrates the presence of yohimbine, ajmaline, and ajmalidine (Roja et al. 1985); these cultures may therefore be used as an efficient source of indolic alkaloids (Heble 1985). Yamamoto and Yamada (1986) selected for optimized reserpine production, under certain stress conditions, in a stem-explant-derived cell culture of *R. serpentina*. Ajmaline production has also been investigated by radio immunoassay (RIA) in cell cultures of *R. vomitoria* and *R. serpentina* (Arens et al. 1987). Kunakh and Alkhimova (1989) have shown that, following mutagenesis, strains derived from stem explants had twice the amount of ajmaline than that accumulated in the whole plant.

Cell suspensions of *R. serpentina* are an excellent source of enzymes involved in the biosynthesis of indolic alkaloids (Schübel et al. 1989). Polyneuridine aldehyde esterase, for example, a key enzyme in the biosynthetic chain leading to sarpagine and ajmaline, has been isolated and purified by Pfitzner and Stöckigt (1983). Similarly, strictosidine synthase, which catalyzes the condensation of secologanine and tryptamine into a glucoalkaloid, H-3 alpha (S) strictosidine, has been purified and characterized by Hampp and Zenk (1988). Moreover, optimized production of raucaffricine, another gluco-alkaloid, has been achieved in cell suspensions of *R. serpentina* (Schübel et al. 1989).

1.4 Importance of Somatic Embryogenesis

Among the in vitro techniques, somatic embryogenesis is a valuable alternative to sexual reproduction, especially for trees. It enables "recalcitrant" species, in which in vitro vegetative multiplication proves difficult, to be improved. Somatic embryos may be obtained from differential explants whether or not they have gone through the callus stage. As a method, embryogenesis can be automated, thereby enabling massive quantities of seedlings to be obtained. The encapsulation of somatic embryos has also been proposed for obtaining artificial seeds.

Direct experimental embryogenesis is a rare phenomenon (Stolarz et al. 1991); few authors have observed this occurring from somatic tissue (Stolarz et al. 1991), single cells (Conger et al. 1983), or protoplasts. The latter are a substantial source of genetic diversity and somaclonal variation (Bajaj et al. 1988). Plants have been regenerated from protoplasts in a large numbers of species, whereas regeneration of plantlets from protoplasts by means of direct somatic embryogenesis has been obtained in only a very small number.

Somatic embryogenesis may also prove to be of value for the *Rauwolfia* genus, since germination of seeds can be as low as 10% in *R. serpentina,*

R. vomitoria, R. tetraphylla, and *R. caffra.* This low level of germination may be due to either the stony endocarp in the drupe or, what is more likely, to the lack of embryos, a possible result of parthenocarpy or deferred somatoplastic sterility (Woodson et al. 1957).

2 Somatic Embryogenesis in *Rauwolfia vomitoria*

Protoplasts were isolated from shoot cultures to regenerate plantlets and compare their alkaloid contents with those of the parent line. This strategy has already been used to establish cell lines from leaf protoplasts of *Choisya ternata* (a woody plant of the Rutaceae family) (Trémouillaux-Guiller 1988); Trémouillaux-Guiller et al. 1988; Crèche et al. 1993).

2.1 Leaf-Protoplast Cloning

The following cloning was carried out with *R. vomitoria* shoot-culture protoplasts (Trémouillaux-Guiller and Chénieux 1991).

In the *first set of experiments,* protoplasts were isolated from very small leaves of axenic shoots which had developed from a callus, after transferring it from a 2,4-D supplemented B5 medium to 2,4-D free and BA-supplemented MT medium. The shoots were propagated by routine periodic subcultures, and had very small leaves (Fig 2A).

Fig. 2A,B. Shoots used for isolating protoplasts. **A** shoots developed on root explant-derived callus; **B** protoplast-derived shoot culture. (Trémouillaux-Guiller and Chénieux 1991)

In the *second set of experiments*, protoplasts were produced from leaves of axenic shoots which were regenerated from protoplast-derived callus obtained during the first set of experiments; the latter shoots, bearing well-developed leaves (Fig. 2B) were propagated on the hormone-free and low-glucose content medium G/4. Regardless of the medium used, protoplasts were isolated with good reproducibility without having to strip the lower epidermis of the leaves.

2.2 Direct Embryogenesis

Protoplasts were cultured in liquid media (Fig. 3A) and similar results were obtained in the two sets of experiments. During the first hours of culture, the protoplasts gradually increased in volume. New cell walls, indicated by changes in shape of the cells, were regenerated after 24–48 h (Fig. 3B). About 25% of the protoplasts divided within 5 days (Fig. 3 B,C). Some cells showed unsymmetrical divisions by the daughter cells (Fig. 3C). After 4–6 weeks, two types of cells appeared in the cultures: one type was present in colonies of 10–15 slightly chlorophyllous cells, the second type was present as isolated isodiametric cells with a dense, light yellow or pale green cytoplasm, a characteristic which is usually found in embryogenic cells (Song et al. 1990; Arya et al. 1991). During the first experiments, we observed a third cellular type formed from elongated cells

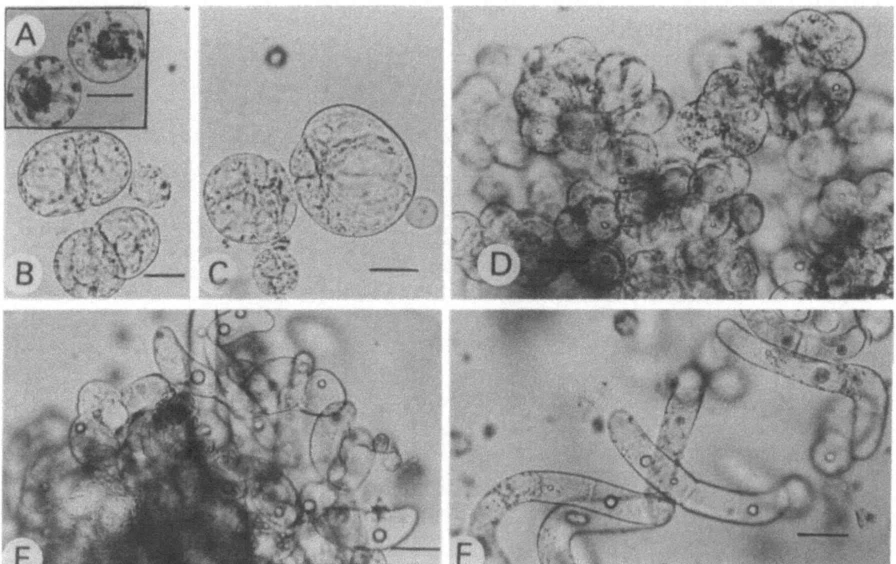

Fig. 3A–F. Culture of protoplasts isolated from *Rauwolfia vomitoria* shoots. **A** (*inset*) Freshly isolated protoplasts; **B** cell divisions after dedifferentiation of leaf protoplasts; **C** unsymmetrical divisions by daughter cells; **D** a protoplast-derived callus formed after 2 months; **E** an embryogenic colony obtained in the first set of experiments after 30–45 days; **F** elongated cells with a structure resembling suspensor; *bar* = 30 μM

(Fig. 3E,F) as suspensor-like cells obtained from embryos of *Picea glauca* (Lu and Thorpe 1987) and *Picea abies* (Boulay et al. 1988; Mo and Von Arnold 1991), but these long cells of *Rauwolfia* were often free in liquid medium.

The cell colonies progressively increased in size, forming green microcalli (Fig. 3D), whereas the isolated cells gave rise to proembryos together with a few elongated cells (suspensor-like structure, Fig. 4A,B). The proembryo is defined as a group of densely compacted cells derived from a single protoplast (Song et al. 1990). The proembyos turned into globular embryos (Fig. 4C), with a central core of tightly packed chlorophyllous cells confined within an epidermic structure, some of which were slightly elongated, similar to the "oblong embryos" (Fig. 4D) described by Schiavone and Cooke (1985). The somatic embryos did not develop further in liquid media and progressively turned brown, however, they were viable for several months, without apparent necrosis.

About 80% embryos and 20% calli were obtained in 3-month-old cultures. However, in a few experiments in which the initial culture medium was not diluted, only a population of calli was obtained. It is known that the relative percentage of embryos and calli obtained from protoplasts may be affected by the composition and dilution of the culture medium (Kobayashi et al. 1985; Miura and Tabata 1986; Song et al. 1990) or by a low and high cell density (Moyne et al. 1988); thus, the embryogenesis of protoplasts is inhibited by a high cell density

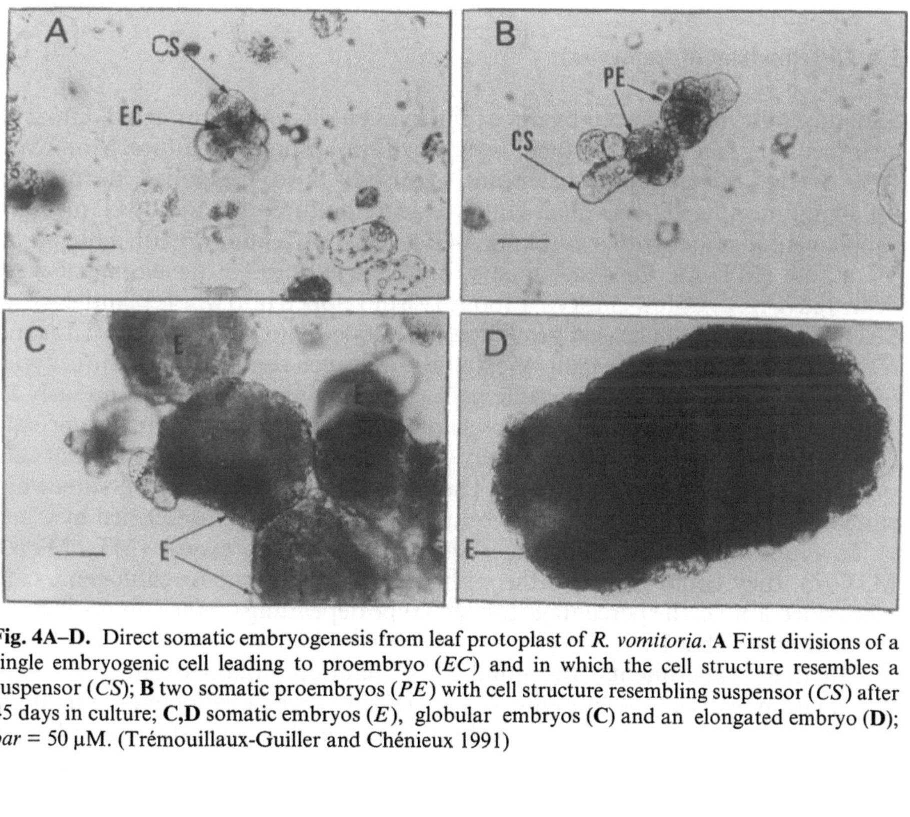

Fig. 4A–D. Direct somatic embryogenesis from leaf protoplast of *R. vomitoria*. **A** First divisions of a single embryogenic cell leading to proembryo (*EC*) and in which the cell structure resembles a suspensor (*CS*); **B** two somatic proembryos (*PE*) with cell structure resembling suspensor (*CS*) after 45 days in culture; **C,D** somatic embryos (*E*), globular embryos (**C**) and an elongated embryo (**D**); *bar* = 50 μM. (Trémouillaux-Guiller and Chénieux 1991)

(Kobayashi et al. 1985; Miura and Tabata 1986). Success of direct embryogenesis is apparently dependent on genotype (Song et al. 1990) and donor tissue (Dijak and Brown 1987). Direct embryogenesis occurred most readily by the formation of small cell colonies (Lu et al. 1982). Dijak and Brown (1987) noted that the embryos developed from a few cells surrounded by aggregates of dying cells. In our study the embryos developed directly from single protoplasts as described by Song et al. (1990).

2.3 Shoot Induction

Microcalli were transferred onto agar medium MT4. They were cultured for 1 month before being transferred again onto MTBA medium. Shoot primordia arose on the calli within 3–5 weeks. Shoots over 1 cm (Fig. 5F) were excised and individually transferred to several media used as rooting inducers by several authors (Kaul 1987; Diaz et al. 1988; Agrawal et al. 1989). However, root formation was never observed. As frequently observed with woody plants, the rooting process in *R. vomitoria* cultures remains obviously a complicated problem (Ochatt and Caso 1986; Rastushnyak et al. 1989). On the other hand, shoots cultured on G/4 medium rapidly elongated, branched out, and developed normal leaves.

2.4 Development of Embryos

Somatic embryos produced by direct embryogenesis were transferred onto MT agar medium. About 10% of them swelled within 1 month of culture. Moreover, 10% of the remaining "recalcitrant" embryos also developed during the following months. The swollen embryos were cultured on solid MT medium supplemented either with a cytokinin (4.5 μM BA or zeatin) or with a high level of sucrose (MT-60). On these media, different patterns of development were observed (Fig. 6). A few embryos (about 1%) developed into whole plantlets (Fig. 5E) but some of them ceased growth at the cotyledonary stage (Fig. 5C,D) and recallused into caulogenic calli. Most of the embryos recallused into three types of calli: (1) yellowish embryogenic calli (Fig. 5A) giving rise continuously to somatic embryos produced by indirect embryogenesis; (2) white rhizogenic calli (Fig. 5B) leading to roots which rapidly ceased development; and (3) nodular calli presenting a chlorophyllous core. These nodules appeared to be somewhat similar to the small sugarcane calli bearing green meristems described by Chen et al. (1988). Upon transferring them onto different media (MT, MT60, MT60Z), they could revert to the embryogenic rhizogenic or caulogenic calli mentioned above, the percentage for each type depending on the composition of the medium (Table 1).

In these experiments, the number of plantlets obtained either through germination of direct somatic embryos or via embryogenic callogenesis remained very low (Table 2). We attempted to enhance the percentage of plantlet

Fig. 5A–F. Shoot regeneration and embryo development from *R. vomitoria* leaf protoplasts. **A** embryogenic calli; **B** rhizogenic calli (*R* root); **C,D** cotyledonary stages; **E** whole plantlet; **F**. shoots developed on microcalli. *Bar* = 1 mm (**C**); 1.5 mm (**D,E**); 2 mm (**A**); 5 mm (**B, F**). (Trémouillaux-Guiller and Chénieux 1991)

formation by increasing the osmolarity of the medium during one passage, since some work has indicated that desiccation may play an important role in the maturation process (Finkelstein and Crouch 1986; Lu and Thorpe 1987). As shown in Table 2, this treatment improved the production of whole plantlets.

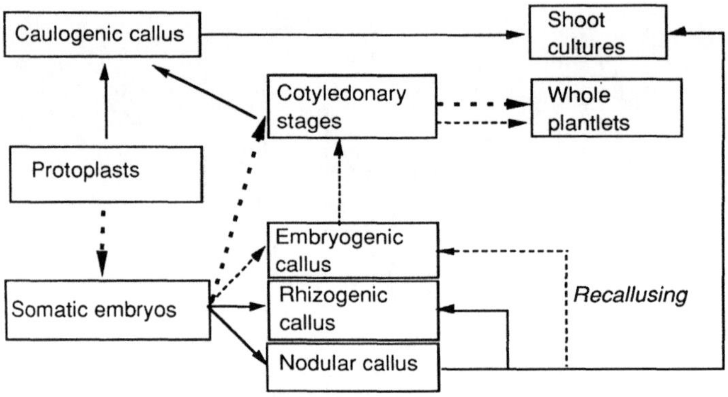

Fig. 6 Morphogenetic response in *R. vomitoria* protoplast cultures. ---→ Direct embryogenesis pathway; – – –→ indirect embryogenesis pathway. (Trémouillaux-Guiller and Chénieux 1991)

Table 1. Morphogenetic responses obtained with nodules grown on different media. (Trémouillaux-Guiller and Chénieux 1991)

Response (%)	Media		
	MT	MT60	MT60Z
Embryogenic calli	10	12	22
Rhizogenic calli	2.5	30	25
Nodular calli	34	24.5	24
Cotyledonary stage	6	4	11
Whole plantlets	0	0.9	3
Shoots	10	13	15
No response	37.5	15.6	0

Total: 875 nodules. Response after 4 weeks in culture; see Table 3 for the formulation of the media

Table 2. Whole plantlets formation from somatic embryos grown on different media. (Trémouillaux-Guiller and Chénieux 1991)

	Media				
	MT	MTBA	MT60	MT60Z	MT60Z[a] MT120Z
Responses (%)	0.7	1.7	1.2	1.7	16.4

Total : 1045 somatic embryos; response after three months in culture
[a] Three transfers to MT60Z, then one transfer to MT 120Z; see Table 3 for the formulation of media

3 Summary

Rauwolfia vomitoria mesophyll protoplasts were isolated from axenic shoot cultures and cultured (10^5–10^6 protoplasts/ml) in Murashige and Tucker (MT) liquid medium containing growth regulators. Within 6–8 weeks, a mixed population of calli and proembryos was obtained and transferred to solid media. Calli produced shoots; however, rooting did not occur. Somatic embryos achieved different patterns of development. In particular, whole plantlets were obtained either directly through germination of primary embryos or via embryogenic calli.

4 Protocol for the Induction of Somatic Embryos

Plant Material

In September 1985, we prepared a callus culture of *Rauwolfia vomitoria* from a seedling root explant placed on B_5 (Gamborg et al. 1976) agar medium supplemented with 4.5 μM 2,4-D (2,4-dichlorophenoxyacetic acid), 0.28 μM K (kinetin), and 0.17 M glucose (pH 5.7). In November 1986, pieces of callus were transferred to MT (Murashige and Tucker 1969) agar medium supplemented with 4.4 μM BA (N^6 benzyladenine) and 0.17 M glucose. Multiple shoots developing on the tissues were subcultured at 4-week intervals and maintained on the latter medium under continuous white light (Philips TL/40 W34 fluorescent tubes, 1000 lx) at 24 ± 1 °C.

Protoplast Isolation

Two sets of experiments were performed. Protoplasts were prepared either from the tiny leaves (1–3 mm length) on the shoots developed on the above-mentioned callus culture (first set of experiments) or from the more developed leaves (10 mm length) excised from a protoplast-derived shoot culture (second set of experiments). In this latter case, the leaves were cut into strips, 0.5–0.8 mm wide. Approximately 0.5 g of leaf material was incubated in 4 ml of enzyme solution (in 55 × 15mm Petri dishes) for 14–16h, in the dark. The enzyme mixture consisted of 2.5 g/l cellulase Onozuka R10, 1.25 g/l macerozyme R 10 (Yakult Honsha Co., Ltd, Japan), 0.25 g/l pectolyase (Sigma-Chimie, France) in 34 mM $CaCl_2 \cdot 2H_2O$ and 0.6 M mannitol (pH 5.5). After digestion, protoplasts were filtered through a 100-mm sieve and collected by centrifugation (80 × g) for 3 min. Supernatants were discarded and pellets resuspended in 0.7 M mannitol. After centrifugation (80 × g, 3 min), protoplasts were collected in the pellet and washed again twice in 0.7 M mannitol. Following washings, protoplasts were resuspended in the liquid MT1 medium (see Table 3 for the formulation of the media). Viability of protoplasts was checked using the induction of fluorescence in living cells with a 1 g/l fluorescein diacetate solution in acetone for 1–5 min: about 20% of the preparation consisted of broken protoplasts and cellular debris, but the survival percentage of the living protoplasts reached 95% after 2 days of culture.

Protoplast Culture

Protoplasts were cultured at concentrations of either 10^5 (first set of experiments) or 10^6 (second set of experiments) per milliliter, in 35-mm Corning dishes containing 0.5 ml MT1 medium. They were initially kept in the dark, at 25 ± 1 °C. After 2 weeks (first experiments) or 1 week (second experiments), the cultures were diluted with 0.2 ml MT2 medium and placed under white light (Philips TDL 18 W/33 fluorescent tubes, 750 lx) with a 12-h photoperiod. Two weeks later, developing cultures

Table 3. Composition of media used in the present work. (Trémouillaux-Guiller and Chénieux 1991)

	MT1	MT2	MT3	MT4	MT	MTBA	MT60	MT60Z	G/4	MT120Z
(M)										
Glucose	0.25	0.25	0.16	0.16	0.16	0.16			0.04	
Mannitol	0.35	0.15	0.05							
Sucrose							0.17	0.17		0.34
(g/l)										
Agar				(9)[a]	9	9	9	9	9	9
(μM)										
2,4-D	4.5	4.5	2.3	2.3						
NAA	5.4	5.4	2.7	2.7						
BA	0.4	0.4	0.2	0.2		4.5				
K	0.5	0.5	0.25	0.25						
Z								4.5	4.5	

Basal medium = mineral salts and organic addition of MT medium
[a] MT4 was used either as liquid or agar medium

were diluted at weekly intervals with 0.2–0.4 ml medium. The first feeding was done with liquid MT2 medium, the second with MT3, and the third with MT4. The cultures were subdivided into two Corning dishes when the liquid culture medium exceeded 1 ml. Within 1 month after the third feeding, protoplast-derived cells gave rise to a mixed population of microcalli and proembryo. Microcalli (1–2 mm in diameter) and proembryos were plated onto an agar medium, MT4, and transferred after 4 weeks onto the various media to be investigated.

References

Adjanohoun EJ (1986) Contribution aux études ethnobotaniques et floristiques en République Populaise du Congo, ACCT, Paris

Agrawal S, Chandra N, Kothari SL (1989) Plant regeration in tissue cultures of pepper (*Capsicum annuum* L.cv. mathania). Plant Cell Tissue Organ Cult 16: 47–55

Arens H, Deus-Neumann D, Zenk MH (1987) Radio immunoassay for the quantitative determination of ajmaline. Planta Med 53 (2): 179–183

Arya S, Liu JR, Eriksson T (1991) Plant regeneration from protoplasts of *Panax ginseng* (C. V. Meyer) through somatic embryogenesis. Plant Cell Rep 10: 277–281

Babcock PA, Carew DP (1962) Tissue culture of the Apocynaceae I. Culture requirements and alkaloid analysis. Lloydia 25: 209–213

Bajaj YPS, Furmanowa M, Olszowska O (1988) Biotechnology of the micropropagation of medicinal and aromatic plants. In: Bajaj YPS (ed) Biotechnology in agriculture and forestry, vol 4. Medicinal and aromatic plants I. Springer, Berlin Heidelberg New York, pp 60–103

Boulay MP, Gupta PK, Krogstrup P, Durzan DJ (1988) Development of somatic embryos from cell suspension cultures of Norway spruce (*Picea abies* Karst). Plant Cell Rep 7: 134–137

Bruneton J (1987) Eléments de phytochimie et de pharmacognosie. Lavoisier, Paris

Chen WH, Davey MR, Power JB, Cocking EC (1988) Control and maintenance of plant regeneration in sugarcane callus culture. J Exp Bot 199: 251–261

Conger BV, Hanning GE, Gray DJ, McDaniel JK (1983) Direct embryogenesis from mesophyll cells of orchardgrass. Science 221: 850–851

Court WE (1983) Alkaloid distribution in some African *Rauwolfia* species. Planta Med 48: 228–233

Créche J, Trémouillaux-Guiller J, Gröger D, Chénieux JC, Rideau M (1993) *Choisya ternata* ("Mexican Orange"): culture in vitro and the production of alkaloids. In: Bajaj YPS (ed)

Biotechnology in agriculture and forestry, vol 21. Medicinal and aromatic plants V. Springer, Berlin Heidelberg New York, pp 107–121

Diaz I, Moreno R, Power JB (1988) Plant regeneration from protoplasts of *Capsicum annuum*. Plant Cell Rep 7: 210–212

Dijak M, Brown CW (1987) Patterns of direct and indirect embryogenesis from mesophyll protoplasts of *Medicage sativa*. Plant Cell Tissue Organ CULT 9: 121–130

Finkelstein RR, Crouch ML (1986) Rapeseed embryo development in culture or high osmoticum is similar to that in seeds. Plant Physiol 81: 907–912

Gamborg OL, Murashige T, Thorpe TA, Vasil IK (1976) Plant tissue culture media. In Vitro 12: 473–478

Habib MS, Court WE (1973) Minor alkaloid of *Rauwolfia caffra*. Phytochemistry 12: 1821

Hampp N, Zenk MH (1988) Homogeneous strictosidine synthase from cell suspension cultures of *Rauwolfia serpentina*. Phytochemistry 27: 3811–3815

Heble MR (1985) Multiple shoot cultures: a viable alternative in vitro system for the production of known and new biologically active plant constituents. In: Neumann KH, Barz W, Reinhard E (eds) Primary and secondary metabolism of plant cell cultures. Springer, Berlin Heidelberg New York, pp 282–289

Heble MR, Benjamin BD, Roja PC, Chadha MS (1983) Studies on shoot organ cultures of *Atropa belladonna* and *Rauwolfia serpentina*. In: Sen SK, Giles KL (eds) Plant cell culture in crop improvement. Plenum Press, New York, pp 57–63

Iwu M (1980) Alkaloids of *Rauwolfia vomitoria* fruits. Planta Med Suppl: 13–16

Iwu M, Court WE (1977) Root alkaloids of *Rauwolfia vomitoria* Afz. Planta Med 32: 88–89

Iwu M, Court WE (1978a) The alkaloids of *Rauwolfia cumminsii*. Planta Med 34: 390–396

Iwu M, Court WE (1978b) Alkaloids of *Rauwolfia cumminsii* stem. Phytochemistry 17: 1651–1654

Iwu M, Court WE (1978c) Leaf alkaloids of *Rauwolfia cumminsii* stapf. Planta Med 33: 360–364

Iwu M, Court WE (1979) Alkaloids of *Rauwolfia monbasiana* stem bark. Planta Med 36: 208–212

Kaul K (1987) Plant regeneration from cotyledon hypocotyl explants of *Pinus strobus* L. Plant Cell Rep 6: 5–7

Kobayashi S, Ikeda I, Uchimiya H (1985) Conditions for high frequency embryogenesis from orange (*Citrus sinensis* Osb.) protoplasts. Plant Cell Tissue Organ Cult 4: 249–259

Kunakh VA, Alkhimova EG (1989) *Rauwolfia serpentina*: in vitro culture and the production of ajmaline. In: Bajaj YPS (ed) Biotechnology in agriculture and forestry, vol 7. Medicinal and aromatic plants II. Springer, Berlin Heidelberg New York, pp 398–416

Lu DY, Thorpe TA (1987) Somatic embryogenesis and plantlet regeneration in cultured immature embryos of *Picea glauca*. J Plant Physiol 128: 297–302

Lu DY, Davey MR, Pental D, Cocking EC (1982) Forage legume protoplasts: somatic embryogenesis from protoplasts of seedling cotyledons and roots of *Medicago sativa*. In: Fujiwara A (ed) Plant tissue culture 1982. Maruzen, Tokyo, pp 597–598

Majumdar SP, Poisson J, Potier P (1973) Alcaloïdes de *Rauwolfia suaveolens*. Phytochemistry 12: 1167–1169

Mitra GC, Kaul KN (1964) In vitro culture of root and stem callus of *Rauwolfia serpentina* Benth. for reserpine. Indian J Exp Biol 2: 49–51

Miura Y, Tabata M (1986) Direct somatic embryogenesis from protoplasts of *Foeniculum vulgare*. Plant Cell Rep 5: 310–313

Mo LH, Von Arnold S (1991) Origin and development of embryogenic cultures from seedlings of norway spruce (*Picea abies*). J Plant Physiol 138: 223–230

Moyne AL, Thor V, Pelissier B, Bergounioux C, Freyssinet G, Gadal P (1988) Callus and embryoids from protoplasts of *Helianthus annuus*. Plant Cell Rep 7: 437–440

Müller JM, Schlittler E, Bein HJ (1952) Reserpin, der sedative Wirkstoff au *Rauwolfia serpentina* Benth. Experientia 8/9: 338

Murashige T, Tucker DPH (1969) Growth factor requirements of *Citrus* tissue cultures. In: Chapmann JV (ed) First Int Citrus Symp. Riverside, (A)pp 1155–1161

Ochatt SJ, Caso OH (1986) Shoot regeneration from leaf mesophyll protoplasts of wild pear (*Pyrus communis* var. pyraster L.). J Plant Physiol 122: 243–246

Ohta S, Yatazawa M (1979) Growth and alkaloid production in callus tissues of *Rauwolfia serpentina*. Agric Biol Chem 11: 2297–2303

Paris RR, Moyse H (1971) Précis de Matière Médicale. Masson, Paris

Pfitzner A, Stöckigt J (1983) Characterization of polyneuridine aldehyde esterase, a key enzyme in the biosynthesis of sarpagine/ajmaline type alkaloids. Planta Med 48: 221–227

Poisson J (1959) Recherches sur les alcaloïdes des racines du *Rauwolfia vomitoria* Afz (Apocynacées). Thèse Doct Etat ès Sci Nat, Paris

Rao AS (1956) A revision of *Rauwolfia* with particular reference to the American species. Ann Mo Bot Gard 43: 253–354

Rastushnyak YI, Piven NM, Rudas VA (1989) Protoplast culture and plant regeneration in *Lycium barbarum* L. Plant Cell Tissue Organ Culture 17: 183–190

Roja PC, Benjamin BD, Heble MR, Chadha MS (1985) Indole alkaloids from multiple shoot cultures of *Rauwolfia serpentina*. Planta Med 1: 73

Sabri NN, Court WE (1978) Stem alkaloids of *Rauwolfia vomitoria*. Phytochemistry 17: 2023–2026

Schiavone FM, Cooke TJ (1985) A geometric analysis of somatic embryo formation in carrot cell cultures. Can J Bot 63: 1573–1578

Schübel H, Ruyter CM, Stöckigt J (1989) Improved production of raucaffricine by cultivated *Rauwolfia* cells. Phytochemistry 28 (2): 491–494

Servier J (1957) Contribution à l'étude du *Rauwolfia serpentina* et aperçus sur quelques espèces voisines. Thèse Doct, Univ Lille

Song J, Sorensen EL, Liang GH (1990) Direct embryogenesis from mesophyll protoplasts in alfalfa (*Medicago sativa* L). Plant Cell Rep 9: 21–25

Stöckigt J, Pfitzner A, Firl J (1981) Indole alkaloïds from cell suspension cultures of *Rauwolfia serpentina* Benth Plant Cell Rep 1: 36–39

Stolarz A, Macewicz J, Lörz H (1991) Direct somatic embryogenesis and plant regeneration from leaf explants of *Nicotiana tabacum* L. J. Plant Physiol 137: 347–357

Takenaka Y (1970) Sur la Rauvanine, alcaloïde du *Rauwolfia vomitoria* Afz. (Apocynacées). Doct Etat ès Sci Pharm, Paris

Timmins P, Court WE (1974a) Alkaloids of *Rauwolfia macrophylla*. Planta Med 13: 281–282

Timmins P, Court WE (1974b) Root bark alkaloids of *Rauwolfia obscura*. Phytochemistry 3: 1997

Timmins P, Court WE (1976) Further alkaloids from the roots of *Rauwolfia obscura*. Planta Med 29: 283–288

Trémouillaux-Guiller J (1988) Etude comparative des méthodologies de sélection de cultures cellulaires végétales à haute capacité d'accumulation. Application à des souches et lignées clonales biosynthétisant des alcaloïdes dihydro-furoquinoléïques. Thèse Doct, Univ Tours

Trémouillaux-Guiller J, Chénieux JC (1991) Somatic embryogenesis from leaf protoplasts of *Rauwolfia vomitoria* shoot cultures. Plant Cell Rep 10: 102–105

Trémouillaux-Guiller J, Kodja H, Andreu F, Crèche J, Chénieux JC, Rideau M (1988) Variability in tissue cultures of *Choisya ternata* III. Comparing alkaloid production in cell lines obtained by various strategies. Plant Cell Rep 7: 456–458

Woodson RE Jr, Youngken HW, Schlittler E, Schneider JA (1957) *Rauwolfia*: botany, pharmacognosy, chemistry and pharmacology. Little, Brown and Co, Boston

Yamamoto O, Yamada Y (1986) Production of reserpine and its optimization in cultured *Rauwolfia serpentina* Benth. cells. Plant Cell Rep 5: 50–53

IV.8 Somatic Embryogenesis in *Arabidopsis thaliana* L

B. HUANG[1] and M.M. YEOMAN[2]

1 Introduction

Arabidopsis thaliana is a small cruciferous plant which is widely distributed in many parts of the world (Kirchheim and Kranz 1981). This species has been used as an important experimental system in studies of classical genetics for nearly half a century (Rédei 1975), and recently also in molecular genetics (for reviews, see Meyerowitz 1987). *Arabidopsis* possesses a number of unique features which render it an extremely suitable object for experiments. It has an extraordinarily short life cycle of 5–6 weeks (seed–seed). The plant is small in size and easy to grow both in soil and in a defined aseptic medium. At maturity, each individual plant may produce over 10 000 seeds, and therfore it is possible to handle large populations of plants in the laboratory at low cost. The haploid genome size is one of the smallest in higher plants, comprising only 7×10^4 kb of DNA (Pruitt and Meyerowitz 1986). The small genome allows the manipulation of DNA such as the construction and screening of genomic libraries, gene cloning by chromosome walking, etc. with minimum effort (Meyerowitz 1987). Moreover, large numbers of different ecological variants (Röbbelen 1965) and various well-characterized morphological, physiological, and biochemical mutants (McKelvie 1962; Koornneef et al. 1983) are available for genetic studies. Based on extensive mutation studies, a rather detailed linkage map of *Arabidopsis thaliana* has been established (Koornneef 1987). More recently, an RFLP map containing nearly 100 restriction fragment length polymorphic markers has also been constructed (Chang et al. 1988). Molecular cloning of *Arabidopsis* genes has also been attempted, and up to 1989, more than 20 genes had been cloned and characterized (Meyerowitz 1989). These studies provide basic knowledge of the organization and function of the genetic background in this species.

The successful tissue culture of *A. thaliana* was reported as early as 1965 (Yokoyama and Jones 1965). It has been shown that the formation of roots and shoots, as well as the regeneration of plants from both callus and suspension cultures, is possible (Negrutiu et al. 1978). Protoplast culture and plant regeneration have also been reported (Karesch et al. 1991). Protoplast fusion between *Arabidopsis* and *Brassica*, giving rise to the somatic hybrid *Arabido-*

[1] Institute of Genetics and Cytology, Northeast Normal University, Changchun 130024, China
[2] Institute of Cell and Molecular Biology, The University of Edinburgh, The King's Buildings, Edinburgh EH9 3JH, UK

Biotechnology in Agriculture and Forestry, Vol. 31
Somatic Embryogenesis and Synthetic Seed II (ed. by Y.P.S. Bajaj)

brassica, was reported by Gleba and Hoffmann (1980). Recently, tissues of *A. thaliana* have been used for genetic transformation through *Agrobacterium* infection (Valvekens et al. 1988). Various other aspects of tissue culture and genetic transformation in *A. thaliana* have been recently reviewed (see Feldmann 1992; Bilang and Potrykus 1993).

In this chapter our work on somatic embryogenesis in *A. thaliana* with emphasis on the protocol together with an account of the structural and biochemical changes which accompany the induction and development of somatic embryos is detailed.

2 Somatic Embryogenesis in *Arabidopsis thaliana* L.

2.1 Pathways of Somatic Embryogenesis

The callus initiated from seedlings of *A. thaliana* on B_5 (Gamborg et al. 1968) agar medium containing 0.5 mg/l 2,4-dichlorophenoxyacetic acid (2,4-D) and 0.05 mg/l kinetin was composed of numerous nodules (Fig. 1A). When the nodular callus of seedling origin was transferred onto a medium containing 0.1 mg/l indole-3-acetic acid (IAA) and 1–2 mg/l kinetin, many green nodules developed which differentiated to form shoots at a later stage on this high kinetin medium (Huang and Yeoman 1984). If the green nodules were placed on an auxin-free medium with a low level of kinetin (0.1 mg/l), some of them differentiated into clumps of embryo-like masses (referred to as embryoidal masses) composed of numerous, tightly packed embryoidal units, most of which showed an abnormal morphology with cotyledon numbers ranging from one to five (Fig. 1B). At a later stage these embryoids started to "germinate" to produce young leaves (Fig.1C). Transfer of the embryoidal masses to a fresh medium without any added growth regulators ensured the "germination" and development of embryoids into plantlets (Fig. 1D). However, these young plantlets often failed to form roots. An important feature arising from this experiment was that the embryoidal masses appeared to be initiated as secondary adventitious embryos from the previously formed leafy structures.

Besides the embryoidal masses, discrete individual somatic embryos may be induced in callus cultures established from explants of zygotic embryos. In this culture system, zygotic embryos were dissected from the seeds of *A. thaliana*, and were placed on either B_5 or SH (Schenk and Hildebrandt 1972) medium supplemented with 0.5 mg/l 2,4-D and 0.05 mg/l kinetin. A nodular callus was formed within 2 weeks. If this callus was kept in the same medium for 4 weeks, or alternatively, transferred to a medium without 2,4-D, individual somatic embryos were initiated on the surface of the callus, which became clearly visible when the early cotyledonary stage had been reached. The embryos induced in this way appeared more or less abnormal in morphology, often possessing swollen hypocotyls (Fig. 2A) and sometimes with only one cotyledon (Fig. 2B), or fasciculr cotyledons (Fig. 2C). When the embryos were detached from the callus

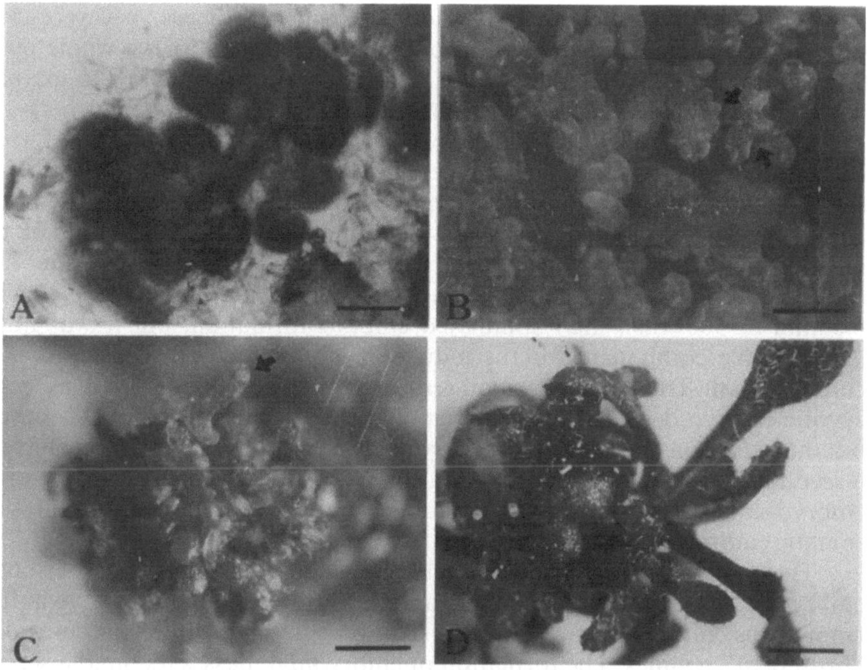

Fig. 1A–D. Plant regeneration via embryoidal masses induced from seedling-derived callus. **A** Nodular callus cleared with lactophenol and stained with cotton blue. **B** Tightly packed, fascicular embryoidal masses formed on callus following a series of subcultures on media of different growth regulator compositions (see text for details). Note the fused cotyledons (*arrows*). **C** Young leaves "germinating" from embryoidal masses. **D** Young plantlets developed from embryoidal masses. *Bar* in **A** = 1 mm; *bars* in **B–D** = 2 mm

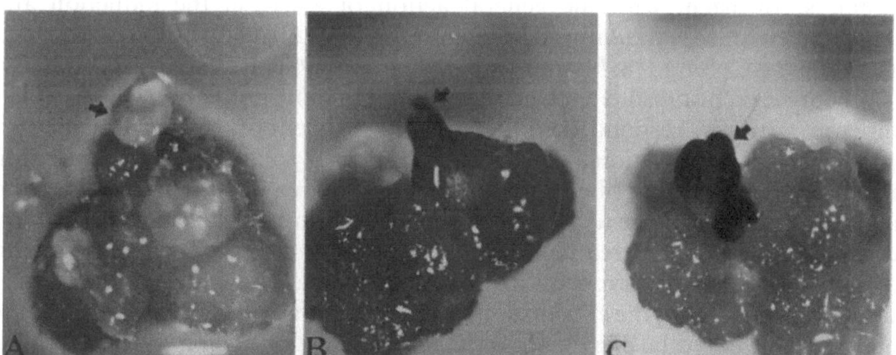

Fig. 2A–C. Individual somatic embryos initiated from zygotic embryo-derived callus. Note the abnormality of the embryos with swollen hypocotyl (**A**, *arrow*), a single cotyledon (**B**, *arrow*) and fascicular cotyledons (**C**, *arrow*)

and placed on a basal medium free of added growth substances, they were able to develop into normal seedlings with roots, and subsequently into whole plants. Some of the regenerated plants were transplanted into soil, grew to maturity, flowered, and set seed under greenhouse conditions.

The auxin 2,4-D when added to the culture medium was critical to the induction of somatic embryos. The highest induction efficiency of 10.7% (percentage of calli which produced embryos) was achieved with a callus initiated and kept over 4 weeks on a medium containing 0.5 mg/l 2,4-D and 0.05 mg/l kinetin, while the embryogenic potency of the callus was lost if the 2,4-D concentration was increased to 4 mg/l. Addition of yeast extract (YE) to culture medium at 500–1000 mg/l stimulated embryo formation, but it was not an indispensible component for the induction of somatic embryogenesis in this culture system. The extent of somatic embryo formation was on average 7% if the zygotic embryo-derived callus was kept in primary culture without subculture, but the induction efficiency was greatly improved if the callus was transferred to a secondary culture devoid of 2,4-D (2,4-D^-). The high reproducibility of this embryogenic pathway ensured a regular supply of embryogenic cultures and somatic embryos for further studies.

The data presented above clearly show that the two forms of somatic embryos, i.e., the embryoidal masses and the individual embryos, were induced from callus cultures via two different pathways involving different culture and "hormonal" regimes. The formation of embryoidal masses required a series of subcultures involving the transfer of seedling-derived callus from the 2,4-D-containing (2,4-D^+) primary medium to a high cytokinin medium (1 mg/l kinetin, 0.1 mg/l IAA), followed by another transfer to an auxin-free medium. In contrast, the formation of individual somatic embryos from zygotic embryo-derived callus occurred either when it was kept in the 2,4-D^+ primary culture for a prolonged period of time, or when it was transferred to a 2,4-D^- secondary culture. It seems clear that in both pathways, the inclusion of 2,4-D in primary culture was essential to the production of potentially embryogenic callus, and the absence of 2,4-D in secondary culture, or the exhaustion of 2,4-D from the primary culture, was necessary for the subsquent development of embryos. This result is consistent with the general action of 2,4-D in the induction and development of somatic embryos as demonstrated with many other species (e.g., Al-Abta and Collin 1978; Nomura and Komanine 1986; El Hadrami et al. 1991).

Moreover, our results showed that, although morphologically alike, calli of different origins responded differently to exogenous growth regulators, particularly to the deletion of 2,4-D from the primary culture in which initiation had taken place. As described above, the transfer of seedling-derived callus from a 2,4-D^+ primary culture to a 2,4-D^- secondary culture resulted in the formation of roots, while the same treatment with zygotic embryo-derived callus gave rise to a compact green embryogenic callus from which individual embryos may develop. Apparently, the explant source was an important factor influencing the embryogenic potential of callus in *A. thaliana,* the zygotic embryo-derived callus being more embryogenic than that of seedling origin. This fact implies that the intrinsic physiological state of callus, which may have been determined at the point when it was induced, plays a key role in deciding the manner in which it responds to external stimuli, e.g., the growth regulators.

2.2 Histological and Structural Aspects

The structure and patterns of callus proliferation and various forms of in vitro morphogenesis in *A. thaliana* were studied. The cells that constitute the callus were heterogeneous in both their morphology and size. The two distinct cell types in a callus were typically small, actively dividing cells with dense cytoplasm, and large cells with an extensive central vacuole and a thin layer of peripheral cytoplasm. During callus proliferation, the small cells divide to produce nodules forming a typical nodular callus. Figure 3A shows a semi-thin section across a nodule of the seedling-derived callus. A conspicuous feature revealed by this section is the presence of groups of small, densely cytoplasmic cells situated at the periphery of the nodule. These cell groups were capable of continued division and growth or of further differentiation when placed under appropriate conditions. They have been referred to as meristematic centers (MCs), and may act like organized and relatively undifferentiated meristems as in the whole plant. In the presence of 2,4-D, cells in MCs undergo active division leading to the formation of new nodules. The MCs in nodular callus were organized into root primordia

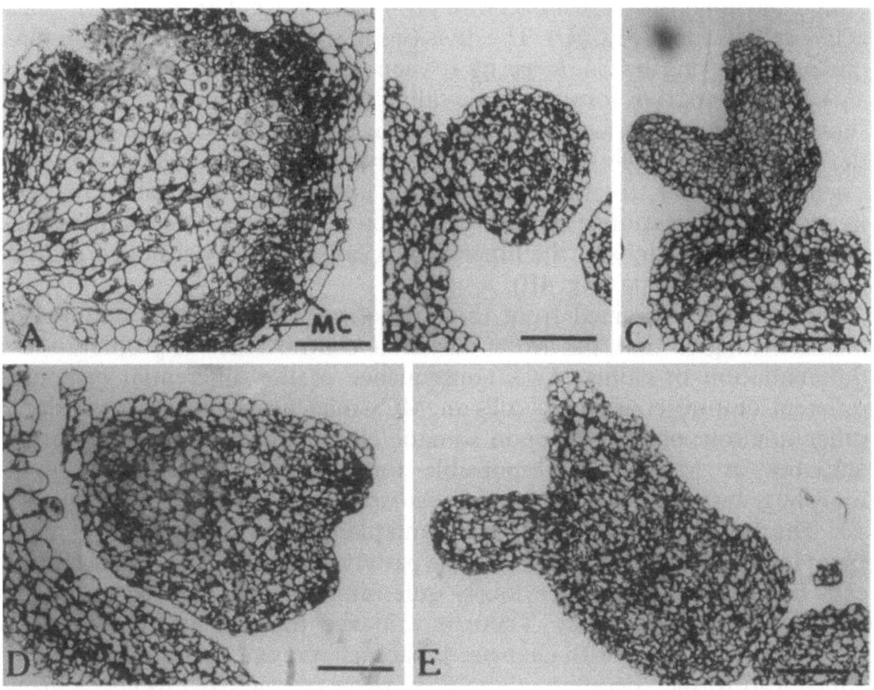

Fig. 3A–E. Toluidine blue-stained semi-thin sections of nodular callus and developing embryoids. **A** Cross section of a nodule taken from callus cultured on 2,4-D$^-$ medium, showing the meristematic centres (*MC*) at the periphery. **B–E** Sections of embryoids as shown in Fig. 1B, showing developmental stages from globular (**B**), early heart-shaped (**C**), late heart-shaped (**D**), and torpedo-shaped (**E**) embryos. Note the histological and structural resemblance between the embryoids and the leafy structures from which they had arisen. *Bar* in *A* = 30 μm, *bars* in **B–E** = 50 μm which subsequently developed into roots when placed on a root-inducing medium (2,4-D$^-$ or NAA$^+$), or differentiated into shoot primordia from which the shoots and plantlets could be regenerated on a high kinetin medium (data not shown). Figure 3B–E shows semi-thin sections of the embryoidal masses at different developmental stages from globular, early heart-shaped, late heart-shaped, and torpedo-shaped embryoids. These sections show that the developmental patterns of embryoidal masses are similar to those followed by zygotic embryos of the same plant. Another striking point revealed by these sections is the secondary nature of the embryoids as they apparently arose from leafy structures which bore close histological and structural resemblance to the embryoids (Fig. 3B–E). Other data indicated that the leafy structures developed from the MCs in compact green callus (not shown).

The ontogeny and segmentation pattern of embryogenic cells were not precisely followed in this study. However, some aspects of the structure of embryogenic callus and somatic embryos have been examined using histological techniques. Like the seedling-derived callus, the zygotic embryo-derived callus also contained groups of meristematic cells near the surface (Fig. 4A). Upon transfer to 2,4-D$^-$ medium, some embryogenic cells were formed at the surface (Fig. 4B,C) which probably originated from the cells in MCs. These embryogenic cells are distinct from the surrounding cells because of their dense cytoplasm, low level of vacuolation, large nuclei with prominent nucleoli (Figs. 4B,C), and their active cell division (Fig. 4C). The divisions in some of the cells were apparently unequal, resulting in some large, more vacuolated cells (Fig. 4C). At a later stage globular proembryos formed on the callus, as can be seen in Fig. 4D. Figure 4E shows a median longitudinal section through an individual somatic embryo at the cotyledonary stage, from which it can be seen that the embryo was a highly organized, bipolar structure with a radical apex at the basal end of the swollen hypocotyl still in contact with the maternal callus, and cotyledons associated with the plumule apex at the other end. A layer of well-developed protoderm tissue was also visible (Fig. 4E).

The results obtained from these experiments with tissue cultures of *A. thaliana* suggest that the MCs played a central role in proliferation and differentiation of callus. As a consequence of the differential responses to different culture conditions, cells in MCs may undergo active division and differentiation, or embark upon somatic embryogenesis. In other words, the activities of MCs were responsible for virtually all forms of in vitro morphogenesis in callus tissues in *A. thaliana.*

The de novo formation of organs and plants from somatic cells of a mature plant must involve an alteration to the pattern of gene expression of specialized cells in such a way that previously quiescent cells are directed into different pathways of development. There is evidence that the process of de novo organogenesis begins with changes in parenchyma cells of an explant or callus which lead to the "dedifferentiation" of these parenchyma to give a group of dividing cells. According to Thorpe (1980), the degree of "differentiation" in meristematic cells determines whether organs or embryos are formed, i.e., if dedifferentiation is complete, embryos will be formed, whereas partial "dedifferentiation" will lead to the formation of organ primordium. Con-

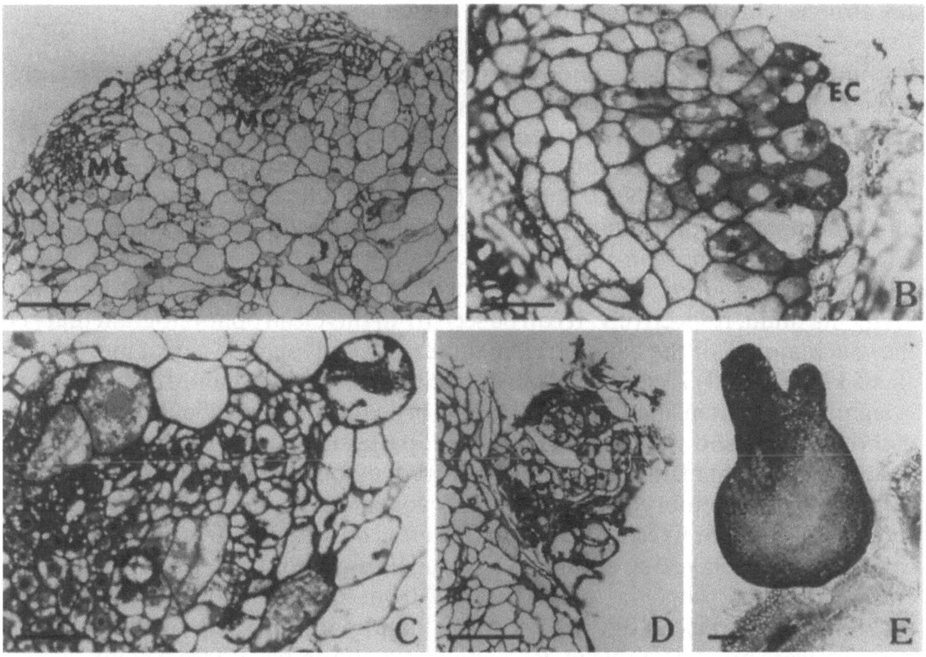

Fig. 4A–E. Semi-thin sections of zygotic embryo-derived embryogenic callus after transfer onto 2,4-D⁻ medium. **A** Meristematic centres (*MC*) in compact green callus. **B,C** Embryogenic cells (*EC*) with dense cytoplasm, large nuclei, and prominent nucleoli at the surface of the callus. **C** The embryogenic cells are actively dividing and some divisions are unequal. **D** A globular proembryo formed on the callus surface. **E** A median longitudinal section of an individual embryo. Note the bipolar nature of this structure. *Bars* in **A–D** = 30 μm; *bar* in **E** = 20 μm

ceivably, the meristematic cells have a basic pattern of gene expression and are in such a state that if they receive the appropriate signals, they can embark on different pathways of differentiation, and each pathway involves a shift to a new specific pattern of gene expression. In tissue cultures of *A. thaliana* the MCs were established as soon as the nodular callus was formed, and this was a result of the dedifferentiation of explanted tissues in response to added 2,4-D during callus induction. If it is the degree of dedifferentiation of cells that decides the pathway of morphogenesis, then the regenerative differences between the seedling-derived and embryo-derived calluses must be a consequence of the differential responses of explanted cells to the same "hormonal" treatment (0.5 mg/l 2,4-D) in primary medium. This is possible since the calli originated from cell types with different degrees of differentiation. In the former case, callus arose from hypocotyls of germinating seedlings, whereas in the latter, it was formed from every part of the isolated zygotic embryos. Therefore, differences in degrees of dedifferentiation can be expected even when these calli were initiated under the same conditions.

2.3 Protein Markers for Somatic Embryogenesis

It is important to recognize molecular changes that take place during the early phase of induction. For this purpose the protein composition of callus maintained on 2, 4-D$^+$ primary medium, and callus that had grown on the 2,4-D$^-$ secondary medium for various periods of time, were analyzed and compared using polyacylamide gel electrophoresis (PAGE). Figure 5 shows an SDS-PAGE separation pattern of proteins in extracts of calli from different cultures. Tracks 1, 2, and 3 are the protein composition of calli grown on 2,4-D$^+$ medium with 2 mg/l kinetin, 2,4-D$^+$ medium without kinetin, and callus 2 weeks after transfer to 2,4-D$^-$ medium, respectively. It can be seen from this electrophoretogram that (1) the vast majority of the bands are the same; (2) a polypeptide with a mol. wt. of ca. 65 kDa (band I) is clearly visible in track 3 (2,4-D$^-$) ; (3) band II (ca. 58 kDa) is significantly intensified in track 3; and (4) calli of primary cultures with or without kinetin show the same SDS-PAGE pattern. Whether the appearance of band I and the intensification of band II were associated with the transition of callus from a nonembryogenic to an embryogenic stage requires further study. Indeed, this may be likely since this pattern was only observed in compact green callus which displayed an embryogenic tendency. In contrast, calli from 2,4-D$^-$ culture (nonembryogenic, track 1), and from high kinetin culture (shoot-differentiating, track 2) did not show this band pattern. Also, these two polypeptides were not detectable in developing somatic embryos (see below).

Meanwhile, the IEF/SDS two-dimensional PAGE coupled with a sensitive silver staining procedure was used to resolve the minor changes in protein pattern of callus at the very early phase of the transition. Figure 6A, B shows the 2-D PAGE patterns of soluble proteins of the callus cultured on 2,4-D$^+$ primary

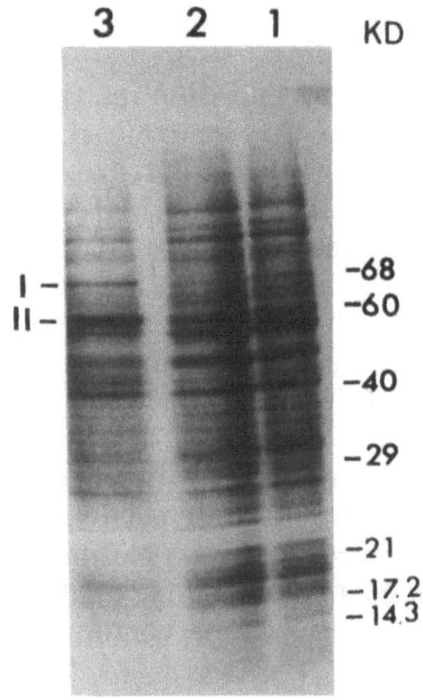

Fig. 5. SDS-PAGE patterns of proteins extracted from calluses of *A. thaliana* grown on different media. *Track 1* 0.5 mg/l 2,4-D + 2 mg/l kinetin (shoot differentiating callus); *track 2* 0.5 mg/l 2,4-D (nonembryogenic callus); *track 3* no added growth regulators (embryogenic callus). Note the intense *bands I* and *II* in embryogenic callus (*track 3*). The gel was stained with Coomassie blue

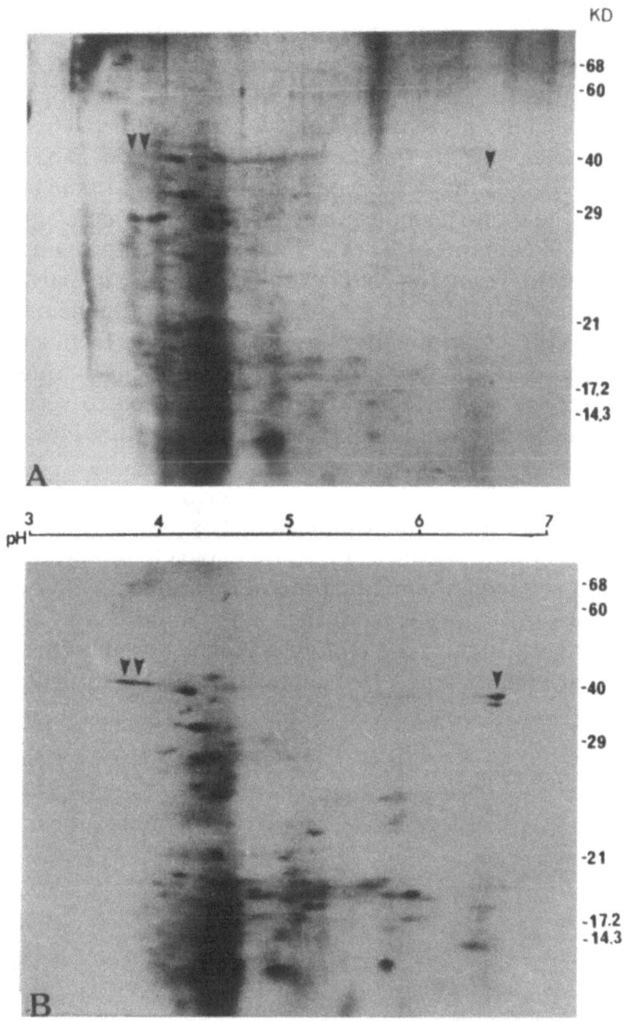

Fig. 6.A,B IEF/SDS two-dimensional PAGE patterns of soluble proteins of nonembryogenic (**A**) and embryogenic (**B**) calli. The positions occupied by the putative "embryogenic proteins" are indicated by *arrow heads*. Gels were stained with a Coomassie blue/sliver double-staining procedure

medium, and the callus which had been transferred to 2,4-D$^-$ secondary culture for about 1 week, respectively. The most evident difference between these two 2-D patterns was the emergence of a few polypeptides unique to embryogenic callus (2,4-D$^-$). Two closely aligned doublet spots at the position of approximately pI 6.4, mol. wt. 37–38 kDa (arrow), and another spot at pI 3.5, mol. wt. 38 kDa (double arrows) can be seen in embryogenic callus (Fig. 6B), but they were absent in nonembryogenic callus (Fig. 6A). Moreover, these spots displayed variable life spans during secondary culture on 2,4-D$^-$ medium. The doublet spots were "long-lived" peptides as they were present in callus which had been continuously kept in secondary culture for up to 1 month, while the more acidic peptide was detectable only in the early phase of secondary culture and had vanished after 2 weeks in culute. Although the identity and biological functions of these polypeptides are unknown, they may well be related to the special functions and metabolism of the initiation of somatic embryogenesis, since they were only specifically present in embryogenic callus. Therefore, these polypeptides, together with the specific bands revealed by 1-D SDS-PAGE, can be putatively regarded as the "embryogenic proteins" in tissue cultures of *A. thaliana.* They may be useful as molecular markers for somatic embryogenesis in this plant. It should be emphasized that the embryogenic proteins" in callus of *A. thaliana* are different from the "embryonic proteins" in carrot described by Sung and Okimoto (1981), because the former is a biochemical change prior to embryo formation marking the conversion of callus from a nonembryogenic state to an embryogenic state, while the latter represent unique components in developing embryos. Much more work will be needed to verify that these "embryogenic proteins" are functionally related to, or required for, in vitro embryogenesis in *A. thaliana.*

In addition, the protein patterns of somatic and zygotic embryos of *A. thaliana* were studied and compared using SDS-PAGE. It is clear from the electrophoretogram in Fig. 7 that apart from certain small differences involving minor bands, the major protein components of somatic (track S) and zygotic (track Z) embryos were the same. As already characterized, the major seed storage proteins of *A. thaliana* were the 12s globulin and the 1.7 s globulin (Heath et al. 1986; Huang 1986) comprising six constituent polypeptides of 19–34 kDa

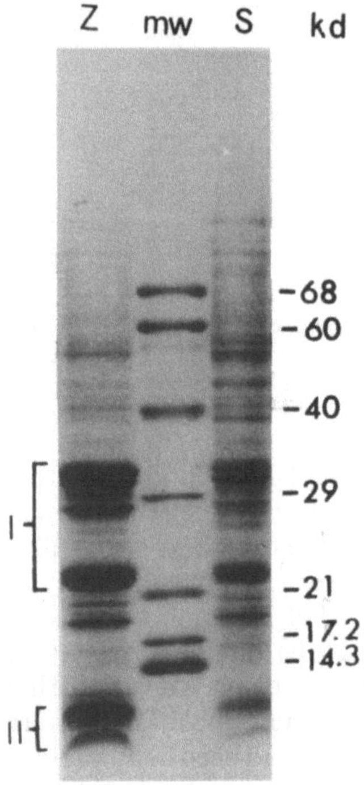

Fig. 7. SDS electrophoretograms of proteins of somatic embryos at the cotyledonary stage (*track S*) and mature zygotic embryos (*track Z*). The 6 characteristic SDS bands for the 12s storage protein (*group I*) and the 3 low mol. wt. bands for the 1.7s storage protein (*group II*) are present in both somatic and zygotic embryos. The gel was stained with Coomassie blue

and three polypeptide of 10–12 kDa, respectively (Huang 1986). Both the somatic and zygotic embryos possessed the characteristic SDS bands of the constituent polypeptides of the major seed storage proteins as shown in Fig. 7. This result indicated that, in *A. thaliana*, somatic embryos synthesized and accumulated the same storage proteins in vitro as zygotic embryos did in vivo. Experiments employing Western blotting and Protein A/gold immuno-electron microscope localization, using the anti-12s IgGs as a probe, provided further evidence that the 12s storage globulin was accumulated in somatic embryos at a similar developmental stage (cotyledonary stage) to that in zygotic embryos, and that the storage protein was deposited within the vacuole-originated protein bodies (Huang 1986). So far, little research attention has been paid to the storage proteins in somatic embryogenesis. The limited number of reports include storage proteins in somatic embryogenesis in *Zea mays* (Cross et al. 1981) and cotton (Shoemaker et al. 1987). As storage proteins are suitable markers for studying biochemical changes and differential gene expression in seed development, they may also be used as markers in similar studies of somatic embryogenesis.

3 Summary and Conclusions

1. Somatic embryos can be induced from callus cultures of *A. thaliana* by two routes, arising either as embryoidal masses from seedling-derived callus, or as individual embryos developing from zygotic embryo-derived callus. These two forms of embryogenesis involve different pathways and "hormonal" regulation but, as a general rule, the removal or exhaustion of 2,4-D from primary culture was essential to both embryogenic pathways.
2. Somatic embryos mimic some of the developmental stages of zygotic embryogenesis, and can grow into normal plants. However, they show certain abnormalities in growth pattern, morphology, and structure when compared with zygotic embryos. Besides, the embryoidal masses seem to arise as secondary adventitious embryoids from the preformed leafy structures.
3. The meristematic centers (MCs) play a central role in callus proliferation and differentiation. These MCs are able to undergo active division to give rise to new nodules, or become organized into root or shoot primordia. They are also actively involved in the induction of somatic embryogenesis. These results support the concept of the "single organogenetic pattern" proposed by Wicart et al. (1984). This hypothesis holds that proliferating MCs or meristemoids in callus are responsible for all regenerated structures such as roots, shoots, and somatic embryos.
4. A few polypeptides specifically present in potentially embryogenic callus were revealed by both 1-D SDS and 2-D IEF/SDS-PAGE. They have been putatively recognized as "embryogenic proteins". Meanwhile, somatic embryos were shown to synthesize and accumulate the same seed storage proteins as zygotic embryos. All these proteins may be useful as biochemical markers for somatic embryogenesis in this plant.

4 Protocols

The tissue culture procedures and the pathways of in vitro morphogenesis in callus cultures of *A. thaliana* are summarized diagrammatically in Fig. 8.

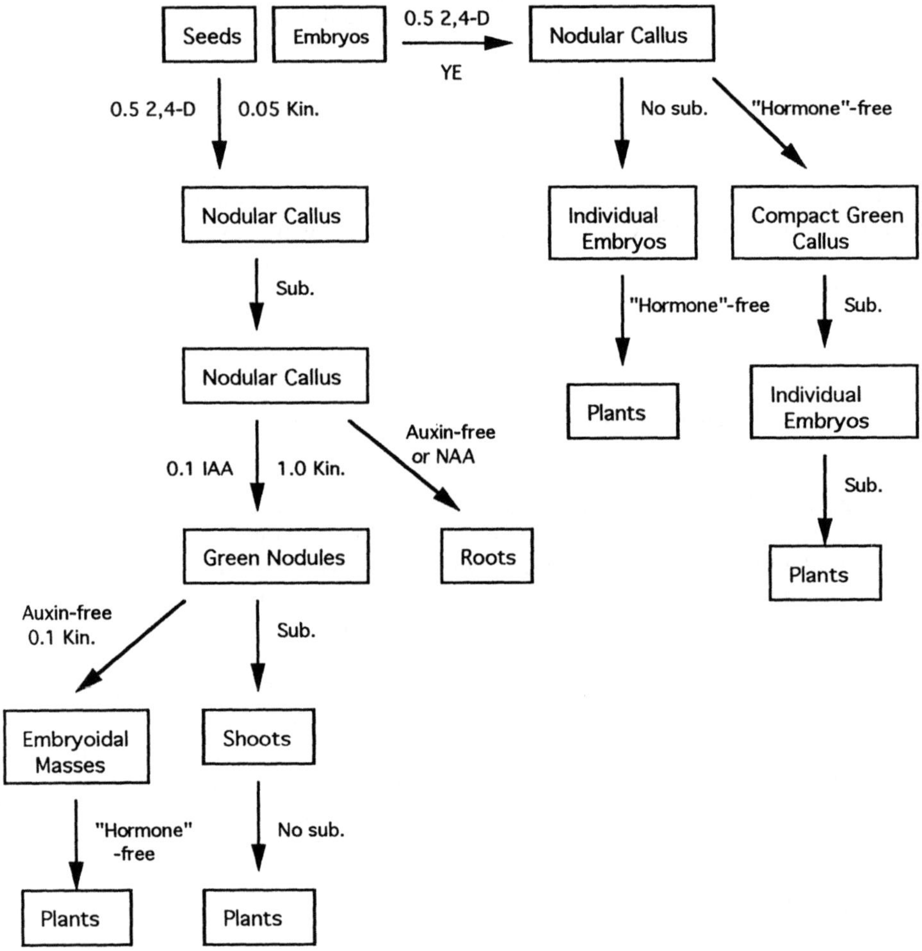

Fig. 8. Diagrammatic representation of the protocols and pathways for in vitro morphogenesis in callus cultures of *A. thaliana*. The basal medium used was either B_5 or SH. *YE* yeast extract. All the growth regulator concentrations are in mg/l

References

Al-Abta S, Collin HA (1978) Control of embryoid development in tissue cultures of celery. Ann Bot 42: 773–782

Bilang R, Potrykus I (1993) Regeneration of *Arabidopsis thaliana* transgenic plants after PEG-mediated direct gene transfer to protoplasts, In: Bajaj YPS (ed) Biotechnology in agriculture and forestry, vol 22. Plant protoplasts and genetic engineering III. Springer, Berlin Heidelberg New York Tokyo, pp 123–134

Chang C. Bowman JL. DeJohn AW. Lander ES, Meyerowitz EM (1988) Restriction fragment length polymorphism map for *Arabidopsis thaliana*. Proc Natl Acad Sci USA 85: 6856–6860

Cross JW, Adams W, Rice TB (1981) Embryo storage proteins from embryogenic tissue cultures of *Zea mays*. Plant Physiol (Suppl 4) 67: 154

El Hadrami I, Carron MP, d'Auzac J (1991) Influence of exogenous hormones on somatic embryogenesis in *Hevea brasiliensis*. Ann Bot 67: 511–515

Feldmann KA (1992) In vitro culture of *Arabidopsis thaliana*. In: Bajaj YPS (ed) Biotechnology in agriculture and forestry, vol 20, High-tech and micropropagation IV. Springer, Berlin Heidelberg New York, pp 471–483

Gamborg OL, Miller RA, Ojima K (1968) Nutrient requirements of suspension cultures of soybean root cells. Exp cell Res 50: 151–158

Gleba YY, Hoffmann F (1980) "Arabidobrassica": a novel plant obtained by protoplast fusion. Planta 149: 112–117

Heath JD, Weldon R, Monnot C, Meinke DW (1986) Analysis of storage proteins in normal and aborted seeds from embryo-lethal mutants of *Arabidopsis thaliana*. Planta 169: 304–312

Huang BC (1986) A comparative study of in vivo and in vitro embryogenesis in *Arabidopsis thaliana* L. PhD thesis, Edinburgh University

Huang BC, Yeoman MM (1983) Formation of somatic embryos in tissue cultures of *Arabidopsis thaliana*. Arab Inform Serv 20: 73–77

Huang BC, Yeoman MM (1984) Callus proliferation and morphogenesis in tissue cultures of *Arabidopsis thaliana* L. Plant Sci Lett 33: 353–363

Karesch H, Bilang R, Potrykus I (1991) *Arabidopsis thaliana*: protocol for plant regeneration from protoplasts. Plant Cell Rep 9: 575–578

Kirchheim B, Kranz AR (1981) New population samples of the AIS-seed bank. Arab Inform Serv 18: 173–176

Koornneef M (1987) Linkage map of *Arabidopsis thaliana* (2n = 10). In: O'Brien SJ (ed) Genetic maps 1987: a compilation of linkage and restriction maps of genetically studied organisms. Cold Spring Harbor Laboratory, Cold Spring Harbor, New York pp 468–496

Koornneef M, van Eden J, Hanhart CJ, Stam P, Braaksma FJ, Feenstra WJ (1983) Linkage map of *Arbidopsis thaliana*. J Hered 74: 265–272

McKelvie AD (1962) A list of mutant genes in *Arabidopsis thaliana* (L.) Heynh. Radiat Bot 1: 233–241

Meyerowitz EM (1987) *Arabidopsis thaliana*. Annu Rev Genet 21: 93–111

Meyerowitz EM (1989) *Arabidopsis*, a useful weed. Cell 56: 263–269

Negrutiu I, Jacobs M, Cachita D (1978) Some factors controlling in vitro morphogenesis of *Arabidopsis thaliana*. Z Pflanzenphysiol 86: 113–124

Nomura K, Komamine A (1986) Somatic embryogenesis in cultured carrot cells. Growth Differ 28: 511–517

Pruitt RE, Meyerowitz EM (1986) Characterization of the genome of *Arabidopsis thaliana*. J Mol Biol 187: 169–183

Rédei GP (1975) *Arabidopsis* as a genetic tool. Annu Rev Genet 9: 111–127

Röbbelen G (1965) The Laibach standard collection of natural races. Arab Inform Serv 2: 36–47

Schenk RU, Hildebrandt AC (1972) Medium and techniques for induction and growth of monocotyledonous and dicotyledonous plant cell cultures. Can J Bot 50: 199–204

Shoemaker RC, Christofferson SE, Galbraith DW (1987) Storage protein accumulation patterns in somatic embryos of cotton (*Gossypium hirsutum* L.). Plant Cell Rep 6: 12–15

Sung ZR, Okimoto R (1981) Embryogenic proteins in somatic embryos of carrot. Proc Natl Acad Sci USA 78: 3683–3687

Thorpe TA (1980) Organogenesis in vitro: structural, physiological and biochemical aspects. Int Rev Cytol (Suppl) 11A: 71–111

Valvekens D, van Montagu M, van Lijsebettens M (1988) *Agrobacterium tumefaciens*-mediated transformation of *Arabidopsis thaliana* root explants by using kanamycin selection. Proc Natl Acad Sci USA 85: 5536–5540

Wicart G, Mouras A, Kutz A (1984) Histological study of organogenesis and embryogenesis in *Cyclamen persicum* Mill. tissue cultures: evidence for single organogenetic pattern. Protoplasma 119: 159–167

Yokoyama K, Jones WH (1965) Tissue cultures of *Arabidopsis thaliana*. Plant Physiol 40 (Suppl): LXXVII (Abstr)

IV.9 Somatic Embryogenesis in *Bellevalia romana* (L.) Rchb. and *Brimeura amethystina* (L.) Salisb. (Liliaceae)

A. Cavallini and L. Natali[1]

1 Introduction

1.1 Botany, Distribution, and Importance

The genera *Bellevalia* and *Brimeura* belong to the *Liliaceae* family. *Bellevalia* was established in 1808 by De Lapeyrouse on *B. operculata* in France, a plant which was, however, identical with *Hyacinthus romanus* described by Linnaeus. Other species of *Bellevalia* were also described first as *Hyacinthus* and *Muscari*. However, since the first half of the nineteenth century, the genus *Bellevalia* has been accepted by taxonomists and rather well distinguished from allied genera (for example, see Boissier 1884).

Brimeura Salisb. is an archaic, pre-tertiary genus, possibly of hybrid origin. Also, *Brimeura amethystina* has often been assigned to other genera, particularly to *Hyacinthus* L., but morphological, karyological, ecological, and chorological analyses have confirmed the validity and the extension of the genus *Brimeura* (Garbari 1970).

Bellevalia romana (L.) Rchb. and *Brimeura amethystina* (L.) Salisb. are perennial species; in both species leaves are linear and canaliculate, shorter than space in *Brimeura*, longer in *Bellevalia*; the inflorescence is a raceme, composed of few flowers, pendulous at anthesis, 3–4 mm wide. The perianth is usually white, campanulate. Flowering occurs in March–April; the fruits are capsules, seeds are small and globular, black. In the adult plant, the bulb is 1.5–3 cm in *Bellevalia*, and 3–4 cm in *Brimeura*.

From an ecological point of view, the two species are mesophyllous-sciophilous found mostly in mountainous or subalpine area; they show a very broad ecological range and are nearly irresponsive to the soil nature. The geographical area of these species is the west Mediterranean basin, including southern France, Corsica, Italy, Malta, Dalmatia, and Greece (Feinbrun 1940; Garbari 1970).

Bellevalia romana may be of some economic interest because of its content of homoisoflavanones (Adinolfi et al. 1989). *Brimeura* is a weed, economically insignificant.

[1] Department of Agricultural Plant Biology, Genetics Section, University of Pisa, Via Matteotti 1/B, 56124 Pisa, Italy

Biotechnology in Agriculture and Forestry, Vol. 31
Somatic Embryogenesis and Synthetic Seed II (ed. by Y.P.S. Bajaj)

1.2 Cytology

Bellevalia romana chromosomes (2n=8) have been distinguished in four pairs: the chromosome pair A (I), the longest, near metacentric, the chromosome pair B (II), subtelocentric, with the long arm longer than either A chromosome arms, the chromosome pair C (III), shorter than A and B, submetacentric, and the chromosome pair D (IV), the shortest, subtelocentric. Small satellites have been observed on both the short arm of chromosome I and the long arm of chromosome IV, with differences in structure and localization depending on the biotype (D'Amato 1948).

The karyotype of *Brimeura amethystina* (2n=28) is typically bimodal, with 10 large subtelocentric and 18 small submetacentric chromosomes; these small chromosomes are also divided into two groups: 6 larger and 12 very small (Garbari 1970). Quinacrine staining shows 16 distinct bands in 8 large subterminal chromosomes (Vosa 1979): of these bands, 12 are evident also in interphase nuclei as large chromocenters (Zolfino et al. 1984).

1.3 Significance of Somatic Embryogenesis in *Bellevalia* and *Brimeura*

Genetic and chromosome variations occur in practically all calli and suspension cultures (D'Amato 1985; Bajaj 1990; Karp 1991). This variability can be useful for the production of plants with novel genotypes or chromosome aberrations (D'Amato 1990). In this context, it is of great importance to know the origin of the regenerates, whether they are formed via adventitious buds (multicellular origin) or via somatic embryos (unicellular origin), since this might determine different chromosomal conditions in the plants (possible mosaicism) with consequences on the segregation rates of the regenerates (see D'Amato 1985).

In our opinion, there is little information on the karyological status of plants obtained via somatic embryogenesis. In particular, little attention is paid to the chromosomal conditions during the different stages of somatic embryogenesis. On the other hand, while karyological changes in response to tissue culture have been amply documented (D'Amato 1985), more cryptic gene alterations, such as amplification or loss of DNA and rearrangements of nuclear DNA, have now been shown to be affected by tissue culture stress (Brown and Lorz 1986; Nagl 1990a; Cecchini et al. 1992).

With the aim to acquire deeper insight into these problems, we initiated a study on the dedifferentiating ability and morphogenetic potential of tissues from different organs of plants of these two species.

2 In Vitro Culture

2.1 *Bellevalia romana*

2.1.1 Callus Induction

In vitro culture of *Bellevalia romana* was established using explants derived from seedlings. After germination, 0.5-cm-long explants of root, cotyledon, and mesocotyl, excised from 5–10 cm tall seedlings, were placed horizontally in 10 cm diameter plastic Petri dishes, on the surface of the MS agar medium containing 4.5, 13.6, and 22.6 μM. 2,4-dichloro-phenoxy-acetic acid (2,4-D), 10.8 μM naphthaleneacetic acid (NAA) plus 2.2 μM 6-benzylaminopurine (BAP).

Callus tissue was formed at the cut ends of the explants after 2 weeks. Among the various explants, the best response was obtained using mesocotyls on MS medium containing 10.8 μM NAA plus 2.2 μM BAP, where a white, friable and fast-growing callus originated (Fig. 1a). Callus induction was also achieved from 100% mesocotyl explants when MS was supplemented with various concentrations of 2,4-D. In these cases, callus cultures were semifriable, nodular, and grew slowly. Cotyledon explants formed calli with a good frequency (49%) when treated with NAA plus BAP, whereas 2,4-D was less effective. Root explants formed calli only rarely in the presence of NAA plus BAP. Both cotyledon- and root-derived calli showed a low growth rate (Lupi et al. 1985).

2.1.2 Somatic Embryogenesis

Cell suspension cultures were established from both mesocotyl- and cotyledon-derived calli cultured for 2 months on solid MS containing either 2,4-D or NAA plus BAP (see Sect. 2.1.1). Small callus pieces were placed into 7 ml liquid MS in 50-ml Erlenmeyer flasks: calli previously induced in the 2,4-D media were transferred to 0.9 μM 2,4-D containing MS, those induced in NAA plus BAP were maintained in the same medium. The suspensions were incubated on a gyratory shaker (80 rpm) at 24±1 °C under continuous light of 1000 lx. For the subcultures of this material (0.08 g/ml of medium), the inoculum was obtained by centrifugation at 1300 rpm for 10 min, of the suspension containing single cells and cell clumps (Lupi et al. 1985; Cavallini et al. 1987).

Somatic embryogenesis was observed after 6 weeks of culture in liquid medium supplemented with 0.9 μM 2,4-D (Fig. 2.). By contrast, cell suspension cultures in NAA plus BAP-containing medium did not produce embryos. Somatic embryos showed the typical morphological organization of monocotyledonous embryos.

The positive role of 2,4-D in inducing the development of embryos in callus of suspension cultures has been demonstrated in a number of species (Thorpe 1982; Terzi and Lo Schiavo 1990; Tran Thanh Van and Trinh 1990). In *Bellevalia* also, embryogenesis occurred only in callus which was precultured at high 2,4-D concentrations. A reduction of the auxin level in the medium, however, was necessary for development of somatic embryos; moreover, only embryos transferred to auxin-free MS produced plantlets (Cavallini et al. 1987).

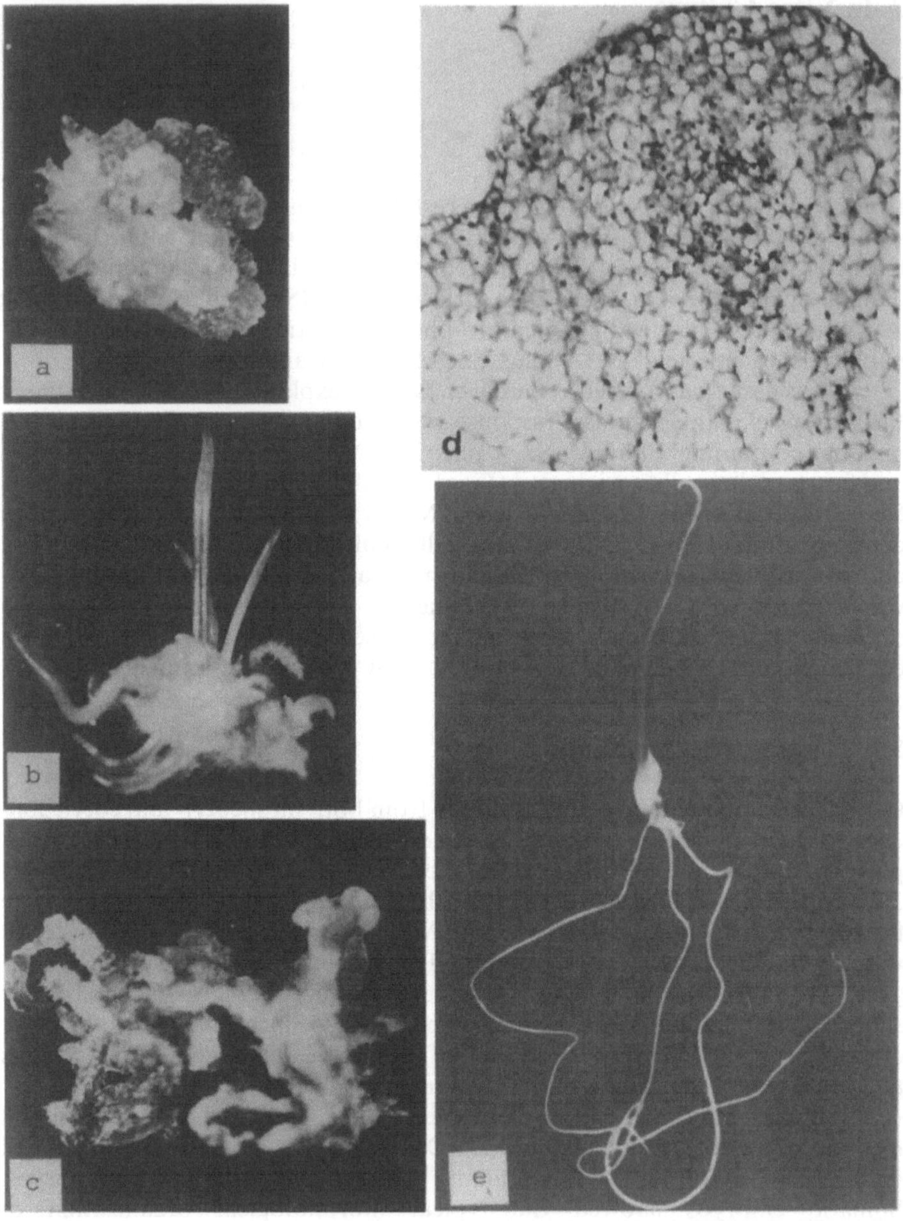

Fig. 1a–e. Different phases of in vitro culture of *Bellevalia romana*. **a** Mesocotyl callus after 50 days of culture on MS plus 10.8 μM NAA plus 2.2 μM BAP; **b** mesocotyl callus with regenerated shoots; **c** cotyledon callus with regenerated callusing roots; **d** historical section of callus showing an initial bud regenerating through organogenesis; **e** one rooted, regenerated plantlet. (Lupi et al. 1985)

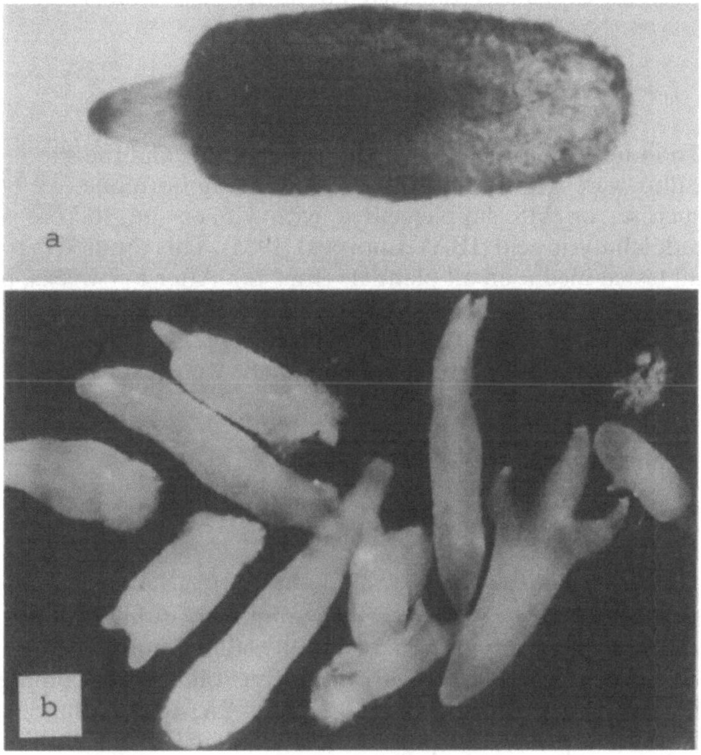

Fig.2 a,b. Somatic embryos of *Bellevalia romana* from liquid culture in MS plus 0.9 μM 2,4-D. **a** One isolated embryo showing a nascent root apex; **b** somatic embryos with one or more developing roots. (Lupi et al. 1985)

2.1.3 Organogenesis

Organogenesis was easily obtained only from mesocotyl-derived calli (Lupi et al. 1985); in particular, in NAA plus BAP-containing MS, 100% mesocotyl calli, after 2–3 months of culture, regenerated shoots (5 to 10 shoots per callus; Fig. 1b) via organogenesis, as revealed by histological sections (Fig. 1d). Shoot regeneration was maintained, for practically all these calli, for at least 2–3 years of culture. In contrast, none of the cotyledon-derived calli in NAA plus BAP-containing MS showed shoot regeneration, but only formation of aggregates of regenerated, callusing roots (Fig. 1c). Calli grown in the presence of different 2,4-D concentrations showed no morphogenetic responses. It has been hypothesized that the inhibitory effect of 2,4-D upon adventitious shoot formation should depend on the fact that this potent auxin stimulates callus cultures, but antagonizes organized development (Murashige 1974). It was observed that by decreasing 2,4-D concentrations in the medium, the inhibitory effect of this hormone could be minimized, thus allowing organogenesis (Groenewald et al. 1977). However, in *Bellevalia romana*, no shoots (but only abnormally growing roots) developed when the callus was transferred to hormone-free medium (Lupi et al. 1985).

2.1.4 Plant Rooting and Acclimatization

To induce root formation, both the embryos and the shoots regenerated from callus were transferred to tubes containing hormone-free MS with 1.5 or 3% sucrose, or MS supplemented with 1.5 or 3% sucrose and 0.5 or 1 μM indolebutyric acid (IBA) (Lupi et al. 1985). This resulted in root development in all tested media on all plantlets (Fig. 1e). After 6–7 weeks, plantlets with well-developed roots were transferred to pots containing garden soil compost and kept under colorless, transparent plastic to prevent desiccation. After 1–2 weeks, the plastic bags were removed and plants were grown in the greenhouse.

2.2 *Brimeura amethystina*

2.2.1 Callus Induction

Brimeura amethystina was cultured in vitro using explants derived from both seedlings and bulbs. After germination, 0.5-cm-long root, leaf, and mesocotyl explants were excised from 5-cm-tall seedlings and placed, to produce callus, in 10 cm diameter plastic Petri dishes on solid media containing the following hormones: 2.2 μM BAP plus 10.8 μM NAA, 4.4 μM plus 5.4 μM NAA, 4.5 μM 2,4-D, 2.3 μM 2,4-D plus 2.2 μM BAP. After sterilization in 0.6% sodium hypochlorite, bulb scales were also placed on the same media. All cultures were routinely maintained in a temperature-controlled room at 24 + 1 °C under continuous fluorescent light of 3000 lx.

Callus formation was recorded 4–6 weeks after the beginning of the culture (Table 1; Cavallini and Natali 1989). A whitish, friable callus tissue was produced first at the cut edges of the explants, with different frequencies depending on the type of explant (Table 1); all calli grew very slowly; the best response was obtained using either mesocotyls or leaves on MS medium containing 2.2 μM BAP plus 10.8 μM NAA (Table 1). As shown in Table 1, very poor callus induction was achieved when MS was supplemented with 2,4-D, in particular when roots or bulb scales were used as explants.

2.2.2 Induction of Embryogenesis

Calli produced by different explants and media were cultured onto fresh solid medium of the same composition after 30 days of culture, and transferred every month. Embryoids developed on the callus surface after 2 weeks of callus culture (Cavallini and Natali 1989; Fig. 3a). They showed the typical morphological organization of monocotyledonous embryos (Fig. 3b). Embryoids could be

Table 1. Percentage of callus formation and number of adventitious somatic embryos from different explants of *Brimeura amethystina* cultured on MS medium supplemented with different combinations of hormones. (Cavallini and Natali 1989)

Type of explant	Culture medium											
	2.2 μM BAP + 10.8 μM NAA			4.4 μM BAP + 5.4 μM NAA			4.5 μM 2,4-D			2.3 μM 2,4-D+ 2.2 μM BAP		
	No. of explants	No. of calli	No. of individual embryos[a]	No.of explants	No. of calli	No. of individual embryos[a]	No. of explants	No. of calli	No. of individual embryos[a]	No. of explants	No. of calli	No. of individual embryos[a]
Root	42	2	12	36	0	0	15	0	0	21	1	1
Leaf	45	26	48	39	13	27	15	4	8	21	5	14
Mesocotyl	27	24	68	21	17	42	12	3	15	15	6	16
Scale	30	18	35	24	7	21	12	0	0	12	0	0

[a]After 3 subcultures

easily recorded and detached to be cultured separately on hormone-free MS medium until a complete plantlet was achieved. Sometimes, particulary on leaf segments cultured on MS medium containing 2.2 μM BAP plus 10.8 μM NAA, direct embryogenesis was observed from the outer cell layer of the explant, without callus formation (Fig. 3c).

The highest frequency of embryo formation (after three subcultures) was achieved by culturing mesocotyls as explants onto MS medium containing 2.2 μM BAP plus 10.8 μM NAA (Table 1). As observed in *Bellevalia* (see Sect. 2.1.3),

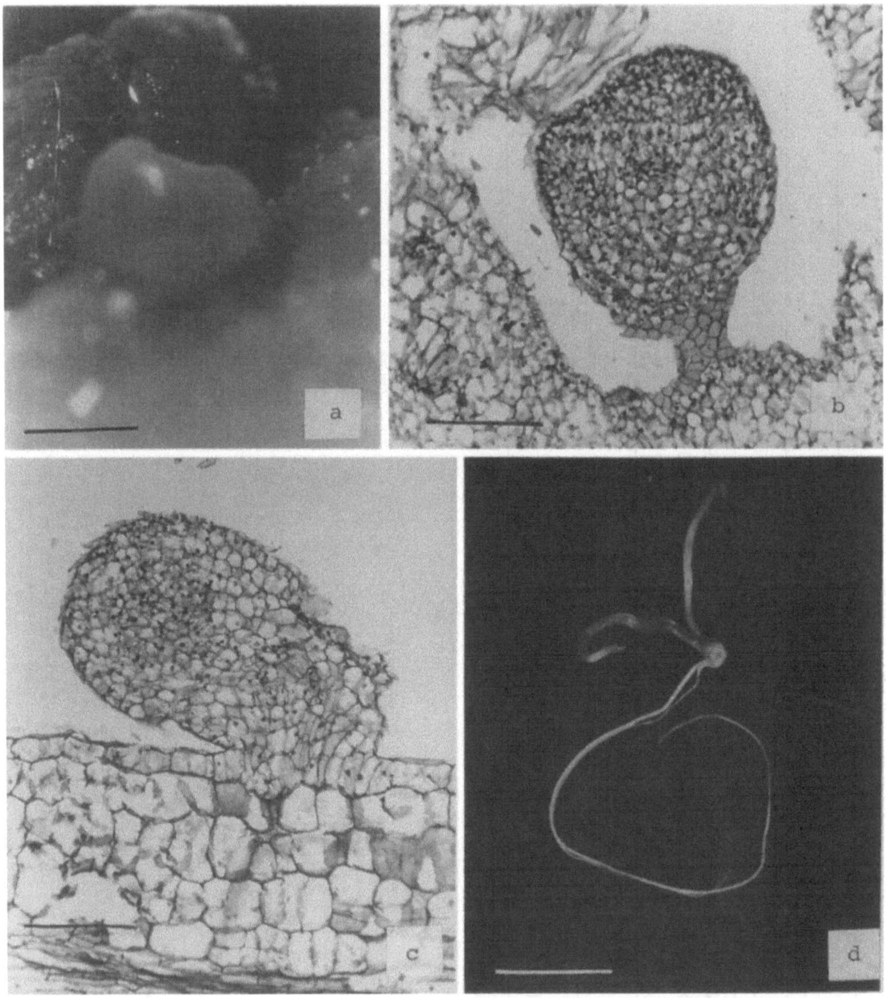

Fig. 3a–d. Different phases of in vitro culture of *Brimeura amethystina.* **a** Globular embryo on the surface of mesocotyl callus after 30 days of culture on MS plus 2.2 μM BAP plus 10.8 μM NAA (*bar*=0.3 mm); **b** histological section of a somatic embryo developing on the callus surface (*bar*=90 um); **c** section of a somatic embryo developing directly on the explant (leaf; *bar*=90 μm); **d** one rooted, regenerated plantlet (*bar*=26 mm). (Cavallini and Natali 1989)

mesocotyl explants of *Brimeura* allow one to obtain the highest frequencies of callus and bud formation; however, in this species, regeneration was achieved also from different types of explants (Tables 1 and 2), as already seen for other liliaceous species.

All calli retained high embryogenetic ability for at least six to seven subcultures on the same medium (Table 2). In particular, this was observed in mesocotyl calli cultured on NAA plus BAP-containing MS. Similar results were reported for other *Liliaceae*, i.e., *Lilium*, *Freesia*, and *Bellevalia* (Hussey 1982; Lupi et al. 1985).

Contrary to expectations (see Sect. 2.1.2), in *Brimeura*, the presence of 2,4-D produces a lower frequency of regenerated plants and a delay in embryogenesis (Table 2). As in *Bellevalia*, a reduction of the auxin level in the medium was necessary for development of *Brimeura* somatic embryos: only embryos transferred to auxin-free MS produced plantlets.

Bulblet formation and adventitious root development (after 4–6 weeks; Fig. 3d) were achieved only when embryoids were transferred to hormone-free MS medium supplemented with activated charcoal. Similar results were reported in a closely related liliaceous species, *Muscari armeniacum* (Peck and Cummings 1986).

After root development, plantlets and bulblets were transferred into earthen pots containing garden soil compost in the greenhouse and kept under colorless, transparent plastic bags to prevent desiccation during the first 3 weeks. After 1 year, the survival percentages was 68.5%.

3 Nuclear Cytology of Callus and Regenerated Plants

3.1 *Bellevalia romana*

As far as the explant is concerned, no chromosomal variability was found: all metaphases analyzed in control seedlings were diploid. In contrast, from the first phases of culture, chromosome counts in callus revealed hypohaploid, haploid, hypodiploid, hyperdiploid, and tetraploid nuclei, in addition to a majority of diploid cells (Cavallini et al. 1986). A higher frequency of aneuploid, haploid, and sometimes tetraploid mitoses was observed throughout the culture period in callus growing in the presence of 2,4-D with respect to callus maintained in NAA plus BAP; the frequency of abnormal mitoses was correlated with higher mitotic rates. Moreover, non-euploid nuclei increased with increasing period of culture. These results were confirmed by the distribution of DNA values determined by Feulgen cytophotometry (Cavallini et al. 1986).

Contrary to other species, the in vitro origin of aneuploid cell lines in *Bellevalia romana* is not clear; probably, aneuploidy is produced during the very initial events of callus induction because obvious abnormalities in nuclear division have not been observed after 2 days of culture (Cavallini et al. 1986). Once produced, aneuploid cell lines may propagate mitotically even when the

Table 2. Embryogenetic ability of *Brimeura amethystina* calli cultured on different media

Culture medium (μM)	Number of calli	Number of somatic embryos						
		After 1 subculture	After 2 subcultures	After 3 subcultures	After 4 subcultures	After 5 subcultures	After 6 subcultures	After 7 subcultures
MS + 2.2 BAP + 10.8 NAA	30	33	51	61	74	79	89	94
MS + 4.4 BAP + 5.4 NAA	30	21	31	45	52	61	64	66
MS + 4.5 2,4-D	12	0	0	13	16	20	21	27
MS + 2.3 2,4-D + 2.2 BAP	12	0	0	11	13	18	20	22

chromosome number is lower than the haploid level (Singh et al. 1972; D'Amato 1985). Metaphase figures of *B. romana* callus cells revealed the completely random distribution of chromosomes in all aneuploid nuclei analyzed.

Since the mesocotyl explants did not contain polyploid nuclei, the dividing tetraploid cells found in *Bellevalia* callus (Fig. 4a) should result from tetraploidization of diploid cells. Polyploidization during callus induction is a very common phenomenon (Pijnacker et al. 1986; Castorena-Sanchez et al. 1988). On the other hand, in these calli, true haploid cells (n=4) have also been found: these cells probably originated by homologous chromosome segregation during diploid mitosis (Fig. 4b), as observed in other species both in vivo (Cavallini et al. 1988) and in vitro (Wen et al. 1989). It is worth noting that 80.4% of regenerating shoots via organogenesis were diploid, and 19.6% were chromosome number mosaics, i.e., aneusomatics or mixoploids, consisting of diploid, hypodiploid, haploid, hyperdiploid, and tetraploid cells (Cavallini et al. 1986). As pointed out by D'Amato (1985), such chimerism in regenerated plants is the best evidence for the participation of many cells in the vitro organization of adventitious buds.

As to somatic embryogenesis, all globular proembryos were diploid (Table 3; Cavallini et al. 1987). In contrast, chromosome number mosaics occurred in the embryos 0.5 to 6 mm in length. Chromosome counts in the embryos of different sizes (Table 3) showed a large increase in chromosome number mosaics during embryo growth: the number of (true) haploid, aneuploid (hypo- and hyper-

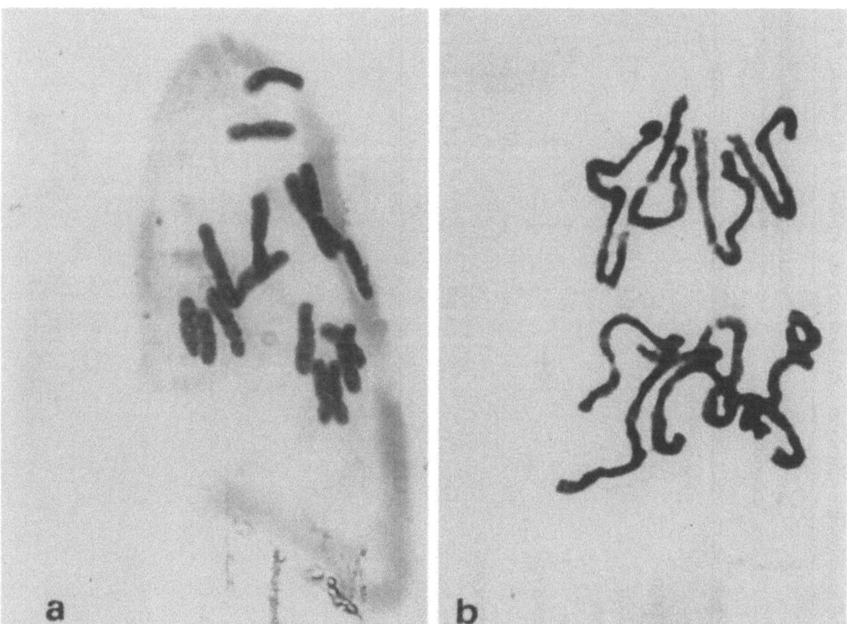

Fig. 4. Tetraploid metaphase (**a**) and homologous somatic chromosome segregation (**b**) in callus cells of *Bellevalia romana*

Table 3. Chromosome counts in somatic embryos of *Bellevalia romana*. At least 20 metaphases per embryo were counted. (Compiled from Cavallini et al. 1987)

Embryo stage	No. of embryos examined	Percentage of cells in each chromosome number class												
		4	5	6	7	8	9	10	11	12	13	14	15	16
Globular proembryos	12	–	–	–	–	100.0	–	–	–	–	–	–	–	–
0.5-mm-long embryos	10	1.4	–	0.5	0.8	94.9	0.3	–	–	–	–	–	–	2.1
0.5-1-mm-long embryos	12	0.9	–	–	1.0	88.0	1.2	–	–	0.4	–	–	–	8.5
2-3-mm-long embryos	10	0.3	–	2.0	1.4	87.6	0.2	1.2	0.8	0.5	–	1.9	–	4.1
4-6-mm-long embryos	14	1.0	1.2	8.4	8.6	67.0	1.2	0.6	–	1.4	–	0.8	0.5	9.3

diploid), and tetraploid mitoses in smaller embryos was significantly lower than in the more developed ones, and complete diploidy was never observed in the 4–6 mm long embryos, either in the root or in the shoot apex.

Our results indicated that in *Bellevalia romana* only diploid cells have the competence to form somatic embryos as reported by Vasil (1986) and Feher et al. (1989) for other species. Though in the callus, from which the cell suspension derived, different chromosome numbers in addition to diploid were present, all globular proembryos were diploid, indicating that somatic embryos are of unicellular origin, in agreement with other studies (D'Amato 1985).

On the other hand, Table 3 shows that, during somatic embryogenesis, chromosome variation is produced, probably due to the culture conditions and, in particular, to the longer presence of 2,4-D in the medium (see Sect. 2.1.2). This aspect seems very important, because 2,4-D has been shown, in many cases, strictly necessary to stimulate the embryogenetic competence in either callus or suspension cultures (Thorpe 1982).

Another aspect to be noted in *Bellevalia romana* somatic embryos is that they are often unable to differentiate into plants, probably due to the high frequency of abnormal chromosome numbers observed in the more mature embryos. Similar inability is reported in the literature for somatic embryos of other species (reported in Cavallini and Natali 1989). However, in cases where the karyological heterogeneity in somatic embryos is lower, it is possible that conditions prevail which favor the selective advantage of diploid over aneuploid cells (diplontic selection), allowing the attainment of an adult plant stage. For example, in *Triticum durum* (Lupi et al. 1981), *Pisum sativum* (Natali and Cavallini 1987), and *Helianthus annuus* (Cavallini and Lupi 1987) aneusomaty was reduced in subsequent growth stages and only diploid cells entered meiosis. Probably, the genetic structure of *Bellevalia romana* (a true diploid) does not allow the survival of the haploid and hypodiploid cells present in high numbers in the somatic embryos (for a discussion on this aspect, see D'Amato 1985).

From the above discussion it is apparent that the aneusomatic condition is common in regenerated plants of *Bellevalia romana*. However, it is worth noting that, when regeneration is achieved via organogenesis, chromosomal mosaicism is immediately present (it reflects the cytological conditions of the callus and demonstrates the multicellular origin of the buds), though it may decrease during plant development (due to diplontic selection). When a somatic embryo is formed, its genetic structure is usually diploid and uniform (showing the unicellular origin of somatic embryos) and the mosaicism is achieved during the first phases of plant development (with probable alterations in segregation rates in the subsequent generation, though a diplontic selection may again take place in later developmental stages).

3.2 *Brimeura amethystina*

In *Brimeura amethystina*, analysis of globular embryos (either from calli or from explant tissues) revealed that nearly all mitoses observed were diploid (Cavallini and Natali 1989); stability of chromosome number was maintained also during subsequent stages of embryo development. On the other hand, somatic embryos can be distinguished into two classes depending on their cell division rate: the former showing many actively dividing cells (mitotic index=6.42), the latter showing less dividing cells (m.i.=0.25). Chromosome counts in nongrowing embryos, even if limited to a few cells, showed that, as in actively growing embryos, chromosome number is mainly euploid and stable (Cavallini and Natali 1989). Hence, as in *Bellevalia*, only diploid cells seem to have the competence to form *Brimeura* somatic embryos.

It is worth noting that *Brimeura amethystina* somatic embryos are often unable to differentiate into plants. Complete development and germination of somatic embryos in monocotyledons have been often proven difficult, in connection with hormone and/or nutrient imbalance in the culture medium (Lu et al. 1983; Rybczynski and Zdunczyk 1986) and chromosome number alterations occurring during embryo development (see Sect. 3.1). In *Brimeura*, we have assayed numerous culture conditions and phytohormones, in order to increase the frequency of developing embryos, but no positive responses were obtained. On the other hand, failure of embryo development does not seem due to the presence of abnormal chromosome numbers.

In a cytophotometric study, mean 4C DNA contents of embryos and plantlets were compared (Fig. 5; Cavallini and Natali 1989): these values are very stable in normal embryos and very different among nondeveloping embryos: moreover, developing plantlets, also show a great uniformity of 4C values (Fig. 5). Cytophotometric analyses of optical density curves of interphase nuclei indicated that precise chromatin fractions (both euchromatic and heterochromatic) are differently represented in embryos showing high, medium, and low total DNA content (Cavallini and Natali 1989).

The nature of DNA sequences involved in the variations remains unknown. Selective loss and regain of chromatin were described by Deumling and Clermont (1989) in *Scilla siberica*, another liliaceous species. The possible explanation to

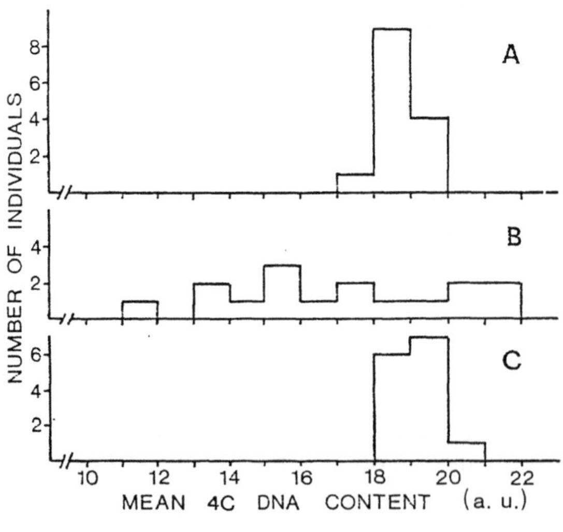

Fig. 5. Mean 4C (G2) DNA contents (arbitrary units) of developing (**A**) and nondeveloping (**B**) embryos, and plantlets (**C**) of *Brimeura amethystina*, regenerated on MS medium containing 2.2 μM BAP plus 10.8 μM NAA. Each value was determined from Feulgen absorption histograms of 200 interphase nuclei per individual. (Cavallini and Natali 1989)

this phenomenon of euchromatic loss is that it may involve structural genes redudant in the plant genome, namely small gene families, without completely losing gene information. It is, however, conceivable that the sequences involved in the variations are mostly noncoding; in this sense, they should belong to the so-called conformational DNA (Nagl 1990b) that determines nuclear general structure and function.

On the other hand, it is worth noting that, in *Brimeura* somatic embryos, these variations are correlated to failure in development, indicating that their function is important for plant life.

4 Summary and Conclusions

Bellevalia romana is a species whose chromosomes are easily recognizable and occur in low number. When cultured in vitro, regenerated plantlets are frequently obtained via either organogenesis or somatic embryogenesis. Hence, this species is very favorable for basic studies on cytology of in vitro culture.

In particular, the results reported in this review seem interesting to explain the onset of somaclonal variation in plants regenerated via somatic embryogenesis. In fact, while Vasil (1986) argues that somaclonal variation is largely confined to organogenetic plant regeneration, a number of reports indicate that variability can occur also in regeneration from somatic embryoids (Orton 1985; Barwale and Windholm 1987; Armstrong and Phillips 1988; Bajaj 1990; Bebeli et al. 1990; Karp 1991), i.e., there is no correlation between variability and regeneration mode. Our studies in *Bellevalia romana* show that, through different developmental pathways, all regenerated plantlets may show the same karyological situation: organogenetic plants are mosaics because aneuploid callus cells participate in bud formation, while embryo-derived plants are mosaics because aneuploid cells originate during embryo development in culture.

In *Brimeura amethystina*, embryogenesis occurred both on the primary explants (roots, leaves, mesocotyls, bulb scales) and after callus formation. Both callus and embryo formation were achieved with high frequencies using BAP plus NAA-containing media. Complete plantlets developed from embryos by inducing adventitious rooting on hormone-free medium containing activated charcoal.

A cytological and cytophotometric study of embryos and plantlets was undertaken to explain the observed frequent failure of embryo development after the globular stage: chromosome number was usually euploid and stable, but a much larger variation of nuclear DNA content was found among non-developing embryos than among normal ones.

As no cytological evidences (i.e., pycnotic nuclei, extrusion of micronuclei) of nuclear DNA changes were observed in the embryos at the stage analyzed, phenomena leading to differences in nuclear DNA content should occur in the callus and/or in early developmental stages of the embryos. Nevertheless, nuclear

DNA changes might be correlated with the lack of further embryo development. Even if some differences in DNA content might also be correlated with rare chromosome alterations, other phenomena may occur, such as DNA underrepresentation, underreplication, or amplification.

5 Protocols

Seeds of either *Bellevalia romana* and *Brimeura amethystina* are scratched to peel the seed coats, then surface sterilized for 10 min in a 2% sodium hypochlorite solution. After three washes in sterile distilled water, the seeds are placed in Erlenmeyer flaks (5 seeds per flask) containing 25 ml of hormone-free MS medium solidified with 0.8% agar, pH 5.6. After germination, 0.5-cm-long mesocotyl explants, excised from 5–10 cm tall seedlings, are placed horizontally on the surface of the MS agar medium (8 explants in 20 ml of medium in 10 cm diameter plastic Petri dishes) supplemented with 4.5, 13.6, or 22.6 μM 2,4-D (for *Bellevalia*) or 2.2 μM BAP plus 10.8 μM NAA (for *Brimeura*).

In *Bellevalia*, cell suspension cultures are established from calli cultured for 2 months. Small callus pieces are placed in 50-ml Erlenmeyer flasks into 7 ml MS liquid medium containing 0.9 μM 2,4-D. The suspensions are incubated on a gyratory shaker (80 rpm). After 14 days of growth, 10 ml of fresh liquid medium is added to the culture flasks. Afterwards, suspension cultures are transferred to fresh medium at 14-day intervals. For the subcultures of this material (0.08 g/ml of medium), the inoculum is obtained by centrifugation, at 1300 rpm for 10 min, of the suspension containing single cells and cell clumps.

Somatic embryos obtained on MS liquid medium develop into plantlets by transferring to hormone-free solid medium, where rooting is also achieved. The plantlets with well-developed roots are moved to pots containing garden soil compost and kept under colorless, transparent plastic to prevent desiccation. After 1 to 2 weeks, the plastic bags are removed and plants are grown in the greenhouse.

In *Brimeura*, after 3–4 weeks of callus culture, somatic embryos formed on the callus surface are detached and transferred to hormone-free MS solid medium containing 1 g/l of activated charcoal to develop bubils and adventitious roots. The plantlets with well-developed roots are moved to pots containing garden soil compost and kept under colorless, transparent plastic to prevent desiccation. After 3 weeks, the plastic bags are removed and plants are grown in the greenhouse.

All cultures are routinely maintained in a temperature-controlled room at 24+1 °C under continuous fluorescent light of 2500 lx (type of lamps Sylvania daylight F30W/ T8/D).

References

Adinolfi M, Aquila T, Barone G, Lanzetta R, Parrilli M (1989) Homoisoflavanones from *Bèllevalia romana*. Phytochemistry 28: 3244–3245

Armstrong CL, Phillips RL (1988) Genetic and cytogenetic variation in plants regenerated from organogenic and friable embryogenic tissue cultures of maize. Crop Sci 28: 363–369

Bajaj YPS (1990) Somaclonal variation-origin, induction, cryopreservation, and implications in plant breeding. In: Bajaj YPS (ed) Biotechnology in agriculture and forestry, vol 11. Somaclonal variation in crop improvement I. Springer, Berlin Heidelberg New York, pp 3–48

Barwale UB, Widholm JM (1987) Somaclonal variation in plants regenerated from cultures of soybean. Plant Cell Rep 6: 365–368

Bebeli P, Karp A, Kaltsikes PJ (1990) Somaclonal variation from cultured immature embryos of rye differing in heterochromatic content. Genome 33: 173–183

Boissier E (1884) Flora orientalis, vol 5. Geneve, Basel.

Brown PTH, Lorz H (1986) Molecular changes and possible origins of somaclonal variation. In: Semal J (ed) Somaclonal variations and crop improvement. Martinus Nijhoff, Dordrecht, pp 148–159

Castorena-Sanchez I, Natali L, Cavallini A (1988) In vitro culture of *Aloe barbadensis* Mill.: morphogenetic ability and nuclear DNA content. Plant Sci 55: 53–59

Cavallini A, Lupi MC (1987) Cytological study of callus and regenerated plants of sunflower (*Helianthus annuus* L.). Plant Breed 99: 203–208

Cavallini A, Natali L (1989) Cytological analyses of in vitro somatic embryogenesis in *Brimeura amethystina* Salisb. (*Liliaceae*). Plant Sci 62: 255–261

Cavallini A, Cremonini R, Lupi MC, Bennici A (1986) In vitro culture of *Bellevalia romana* (L.) Rchb. II. Cytological study of callus and regenerated plantlets. Protoplasma 132: 58–63

Cavallini A, Lupi MC, Cremonini R, Bennici A (1987) In vitro culture of *Bellevalia romana* (L.) Rchb. III. Cytological study of somatic embryos. Protoplasma 139: 66–70

Cavallini A, Cremonini R, Cionini G, Cionini PG (1988) Polysomaty and somatic reduction in *Phaseolus coccineus* L. Genome 30: 671–676

Cecchini E, Natali L, Cavallini A, Durante M (1992) DNA variations in regenearated plants of pea (*Pisum sativum* L.). Theor Appl Genet 84: 874–879

D'Amato F (1948) Contributo alla migliore conoscenza del cariogramma di *Bellevalia romana* Sweet. Nuovo Giorn Bot Ital NS LV: 1–4

D'Amato F (1985) Cytogenetics of plant cell and tissue cultures and their regenerates. CRC Crit Rev, Plant Sci 3: 73–112

D'Amato F (1990) Somatic nuclear mutations in vivo and in vitro in higher plants. Caryologia 43: 191–204

De Lapeyrouse P (1808) *Bellevalia operculata*. J Physiol 67: 425

Deumling B, Clermont L (1989) Changes in DNA content and chromosomal size during cell culture and plant regeneration of *Scilla siberica*: selective chromatin diminution in response to environmental conditions. Chromosoma 97: 439–448

Feher F, Tarczy MH, Bocsa I, Dudits D (1989) Somaclonal chromosome variation in tetraploid alfalfa. Plant Sci 60:91–99

Feinbrun N (1940) A monographic study on the genus *Bellevalia* Lapeyr. Palest J Bot Jerus Ser 1,4: 336–409

Garbari F (1970) II genere *Brimeura* Salisb. (*Liliaceae*). Atti Soc Toscana Sci Nat Mem 3: 12–36

Groenewald EG, Wessels DCJ, Koeleman A (1977) Callus formation and subsequent plant regeneration from seed tissue of an *Agave* species (*Agavaceae*). Z Pflanzenphysiol 81: 369–373

Hussey G (1982) In vitro propagation of monocotyledonous bulbs and corms. In: Fujiwara A (ed) Plant tissue culture 1982. Maruzen, Tokyo, pp 677–680

Karp A (1991) On the current understanding of somaclonal variation. Oxford Surv Plant Mol Cell Biol 7: 1–58

Lu C, Vasil V, Vasil IK (1983) Improved efficiency of somatic embryogenesis and plant regeneration in tissue culture of maize (*Zea mays* L.). Theor Appl Genet 66: 285–289

Lupi MC, Bennici A, Baroncelli S, Gennai D, D'Amato F (1981) In vitro regeneration of *Durum* wheat plants. II. Diplontic selection in aneusomatic plants. Z Pflanzenzücht 87: 167–171

Lupi MC, Bennici A, Gennai D (1985) In vitro culture of *Bellevalia romana* (L.) Rchb. I. Plant regeneration through adventitious shoots and somatic embryos. Protoplasma 125: 185–189

Murashige T, Skoog F (1962)A revised medium for rapid growth and bioassays with tobacco tissue cultures. Physiol Plant 15: 473–497

Nagl W (1990a) Gene amplification and related events. In: Bajaj YPS (ed) Biotechnology in agriculture and forestry. vol 11. Somaclonal variation in crop improvement I. Springer, Berlin Heidelberg New York, pp 153–201

Nagl W (1990b) Polyploidy in differentiation and evolution. Int J Cell Cloning 8: 216–223

Natali L, Cavallini A (1987) Nuclear cytology of callus and plantlets regenerated from pea (*Pisum sativum* L.) meristems. Protoplasma 141: 121–125

Orton TJ (1985) Genetic instability during embryogenic cloning of celery. Plant Cell Tissue Organ Cult 4: 159–169

Peck DE, Cummings BG (1986) Beneficial effects of activated charcoal on bulblet production in tissue culture of *Muscari armeniacum*. Plant Cell Tissue Organ Cult 6: 9–14

Pijnacker LP, Walch K, Ferwerda MA (1986) Behaviour of chromosomes in potato leaf tissue cultured in vitro as studied by BrdU Giemsa labelling. Theor Appl Genet 72: 833–839

Rybczynski JJ, Zdunczyk W (1986) Somatic embryogenesis and plant regeneration in the genus *Secale*. 1. Somatic embryogenesis and organogenesis from cultured immature embryos of five wild species of rye. Theor Appl Genet 73: 267–271

Singh BD, Harvey RL, Kao KN, Miller RA (1972) Selection pressure in cell population of *Vicia hajastana* cultured in vitro. Can J Genet Cytol 14:65–72

Terzi M, Lo Schiavo F (1990) Somatic embryogenesis. In: Bhojwani SS (ed) Plant tissue culture: applications and limitations. Elsevier, Amsterdam, pp 54–66

Thorpe (1982) Callus organization and de novo formation of shoots, roots and embryos in vitro. In: Tomes DT, Ellis BE, Harney PM, Kasha KJ, Peterson RL (eds) Application of plant cell and tissue culture to agriculture and industry. University of Guelph, Guelph, pp 115–138

Tran Thanh Van K, Trinh TH (1990) Organogenic differentiation. In: Bhojwani SS (ed) Plant tissue culture: applications and limitations. Elsevier, Amsterdam, pp 34–53

Vasil IK (1986) Relative stability of embryogenic cultures of the *Graminae* and uniformity of regenerated plants. In: Semal J (ed) Somaclonal variations and crop improvement. Martinus Nijhoff, Dordrecht, pp 108–116

Vosa CG (1979) Heterochromatic banding patterns in the chromosomes of *Brimeura* (*Liliaceae*). Plant Syst Evol 132: 141–148

Wen F, Barnett FL, Liang GH (1989) Somatic pairing and separation of chromosomes in root tip cells of regenerated *Sorghum* plants. J Hered 80: 159–160

Zolfino C, Cavallini A, Cionini PG (1984) Appaiamento dei cromosomi omologhi e riduzione somatica in *Brimeura*. Nuovo G Bot Ital 118 Suppl 2: 158–160

IV.10 Somatic Embryogenesis in a Mistletoe – *Dendrophthoe falcata* (L.f.) Ettingsh.

Y.P.S. Bajaj[1]

1 General Account

Dendrophthoe falcata (Loranthaceae) commonly grows as a semiparasite on the branches of a wide range of trees (Singh 1954), especially *Ficus religiosa* and *Ehretia laevis*. It flowers and fruits during January to April. The tubular/companulate flowers are orange to red in color. Several embryo sacs develop simultaneously in the same ovary and their tips which enclose the egg apparatus extend up to two-thirds the length of the style (Singh 1952). Usually, several zygotes elongate appreciably and divide vertically. This is followed by transverse divisions resulting in biseriate proembryos comprising a terminal embryonal tier and the remaining suspensor tiers. Further transverse divisions of the embryonal and suspensor tiers and elongation of the latter push the proembryos through the style into the lower part of the embryo sac (situated in the ovary) containing a four-seriate cellular endosperm. After reaching the saucer-shaped hypostase, the embryonal tier gives rise to the globular, heart-shaped, and dicot stages. In a mature embryo, the cotyledons (except in the region of the plumule) become closely appressed against each other, giving a pseudomonocotyledonous appearance. The mature embryo is peg-shaped and the cotyledonary portion remains buried in the endosperm.

The fruits are mostly dispersed by birds, and the embryo, deposited on the bark of the host tree, starts to germinate. Due to desiccation of the viscid layer, the endosperm adheres closely to the bark. The radicular end swells, bends sharply at right angles to the cotyledons, and attaches itself to the host forming an irregular pad called the holdfast. The plumule also elongates and produces one or two pairs of leaves.

In vitro culture studies were conducted on globular embryos and mature embryos (with and without endosperm) of various mistletoes to study their growth behavior and mode of germination. During the course of these investigations (Johri and Bajaj 1962, 1963, 1964, 1965; Bajaj 1963, 1966, 1967, 1968, 1970), somatic embryogenesis/polyembryony was observed in vitro, regenerating potentials studied, and the growth of seedlings on synthetic media was observed. This chapter is primarily based on the author's work (Bajaj 1963).

[1] Former Professor of Tissue Culture, Punjab Agricultural University, Ludhiana (Punjab), India. Present address: A–137 New Friends Colony, New Delhi 110065, India

Biotechnology in Agriculture and Forestry, Vol. 31
Somatic Embryogenesis and Synthetic Seed II (ed. by Y.P.S. Bajaj)

2 Material and Methods

The fruits at various stages of development were excised and sterilized either with chlorine water for nearly 15 min, or by dipping them in 90% ethanol and subjecting them to flaming. The latter method proved to be better. The proembryos were dissected out under a microscope under aseptic conditions, whereas mature embryos could simply be squeezed out of the mucilaginous fruit. The embryos were cultured on basal medium (BM) of White (1943) supplemented with various combinations and concentrations of indole-acetic acid (IAA), casein hydrolysate (CH), and coconut milk (CM) and gelled with 0.8% agar. The pH of the medium (5.6–5.8) was adjusted before autoclaving. The cultures were maintained in diffuse daylight at 25 ± 2 °C and 55–65% relative humidity.

The cultured material was fixed in FAA or acetic alcohol. The squashes and whole mounts of callus and somatic embryos were stained with acetocarmine, dehydrated through an acetic acid-n-butyl alcohol series, and mounted in Canada balsam.

3 Experimental Results

3.1 Culture of Globular Proembryos (Figs. 1 and 2; Table 1)

The globular embryos[2] (0.8–1.5 mm long; Fig. 1A), when cultured on BM alone, attained the heart-shaped stage and in a few cases the cotyledons developed within 6 weeks. With the addition of 1 mg/l IAA, the proembryo showed only slight callusing and failed to attain maturity.

On BM + 1 mg/l IAA+ 1000 mg/l CH the proembryo swelled (Fig. 1B) and became green in 2–4 days. During the next week, the proembryo attained the heart-shaped stage in most of the cultures. The elongation of the two cotyledonary primordia was accompanied by the formation of a callus between them. In 2 or 3 weeks the callus covered the entire embryo, and after another 2 to 4 weeks 20–40 somatic embryos (the terms accessory embryos, embryo-like structures, adventitious embryos, accessory embryos, and secondary embryos have been used in previous publications) differentiated from the callused base of the proembryo. Sometimes, after 3 or 4 weeks, embryos differentiated over the entire surface of the proliferated proembryo (Fig. 2A). In still other cultures the proembryo swelled and within 6 weeks gave rise to a number of outgrowths simulating the globular, heart-shaped, and early dicot stages (Fig. 2B).

---

[2] These embryos were invariably attached to the suspensor (length of the suspensor, 0.6–1.1 mm; length of embryonal mass, 0.2–0.4 mm).

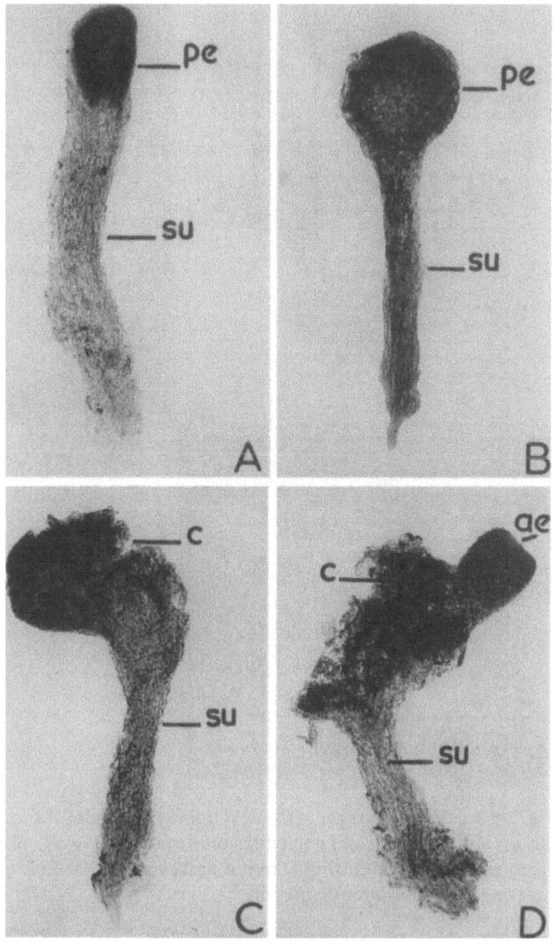

Fig. 1A–D. Various stages in the initiation of callus and the differentiation of adventitious embryos (somatic embryos) from a globular proembryo of *Dendrophthoe falcata* cultured on BM+1 mg/l IAA+1000 mg/l casein hydrolysate. *ae* adventive embryo; *c* callus; *pe* proembryo; *su* suspensor. (Bajaj 1963, Plate 15)

Occasionally, the embryos passed through the normal pregerminal stages and the cotyledons lying in contact with the medium became flattened and leaf-like. Subsequently, the cotyledons proliferated forming a massive callus which differentiated into numerous somatic embryos. The suppression of one cotyledon or the formation of more than two cotyledons was common.

The combined effect of different concentrations of IAA and CH on the response of embryos is shown in Table 1. There was no polyembryony on BM alone or BM containing IAA (even up to 2 mg/l). With the addition of CH to BM the percentage of polyembryony varied from 16 to 40; 25 to 62% polyembryony occurred with 0.25 to 2 mg/l IAA+500 to 2000 mg/l CH (see Table 1).

The globular proembryo grown on BM+1 mg/l IAA developed into a small dicot structure, or formed callus which remained undifferentiated. On BM+CH the proembryos developed to maturity with free cotyledons. In 16 to 40% of the

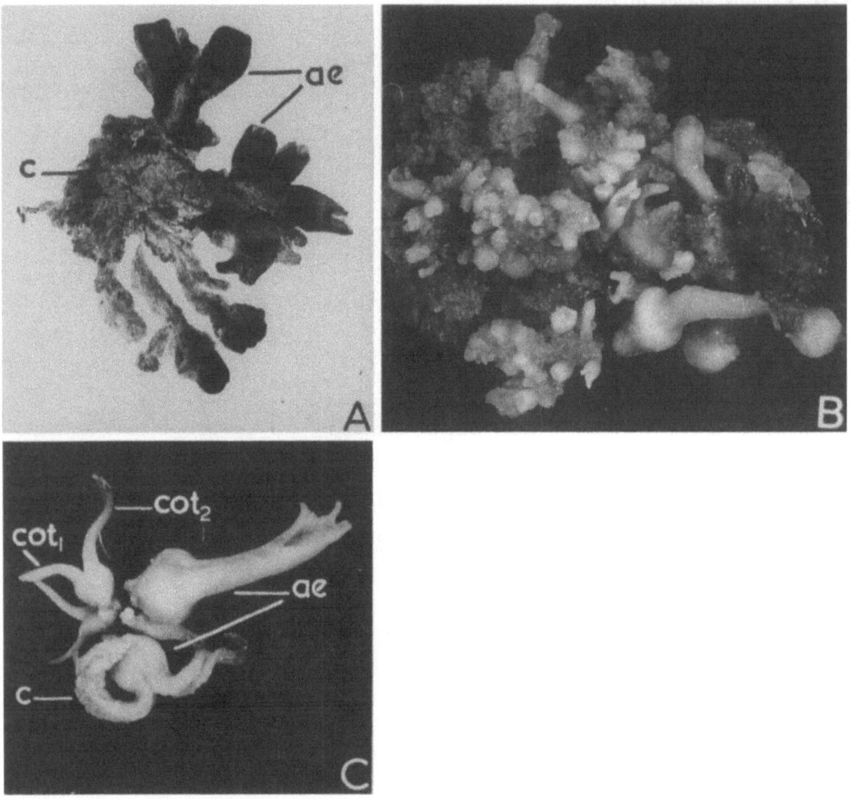

Fig. 2A–C. Various stages in the development and maturation of somatic embryos (Bajaj 1963, Plates 16 and 19). **A** 6-week-old polyembryonal mass; **B** 15-week-old polyembryonal mass of callus showing various stages of maturing somatic embryos; **C** 20-week-old polyembryonal mass with embryos showing well-developed cotyledons

cultures the proembryos proliferated and the callus differentiated into a polyembryonal mass. In each mass there were 20–50 somatic embryos at various stages of development and the length of the mature embryos varied from 6 to 9 mm. In these embryos the cotyledons remained free from each other (Fig. 2C), in contrast to the pseudomonocot condition in vivo.

On BM+1 mg/l IAA+1000 mg/l CH the proembryo either passed through the normal stages of development leading to the formation of mature embryos with free cotyledons, or proliferated extensively to produce a polyembryonal mass (Fig. 2B).

The addition of 10–25% CM to BM+1 mg/l IAA proved equally effective for the growth of the proembryos, and the induction of polyembryony. In 20–40% of the cultures the proembryo proliferated and differentiated into 30–40 accessory embryos in 3–5 weeks.

Even when the suspensor was separated and only the proembryonal mass was grown on BM+1 mg/l IAA+25% CM, it turned dark green, formed callus

Table 1. Interaction of CH + IAA on the induction of polyembryony in *Dendrophthoe falcata.* (Johri and Bajaj 1965)

Indoleacetic acid (mg/l)	Casein hydrolysate (mg/l)					
	0	125	250	500	1000	2000
0	0	16	16	18	27	40
0.125	0	22	20	16	18	30
0.25	0	18	18	40	25	43
0.5	0	43	36	40	50	62
1	0	27	33	44	57	57
2	0	28	20	25	40	50
	Percentage of cultures showing polyembryony					

within 2 weeks, and produced 10–25 accessory embryos in another 2 weeks. In some of the 5-week-old cultures the cotyledons became flattened and leaf-like and proliferated almost everywhere into a callus. Occasionally, some of the accessory embryos in the polyembryonal mass produced one to five shoot apices.

3.2 Culture of Mature Zygotic Embryos (Figs. 3 and 4)

The mature embryo has a broad radicular and a pointed cotyledonary end. The two cotyledons are adpressed to one another, and the whole embryo except the radicular end is enclosed in the endosperm.

The mature embryos, cultured with and without endosperm, showed various modes of development as shown in Fig. 4. The mature embryos without endosperm germinated in situ, formed holdfast, haustorial disks, and plumular leaves. On the other hand, embryos with the intact endosperm showed various patterns of growth, development, and differentiation. In some cases, the embryo along with the endosperm formed a mass of callus (Fig. 3A) which eventually differentiated to form embryo-like structures/buds/shoots (Fig. 3 B,C). A 20-week-old culture with bulbous, callused holdfast, numerous accessory leaves, and a pair of plumular leaves is shown in Fig. 3D, whereas Fig. 3E shows the somatic embryos at the radicular end of the embryo.

3.3 Culture and Germination of Somatic Embryos (Fig. 5)

The growth behavior of the somatic embryos in vitro depended to a large extent on the stage of development at the time of culture. The very young (preglobular, globular, and early heart-shaped) developing somatic embryos, when excised and recultured on the same medium, had a tendency to undergo proliferation to form callus which eventually differentiated and formed a polyembryonal mass of accessory embryos (Fig. 2A). However, in somewhat older embryos (late heart-shaped, early dicot stage) the cotyledons developed and callused, but fully

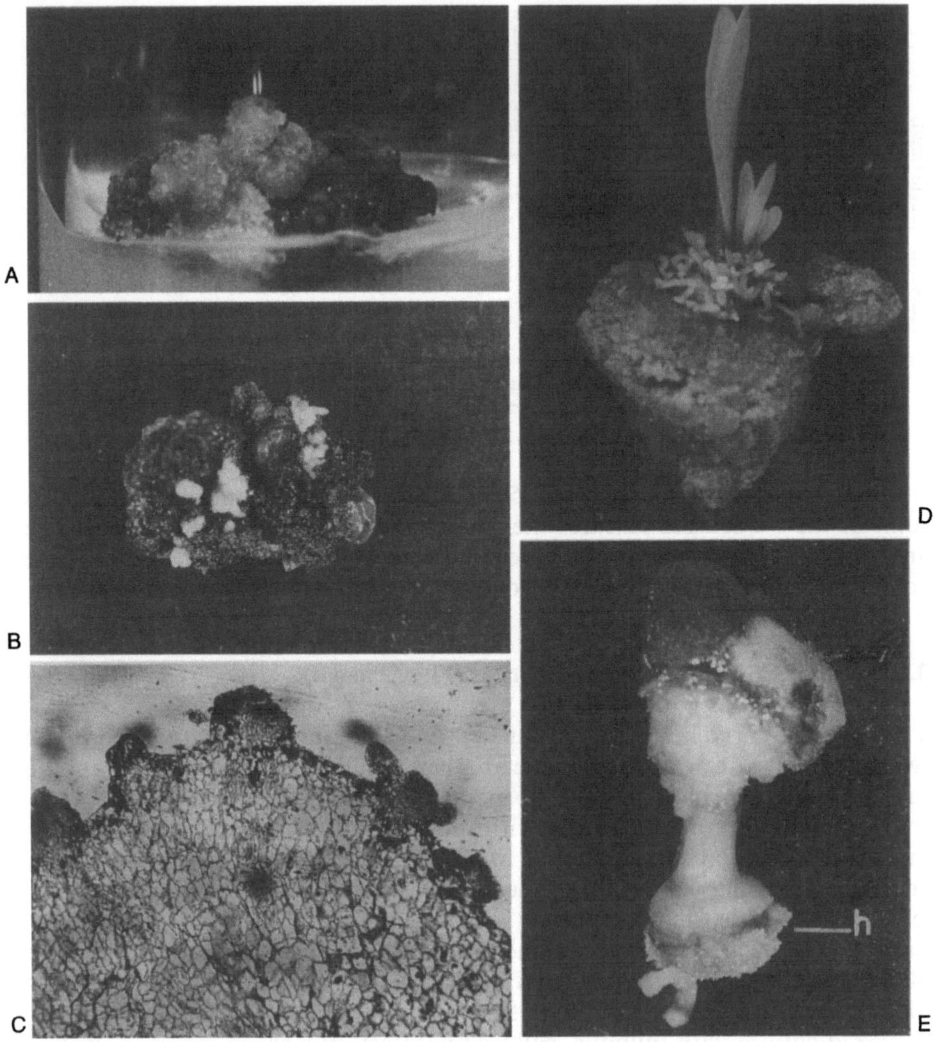

Fig. 3A–E. Organogenesis/somatic embryogenesis in mature embryos with an intact endosperm cultured on BM+0.5 mg/l IAA+500 mg/l CH. **A** 10-week-old culture showing a mass of callus formed as a result of proliferation of embryo/endosperm. **B,C** Endosperm-derived callus showing differentiation of embryo-like structures/buds/shoots. **D** 20-week-old seedling showing the bulbous, callused holdfast with numerous accessory leaves and a pair of plumular leaves. **E** 15-week-old seedling showing somatic embryos at the proliferated radicular end; *h* holdfast

mature embryos were rarely formed. Some embryos formed multiple shoots. The older/mature somatic embryos with well-developed cotyledons underwent germination in situ in 5–25 weeks and formed holdfast and haustorial disks (Fig. 5B,C), like the ones obtained by the in vitro germination of zygotic embryos (Fig. 4).

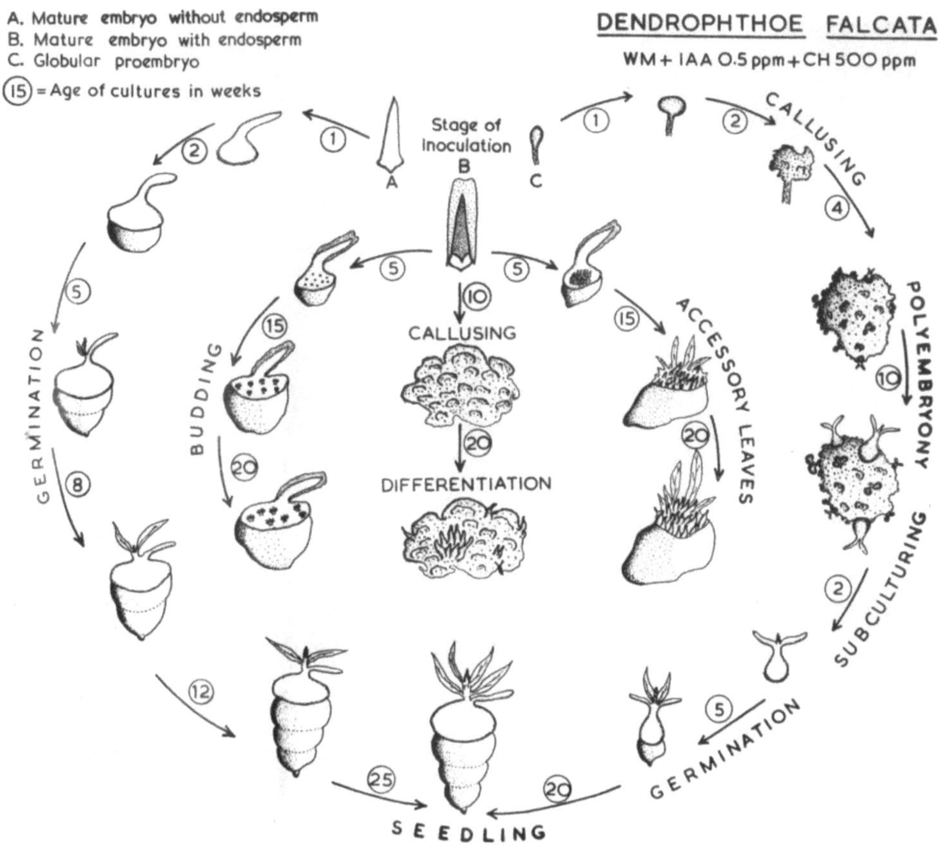

Fig. 4. Diagrammatic representation showing morphogenesis/somatic embryogenesis in cultures derived from proembryos, mature embryos, and embryos cultured with the endosperm; *WM* White's basal medium. (Bajaj 1963, Plate 27)

4 Summary

Embryos of *Dendrophthoe falcata* (a semiparasitic stem mistletoe), cultured at various stages of development on basal synthetic medium supplemented with IAA and casein hydrolysate, showed various modes of development, organogenesis, and somatic embryogenesis (Fig. 4). The young globular and preglobular embryos underwent proliferation to form callus which eventually differentiated into polyembryonal masses of somatic embryos. The mature zygotic embryos, cultured without the endosperm, germinated directly, whereas the mature embryos with an intact endosperm showed overall callusing of the embryo and the endosperm, differentiation of shoot buds, papillate structures, and somatic embryos.

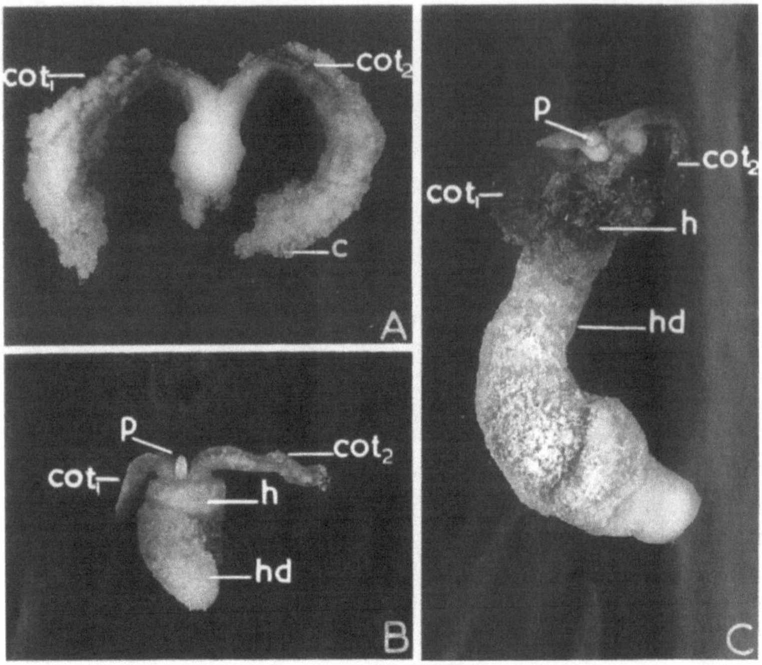

Fig. 5A–C. Various stages in the maturation and germination of somatic embryos (Bajaj 1963, Plate 18). **A** 10-week-old subculture of an embryo showing callused cotyledons (*cot*$_1$, *cot*$_2$); **B,C** 20- and 25-week-old subcultures of somatic embryos showing germination and the development of haustorium and the haustorial disks. *c* Cotyledon; *h* haustorium; *hd* haustorial disk; *p* plumule

The young somatic embryos, when excised from the polyembryonal mass and cultured on the same medium, proliferated and formed accessory embryos, whereas the mature somatic embryos with well-developed cotyledons underwent germination.

References

Bajaj YPS (1963) In vitro studies on the embryos, ovules and ovaries of some angiosperms. PhD Thesis, Univ Delhi

Bajaj YPS (1966) Behavior of embryo segments of *Dendrophthoe falcata* (L.f.) Ettings. in vitro. Can J Bot 44: 1127–1131

Bajaj YPS (1967) In vitro studies on the embryos of two mistletoes, *Amyema pendula* and *Amyema miquelli*. NZ J Bot 5: 49–56

Bajaj YPS (1968) Some factors affecting growth of embryos of *Dendrophthoe falcata* in cultures. Can J Bot 46: 429–433

Bajaj YPS (1970) Growth responses of excised embryos of some mistletoes. Z Pflanzènphysiol 63: 408–415

Johri BM, Bajaj YPS (1962) Behavior of mature embryos of *Dendrophthoe falcata* (L.f.) Ettings. in vitro. Nature (Lond) 193: 194–195

Johri BM, Bajaj YPS (1963) In vitro response of the embryo of *Dendrophthoe falcata* (L.f.) Ettings. In: Maheshwari P, Rangaswamy NS (ed) Plant tissue and organ culture – a symposium. Int Soc Plant Morphol, Delhi pp 292–301

Johri BM, Bajaj YPS (1964) Growth of embryos of *Amyema, Amylotheca* and *Scurrula* on synthetic media. Nature (Lond) 204: 1220–1221

Johri BM, Bajaj YPS (1965) Growth responses of globular proembryos of *Dendrophthoe falcata* (L.f.) Ettings. in culture. Phytomorphology 15: 292–300

Singh B (1952) A contribution to the floral morphology and embryology of *Dendrophthoe falcata* (L.f.) Ettings. J Linn Soc (Bot) 53: 449–473

Singh B (1954) Studies in the family Loranthaceae – list of new hosts of *Dendrophthoe falcata* (L.f.) Ettings. Its relations with hosts. The anatomy of its seedlings and mature haustorium. Agra Univ J Res Sci. 3: 301–315

White PR (1943) A handbook of plant tissue culture. The Jaques Cattal Press, Tempe, AZ

IV.11 Somatic Embryogenesis in Buckwheat (*Fagopyrum* Mill.) and Sorrel (*Rumex* L.), Polygonaceae

M. Nešković[1], Lj. Ćulafić[1], and R. Vujičić[2]

1 Introduction

1.1 Distribution

The genera *Rumex* and *Fagopyrum* are members of the family Polygonaceae, which includes, in addition, *Polygonum* and *Rheum* and a few genera of limited distribution and significance. There are a few hundred species in the genus *Rumex*, divided into three subgenera: *Lapathum*, *Acetosa*, and *Acetosella*. The genus, which probably originated in Europe, displays at present a worldwide distribution. Due to the richness in morphological forms and ecological adaptations, the genus is considered an ideal material for studies on plant taxonomy, cytogenetics, and speciation (Löve 1944).

The genus *Fagopyrum* comprises about 15 species. The mountain regions of central and northeast Asia are believed to be its land of origin. Common buckwheat, *F. esculentum*, is the only widely cultivated species. Tartary buckwheat, *F. tataricum*, and the perennial *F. cymosum* are occasionally cultivated, but they usually grow as wild populations.

1.2 Economic Importance

Most *Rumex* species grow as common weeds in pastures and disturbed soils. *R. acetosa*, comprising several widespread varieties, seems to be the only species cultivated for human consumption. It is a leafy vegetable; young leaves are rich in K, Fe, and other mineral salts, carotene, ascorbic acid, and oxalic acid, which amounts to 10–13% of dry weight.

Buckwheat is a much more important crop. Its use for human consumption was mentioned in some Chinese writings more than a 1000 years ago. According to Pomeranz (1983), common buckwheat is nowadays grown on about 2 million ha all over the world and its annual production amounts to 1 million metric tons. Buckwheat is not a demanding crop and grows well on rather poor soils. Its

[1] Institute of Botany, Faculty of Science, and Institute for Biological Research "Siniša Stanković", University of Belgrade, 29 novembra 142, 11060 Belgrade, Yugoslavia
[2] Center for Multidisciplinary Studies, University of Belgrade, 29 novembra 142, 11060 Belgrade, Yugoslavia

Biotechnology in Agriculture and Forestry, Vol. 31
Somatic Embryogenesis and Synthetic Seed II (ed. by Y.P.S. Bajaj)

nutritive value is, nevertheless, very high (Sure 1955). Buckwheat grains contain 11–15% storage proteins, which are remarkably rich in lysine (6.1%) and other essential amino acids (Pomeranz and Robbins 1972; Javornik 1983).

Both buckwheat and certain sorrel species contain substances of pharmaceutical importance. These include rutin (Bowen and Cubbin 1993), which in buckwheat leaves reaches up to 6% of dry weight. *Rumex* species contain anthraquinones and their derivatives (van den Berg and Labadie 1989). The capacity for anthraquinone synthesis is reportedly retained in callus cultures (Koblitz 1988).

1.3 Botanical Characteristics and Biotechnology

The *Fagopyrum* species are diploid (2x = 2n = 16), but tetraploid varieties occur either spontaneously, or can be induced (De Jong 1972). Buckwheat has for centuries remained a crop with low yield (De Jong 1972), due to certain characteristics which prevent the application of conventional breeding methods (Kreft 1983). The main obstacles in buckwheat breeding include its very strong sporophytic self-incompatibility and its indeterminate type of growth and flowering. As pointed out earlier (Nešković et al. 1986), modern biotechnology may provide means to address these problems in a novel way.

The basic chromosome number in most *Rumex* species is x = 7. There exist diploid, tetraploid, hexaploid, and octoploid varieties. Male plants in the subgenus *Acetosa* have an additional Y chromosome (2n = 15). Bearing in mind the limited use of *Rumex* species, it is conceivable why they did not attract much interest of breeders. All species belonging to the subgenera *Acetosella* and *Acetosa* (section Euacetosae) include only dioecious plants. These plants, being obligatorily cross-pollinated, are highly heterozygous. In order to provide genetically uniform material for studies on sex determination, methods of clonal propagation of *Rumex* species have been elaborated in the author's laboratory.

1.4 Brief Survey of Tissue Culture Work

In vitro studies on species of the family Polygonaceae are rather scarce. A survey of results obtained with *Fagopyrum* and *Rumex* species is presented in Table 1. Reports dealing with somatic embryogenesis are listed in Table 2. Roggemans and Boxus (1988) have reviewed the research on species belonging to the genus *Rheum*.

2 Somatic Embryogenesis

Somatic embryogenesis has been hitherto described in *F. esculentum*, *F. tataricum*, and *R. acetosella* (Table 2). All species displayed the capacity for simultaneous bud formation and embryogenesis.

Table 1. Summary of in vitro studies on *Fagopyrum* and *Rumex*. (See also Bowen and Cubbin 1993)

Species and primary explant	Growth response[a]	Purpose of study	Reference
F. esculentum			
Germinating seeds	C, P	Callus induction, plant regeneration, cytological observation	Yamane (1974)
Cotyledon fragments	C, P	Organogenesis, plant regeneration	Srejović and Nešković (1981)
Hypocotyl sections	Cs	Compartmentation of shikimic acid and phenylalanine	Holländer-Czytko and Amrhein (1983)
Seedling stem tips	S	Micropropagation	Bohanec (1985)
Hypocotyl segments	C,P	Callus formation, plant regeneration, cytological observation	Takahata and Jumonji (1985)
Young leaves	C, P	Plant regeneration	Surikov and Mazur (1985)
Stem tips, inflorescence	P	Micropropagation	Bohanec (1987)
Young inflorescence	C, P	Regeneration capacity	Takahata (1988)
Anthers	C, P	Plant regeneration	Adachi et al. (1988)
Hypocotyl sections	Pr, C, P	Protoplast culture, plant regeneration	Adachi et al. (1989)
Seedling stem tips	P	Micropropagation of determinate plants	Nešković et al. (1990a)
Stem, leaves	C, R	Crown gall and hairy root induction	Nešković et al. (1990b)
Hypocotyl segments	Pr	Protoplast culture and regeneration	Lachmann (1991)
Cotyledon fragments	P	Genetic transformation	Miljuš-Djukić et al. (1992)
Anthers	C, P	Induction of androgenesis, dihaploid plants	Bohanec et al. (1993)
F. tataricum			
Hypocotyl sections	Pr	Protoplast culture and regeneration	Lachmann and Adachi (1990)
Immature embryo	P	Embryo rescue	Samimy (1991)
Hypocotyl sections	C, Pr	Protoplast culture and regeneration	Lachmann (1991)
F. cymosum			
Young inflorescence	C, P	Regeneration capacity	Takahata (1988)

Hypocotyl sections	C, Pr	Protoplast culture and regeneration	Adachi et al. (1989)
Hypocotyl sections	C, Pr	Protoplast culture and regeneration	Lachmann (1991)
R. acetosa			
Virus tumor	C, S, R,	Nutritional requirements in culture	Nickell (1954)[b]
Hypocotyl sections	C, P	Cloning of male and female plants	Ćulafić et al. (1987b)
R. acetosella			
Hypocotyl sections	C, P	Cloning of male and female plants	Ćulafić et al. (1987b)
R. abyssinicus			
Apical and lateral buds	C, P	Vegetative propagation	(unpubl.)
R. patientia			
Root segments	C	Anthraquinone synthesis	Suchy et al. (1973)
R. alpinus			
Root segments	C	Anthraquinone synthesis	Van den Berg and Labadie (1981)

[a] C = callus; P = Plant; S = shoot; R = root; Pr = protoplasts; Cs = cell suspension.
[b] Review article, see also references cited therein.

Table 2. Somatic embryogenesis in *Fagopyrum* and *Rumex*

Species and cultivar	Ploidy	Primary explant	Embryogenic capacity recorded in	Reference
F. esculentum				
Pennquad	4n	Immature embryos	Immature embryos	Nešković et al. (1987)
Kazanskaja krupnozernaja	NS	Immature embryos	Globular callus	Rumyantseva et al. (1989)
Miyazaki-ootsubu	4n	Hypocotyl sections	Protoplast-derived calli	Adachi et al. (1989)
Miyazaki-zairai	2n	Hypocotyl sections	Hypocotyl callus	Lachmann (1991)
Sando-soba	NS	Hypocotyl sections	Hypocotyl callus	Lachmann (1991)
La Harpe	NS	Hypocotyl sections	Hypocotyl callus	Lachmann (1991)
F. tataricum				
Wild growing	2n	Immature embryos	Globular callus	Rumyantseva et al. (1989)
R. acetosella				
Wild growing	6n	Axillary shoot tips	Leaves, calli	Ćulafić et al. (1987a)

NS = not stated.

2.1 Initiation

2.1.1 Explants

Somatic embryogenesis in buckwheat tetraploid variety Pennquad was induced in proliferating tissue of immature zygotic embryos or, less frequently, directly on the cotyledon surface. Immature embryos, 2–3-mm-long, were isolated from unripe, green fruits, in which they were embedded in a gelatinous environment (Nešković et al. 1987). Rumyantseva et al. (1989) stressed the importance of the zygotic embryo developmental stage in common and Tartary buckwheat. They reported that the formation frequency of appropriate callus type and, hence, the subsequent somatic embryogenesis, was highest in embryos 1–3.5 mm long, at the milky ripeness stage. Adachi et al. (1989) noted that calli obtained by protoplast division gave rise to a few embryoids. Lachmann (1991) reported the induction of somatic embryogenesis in calli derived from hypocotyl segments.

Embryo formation in *R. acetosella* occurred on previously established shoot cultures. The primary explants from which these shoots originated consisted of vegetative axillary buds of male and female field-grown adult plants (Ćulafić et al. 1987b). They were propagated in vitro by a multiple bud technique or de novo organogenesis. The tissues at the shoot base, on petioles, and on the leaf base, were capable of producing a soft, friable callus and somatic embryos under certain conditions.

2.1.2 Media

The composition of the media used for somatic embryogenesis in the authors' laboratory is presented in Table 3. The media used by other authors differed mainly in hormone content. Rumyantseva et al. (1989) used three different media containing (1) kinetin, α-naphthalene acetic acid (NAA), and gibberellic acid (GA_3) (0.05 mg/l each); (b) 2,4-dichlorophenoxyacetic acid (2,4-D) (0.1 mg/l); or (3) hormone-free medium. The content of sugars (sucrose and glucose) in the media did not exceed 3%. Lachmann (1991) used 2 mg/l 6-benzylaminopurine (BAP), 0.1 mg/l NAA, and 6% sucrose.

2.1.3 Hormonal and Osmotic Factors in Embryo Induction

Numerous experiments have shown that the procedure elaborated for buckwheat shoot regeneration in cotyledon fragments (Srejović and Nešković 1981) is also suitable for regeneration in immature embryos. The isolated embryos of *F. esculentum* were cultured on medium (A), containing a high 2,4-D concentration, for 5 days, and then transferred to medium (B) (Table 3). The embryos displayed three patterns of development. In 20–30% of the explants a voluminous callus was developed, frequently with numerous roots. The rest of the explants produced buds only (about 30%), or buds and somatic embryos at the same time (about 30–40%). An entirely embryogenic tissue was selected

Table 3. Media for induction and growth of somatic embryos

Basal Medium			Hormones (mg/l)	Reference
Mineral salts	Sugar (g/l)	Vitamins, etc. (mg/l)		
F. esculentum				
B5[a]	Sucrose 60	Thiamine 0.4 Pyridoxine 0.5 Nicotinic acid 0.5 *myo*-Inositol 100 CH 2000	Induction: (A) 2,4-D, Kinetin 0.1 (B) IAA 0.17, BAP 2.2 Plant growth: (C) IBA 0.2–0.5	Nešković et al. (1987)
R. acetosella				
MS[a]	Sucrose 60, or Sucrose 20 + mannitol or sorbitol 21.3	Thiamine 1 Pyridoxine 1 Nicotinic acid 5	Induction: (D) BAP 2.2, IAA 0.17 Maturation: (E) GA_3 10 Plant growth: (F) None	Ćulafić et al. (1987a)

[a] All media were solidified by 0.7% agar, except (C) and (E) which were sometimes liquid. Abbreviations: B5 = Gamborg et al. (1968); MS = Murashige and Skoog (1962); CH = casein hydrolysate; 2,4-D = 2,4-dichlorophenoxyacetic acid; IAA = indole-3-acetic acid; BAP = 6-benzylaminopurine; GA_3 = gibberellic acid.

through several subcultures. When the pretreatment with high 2,4-D content was omitted, the percentage of embryogenic structures fell to about 10%. It is worth mentioning that direct embryo formation on the cotyledon surface, without the formation of buds or the intervening callus, was occasionally observed only if 2,4-D was omitted. However, further attempts to increase the frequency of direct somatic embryogenesis by using different amounts of cytokinin and auxin failed.

Another requirement for somatic embryogenesis in *F. esculentum* was the increased sucrose concentration. All embryogenic cultures were initiated only in the presence of 6% sucrose (Nešković et al. 1987). In subsequent experiments with the diploid varieties, Siva and Darja, embryogenic structures sporadically developed on 3% sucrose, but did not exhibit a high proliferation capacity (unpubl.). Lachmann (1991) reported that somatic embryogenesis was initiated only when hypocotyl explants were grown on media containing 6% sucrose.

The organogenic callus of *R. acetosella*, described above, was grown for about 40 subcultures on shoot induction medium (Ćulafić et al. 1987a). Embryogenesis was induced on the same medium by raising the sucrose concentration from 2 to 6%. The same result was obtained by supplementing 2% sucrose with mannitol or sorbitol (Table 3), thus producing the same osmolality as with 6% sucrose. On media with low osmotic potentials the initiation of buds declined, and the occurrence of somatic embryos greatly prevailed (Ćulafić et al. 1987a).

2.2 Embryo Maturation and Conversion to Plantlets

Embryogenic clones of buckwheat could be maintained on the induction medium for years. The tissue consisted almost entirely of somatic embryos in various developmental stages. Meristematic globules and young embryos were surrounded by unorganized callus tissue (Figs. 1 and 2) and the root pole was well delineated from callus cells. The small embryos were often released together with callus globules at the suspensor site (Fig. 3). Secondary embryo formation (Fig. 4) was the main pattern of proliferation. At cotyledonary stages many embryos were short, with more than two cotyledons, usually fused together. Treatments with cytokinins and auxins, in various combinations and balance, did not improve embryo quality. Low concentrations of GA_3 were ineffective, whereas higher ones were detrimental. Abscisic acid slightly increased the volume of cultures, but did not affect embryo differentiation (unpubl.). Some embryos spontaneously elongated and were transplanted to the hormone-free medium. Embryos could also be converted to plantlets by growing them singly on filter paper bridges in liquid medium-containing test tubes. Root growth could be improved by adding 0.2–0.5 mg/l indole-3-butyric acid (IBA) (Table 3, medium C).

Somatic embryos in *R. acetosella* appeared upon transferring leaves to medium with 6% sucrose (Fig. 5). Histological sections of tissues grown on medium (D) with 6% sucrose (Table 3) show numerous meristematic centers (Fig. 6), and embryos in globular (Fig. 7) to torpedo stages with a clearly discernible root meristem (Fig. 8). Most embryos did not develop spontaneously

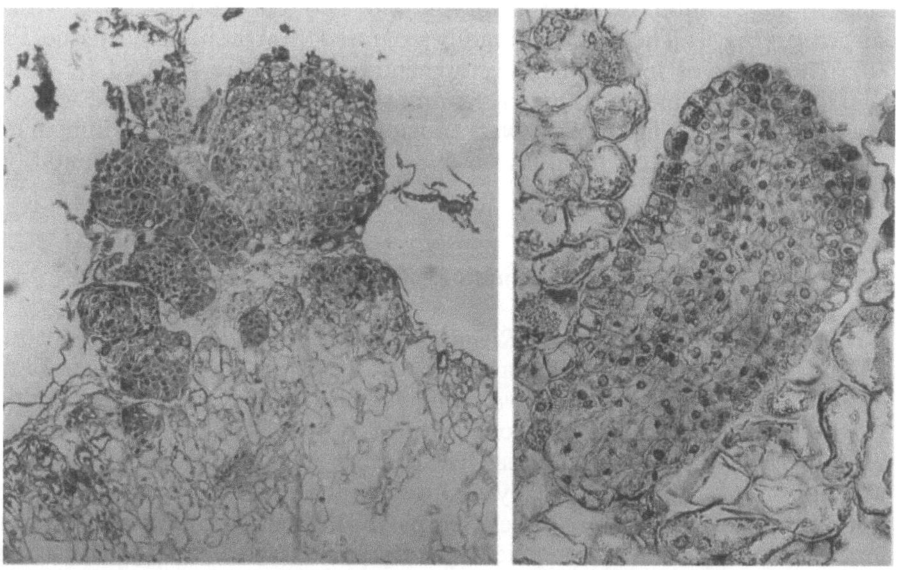

1 2

Figs. 1,2. Histology of somatic embryogenesis in *F. esculentum*. **Fig. 1.** Meristematic globules in an embryogenic callus. **Fig. 2.** An embryo at the torpedo stage. (Nešković et al. 1987)

3 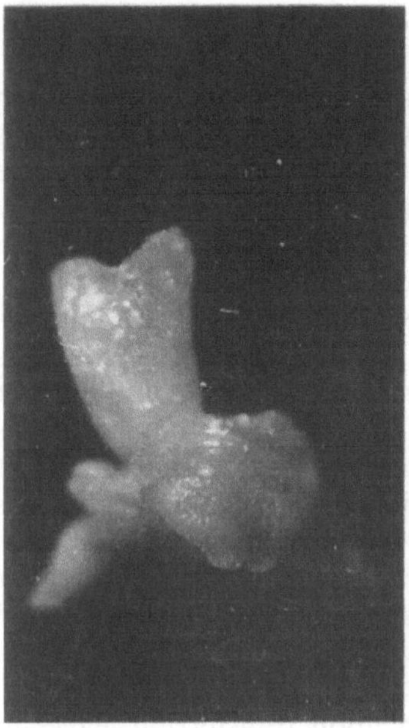 4 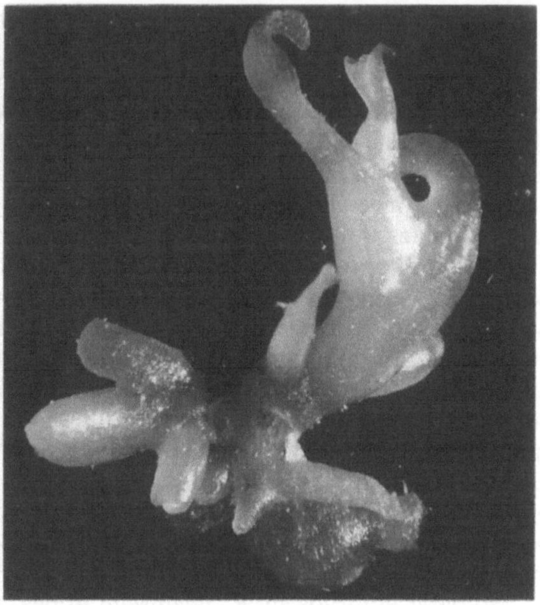

Figs. 3,4. Somatic embryos of *F. esculentum*. **Fig. 3.** A young embryo with a callus globule at the suspensor site, liberated from the mass of embryogenic callus. **Fig. 4.** Secondary embryogenesis; several young adventitious embryos developing on an embryo in advanced stage

into plantlets, their growth being arrested at the torpedo stage. Omission of BAP and IAA, as well as addition of GA_3 (Table 3, medium E) were required for embryo maturation. Mature embryos had two cotyledons and a visible suspensor (Fig. 9). They could be readily separated from the maternal tissue (Fig. 10). The embryos in various developmental stages were released by gently shaking in liquid medium (Fig. 11). Almost all embryos germinated within 2 weeks after GA_3 addition. They were transplanted into the solid medium in test tubes (Table 3,F). Most plantlets survived the transfer to nonsterile conditions fairly well.

2.3 Genotypic Specificity and Somatic Embryogenesis

Given that both *Fagopyrum* and *Rumex* are outbreeding species, each individual seedling may be considered as a particular genotype. Genotypic factors for somatic embryogenesis in *Fagopyrum* seem to be markedly expressed, as judged by the finding that only about one-third of zygotic embryos in each sample exhibited an embryogenic potential. The number of embryogenic explants could not be significantly increased by controlling the medium composition. The authors believe that different hitherto employed media were merely permissive to embryo development in competent genotypes. It is noteworthy that a high genotypic specificity was likewise reported in the formation of buckwheat androgenic calli and embryos (Bohanec et al. 1993). Besides, somatic

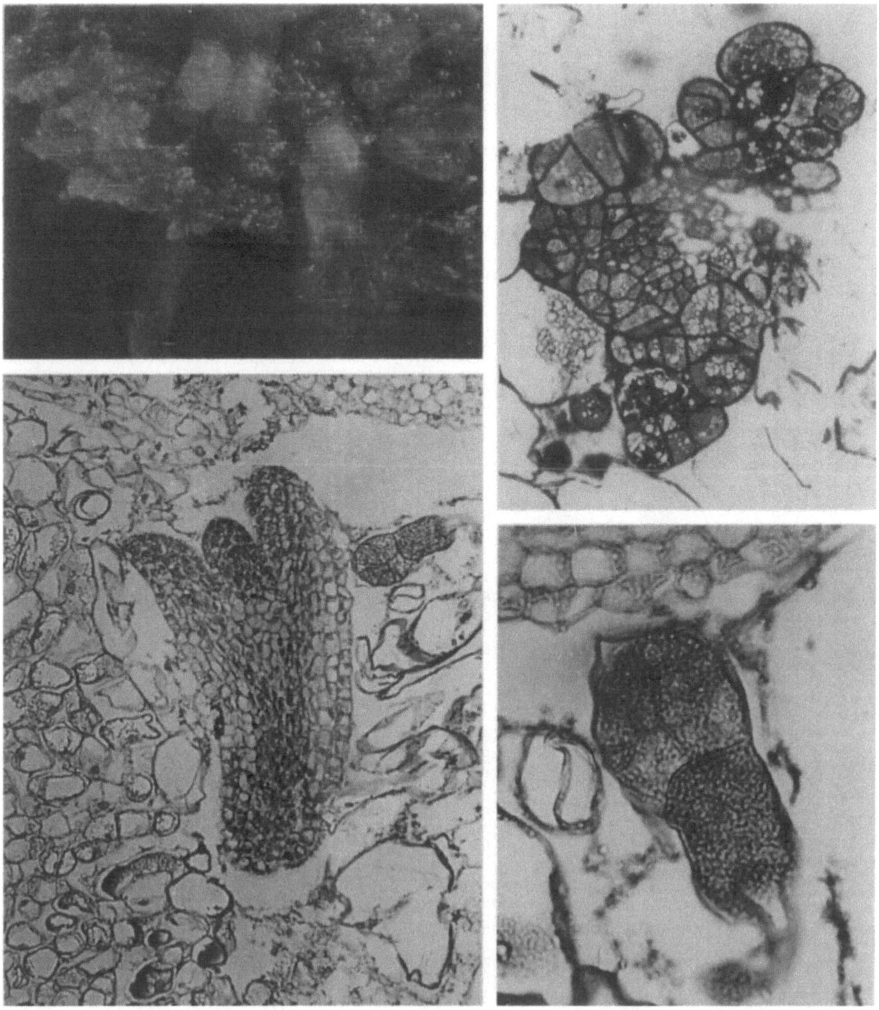

5 6 8 7

Figs. 5–8. Somatic embryogenesis in *R. acetosella.* **Fig. 5.** Somatic embryos formed at the leaf base. **Fig. 6.** Embryogenic callus with meristematic clusters. **Fig. 7.** A proembryo made up of several cells and a suspensor. **Fig. 8.** Embryo at cotyledonary stage with defined polarity. (Ćulafić et al. 1987a)

embryogenesis occurred either during the primary culture period and the first subculture, or it never occurred. Contrary to the findings of other authors (Rumyantseva et al. 1989; Lachmann 1991), embryogenesis was not inducible in established callus or shoot cultures.

Unlike buckwheat, *R. acetosella* seems to possess a high intrinsic capacity for somatic embryogenesis. This pattern of development could be induced in all clones tested so far, regardless of the preceding culture periods. However, under the same conditions *R. acetosa* and *R. abyssinicus*, which have been readily multiplied in vitro, did not display an embryogenic potential.

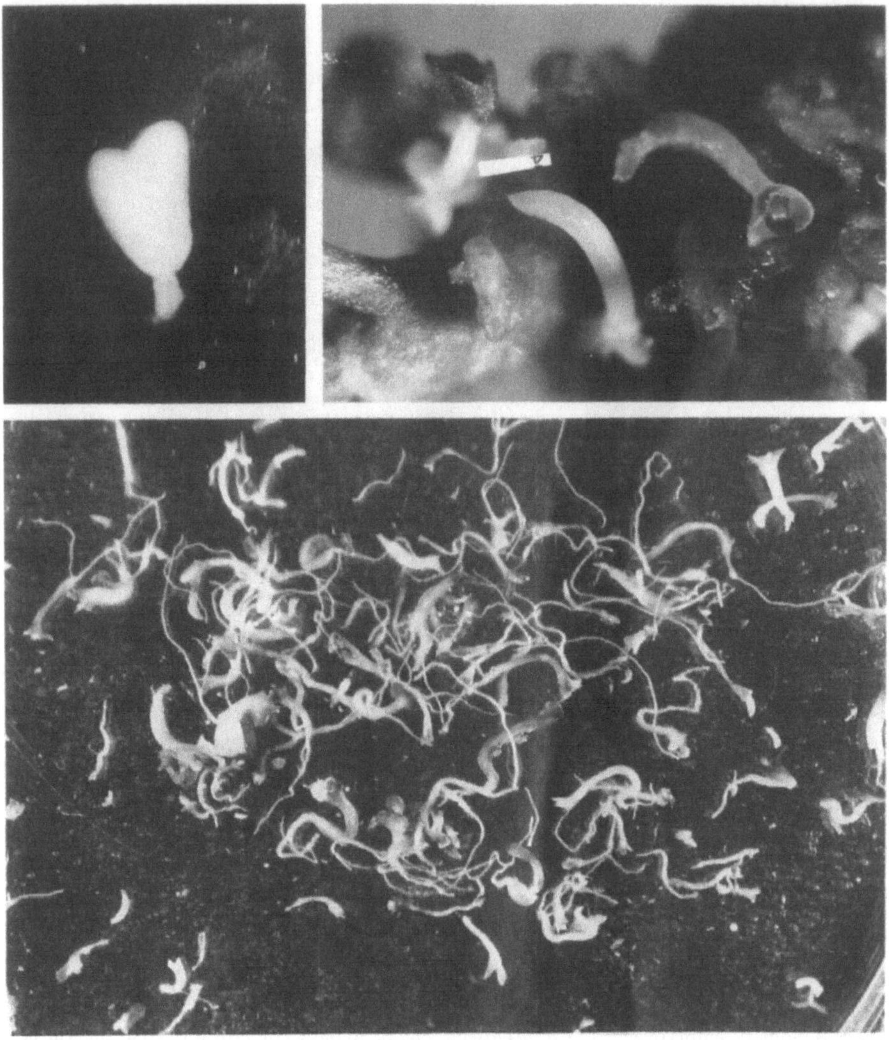

9 10 11

Figs. 9–11. Several stages of *R. acetosella* embryo germination upon addition of GA_3. **Fig. 9.** Cotyledonary stage of an embryo with suspensor. **Fig. 10.** Germinating embryos lying free in a callus mass. **Fig. 11.** Germinating embryos in various developmental stages released by shaking in liquid medium

2.4 Genetic and Phenotypic Stability

Propagation of plant species via somatic embryogenesis may result in genetic changes which are transmitted and reproduced within a clone. The use of hormones required for organ induction may also influence the morphology of regenerated plants and modify their features. These problems were approached with respect to sex determination in *R. acetosella* cultures, because they were exposed consecutively to auxins, cytokinins, and GA_3. In a sample of 166 somatic embryos derived from male and female plants, 66 were successfully grown under nonsterile conditions. Planted in the Botanical Garden in early spring, 1988, they formed large rosettes by the end of the year, but remained vegetative. Flowers were set only in the next season, simultaneously with a wild-growing population. When the regenerates were grown under controlled greenhouse conditions, they displayed a long juvenile period typical of seedlings, even though the primary explants had been taken from adult plants. The photoperiodic reaction was also conserved, since the plants flowered on long days only (Fig. 12). All male and female clones bore only staminate and pistillate flowers, respectively. Therefore, culture conditions, as well as dedifferentiation and redifferentiation processes throughout somatic embryogenesis, did not affect either the photoperiodic sensitivity, or the sex of flowers.

Fig. 12a–c. *R. acetosella* plants regenerated from somatic embryos. **a** Plant in vegetative stage grown on short days of 8/16 h. **b** Flowering female plant grown on long days 16/8 h. **c** Flowering male plant grown on long days 16/8 h

3 Summary and Conclusions

Somatic embryogenesis can be induced in certain species belonging to the genera *Fagopyrum* and *Rumex*. In the most thoroughly studied species, *F. esculentum* and *R. acetosella*, this can be achieved in competent tissues using the experimental methods described above. Immature embryos of buckwheat and parts of micropropagated sorrel shoots respond to a high cytokinin/auxin ratio and to a low osmotic potential by initiating embryogenic structures. Still, the general pattern of somatic embryogenesis in the two species is essentially different. In *R. acetosella* the embryogenic capacity was confirmed in all genotypes employed so far, if minimum requirements were met. These include factors for overcoming the arrest of growth at two points, namely, at the transition from organogenesis to embryogenesis (low osmotic potential) and at the progression of embryo development beyond the torpedo stage (addition of GA_3, omission of BAP/IAA). The reliability of these responses suggests that tissues of *R. acetosella* may provide a suitable model system for further studies on somatic embryogenesis.

In contrast to *R. acetosella*, in *F. esculentum* about 70% of plants represent recalcitrant genotypes, in which genetic factors seem to control somatic embryogenesis. Genotypic specificity is a phenomenon encountered in a great number of species, particularly crops, and no satisfactory explanation has been offerred so far. Studies on molecular mechanisms involved in somatic embryogenesis induction in buckwheat merit further attention.

The question could be raised as to whether somatic embryogenesis in *Fagopyrum* and *Rumex* may be of practical importance. If vegetative cloning of the two species is desirable, this can be easily done by shoot cultures. Both species share the prominent capacity for micropropagation, starting from apical meristems, and the phenotypic and genetic stability seems to be fairly high. Plant regeneration via somatic embryogenesis may perhaps be employed to other ends, e.g., to increase variation of regenerated clones. In *Rumex* this might result in cell lines and regenerates which would be richer in substances of pharmaceutical importance. In buckwheat, some of the traits that depend on single genes might perhaps be modified. These include the self-incompatibility reaction and the indeterminate growth habit. It can be envisaged that spontaneous or induced variation in an embryogenic culture might give rise to altered genotypes. Genetic manipulation may also be applicable. The single cell origin of somatic embryos would alleviate the problem of chimerism, so that the recovery of genotypes with novel traits would be more feasible.

4 Protocols

4.1 *Fagopyrum esculentum*

Plant Material. Immature, green fruits containing embryos 2–3 mm long, which are surrounded by liquid or milky endosperm.

Nutrient Media. Use the basal medium (BM) composed of: B5 mineral salts (Gamborg et al. 1968), 0.7% agar, 60 g/l sucrose, 0.4 mg/l thiamine, 0.5 mg/l pyridoxine, 0.5 mg/l nicotinic acid, 100 mg/l myo-inositol, and 2000 mg/l casein hydrolysate.

–Medium for pretreatment (A) contains 5 mg/l 2,4-D and 0.1 mg/l kinetin.

–Medium for cultivation (B) contains 2.2 mg/l BAP and 0.17 mg/l IAA.

–For embryo growth, add 0.2 mg/l IBA to the BM (medium C).

Procedure. Surface sterilize fruits by washing under running water (30 min), immersing in 95% alcohol with a drop of Tween 20 (1 min), followed by sodium hypochlorite containing 0.4% active chlorine (20 min); rinse several times with sterile water. If the seeds are highly contaminated, use also 0.1% $HgCl_2$ (10 min) after ethanol.

–Remove or cut the pericarp and squeeze out the embryo by gentle pressure. In longer embryos cut the root immediately. In smaller embryos cut both root and stem when sufficiently elongated.

–Place the embryo on medium (A) for 5 days, then transfer the explants to medium (B).

–At the end of the primary culture period, or during the first subculture, examine the explants carefully for the presence of abundant, compact, embryo-like structures. Remove the shoots and subculture the embryo-like groups on medium (B).

–Remove under the dissecting microscope single germinating embryos of about 5 mm, and place them on filter paper bridges in test tubes containing the liquid BM, supplemented with 0.2% mg/l IBA. Plantlets with well-developed roots and shoots are transferred to soil and protected from transpiration.

–Keep all cultures in light in the culture room, preferably at 20 °C.

4.2 *Rumex acetosella*

Plant Material. Lateral vegetative buds from adult plants grown under natural conditions are used as primary explants to obtain embryogenic tissues.

Nutrient Media. Use the following basal medium (BM): MS mineral salts (Murashige and Skoog 1962), 0.7% agar, 20 g/l sucrose, 1 mg/l thiamine, 1 mg/l pyridoxine, and 5 mg/l nicotinic acid.

–Medium for shoot multiplication contains BM, supplemented with 2.2 mg/l BAP and 0.17 mg/l IAA.

–Medium for induction of somatic embryogenesis (D) contains BM with the sucrose content increased to 60 g/l, 2.2 mg/l BAP, and 0.17 mg/l IAA.

–Medium for embryo maturation (E) is BM supplemented with 10 mg/l GA_3.

–For plant growth, use BM with half-strength MS salts, without hormones (F).

Procedure. Surface sterilize the explants in sodium hypochlorite solution containing 0.4% active chlorine (10 min) and rinse several times with sterile water.

–Isolate 5-mm-long shoot tips and place them on medium for shoot multiplication.

–Maintain the shoot cultures by regular transfers to the medium for shoot multiplication.

–Remove the leaves or soft callus and place them on medium (D).

–Transfer the embryogenic callus with proembryogenic cell clusters and early embryo stages to medium (E).

–When most embryos become macroscopically visible, shake the embryogenic tissue in 10 ml of liquid medium for 24 h, to release the embryos, and leave them in the same medium for a few more days.

–Transfer germinating embryos to test tubes containing the medium (F).

–Plant the vigorously growing plantlets in soil and cultivate under normal greenhouse conditions.

Acknowledgement. This work was supported by the Republic of Serbia Research Fund.

References

Adachi T, Suputtitada S, Miike Y (1988) Plant regeneration from anther culture in common buckwheat (*Fagopyrum esculentum*). Fagopyrum 8: 5–9

Adachi T, Yamaguchi A, Miike Y, Hoffmann F (1989) Plant regeneration from protoplasts of common buckwheat (*Fagopyrum esculentum*). Plant Cell Rep 8: 247–250

Bohanec B (1985) A modified method for induction of shoot formation for micropropagation of buckwheat in vitro. Fagopyrum 5: 13–15

Bohanec B (1987) Improvements in buckwheat (*Fagopyrum esculentum* Moench.) micropropagation procedures. Fagopyrum 7: 13–15

Bohanec B, Nešković M, Vujičić R (1993) Anther culture and androgenic plant regeneration in buckwheat (*Fagopyrum esculentum* Moench.). Plant Cell Tissue Organ Cult 35: 259–266

Bowen IH, Cubbin IJ (1993) *Fagopyrum esculentum* Moench. (Buckwheat): In vitro culture, and the production of rutin. In: Bajaj YPS (ed) Biotechnology in agriculture and forestry, vol 24. Medicinal and aromatic plants V. Springer, Berlin Heidelberg New York, pp 202–217

Ćulafić Lj, Budimir S, Vujičić R, Nešković M (1987a) Induction of somatic embryogenesis and embryo development in *Rumex acetosella* L. Plant Cell Tissue Organ Cult 11: 133–139

Ćulafić Lj, Samofalova A, Nešković M (1987b) In vitro organogenesis in two dioecious species, *Rumex acetosella* L. and *R. acetosa* L. (Polygonaceae). Plant Cell Tissue Organ Cult 11: 125–131

De Jong H (1972) Buckwheat. Field Crop Abstr 25: 389–396

Gamborg OL, Miller RA, Ojima K (1968) Nutrient requirements of suspension cultures of soybean root cells. Exp Cell Res 50: 148–151

Holländer-Czytko H, Amrhein N (1983) Subcellular compartmentation of shikimic acid and phenylalanine in buckwheat cell suspension cultures grown in the presence of shikimate pathway inhibitors. Plant Sci Lett 29: 89–96

Javornik B (1983) Nutritional quality and composition of buckwheat proteins. In: Nagatomo T, Adachi T (eds) Buckwheat research (1983). Proc 2nd Int Symp Buckwheat. Miyazaki, Japan, pp 199–212

Koblitz H (1988) Anthraquinones. In: Vasil IK, Constabel F (eds) Cell culture and somatic cell genetics of plants, vol 5. Academic Press, San Diego, pp 113–139

Kreft I (1983) Buckwheat breeding perspectives. In: Nagatomo T, Adachi T (eds) Buckwheat research 1983. Proc 2nd Symp Buckwheat. Miyazaki, Japan, pp 3–12

Lachmann S (1991) Plant cell and tissue culture in buckwheat: an approach towards genetic improvements by means of unconventional breeding techniques. In: Adachi T (ed) Overcoming breeding barriers by means of plant biotechnology. Proc Int Coll, Miyazaki, Japan, pp 145–154

Lachmann S, Adachi T (1990) Callus regeneration from hypocotyl protoplasts of Tartary buckwheat (*Fagopyrum tataricum* Gaertn.). Fagopyrum 10: 62–64

Löve A (1944) Cytogenetic studies on *Rumex* subgenus *Acetosella*. Hereditas 30: 1–136

Miljuš-Djukić J, Nešković M, Ninković S, Crkvenjakov R (1992) *Agrobacterium*-mediated transformation and plant regeneration of buckwheat (*Fagopyrum esculentum* Moench.). Plant Cell Tissue Organ Cult 29: 101–108

Murashige T, Skoog F (1962) A revised medium for rapid growth and bioassays with tobacco tissue cultures. Physiol Plant 15: 473–497

Nešković M, Srejović V, Vujičić R (1986) Buckwheat (*Fagopyrum esculentum* Moench.). In: Bajaj YPS (ed) Biotechnology in agriculture and forestry, vol 2. Crops I. Springer, Berlin Heidelberg New York, pp 579–593

Nešković M, Vujičić R, Budimir S (1987) Somatic embryogenesis and bud formation from immature embryos of buckwheat (*Fagopyrum esculentum* Moench.). Plant Cell Rep 6: 423–426

Nešković M, Vinterhalter B, Miljuš-Djukić J, Ghalawenji N (1990a) Micropropagation of recessive determinate genotypes of buckwheat (*Fagopyrum esculentum* Moench) as an alternate approach to uniform seed production. Plant Breed 105: 377–340

Nešković M, Vinterhalter B, Miljuš-Djukić J, Ninković S, Vinterhalter D, Jovanović V, Kneževič J (1990b) Susceptibility of buckwheat (*Fagopyrum esculentum* Moench.) to *Agrobacterium tumefaciens* and *A. rhizogenes*. Fagopyrum 10: 57–61

Nickell LG (1954) Nutritional aspects of virus-tumor growth. In: Abnormal and pathological plant growth. Brookhaven Symp Biol No 6, US Atomic Energy Commission, Washington, pp 174–186

Pomeranz Y (1983) Buckwheat: structure, composition, and utilization. CRC Crit Rev Food Sci Nutr 19: 213–257

Pomeranz Y, Robbins GS (1972) Amino acid composition of buckwheat proteins. J Agric Food Chem 20: 270–274

Roggemans J, Boxus P (1988) Rhubarb (*Rheum rhaponticum* L.). In: Bajaj YPS (ed) Biotechnology in agriculture and forestry, vol 6. Crops II, Springer, Berlin Heidelberg New York, pp 339–350

Rumyantseva NI, Sergeeva NV, Khakimova LE, Sal'nikov VV, Gumerova EA, Lozovaya VV (1989) Organogenesis and somatic embryogenesis in tissue culture of two buckwheat species. Fiziol Rast 36: 187–194 (in Russian)

Samimy C (1991) Barrier to interspecific crossing of *Fagopyrum esculentum* with *Fagopyrum tataricum*. I. Site of pollen-tube arrest. II. Organogenesis from immature embryos of *F. tataricum*. Euphytica 54: 215–219

Srejović V, Nešković M (1981) Regeneration of plants from cotyledon fragments of buckwheat (*Fagopyrum esculentum* Moench.). Z Pflanzenphysiol 104: 37–42

Suchy C, Duskova J, Masterova I (1973) Pritomnost frangulaemodinu a chrysofanolu v kalusovych tkanovych kulturach *Rumex patientia* L. Farm Obz 42: 168–178 (in Czech)

Sure B (1955) Nutritive value of proteins in buckwheat and their role as supplements to proteins in cereal grains. J Agric Food Chem 3: 793–795

Surikov IM, Mazur VA (1985) Poluchenie regenerantov iz razlichnych tkanei grechihi v kulture in vitro. In: Geneticheskie osnovy selekcii i semenovodstva grechihi. Kishenev, pp 98–100 (in Russian)

Takahata Y (1988) Plant regeneration from cultured immature inflorescence of common buckwheat (*Fagopyrum esculentum* Moench.) and perennial buckwheat (*F. cymosum* Meisn.). Jpn J Breed 38: 409–413

Takahata Y, Jumonji E (1985) Plant regeneration from hypocotyl section and callus in buckwheat (*Fagopyrum esculentum* Moench.). Annu Rep Fac Educ Iwate Univ 45: 137–142

Van den Berg AJJ, Labadie RP (1981) The production of acetate derived hydroxyanthraquinones, -dianthrones, -naphthalenes and -benzenes in tissue cultures from *Rumex alpinus*. Planta Med 41: 169–173

Van den Berg AJJ, Labadie RP (1989) Quinones. In: Dey PM, Harborne JB (eds) Methods in plant biochemistry, vol 1. Academic Press, San Diego, pp 451–491

Yamane Y (1974) Induced differentiation of buckwheat plants from subcultured calluses in vitro. Jpn J Genet 49: 139–146

IV.12 Somatic Embryogenesis in *Ranunculus asiaticus* L

J. MEYNET[1]

1 Introduction

1.1 Importance and Distribution

R. asiaticus L. is the only *Ranunculus* species cultivated for its ornamental attributes. It is native to the eastern Mediterranean basin, from the Greek Islands to Turkestan, and was introduced to western Europe from Turkey as early as the sixteenth century. The Persian buttercup is an excellent plant for spring gardens and a popular cut flower; the dwarf types can be used as potted plants.

All varieties are multiplied by seed, most of them are hybrids, and the parents are often vegetatively propagated by tuberous root divisions or in vitro multiplication. The material cultivated for flower production originates from tuberous roots obtained from seedlings after 1 year of culture (Krabbendam and Baardse 1967).

Flower production is principally located in Western Europe, Israel, and Japan. It is estimated that 30–50 ha are used for bulb production, and 80–100 ha for cut flower cultivation, mainly in Mediterranean countries (personal estimation).

1.2 Some Morphological and Physiological Traits

The commercial types are double-flowered. The full flowers resulting from the petalody of the stamens are mostly male sterile without a stamen. The petalody rate depends on genotype and the effects of nutritional competition: the highest number of stamens is observed in late spring when the growing rate is the fastest, and on axillary buds.

The flower color is very diversified, depending on the amount and distribution in the petals of two anthocyanins, cyanidin and delphinidin, and carotenoids.

This species is self-incompatible and very sensitive to inbreeding. It is well adapted to survive Mediterranean summer drought. Thus, the tuberous roots and seeds are dormant in summer, sprouting only in autumn after the first rain when the soil temperature reaches about 20 °C. The vegetative and flowering

[1] Station d'Amelioration des Plantes Florales, INRA, La Gaudine–Frejus, 83370 Saint-Aygulf, France

Biotechnology in Agriculture and Forestry, Vol. 31
Somatic Embryogenesis and Synthetic Seed II (ed. by Y.P.S. Bajaj)

stages occur in winter and spring under cool temperature and short day length. As soon as the day length exceeds 13 h and soil temperature 22 °C, the plants enter dormancy: the roots enlarge and all aerial organs wilt and disappear.

1.3 Significance of Somatic Embryos in *Ranunculus asiaticus*

The vegetative multiplication of the hybrid parents is difficult due to numerous viruses and diseases that are transmitted by tuberous root divisions. On the other hand, in vitro culture of meristem tips is problematic because of systematic bacterial contaminations (Meynet 1993). When produced on less contaminated floral organs, we believe that somatic embryos should be disease-free and easier to obtain. Thus, in our work, we have used several floral explants: young floral buds, petals, slices of thalamus, and stamens. If successful, the main applications should be:

1. The multiplication of clonal parents of hybrid varieties if the somatic embryos are true to the original types.
2. The production of clonal varieties if secondary somatic embryogenesis can be maintained efficiently to give easily produced, good quality, and inexpensive plants for growers.
3. The induction of genetic variability and the development of a breeding method if somatic embryogenesis reveals or induces a somaclonal variation.

2 Somatic Embryogenesis

In 1965, Konar and Nataraja obtained embryos from different organs of *Ranunculus sceleratus L.* In their first paper (1965a) they reported the establishment of embryos from epidermal cells of the floral stem, particularly on media without coconut milk. In a second article (1965b), they related the production of embryoids on ripe anthers, and showed by histological studies that these embryoids originated from the connective tissue through a callus. Our work on somatic embryogenesis in *Ranunculus asiaticus* is described here.

2.1 Effects of Organs and Tissues

Whole young buds and several floral organs: petals, sepals, stamens, slices of female receptacle (thalamus), and ovaries were tested for somatic embryogenesis. Up to now petals, sepals, and ovaries have never produced organogenetic formations. We obtained neoformations of plants from floral buds and, above all, from stumps of thalamus. In these cases, the regeneration process is long, often requiring two to four subcultures and does not induce somatic embryogenesis (Meynet and Duclos 1990a). Only in vitro culture of anthers produced typical somatic embryos (Fig.1) from epidermal tissue (Meynet and Duclos 1990b). The first globular embryos occur generally after 2 months, developing into heart-shaped and torpedo-stage embryos in about 2 weeks. Most of the embryos originate from anther, sometimes from the connective, but rarely from the filament (Fig. 2). When they are numerous, the development of the embryos is not synchronous. If placed on adequate medium and in full light, they become green (Fig. 3), and root (Fig. 4) quickly. Then they are able to grow under horticultural conditions.

Beside these typical cases, it was observed that organogenetic formations occurred inside the anthers through their line of dehiscence. These formations appear after 2–3 months as yellowish, compact callus giving (Fig. 5) young green plants or bunches of plants.

Cytogenetic and morphological studies of these last plants show that they originate from either tapetal tissue or pollen, but regardless of the origin, the organogenetic process is the same in both cases and it is very difficult to distinguish haploid from somatic plants during the in vitro culture stage. On the other hand, the plants originating from pollen are often spontaneously double-haploid, so the chromosome number (Fig. 6A,B) does not prove that embryos developing inside the anther have a somatic origin.

Fig. 1. Somatic embryos originating from the epidermis of the anther. Their developmental stage is not synchronous

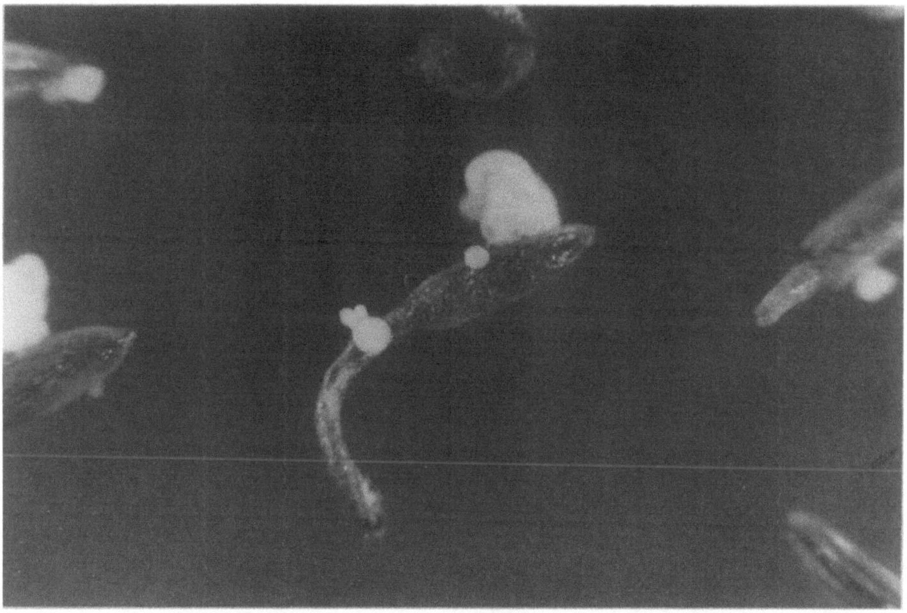

Fig. 2. Some somatic embryos can originate from the filament

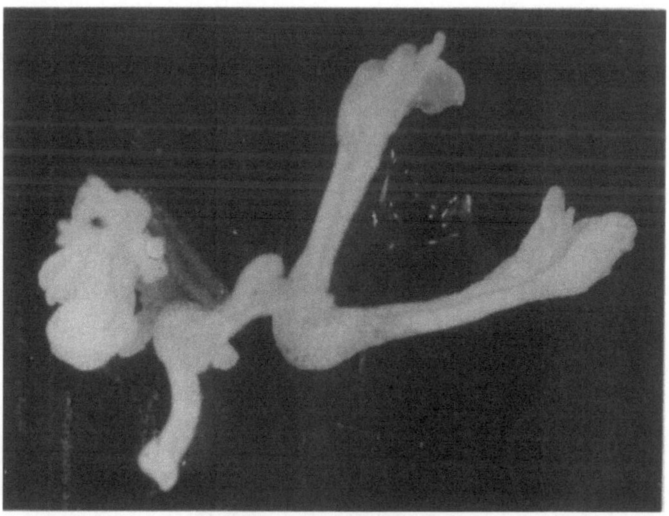

Fig. 3. Maturation of the somatic embryos, 3 weeks after the torpedo stage

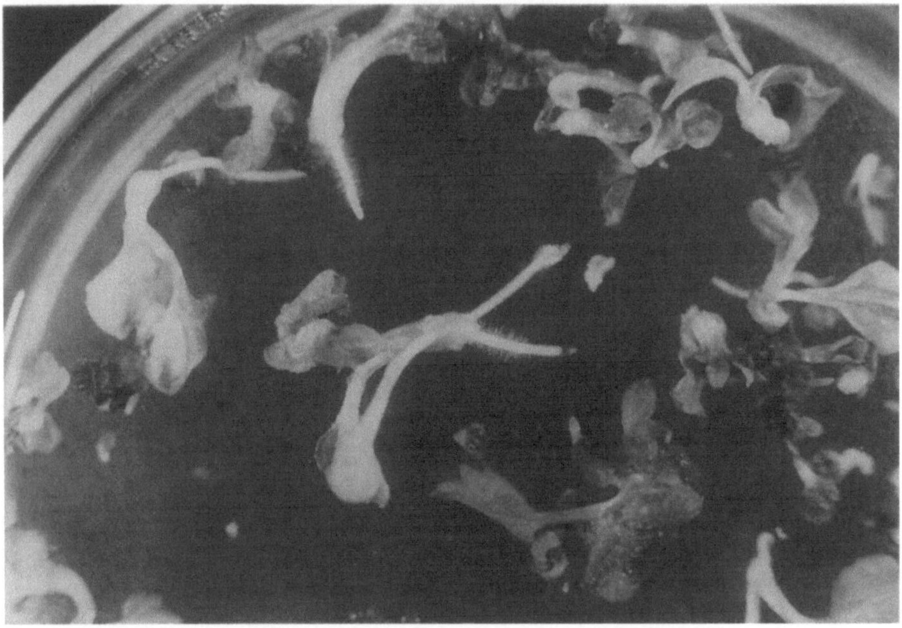

Fig. 4. Direct rooting of plants produced from somatic embryos when placed on a medium without amino acid

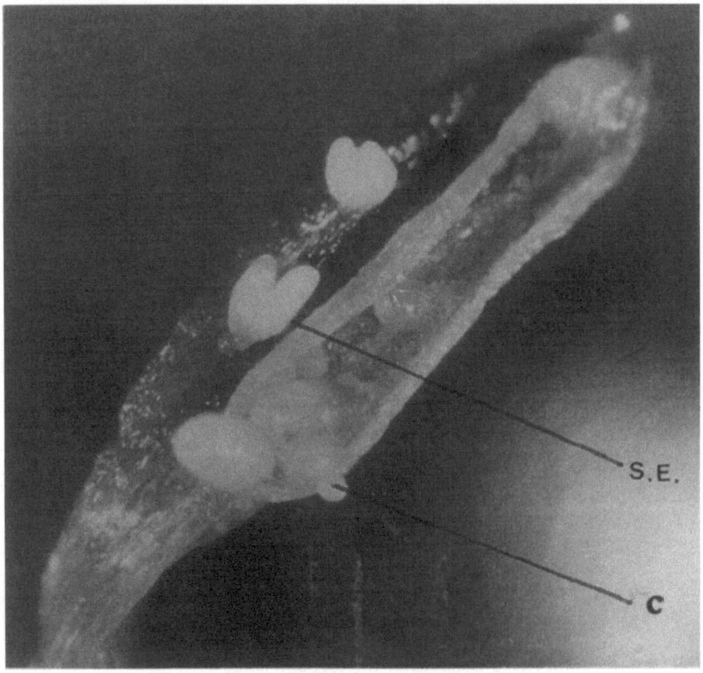

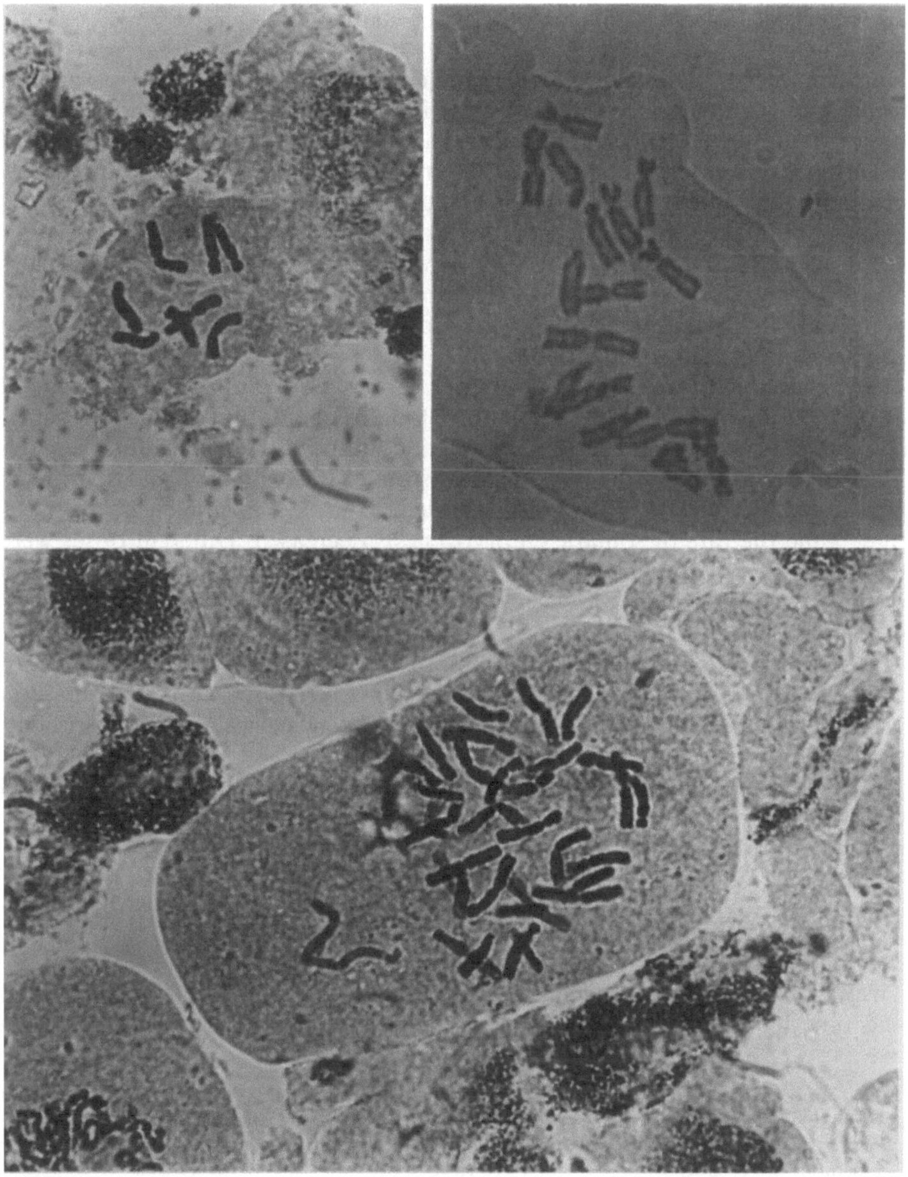

A B C

Fig. 6A–C. Karyograms of root tips produced by anther culture. Different ploidy levels are observed: **A** haploid (n = 8) from pollen; **B** diploid from different anther tissues; **C** tetraploid: a variant somatic embryo

Fig. 5. Somatic embryos (*S.E.*) from epidermal tissue and embryogenic callus (*C*) from inside the anther. These last formations, appearing through the line of dehiscence, can originate from either pollen (haploid or spontaneously double-haploid) or tapetal tissue

2.2 Effects of Culture Media and Physical Conditions

Numerous experiments were carried out in order to optimize the in vitro culture techniques. In first trials, different cytokinins (N-6 furfurylaminopurine, N-6 benzylaminopurine (BA), and 2 iso-pentyladenine) were tested at different concentrations (0, 0.1, 1, and 2 mg/l). Somatic embryos were induced in every medium even without growth regulator, but the maximum number of plants was always obtained with BA. All combinations of BA and 2,4-dichlorophenoxy-acetic acid (2,4-D) at 0, 0.1, 1, and 2 mg/l were also tested. The best results were observed when 0.1 mg/l 2,4-D was combined with 1 to 2 mg/l BA. Different minerals salts (Murashige and Skoog, Linsmaier and Skoog, Tulecke, Nitsch) had no significant effects.

High temperature, 28 °C (optimum is about 18 °C), during the first stage of in vitro culture enhanced the induction of embryogenesis (Table 1). However, cold pretreatments at 2 °C or different light conditions had little influence.

In order to modify the physiological state of the explants some treatments were tested. For instance, silver nitrate added to the medium influenced the callus rate, while the rate of somatic embryogenesis decreased. Different doses were tested on the 11095 hybrid and are reported in Table 2. This effect of 20 μM/l $AgNO_3$ was confirmed on several donors, however, silver thiosulfate was ineffective.

Recent experiments have shown important effects of an elevated CO_2 concentration, e.g., 2% CO_2 hastened and increased the somatic embryo production of several donors. Similar results were also observed when an ethylene adsorbant[2] was used, but they must be confirmed.

These last results emphasize the importance of controlling the physiological conditions, especially through ethylene synthesis with respect to the senescence process of the explants.

2.3 Genotypic Effects

Twelve inbred genotypes (originating from two to five sister × brother crossings) and nine hybrid families were used as anther donors. The first experiments

Table 1. Effect of different temperatures under dark conditions at the beginning of in vitro culture of anthers. (F. Chirat and J. Meynet, unpubl. data 1990)

Temperature	5 °C		18 °C		28 °C	
Duration	1day	14 days	1 day	14 days	1 day	14 days
Embryos/ anthers	5/805	3/952	3/851	4/815	7/795	31/742
Percent	0.62NS	0.32NS	0.35NS	0.49NS	0.88NS	4.18**

NS, ** Nonsignificant or significant at $P \leq 0.01$

[2] Activated alumina XMA 99 provided by Procatalyse, 212, Av. Paul Doumer, 92500 Rueil Malmaison, France

Table 2. Effect of the addition of silver nitrate to the medium on the production rate of somatic embryos (S.E.) and organogenous callus (C) in the 11095 hybrid. (J. Meynet, unpubl. result 1991)

$AgNO_3$ µM/l	0		10		20		50	
Anthers cultivated	440		358		367		396	
Organogenetic process	S.E.	C	S.E.	C	S.E.	C	S.E.	C
Number of formations	21	8	20	19	10	59	7	45
Percent	4.8	1.8	5.6	5.3	2.7	16.1	1.7	11.4

allowed one to distinguish two types of donors (Table 3): either they produced only callus inside the anthers (type C) or they produced typical somatic embryos only or somatic embryos with callus (type S.E). The experiments were replicated several times during the flowering period from January to April. The two types of behavior, depending on the genotype, were always confirmed, but the rate of plants produced and the proportion of callus obtained in donors suitable for somatic embryogenesis were highly variable. Generally, the maximum number of somatic embryos was obtained when the samples were taken in March during the peak of flowering.

Table 3. Effect of genotypes on the production of somatic embryos from epidermis or callus from inside the anthers. (J. Meynet and A. Duclos, unpubl. data 1991)

[a] Number of embryogenic formations/number of anthers in %.

Genotypes		Replication number	Somatic embryos[a]	Callus[a]
Inbred	R2	1	0	0.7
	R4	2	0	0.5
	R13	5	0	1.5
	R14	2	0	1.1
	R15	4	0	2.9
	R16	3	0	2.3
	R20	6	5.3	4.1
	R22	1	3.2	0.8
	R29	2	0	0
	R30	4	0	5.4
	R32	3	4.8	0
	R34	2	2.2	0
Hybrids	2059	1	0	2.3
	4129	2	0	1.8
	11095	6	6.8	2.1
	12196	2	0	0.7
	13059	2	2.5	0.6
	13065	5	0	4.6
	13078	3	0	0.4
	13111	4	0	0.9
	14003	1	0	0.3

Modifications of in vitro media and the physical conditions can influence the production rates of somatic embryos or embryogenic calli, but the genotype remains the essential factor which determines the organogenetic process producing either S.E (from epidermal tissue) or C (from tapetal tissue or pollen) types. Thus, the genetic potential for somatic embryogenesis was examined. The analysis of the first crossings is reported in Table 4.

In spite of the few plants examined (due to the full flowers of a majority of offspring and the work required for an exhaustive study) and the lack of precision regarding the C types (masked by S.E. types ?), the first results seem well established. Up to now, when C genotypes are combined, their offsprings are only C types, while the S.E. plants used in crossing may generate segregation without a maternal effect. These results suggest that the ability to undergo somatic embryogenesis has a Mendelian dominant heredity.

Table 4. Analysis of combinations of two parental types: C genotypes produce only calli from inside the anthers, S.E. genotypes produce somatic embryos from stamen epidermis (with or without callus of tapetal or pollen origin). (F. Chirat and J. Meynet, unpubl. data 1990)

Crossings		Offspring		
Code	Type[a]	Number observed	S.E. type	C. type
0087	S.E. × S.E.	4	4	0
0089	S.E. × S.E.	3	2	1
0084	S.E. × S.E.	4	3	1
9041	S.E. × C	4	3	0
9030	S.E. × C	3	3	0
9031	S.E. × C	4	1	2
9035	S.E. × C	2	0	1
9055	*C × S.E.	4	3	1
9080	*C × S.E.	2	2	0
0048	C × S.E.	5	3	2
9050	*C × C	3	0	3
9061	*C × C	4	0	3
9064	*C × C	2	0	2
9096	*C × C	2	0	2
9105	C × C	2	0	2
0029	*C × C	4	0	4
0042	*C × *C	2	0	2
0045	*C × *C	2	0	2
0051	*C × *C	3	0	3

[a] *C parents are homozygous, double-haploid

2.4 Study of Plants Obtained from the Anthers by Somatic Embryogenesis

Numerous plants obtained through somatic embryogenesis from 21 different donors were examined; the color, double flower characteristics and sometimes flavonoid components of the petals, chromosome number, and self-fertility were noted. With regard to the conformity of the somatic derivatives to their original genotypes, two types of donors were detected. The donors generating no in vitro variation were always homogeneously colored on the petal surfaces. The others showed a very important somaclonal variation; these donors were always bicolored, especially yellow or white, with red or purple fringe petals. In these last cases, the rate of color variation, especially distribution and intensity of anthocyanic synthesis, reached more than 90%. The variant flowers always produced higher concentrations of anthocyanins with larger, colored surfaces of the petals than in the original donors.

Several other variant characters were observed:

–The glycosylation pattern of the flavonoids was sometimes modified; 2 out of 136 plants contained monoglycosides, while the donors contained diglycosides.

–In a first study, we counted (out of 25 somatic derivatives) one tetraploid, (Fig. 6C) one mixoploid (4n/2n), and one trisomic plant (Meynet and Duclos 1990b).

–One plant was found to be self-fertile, while the donor (as all the other original genotypes), was self-incompatible.

Numerous other traits were observed, but are not reported here since they depend too much on environmental conditions.

All these new characteristics were stable after more than eight in vitro subcultures and several cycles of tuberous root division under horticultural conditions.

The genetic studies of self-fertility and increasing anthocyanic synthesis, which appeared in somaclonal variants, were carried out and showed a very good transmissibility from one generation to the next without affecting reciprocal crosses (Chirat 1990; Meynet and Duclos 1990c). New variants should be considered as recessive nuclear mutants.

3 Summary and Conclusion

The somatic embryogenesis in anthers of *Ranunculus asiaticus* can be of great practical and scientific interest. From a practical point of view, it is a good method to obtain healthy material in a species which is very sensitive to numerous viruses, vascular fungi, and endophytic bacteria. Indeed, the plants obtained were always pathogen-free.

When the donor is homogeneously colored, somatic embryogenesis can be employed to obtain a true-to-original-type multiplication. Up to now, the culture trials in shaken liquid medium have failed and an efficient secondary embryogenesis for the mass production of plants has not been achieved. Thus, the healthy plants induced are propagated by classical in vitro culture. This expensive method is only applicable for the multiplication of hybrid parents.

The genetic variation obtained when the donor is bicolored may be very useful in breeding programs, especially to create varietal assortments of different colors with a common genetic background, having identical morphological, physiological, and agronomic characteristics. Contrary to classical mutagenetic methods, many solid mutants (especially for color characteristics) can be obtained directly and are readily available.

However, this variation presents very interesting scientific questions. All the plants originating from the entire somatic staminal territory (epidermal tissue of the anther and filament as well as tapetal tissue) showed a very important variation. On the one hand, the pollen-derived plants, through androgenesis, conformed to the expected segregation, while on the other hand, the plants regenerated from other organs (thalamus, leaves) were true to donor types. This variation appeared regardless of the vitro culture medium (even without growth regulators). These facts suggest that this variation should preexist in situ before somatic embryogenesis, which only reveals it, but should not induce it. However, no explanatory mechanism has been put forward up to now.

Only bicolored donors gave rise to a large variation. The genetic determination of the bicolor character is uncertain. It is very difficult to establish by inbreeding and often results from hetergamous crossings between cyanic and acyanic parents. The rate and the nature of the variation observed suggest the presence of a transposon associated with (or causing) the bicolor character, which should only be activated in the somatic tissues of the anthers.

4 Protocol for the Induction of Somatic Embryos

The anthers are taken from the first closed axillary bud as the main terminal flower just opens. The development stage of anthers is heterogeneous, the inner stamens being the youngest. Generally, the optimal sample is defined when the longitudinal axis of the anther is slightly curved, the yellowish pollen sacs just cover the connective, and the pollen exine starts to thicken. The buds are used immediately or can be kept at 2 °C for 2 or 3 days. They are disinfected by immersion in 70% ethanol for 30 s, then in 15 g/l calcium hypochlorite for 20 min. In spite of these treatments many stamens are contaminated by endophytic bacteria, especially in vegetatively multiplied donors after several tuberous root divisions.

Sixteen anthers are placed in a 55 mm diameter Petri dish on a solid induction medium containing full strength microelements and half-strength Murashige and Skoog macroelements, 0.5 μM/l Fe EDTA, 100 mg/l inositol, 1 mg/l thiamine, 1 mg/l pyridoxine, 2 mg/l Ca pantothenate, 5 mg/l nicotinic acid, 50 mg/l glutamine, 50 mg/l glycine, 3% sucrose, 1 mg/l BA, 0.1 mg/l 2,4-D; pH is adjusted to 5.8 before autoclaving for 20 min at 120 °C. The anthers, placed at 28 °C in darkness for 14 days, are then placed under light with a 8-h photoperiod of 12 W/m^2 provided by cool white fluorescent lamps at 18 °C until the emergence of the somatic embryos. They are then transferred to a new medium without 2,4-D for the maturation of the embryos. Rooting is easily obtained by removing amino acids and BA and, if necessary, 1 mg/l indole-3-butyric acid is added.

When the anthers are taken in March, most of the plants produced through somatic embryogenesis are rooted in late September and can be transferred to horticultural conditions in November. They flower in the greenhouse in February/March and can be used again.

References

Chirat F (1990) Etudes des différents processus embryogènes somatiques chez la renoncule, Mémoire de fin d' études ENSAIA, BP 172–54505, Vandoeuvre-less-Nancy (Libraire)

Johansson L, Andersson B, Eriksson T (1982) Improvement of anther culture technique: activated charcoal bound in agar medium in combination with liquid medium and elevated CO_2 concentration. Physiol Plant 54: 24–30

Konar RN, Nataraja K (1965a) Experimental studies in *Ranunculus sceleratus L.* Development of embryos from the stem epidermis. Phytomorphology 15: 134–137

Konar RN, Nataraja K (1965b) Production of embryoids from the anthers of *Ranunculus sceleratus L.* Phytomorphology 15: 245–248

Krabbendam P, Baardse AA (1967) Bijgoed Bloembollenteelt. In: Tjeenk Willink WEJ (eds) Zwolle, The Netherlands, pp 349–360

Meynet J (1993) *Ranunculus.* In: de Hertogh AA, Le Nard M (eds) The physiology of flower bulbs. Elsevier, Amsterdam, pp 603–610

Meynet J, Duclos A (1990a) Culture in vitro de la renoncule des fleuristes (*Ranunculus asiaticus* L.) I Néoformation et multiplication végétative in vitro de plantes à partir de tronçons de thalamus. Agronomie 10: 157–162

Meynet J, Duclos A (1990b) Production de plants par cultures d'anthères in vitro. Agronomie 10: 213–218

Meynet J, Duclos A (1990c) Etude des plantes produites par embryogenèse somatique à partir de tissus superficiels de l'anthère. Agronomie 10: 285–290

Subject Index

MIX
Papier aus verantwortungsvollen Quellen
Paper from responsible sources
FSC® C105338

If you have any concerns about our products,
you can contact us on
ProductSafety@springernature.com

In case Publisher is established outside the EU,
the EU authorized representative is:
Springer Nature Customer Service Center GmbH
Europaplatz 3, 69115 Heidelberg, Germany

Printed by Libri Plureos GmbH
in Hamburg, Germany